Rainer Kimmich

NMR – Tomography, Diffusometry, Relaxometry

Springer

Berlin
Heidelberg
New York
Barcelona
Budapest
Hong Kong
London
Milan
Paris
Santa Clara
Singapore
Tokyo

Rainer Kimmich

NMR

Tomography, Diffusometry, Relaxometry

With 127 Figures

 Springer

Prof. Dr. Rainer Kimmich

Universität Ulm
Sektion Kernresonanzspektroskopie
Albert-Einstein-Allee 11
D – 89069 Ulm

ISBN-13: 978-3-642-64465-8   e-ISBN-13: 978-3-642-60582-6
DOI: 10.1007/ 978-3-642-60582-6

Library of Contress Cataloging-in-Publication Data

Kimmich, R. (Rainer), 1941–
NMR : tomography, diffusometry, relaxometry / R. Kimmich.
p. cm. Includes bibliographical references.

1. Nuclear magnetic resonance–Methodology. 2. Tomography. 3. Nuclear magnetic resonance–
Industrial applications. I. Title
YC762.K56 1997
538'.362–dc21

© Springer-Verlag Berlin Heidelberg 1997

Softcover reprint of the hardcover 1st edition 1997

Typeset with LaTeX: Danny Lewis Book Production, Berlin. Keyboarding by the author.
Cover: de'blick, Berlin
SPIN: 10129018   51/3020 - 5 4 3 2 1 0 - Printed on acid-free paper

Preface

When I planned this book seven years ago I had my graduate students at the University of Ulm in mind, diploma as well as doctoral students, who often asked me what literature they should work with. I used to suggest a list of ten to twenty (for my taste: excellent) treatises on NMR. Apparently this did not make them entirely happy. The difficulty which newcomers to the field face is to practise and to apply theoretical formalisms from different sources while still learning the principles of NMR and being actively engaged in NMR research. Although the text presented here is largely based on my lecture notes, the result is a "working book" rather than an introduction. It is intended to provide direct access to the basic information one needs for NMR diffusometry, relaxometry, and tomography applications.

A "working book" is certainly not suitable to be read starting on page one and then carrying on until the last page. Boldly extrapolating my own reading habits to those of the typical scientist I am sure that this is not the way in which monographs of this kind are read nowadays. So my aim was to produce a treatise that offers easy and quick access for the reader to relevant matters of interest. I tried hard to ease the comprehension of NMR principles by extensive cross-referencing among the sections and chapters.

Tomography, diffusometry and relaxometry are fields based on common physical principles. The combined use of such techniques provides synergistic insight into physicochemical material properties of an object. Therefore I focused on the methodology offered by NMR in this respect. On the other hand, I virtually ignored conventional NMR spectroscopy apart from what is needed for spectroscopic tomography. This appears to be justified because liquid and solid-state NMR spectroscopy is represented in the vast NMR literature to a much greater extent than are the three fields to which this book is devoted.

The two middle parts, "Molecular Motion" and "Localization and Imaging," are mainly of a methodological nature. They are framed by representations of a more axiomatic character. The first part "Spin Coherences and Echoes" gives full-length treatments of elementary pulse sequences where relaxation, diffusion, and localization effects are not yet accounted for. The fourth part offers a look-up compendium of the formalisms needed in basic NMR theories. The working book character is particularly obvious in this latter part. Researchers and PhD students are busy people and do not have much time for dealing with background formalisms. My aim therefore was to supply a compilation of the theoretical framework relevant for NMR expressed in a consistent way. In particular, I carefully avoided "burying" important formalisms in the text. The intention is to permit straightforward access to what is needed while working analytically with NMR problems. Rules, basic principles, and algebraic techniques should be at hand and ready for use at any time. This part of the book is intended to "accompany" the reader while browsing in the text.

Part IV certainly cannot replace an introductory treatise on NMR. I avoided any "handwaving arguments" which can be so helpful for the beginner (although they are almost never really convincing!). On the other hand, complex formalisms and implicit definitions are circumvented as far as possible. The intention was to provide reasonably tutorial explanations one can "live" with.

The part on "Spin Coherences and Echoes" pays tribute to the fact that – possibly with the exception of more recent force detection experiments – almost everything observable in magnetic resonance refers to coherences or is based on its conversion into coherences. Fundamental spin-coherence phenomena are treated in a closed form. Coherence manipulation is the magic formula disclosing the unmatched beauty of NMR.

The representations in Parts II and III, "Molecular Motion" and "Localization and Imaging," are the logic follow-up to coherence and echo methods utilized for practical measuring techniques. Although I attempted to give an idea of the vast variability of NMR methodology, the primary aim was to emphasize the principles of representative experimental protocols. How can we study the manifold ways morphology, texture, or heterogeneity reveal themselves? What is the type of molecular dynamics in a structured environment? The central two parts of the book framed by the elementary coherence and echo formalisms on the one hand, and by the analytical NMR toolbox on the other, are likely to tone in with the fields of interest of those who felt attracted by the book title.

The organization of the main text can also be viewed in a more hierarchical way. The first part demonstrates pulse sequences suitable for the production of coherences in the form of spin echoes using different coherence pathways. Relaxation and diffusion are not yet considered. The same sort of pulse schemes are more or less revisited in the second part, but with thermal equilibration now taken into account. Irreversible processes such as relaxation, self-diffusion and exchange are treated. In the third part, localization comes into play. Practically all principles outlined in the first two parts are now combined with imaging and volume selection procedures.

This book is meant to be neither a literature review nor a historic survey! Even a moderate attempt of this sort would unavoidably result in an oeuvre such as the "Encyclopedia of NMR" [163] which has a range one order of magnitude larger. The literature references I have listed in the bibliography are intended as a suggestion for further reading. They were not consistently selected according to any original priority. On the contrary, I have preferred references where further citations of previous development are listed. In particular, there are many citations of monographs and review articles.

The above outline of the philosophy of this treatise will, hopefully, convince the reader that my aim was to provide a book for the researcher's desk rather than for his or her bookshelf! I would be happy if I should have succeeded in achieving this goal.

During my struggle with the preparation of this book, I remembered the words of Jim Brown whose office I had the pleasure to share while I spent my sabbatical at the University of Kent at Canterbury two years ago. He once tried to reassure me by stating that "one can never finish writing a book, one can merely abandon it." Although I do not really have the feeling of doing so at this instant, I must admit that it was a long, hard slog with many temptations to break off. This is the time when I must thank all my friends, colleagues and coworkers for supporting, teaching, inspiring and encouraging me in the course of this work and earlier. I enjoyed much fruitful cooperation and discussion, helping me to advance my view of the subject of this book. Representative to many others I would like to mention Jerzy Blicharski, Myer Bloom, Bernhard Blümich, Cesare Borgia, Robert Botta, Paul Callaghan, N. Chandrakumar, Dan Demco, R. R. Ernst, Nail Fatkullin, Franz Fujara, Eichii Fukushima, Farida Grinberg, Geneviève Guillot, Ulrich Haeberlen, Siegfried Hafner, Morley Halse, Jukka Jokisaari, Stefan Jurga, J. Kärger, Joseph Klafter, Winfried Kuhn, Bruno Maraviglia, Peter McDonald, Robert Muller, Narczys Pislewski, Jack Powles, Daniel Pusiol, H. W. Spiess, John Strange, Jan Weis. My thanks are also due to the numerous students who have passed through this laboratory, and, last but not least, to Hans Wiringer, Klaus Gille, and David Tomalin who so kindly assisted me in the computer work involved with this book.

Ulm, August 1996   Rainer Kimmich

Principles of magnetic resonance? Great!
Principles of science? More and more!
What about principles of humanity?
Dedicated to W. K., B. S., and all those who never cease to explore our
determination, examine the facts, and take action with social devotedness.

Contents

Frequent Symbols and Abbreviations

Symbols meant to represent vectors, matrices, or tensors as whole arrays are typed in bold face. No distinct printing style is normally used for operators.

A	attenuation factor
ADLF	adiabatic demagnetization in the laboratory frame
AJCP	adiabatic J cross polarization
ADRF	adiabatic demagnetization in the rotating frame
α, β	eigenfunctions for spin "up" and "down"
$\tilde{\alpha}, \tilde{\beta}, \tilde{\gamma}$	Euler angles
$(\alpha)_x, (\beta)_x, (\gamma)_x$	RF pulses (moduli of the flip angles: α, β, γ; rotating-frame phase direction: x)
AQ	acquisition
AW	Anderson/Weiss (theory)
$\boldsymbol{B}$	magnetic-flux density
$\boldsymbol{B}_d$	magnetic-flux density in the detection interval
$\boldsymbol{B}_e$	magnetic-flux density effective in the rotating frame
B_{loc}	local magnetic-flux density contribution by secular interactions
B_p	magnetic-flux density in the polarization interval
BPP	Bloembergen/Purcell/Pound (theory)
B_r	magnetic-flux density in the relaxation interval
$\boldsymbol{B}_{rf}$	magnetic-flux density of the (linearly polarized) RF field
BWR	Bloch/Wangsness/Redfield (theory)
$\boldsymbol{B}_0$	stationary main magnetic-flux density in z direction
$\boldsymbol{B}_1'$	stationary component of the RF flux density in the rotating frame (modulus: B_1)
$B_1^{(sl)}$	rotating-frame amplitude of the spin-lock RF pulse
cc	complex conjugate
COSY	correlated spectroscopy
CP	cross polarization
CPMG	Carr/Purcell/Meiboom/Gill method
CSA	chemical-shift anisotropy
cw	continuous wave
CYCLCROP	cyclic cross polarization
CYCLPOT	cyclic polarization transfer

D	self-diffusion coefficient
d.c.	direct current
DO (subscripts: do)	dipolar order
DOPT	dipolar order polarization transfer
δ_{lm}, $\delta_{l,m}$	Kronecker symbol
e	(positive) elementary charge if not Euler's number
E	echo amplitude if not electric field strength
$\mathcal{E}$	unity operator
EFG	electric field gradient
ϵ_0	electric field constant
$\mathrm{erf}(x)$	error function: $\mathrm{erf}(x) \equiv (2/\sqrt{\pi}) \int\limits_0^x \exp\{-u^2\}\,\mathrm{d}u$
$\mathrm{erfc}(x)$	complementary error function: $\mathrm{erfc}(x) \equiv 1 - \mathrm{erf}(x)$
ESR	electron spin resonance
η	asymmetry parameter (or anisotropy constant)
EXSY	exchange spectroscopy
F	quantum number of the total spin
$\mathbf{F}$	total spin vector operator (in units $\hbar$)
$\mathcal{F}\{\ \}$	Fourier transform
FC	field cycling
FID	free-induction decay
FFT	fast Fourier transformation
FT	Fourier transform
FWHM	full (line) width at half maximum
φ	azimuthal angle; phase shift; eigenfunction
ϕ	wavefunction; phase shift
Φ	electrostatic potential
$g(t), G(t)$	(reduced) time (auto)correlation function
$g(\sigma_{zz})$	chemical-shift anisotropy lineshape function
$g(z)$	slice profile along the z direction of the laboratory frame
G	gradient of the main magnetic-flux density
γ_n	gyromagnetic ratio
$\Gamma_{11}, \Gamma_{22}, \Gamma_{33}$	principal axes values of the EFG tensor
h $(\hbar)$	Planck's constant (divided by 2π)
$\mathcal{H}$	Hamilton operator (in energy units)
$\mathcal{H}_{cs}$	Hamilton operator of chemical shifts
$\mathcal{H}_d$	Hamilton operator of dipole-dipole interaction
$\mathcal{H}_d^{(0)}$	secular part of the Hamilton operator of dipole-dipole interaction
$\mathcal{H}_d^{(tr)}$	truncated Hamilton operator of dipole-dipole interaction
$\mathcal{H}_e$	Hamilton operator effective in the rotating frame
$\mathcal{H}_i$	Hamilton operator of spin interactions in general
$\mathcal{H}_J$	Hamilton operator of J coupling
$\mathcal{H}_J^{(0)}$	secular part of the Hamilton operator of J coupling
$\mathcal{H}_q$	Hamilton operator of quadrupole coupling

$\mathcal{H}_q^{(0)}$	secular part of the Hamilton operator of quadrupole coupling
$\mathcal{H}_{rf}$	Hamilton operator of the interaction to the RF field
$\mathcal{H}_0$	Hamilton operator of the Zeeman interaction
i	imaginary unit
I	spin quantum number
$\mathcal{I}(\omega)$	reduced intensity function (spectral density)
I	spin vector operator (in units $\hbar$)
I_x, I_y, I_z	Cartesian components of the spin vector operator (in units $\hbar$); the same symbols are used in laboratory and rotating reference frames
I^+, I^-	raising and lowering operators (in units $\hbar$); the same symbols are used in laboratory and rotating reference frames
$\Im\{\ \}$	imaginary part
INEPT	insensitive nuclei enhanced by polarization transfer
IST	irreducible spherical tensor
J	(indirect) spin-spin coupling constant (in ν units)
JCP	J cross-polarization
k_B	Boltzmann's constant
LAS	laboratory axes system
LOSY	localized spectroscopy
m	complex transverse magnetization if not a magnetic quantum number
MAGROFI	magnetization grid rotating-frame imaging
M	magnetization in the laboratory frame
M'	magnetization in the rotating frame
M_x', M_y', M_z'	magnetization components in the rotating frame along the x', y', and z' axes, respectively
M_0	equilibrium (Curie) magnetization
MQC (subscripts: mqc)	multiple-quantum coherences
MRI	magnetic resonance imaging
MRSI	magnetic resonance spectroscopic imaging
μ	magnetic dipole moment
μ_0	magnetic field constant
n	(spin) number density if not a count number
NMR	nuclear magnetic resonance
NOE	nuclear Overhauser effect
NOESY	nuclear Overhauser effect spectroscopy
nQC	n-quantum coherence
NQR	nuclear quadrupole resonance
ω_c	carrier (angular) frequency
ω_e	Larmor (angular) frequency in B_e
ω_L	Larmor (angular) frequency
ω_0	Larmor or resonance (angular) frequency in B_0

ω_1	Larmor (angular) frequency in B_1
Ω	(angular) frequency offset or solid angle
p, P	probability or probability density
PAS	principal axes system
PGSE	pulsed-gradient spin-echo (diffusometry)
$', '', \ldots$	primes indicate quantities and operators referring to rotating, rotated or tilted coordinate frames; in context with spin operators, primes are only used if needed for unambiguity
ψ, Ψ	wavefunction
q	electric field gradient in 33 direction (of the PAS) divided by e
Q	nuclear quadrupole moment
r	position vector
$\Re\{\ \}$	real part
RF (subscripts: rf)	radio frequency
RMTD	reorientation mediated by translational displacements
ρ	density operator
ρ_c	charge density
ρ_s	spin density
ρ_0	equilibrium density operator
RODI	rotating-frame relaxation dispersion imaging
R_1	spin-lattice relaxation rate
R_2	transverse relaxation rate
S	spin quantum number if not signal intensity
S	spin vector operator (in units $\hbar$)
SECSY	spin echo correlated spectroscopy
S_x, S_y, S_z	Cartesian components of the spin vector operator (in units $\hbar$); the same symbols are used in laboratory and rotating reference frames
S^+, S^-	raising and lowering operators (in units $\hbar$); the same symbols are used in laboratory and rotating reference frames
SGSE	steady-gradient spin-echo (diffusometry)
$\mathrm{sinc}(x)$	sinc function $= \sin(\pi x)/(\pi x)$
SL (subscripts: sl)	spin-lock pulse
SLOPT	spin-locking polarization transfer
SO (subscripts: so)	scalar (or J) order
SQC (subscripts: sqc)	single-quantum coherence
σ	reduced density operator
σ	chemical shift shielding tensor
σ_0	reduced equilibrium density operator
t	time
T_E, t_e	echo time
T_R, t_r	repetition time

$t-$	time just before an RF pulse
$t+$	time immediately after an RF pulse
T	absolute temperature
$\mathcal{T}$	Dyson time ordering operator
$T^{(l)}$	irreducible spherical tensor operator of rank l
T_d	dipolar-order relaxation time
$T_{i,j}$	element of a second-rank Cartesian tensor
$T_{l,m}$	component of an irreducible spherical tensor operator of rank l
T_q	quadrupolar-order relaxation time
t_{SL}	duration of a spin-lock pulse
$t_1, t_2, \ldots$	time domains in multi-dimensional experiments
T_1	spin-lattice relaxation time
$T_{1\rho}$	spin-lattice relaxation time during a spin-lock pulse
T_2	transverse relaxation time
T_2^*	time constant of the FID in the presence of B_0 inhomogeneities
T_{2e}	transverse relaxation time effective under multiple-pulse irradiation
$T_{2\rho}$	transverse relaxation time during a spin-lock pulse
t_p	length of a preparation pulse
t_w, τ_w	pulse width
τ_c	correlation time
$\tau, \tau_1, \tau_2, \ldots$	intervals in pulse sequences
u	unit vector with the polar coordinates $\varphi, \vartheta, 1$
u_x, u_y, u_z	unit vectors along the x, y, z axes
ϑ	polar angle
Θ	tilt angle of the effective magnetic-flux density
VOSING	volume-selective spectral editing
VOSY	volume-selective spectroscopy
x, y, z	Cartesian laboratory frame coordinates; as subscripts of spin operators also referring to rotating-frame coordinates
$Y_{l,m}$	spherical harmonics

I

Spin Coherences and Echoes

CHAPTER 1

Introductory Remarks

Time-domain magnetic-resonance signals normally are a result of a finite transverse magnetization precessing about the main magnetic field [154].[1] As formulated in more detail in Sect. 47.1, a finite transverse magnetization can only appear if the spin states, i.e., primarily the wave functions, are in-phase. Such **coherences** in turn are readily generated by radio-frequency (RF) pulses exciting the whole sample in a more or less homogeneous, but nevertheless perfectly phase-coherent, manner.

One of the most fundamental phenomena in the field of magnetic resonance is the occurrence of **spin echoes**. In the course of an RF pulse experiment the transverse magnetization appears to fade away irretrievably but reappears, for instance, after some manipulation with the aid of one or more other pulses.

The possibility of reviving spin coherences in the form of a spin echo indicates that the preceding coherence loss was, in a sense, of an apparent nature. On the one hand, we are dealing here with the average magnetization of the whole spin ensemble from which the magnetic-resonance signal originates. Fading induction signals definitely indicate that the average transverse magnetization declines. However, the object under investigation may consist of an ensemble of "isochromats." This term again stands for an ensemble, but now with respect to all sorts of interactions. An isochromat is thought to comprise all spins or spin systems with identical Zeeman, chemical-shift, quadrupolar, dipolar, and scalar spin interactions. That is, the quantum-mechanical coherence within a selected isochromat goes on the whole time an echo can be generated. In this sense, echo formation stands for refocusing of isochromat coherences where the ensemble of isochromats is subject to a distribution of spin interactions, spatially, orientationally, structurally, or chemically.

The formation of a spin echo stipulates that the isochromat coherences still exist at the time the echo is to occur. The limit is given by **spin relaxation** which is extensively treated in Part II of this book. Complete relaxation in the proper sense means that coherences *on any level* have evanesced.

Spin-bearing particles are certainly not the only quantum-mechanical objects which can be subject to echo effects. It is more a question of the lifetime of the quantum-mechanical states and the experimenter's ability to generate these states in a coherent way before manipulating them within their lifetime. For instance, the development of lasers producing ultrashort light pulses made it possible to

---

[1]This statement does not refer to magnetic-resonance experiments based on β-**radiation**, **optical** or **force microscopy** detection schemes (see Sections 32.3.2 and 35.3).

excite optical transitions coherently by a first pulse. It then becomes feasible to intervene in the evolution of the light emission process by a second pulse before the emission of the photon is completed and before the photon has left the experimental setup [18, 190]. Echo phenomena may thus be classified generally as refocusing processes of quantum-mechanical coherences by pulsed manipulation of a system which underlies a certain distribution of (defocusing) interactions.

Spin echoes are nevertheless peculiar effects: there is an unparalleled diversity of phenomena of this sort. The interactions leading to defocusing of coherences can be of very different natures. What began with Hahn's pioneering work in 1950 [186] dealing with "Hahn echoes," as they are known nowadays, led to a whole class of experiments. The first part of this book is devoted to the description and treatment of the basic principles of spin echoes.

Spin echoes appear if spin coherences, which have been spoiled by the influence of a distribution of interactions within a spin ensemble, are refocused. We distinguish **isolated spins** from spins forming a **spin system** due to spin interactions. Isolated spins adopt states which can be suitably described by the single-spin eigenfunctions or linear combinations thereof.

By contrast, the states of spins forming a system are defined as global states.[2] A **spin system** consists of all particles interacting with each other via dipole-dipole, scalar or J couplings (see Chap. 46). In a solid the interaction network may span the whole sample so that the spin system more or less comprises all spin-bearing particles. In a non-viscous isotropic liquid where molecular motions are fast enough, the anisotropic parts of the interactions average out over the time scale of magnetic resonance so that the size of the spin systems may be restricted to a molecule or a molecular group with a few spin-bearing nuclei.

Isolated spins underlie Zeeman interaction to the external (but potentially internally modified or partly shielded) magnetic field. Another coupling to which isolated spins $I > 1/2$ can be subject is interaction of the electric quadrupole moment of the spin-bearing nucleus with electric field gradients originating from the surrounding charge distribution (see Table 46.1 on page 420).

Although occurring with isolated spins, quadrupolar echoes of spins $I = 1$ are of a similar nature to those of dipolar coupled equivalent spins 1/2. The reason is that the interaction Hamiltonians have a common analytical structure as concerns the operator composition (see Sect. 46.3). Moreover, note that dipolar coupled spin 1/2 pairs form triplet states with a total spin of $F = 1$ and singlet states with a total spin $F = 0$. Only the triplet states contribute to NMR signals and are therefore responsible for any visible phenomena such as spin echoes. Thus the effective spin quantum numbers are the same.

In all cases considered here, we assume that the **quantization direction** is that of the external magnetic field, i.e., we are dealing with "high-field cases" only.[3]

---

[2] In this context, one differentiates **equivalent** from **inequivalent** spins. Equivalent spins refer to quantum-mechanically indistinguishable particles. In particular this means that the magnetic quantum numbers of the individual system members cannot be "good."

[3] In low-field or even zero-field situations, the axis of quantization is determined by the coupling tensor itself. The dominating Hamiltonian may then be due to dipole-dipole or quadrupo-

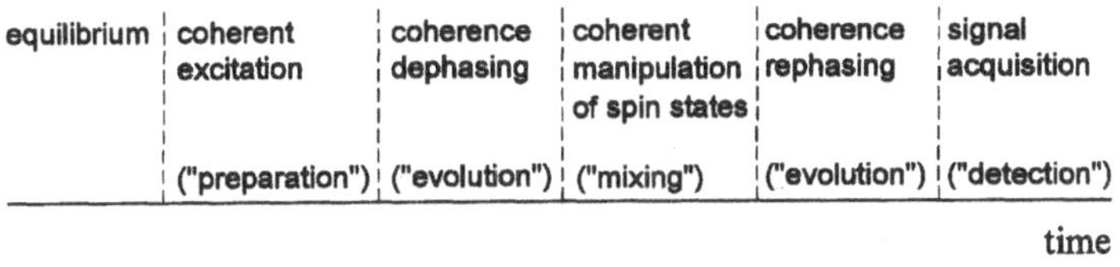

equilibrium	coherent excitation	coherence dephasing	coherent manipulation of spin states	coherence rephasing	signal acquisition
	("preparation")	("evolution")	("mixing")	("evolution")	("detection")

time

Fig. 1.1. Scheme of the intervals of spin manipulation in a typical spin-echo experiment

A typical spin-echo experiment consists of a series of different intervals of spin manipulations. Figure 1.1 shows a simple scheme which represents most species of spin echoes.[4]

The experiment usually begins with the spins in thermal equilibrium. An RF pulse sequence is applied, the first pulse(s) of which coherently excite(s) the spins. RF pulses applied in the "preparation interval" thus serve the generation of appropriate spin coherences. This does not necessarily mean single-quantum coherences (or transverse magnetization). It may also be of interest that, with coupled spins, multiple-quantum coherences can be produced by appropriate pulse sequences within the preparation interval.

The next interval is usually characterized by the absence of any RF pulses, so that the spin coherences evolve freely. This means that the coherences are dephased by distributions of the spin interactions. The purpose of the subsequent "mixing interval" is to initiate the refocusing process, eventually leading to an echo in the free-evolution interval thereafter.

There are two ways of spin manipulation serving this end:

- the phases of the **spin states** can be changed coherently by one or more RF pulses so that coherences lagging behind are promoted and coherences hurrying on ahead are placed back.
- the sign of the **Hamiltonian** responsible for dephasing of the coherences can be reversed so that the coherences evolve "backwards."

---

lar interactions. Spin echoes can also be produced under such circumstances. A quite conventional experimental procedure employing the usual RF pulse sequences is feasible in **pure nuclear quadrupole resonance (NQR)** [108], for example, provided that the NQR frequencies are high enough for direct signal detection. More recently **zero-field experiments** with quadrupolar [29] as well as with dipolar [513] coupled nuclei have been suggested on the basis of the field-cycling technique (see Chap. 15). In this case, coherent zero-field states are produced by non-adiabatically switching the magnetic field off or by applying a d.c. "90° pulse" while the external magnetic field is off. The evolution of the zero-field coherences is then probed by non-adiabatically switching back to a high magnetic field after a certain interval, or by another d.c. "90° pulse. " In the high detection field, the magnetization into which the zero-field coherences have been transferred can be detected conventionally. Zero-field spin echoes can be generated by applying weak d.c. pulses of the magnetic field in the zero-field interval. That is, the RF pulses of conventional echo experiments are substituted by d.c. field pulses corresponding to the "zero carrier frequency" of the experiment.

[4]With rotary echoes (see Chap. 3), for instance, the terminology used for the intervals must be understood in a more figurative sense.

Examples of the former category are all sorts of Hahn spin echoes, whereas gradient-recalled echoes or magic echoes are representative of the latter.

After the refocusing interval, the spin echo arises and the spin coherences are reestablished. If the coherences are of the single-quantum type, an echo signal can be acquired in the "detection interval." Echoes of multiple-quantum coherences remain invisible and must first be transferred to single-quantum coherences by further RF pulses before they can be detected as an indirect signal.

CHAPTER 2

Isolated Spins in Inhomogeneous Fields

In this chapter, the evolution of coherences of uncoupled spins in inhomogeneous magnetic fields is treated. The attenuation of echo signals by irreversible effects such as relaxation or translational diffusion is not yet considered, but will be discussed in detail in Part II.

2.1
Two-Pulse Hahn Echo

Hahn spin echoes [186] are the response of primarily isolated spins to RF pulse sequences of the type displayed in Figs. 2.1 and 2.2. For simplicity, all couplings, except the Zeeman interaction to an (inhomogeneous) magnetic field, are assumed to be negligible or are ignored, i.e., we concentrate on the basic principle of this sort of echo and defer the potential modification of its appearance by other interactions until later.

The coherences prepared by an initial RF pulse are dephased by magnetic field inhomogeneities so that the total FID disappears faster than expected from the attenuation by relaxation or translational diffusion alone. It can easily be perceived that dephasing of coherences by a (stationary) distribution of local Larmor frequencies is a **reversible** process, whereas attenuation by relaxation or translational diffusion is **irreversible**.[1] In this chapter we merely consider the reversible atten-

---

[1] If the magnetic field is particularly homogeneous so that coherences exist for a relatively long time without getting spoiled by inhomogeneities, so-called **"radiation damping"** [1, 45] comes into play as a third sort of irreversible attenuation mechanisms: Coherences manifest themselves as a precessing magnetization which again produces a rotating magnetic field. Hence, an oscillating voltage is induced in the pick-up RF coil. The corresponding current in the resonance circuit causes a secondary oscillating magnetic field which tends to drive the magnetization vector towards the z direction. The term "damping" is justified because Zeeman energy is dissipated in the resistive elements of the probehead circuit. One should, however, bear in mind that the magnitude of the magnetization is not changed, and the transverse component may even rise if the initial flip angle was greater than 90°.

It is also known that in probeheads with high quality factor Q, back action of the coil on the evolution of spin coherences can arise [310] leading to "artifacts" in high-field high-resolution NMR. An operational remedy is the application of dephasing/rephasing field-gradient pairs in the non-acquisition intervals of the RF pulse sequence so that coherences are spoiled when not needed for acquisition. Other possibilities are (i) gated tuning of the probehead during the RF pulses and acquisition only [307]; (ii) feedback suppression of high currents in the RF coil except during the excitation pulses [64]; (iii) the use of composite RF pulses [504].

uation of coherences. Irreversible phenomena will be covered in Part II of this book.

Let us now consider an ensemble of isolated spins I in a magnetic field

$$B(r) = B_0 + G \cdot r \tag{2.1}$$

where $G = \nabla B(r)$ is the field gradient at position r. The field gradient is to stand for all inhomogeneities of the field irrespective of the provenance. All field components perpendicular to the z axis can safely be neglected. The Hamiltonian during RF irradiation can then be written as (see Chap. 46)

$$\mathcal{H} = -\gamma_n \hbar I_z (B_0 + G \cdot r) - 2\gamma_n \hbar B_1 I_x \cos \omega_c t \tag{2.2}$$

where we have assumed that the RF flux density oscillates along the x axis of the laboratory frame. The carrier frequency is assumed to be resonant at the position $r = 0$ so that $\omega_c = \omega_0 = \gamma_n B_0$. As outlined in Sect. 48.8, the transformation to the reference frame rotating with ω_c about the z axis leads to the effective Hamiltonian

$$\mathcal{H}_e = \underbrace{-\gamma_n \hbar I_z (G \cdot r)}_{\mathcal{H}_0'} \underbrace{-\gamma_n \hbar B_1 I_x}_{\mathcal{H}_{rf}'} \tag{2.3}$$

where the spin operator component I_x now refers to the rotating frame. The rapidly oscillating component of the RF field practically averages out and has therefore been discarded. In the high-temperature approximation, the equilibrium density operator at position r is according to Eq. 47.24

$$\rho_0 = \frac{1 + \gamma_n \hbar B(r) I_z / (k_B T)}{\mathrm{Tr}\{\mathcal{E}\}} \approx \frac{1 + \gamma_n \hbar B_0 I_z / (k_B T)}{\mathrm{Tr}\{\mathcal{E}\}}$$
$$\equiv a + b I_z \tag{2.4}$$

where $a = 1/\mathrm{Tr}\{\mathcal{E}\}$ and $b = \gamma_n \hbar B_0 / (k_B T \mathrm{Tr}\{\mathcal{E}\})$.

Figure 2.1 illustrates the sequence producing two-pulse Hahn spin echoes. Symbolically it may be represented by

$$(\pi/2)_x - \tau_1 - (\beta)_y - \tau - \text{(acquisition)} \tag{2.5}$$

The pulse amplitude B_1 is assumed to be much larger than the offset field $G \cdot r$ valid in the rotating frame defined above.[2] The first term in Eq. 2.3 can therefore be neglected as long as the pulses are on. The solution of the rotating-frame Liouville/von Neumann equation, Eq. 48.75, for the first 90° pulse of length $\tau_w = \pi/(2|\gamma_n|B_1)$ is (Sect. 48.5.2)[3]

$$\rho(0+) = e^{-(i/\hbar)\mathcal{H}_{rf}' \tau_w} \rho_0 e^{(i/\hbar)\mathcal{H}_{rf}' \tau_w}$$
$$= a + b I_y \tag{2.6}$$

---

[2] Such pulses are usually designated as "hard."
[3] Instants just before and immediately after an RF pulse are specified by the time at which the pulse occurs supplemented by − and + signs, respectively.

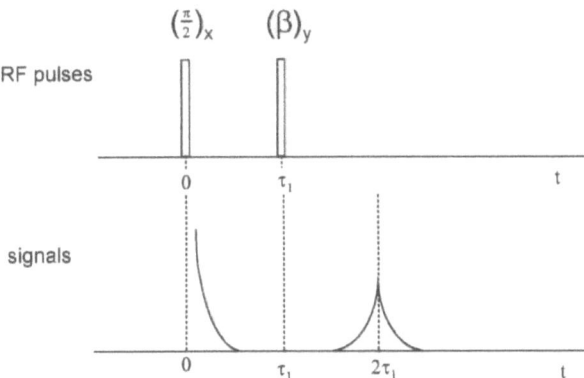

Fig. 2.1. Sequence of two RF pulses generating a Hahn spin echo in inhomogeneous magnetic fields. The scheme at the bottom is to represent an envelope of the expected NMR signals. The flip angle of the second pulse preferably (but not necessarily) is 180°. With coupled spin systems, the second pulse may lead to coherence transfer in addition to the refocusing effect. The pulse phases have been chosen arbitrarily and refer to the treatment in the text. The 90° phase shift between the pulses is assumed in order to have in-phase initial FID and spin-echo coherences.

where we have assumed that the circularly polarized B_1 RF field component rotating synchronously with the rotating frame is aligned along the $+x'$ axis. Equation 2.6 represents coherences with a magnetization aligned along the $+y'$ axis of the rotating frame for positive gyromagnetic ratios (compare Sect. 48.5.3).

Free evolution in the inhomogeneous offset field represented by the first term in Eq. 2.3 leads to the Liouville/von Neumann solution at time $t = \tau_1-$

$$\rho(\tau_1-) = e^{-(i/\hbar)\mathcal{H}'_0\tau_1}\,\rho(0+)\,e^{(i/\hbar)\mathcal{H}'_0\tau_1}$$
$$= a + b\,e^{i\varphi_1 I_z}\,I_y\,e^{-i\varphi_1 I_z}$$
$$= a + b\,(I_y \cos\varphi_1 + I_x \sin\varphi_1) \qquad (2.7)$$

where the local precession angle in the rotating frame is

$$\varphi_1 = \gamma_n\,(\mathbf{G}\cdot\mathbf{r})\,\tau_1 = \Omega\tau_1 \qquad (2.8)$$

The angular frequency $\Omega = \Omega(\mathbf{r})$ represents the local offset from the resonance frequency ω_0.

In order to derive the transverse magnetization we must form the average of the density operator over all offsets $\Omega(\mathbf{r})$ occurring in the spin ensemble:

$$\langle\rho(\tau_1-)\rangle_\Omega = a + b\,\left(I_y\langle\cos\varphi_1\rangle_\Omega + I_x\langle\sin\varphi_1\rangle_\Omega\right) \qquad (2.9)$$

Using Eq. 47.6, we obtain for the complex transverse rotating-frame magnetization

$$m(\tau_1-) = M'_x + iM'_y$$
$$= n\gamma_n\hbar\,\mathrm{Tr}\{\langle\rho(\tau_1-)\rangle_\Omega\,(I_x + iI_y)\}$$
$$= M_0\,\{\langle\sin\varphi_1\rangle_\Omega + i\,\langle\cos\varphi_1\rangle_\Omega\}$$
$$= m(0+)\,\langle e^{-i\varphi_1}\rangle_\Omega \qquad (2.10)$$

where n is the number density of the spins, $m(0+) = iM_0$, and M_0 is the Curie magnetization (see Eq. 47.31).[4] Depending on the distribution of the offset frequencies at different positions the superposition of the local coherences tends to be destructive so that the transverse magnetization cancels.

After the interval τ_1, a second RF pulse is applied with the rotating RF field component arbitrarily chosen along the $+y'$ axis. Assume an unspecified flip angle β for the time being.[5] The density operator immediately after this pulse at the time $t = \tau_1+$ is then

$$
\begin{aligned}
\rho(\tau_1+) &= a + b\,(I_y \cos\varphi_1 + e^{i\beta I_y}\, I_x\, e^{-i\beta I_y}\, \sin\varphi_1) \\
&= a + b\,[I_y \cos\varphi_1 + (I_x \cos\beta + I_z \sin\beta) \sin\varphi_1]
\end{aligned}
\tag{2.11}
$$

At an evolution interval τ later, when time $t = \tau_1 + \tau$, we have

$$
\begin{aligned}
\rho(\tau_1 + \tau) &= a + b\,\big[\, e^{i\varphi I_z}\, I_y\, e^{-i\varphi I_z} \cos\varphi_1 \\
&\quad + \big(e^{i\varphi I_z}\, I_x\, e^{-i\varphi I_z} \cos\beta + I_z \sin\beta\big) \sin\varphi_1 \,\big] \\
&= a + b\,\{\,[I_y \cos\varphi + I_x \sin\varphi]\cos\varphi_1 \\
&\quad + [\,(I_x \cos\varphi - I_y \sin\varphi)\cos\beta + I_z \sin\beta\,]\sin\varphi_1\,\} \\
&= a + b\,\{\,I_x[\sin\varphi \cos\varphi_1 + \cos\varphi \sin\varphi_1 \cos\beta] \\
&\quad + I_y\,[\cos\varphi \cos\varphi_1 - \sin\varphi \sin\varphi_1 \cos\beta] \\
&\quad + I_z \sin\varphi_1 \sin\beta\,\}
\end{aligned}
\tag{2.12}
$$

where

$$
\varphi = \gamma_n\,(\mathbf{G}\cdot\mathbf{r})\,\tau = \Omega\tau
\tag{2.13}
$$

The average over all offsets $\Omega(\mathbf{r})$ (i.e., over all "isochromats") in the spin ensemble,

$$
\begin{aligned}
\langle\rho(\tau_1 + \tau)\rangle_\Omega &= a + b\,\{\,I_x[\langle\sin\varphi \cos\varphi_1\rangle_\Omega + \langle\cos\varphi \sin\varphi_1\rangle_\Omega \cos\beta] \\
&\quad + I_y[\langle\cos\varphi \cos\varphi_1\rangle_\Omega - \langle\sin\varphi \sin\varphi_1\rangle_\Omega \cos\beta] \\
&\quad + I_z\,\langle\sin\varphi_1\rangle_\Omega \sin\beta\,\}
\end{aligned}
\tag{2.14}
$$

leads to the complex transverse magnetization in the rotating frame (see Eq. 47.6)

$$
m(\tau_1 + \tau) = n\gamma_n\hbar\,\mathrm{Tr}\{\langle\rho(\tau_1 + \tau)\rangle_\Omega(I_x + iI_y)\}
$$

---

[4]Generalizing this result leads to a relation of the local complex transverse magnetization after a free evolution interval t to the initial complex transverse magnetization:

$$
m(r, t) = m(r, 0)e^{-i\varphi(r,t)}
$$

where $\varphi(r, t) = \int_0^t \Omega(r, t')\, dt'$.

[5]Flip angles are given throughout as magnitude values of the actual precession angle about the effective field in the rotating frame.

$$
\begin{aligned}
= \;& \frac{M_0}{2}\{\langle\sin(\varphi-\varphi_1)\rangle_\Omega + \langle\sin(\varphi+\varphi_1)\rangle_\Omega \\
& + \langle\sin(\varphi_1-\varphi)\rangle_\Omega \cos\beta + \langle\sin(\varphi_1+\varphi)\rangle_\Omega \cos\beta \\
& + i\,(\langle\cos(\varphi-\varphi_1)\rangle_\Omega + \langle\cos(\varphi+\varphi_1)\rangle_\Omega \\
& - \langle\cos(\varphi-\varphi_1)\rangle_\Omega \cos\beta + \langle\cos(\varphi+\varphi_1)\rangle_\Omega \cos\beta)\}
\end{aligned}
\tag{2.15}
$$

The **two-pulse Hahn spin echo** reveals itself as the maximum of the $\langle\cos(\varphi-\varphi_1)\rangle_\Omega$ terms. The condition for the maximum is obviously

$$
\varphi_1 = \varphi \qquad\qquad \text{or} \qquad\qquad \tau = \tau_1 \tag{2.16}
$$

If the B_0 gradient is strong enough, a broad distribution of the phase angles can be expected, so that $\langle\cos(\varphi+\varphi_1)\rangle_\Omega \approx \langle\sin(\varphi+\varphi_1)\rangle_\Omega \approx 0$. The amplitude of the two-pulse Hahn spin echo is thus

$$
m(2\tau_1) = i\,\frac{M_0}{2}\,(1-\cos\beta) = i\,M_0 \sin^2\frac{\beta}{2} \tag{2.17}
$$

With the RF pulse phases assumed here, the transverse magnetization is aligned along the y axis of the rotating frame, that is in-phase to the initial FID.

If the flip angle β of the second RF pulse is adjusted equal to 180° the initial coherences are fully refocused (apart from potential relaxation and diffusion losses). If $\beta = 90°$, only half the initial transverse magnetization is recovered. The other half is transferred to the z' direction by the second RF pulse as discussed in the following section.

Note that in principle two-pulse spin echoes appear with any combination of flip angles with the exception of integer multiples of 180° for the first and integer multiples of 360° for the second pulse. If the first pulse is not a 90° pulse as assumed above, the transverse magnetization is reduced by a factor of $\sin\alpha$ where α is the flip angle.

The phases of the RF pulses are also inessential for the echo formation process. The coherences tend to be dephased already when the second RF pulse is applied so that all coherence phases relative to the pulse phase occur anyway. If the two pulses were chosen in-phase instead of in quadrature, the initial FID and the spin echo were antiphase in contrast to the above result, but the signal amplitudes and the flip angle dependences were the same.

2.2
Three-Pulse Hahn Echoes

Let us now turn to Fig. 2.2 illustrating the pulse sequence

$$
(\pi/2)_x \;-\; \tau_1 \;-\; (\pi/2)_y \;-\; \tau_2 \;-\; (\pi/2)_y \;-\; \tau \;-\; \text{(acquisition)} \tag{2.18}
$$

Continuing the treatment with Eq. 2.12 and setting the flip angle of the second pulse $\beta = \pi/2$, we obtain at the end of the second interval, at the time $t = \tau_1 + \tau_2-$

$$
\rho(\tau_1 + \tau_2-) = a + b\,\{I_x \cos\varphi_1 \sin\varphi_2 + I_y \cos\varphi_1 \cos\varphi_2 + I_z \sin\varphi_1\} \tag{2.19}
$$

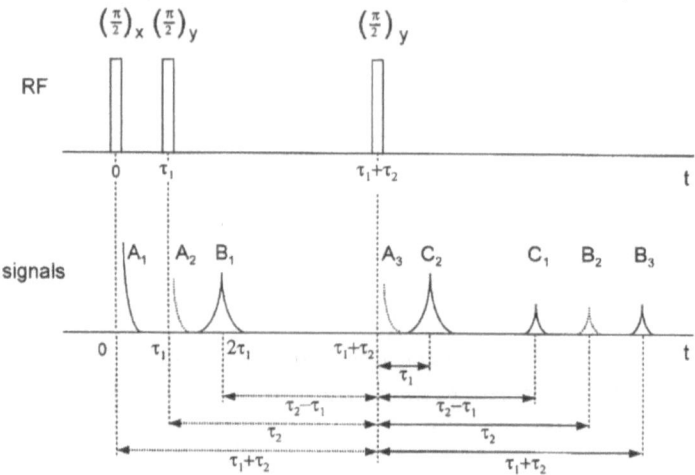

Fig. 2.2. Hahn spin echo envelopes produced by three RF pulses. The flip angles preferably (but not necessarily) are 90°. The pulse phases were arbitrarily chosen in order to get the FIDs and echoes in-phase. A prerequisite for the production of distinct echo signals is a sufficiently inhomogeneous magnetic field. The nomenclature of the indicated signals is: single-pulse signals are labeled by the letter A, two-pulse (or primary) echoes by the letter B, and three-pulse echoes by the letter C. That is, A_1, A_2, A_3 are **free-induction decays** immediately after the RF pulses; B_1, B_2, B_3 are **primary spin echoes** formed by the second or third RF pulses from any of the preceding free-induction decays; C_1 is the **secondary echo** formed by the third RF pulse from the primary echo B_1; C_2 is the **stimulated echo**. With ideal 90° RF pulses, the free-induction decays A_2 and A_3 and – consequently – the echo B_2, refer solely to z magnetization restored by spin-lattice relaxation in the pulse intervals. Otherwise, these signals (dotted lines) may also be indicative of imperfections of the 90° flip angles.

The local coherence phase is shifted in the second interval by the angle

$$\varphi_2 = \gamma_n \, (G \cdot r) \, \tau_2 = \Omega \tau_2 \tag{2.20}$$

The third 90° pulse produces

$$\rho(\tau_1 + \tau_2 +) = a + b\{- I_x \sin \varphi_1 + I_y \cos \varphi_1 \cos \varphi_2 + I_z \cos \varphi_1 \sin \varphi_2\} \tag{2.21}$$

The subsequent free evolution during a period τ leads to

$$\begin{aligned}
\rho(\tau_1 + \tau_2 + \tau) = a + b\{&-(I_x \cos \varphi - I_y \sin \varphi) \sin \varphi_1 \\
&+ (I_y \cos \varphi + I_x \sin \varphi) \cos \varphi_1 \cos \varphi_2 \\
&+ I_z \cos \varphi_1 \sin \varphi_2\}
\end{aligned} \tag{2.22}$$

where the coherences are subject to further local phase shifts

$$\varphi = \gamma_n \, (G \cdot r) \, \tau = \Omega \tau \tag{2.23}$$

Forming the average over all offsets $\Omega(r)$,

$$\langle \rho(\tau_1 + \tau_2 + \tau) \rangle_\Omega = a + b \; \{ I_x \, \langle \sin \varphi \cos \varphi_1 \cos \varphi_2 - \cos \varphi \sin \varphi_1 \rangle_\Omega$$
$$+ I_y \, \langle \sin \varphi \sin \varphi_1 + \cos \varphi \cos \varphi_1 \cos \varphi_2 \rangle_\Omega$$
$$+ I_z \, \langle \cos \varphi_1 \sin \varphi_2 \rangle_\Omega \} \tag{2.24}$$

we obtain for the complex transverse magnetization in the rotating frame (see Eq. 47.6)

$$m(\tau_1 + \tau_2 + \tau) = n \gamma_n \hbar \, \mathrm{Tr}\{ \langle \rho \rangle_\Omega (\tau_1 + \tau_2 + \tau)(I_x + iI_y) \}$$
$$= \frac{M_0}{4} \{ \langle \sin(\varphi + \varphi_1 - \varphi_2) \rangle_\Omega - \langle \sin(\varphi_1 + \varphi_2 - \varphi) \rangle_\Omega$$
$$+ \langle \sin(\varphi + \varphi_2 - \varphi_1) \rangle_\Omega + \langle \sin(\varphi + \varphi_1 + \varphi_2) \rangle_\Omega$$
$$-2 \, \langle \sin(\varphi + \varphi_1) \rangle_\Omega + 2 \langle \sin(\varphi - \varphi_1) \rangle_\Omega$$
$$+ i \, [2 \langle \cos(\varphi - \varphi_1) \rangle_\Omega - 2 \langle \cos(\varphi + \varphi_1) \rangle_\Omega$$

$$\underbrace{\qquad}_{\rightsquigarrow \, C_2}$$

$$+ \underbrace{\langle \cos(\varphi + \varphi_1 - \varphi_2) \rangle_\Omega}_{\rightsquigarrow \, C_1 \; \text{if} \; \varphi_2 > \varphi_1} + \underbrace{\langle \cos(\varphi_1 + \varphi_2 - \varphi) \rangle_\Omega}_{\rightsquigarrow \, B_3}$$

$$+ \underbrace{\langle \cos(\varphi + \varphi_2 - \varphi_1) \rangle_\Omega}_{\rightsquigarrow \, C_1 \; \text{if} \; \varphi_2 < \varphi_1} + \langle \cos(\varphi + \varphi_1 + \varphi_2) \rangle_\Omega] \} \tag{2.25}$$

The terms underbraced in the expression above indicate non-destructive superpositions of coherences for vanishing arguments of the cosine functions. At the corresponding evolution times additional spin echoes appear.

Altogether a three-pulse sequence can produce five spin echoes.

a) A **primary echo** (B_1 in Fig. 2.2) after the first two pulses at time $t = 2\tau_1$, which we already know from Sect. 2.1. The only prerequisite is that $\tau_2 > \tau_1$, so that the third RF pulse does not interfere with the echo formation. The amplitude of the unrelaxed complex transverse magnetization is (Eq. 2.17)

$$m(2\tau_1) = \frac{i}{2} M_0 \tag{2.26}$$

The other four echoes appear as a result of the action of the third RF pulse.

b) The **stimulated echo**[6] is marked by the symbol C_2 in Fig. 2.2 and occurs under the condition that

$$\varphi = \varphi_1 \qquad \text{or} \qquad \tau = \tau_1 \tag{2.27}$$

Its amplitude in terms of the unrelaxed complex transverse magnetization is

$$m(2\tau_1 + \tau_2) = \frac{i}{2} M_0 \tag{2.28}$$

---

[6]Note that we comply with Hahn's original definition of the stimulated echo [186] which refers to isolated spins in inhomogeneous magnetic fields. Coherence-transfer or spin-order transfer effects generated in coupled spin systems are therefore considered as phenomena of a different physical origin even if the echo-time schedules coincide (see Sect. 7.2, for instance). In order to avoid confusion, the term "stimulated echo" should therefore not be used under such circumstances.

The stimulated echo represents components of transverse-magnetization iso-chromats "stored" by the second pulse in z' direction[7] during the interval τ_2 (last density operator term in Eq. 2.19, $I_z \sin \varphi_1$). These magnetization isochromat components are "read out" and refocused by the third RF pulse as the stimulated echo.

c) The **secondary echo** (C_1 in Fig. 2.2) appears for

$$\varphi = |\varphi_2 - \varphi_1| \quad \text{i.e.,} \quad \tau = \tau_2 - \tau_1 > 0 \tag{2.29}$$

The unrelaxed echo amplitude is

$$m(2\tau_2) = \frac{i}{4} M_0 \tag{2.30}$$

The corresponding signal arises after the coherences of the primary echo (B_1 in Fig. 2.2) have been refocused by the third RF pulse. The secondary echo can be traced back to the density operator term $I_y \cos \varphi_2 \cos \varphi_1$ in Eq. 2.19.

d) Another **primary echo** (B_3 in Fig. 2.2) emerges for

$$\varphi = \varphi_1 + \varphi_2 \quad\quad \text{or} \quad\quad \tau = \tau_1 + \tau_2 \tag{2.31}$$

The amplitude of the unrelaxed complex transverse magnetization at this instant is

$$m(2\tau_1 + 2\tau_2) = \frac{i}{4} M_0 \tag{2.32}$$

This echo is formed by refocusing the part of the initial coherences (A_1 in Fig. 2.2) that has neither been "stored" in z' direction nor refocused by the second RF pulse. It originates from the density operator term $I_x \sin \varphi_2 \cos \varphi_1$ in Eq. 2.19.

e) The fifth echo phenomenon that may appear in this context, the **primary echo** B_2 in Fig. 2.2, is not included in the above treatment because ideal 90° pulses were assumed, and spin-lattice relaxation was neglected. In reality none of these conditions is fulfilled exactly. As a consequence, an FID arises after the second RF pulse (A_2 in Fig. 2.2). These coherences are refocused by the third RF pulse for

$$\varphi = \varphi_2 \quad\quad \text{or} \quad\quad \tau = \tau_2 \tag{2.33}$$

leading to a further echo of primary nature (B_2 in Fig. 2.2) at time $t = \tau_1 + 2\tau_2$. The echo amplitude depends on the degree of the pulse imperfections and relaxational recovery during the first pulse interval.

2.3
Gradient-Recalled Echo

The gradient recalled echo is of particular interest in context with soft-pulse slice excitation (Sect. 25.1.3) and fast imaging techniques (section 25.4.3). The coherences

---

[7]This does not mean that a net z magnetization arises!

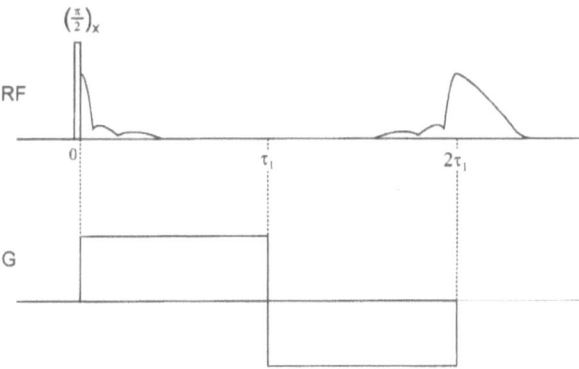

Fig. 2.3. Typical RF and field-gradient pulse sequence for the formation of a gradient echo (drawn as a schematic envelope).

are dephased by field gradients which are deliberately applied with these methods. The application of a subsequent gradient pulse with opposite sign (see Figs. 2.3 and 25.1) reverses the dephasing effect, that is, the dephasing Hamiltonian, so that an echo appears.

The initial 90° RF pulse is assumed to excite the sample homogeneously. Field gradients due to magnet inhomogeneities are neglected relative to the externally controlled gradient pulses. At the time $t = \tau_1$ the density operator thus takes the form of Eq. 2.7 where the field gradient G defining the dephasing angle φ_1, Eq. 2.8, is equated with the externally controlled, spatially constant gradient G:

$$\rho(z, \tau_1) = e^{i\varphi_1 I_z}\, \rho(z, 0+)\, e^{-i\varphi_1 I_z} \tag{2.34}$$

The gradient direction is arbitrarily assumed parallel to the z axis, so that the dephasing angle at a position z is

$$\varphi_1 = \varphi_1(z, \tau_1) = \gamma_n\, G\, z\, \tau_1 \tag{2.35}$$

At the time $t = \tau_1$ the gradient direction is reversed[8] so that the evolution of the spin coherences is also reversed, and the coherences begin to rephase in a subsequent interval τ by a precession phase angle

$$\varphi = \varphi(z, \tau) = -\gamma_n\, G\, z\, \tau \tag{2.36}$$

The local density operator evolves according to

$$\begin{aligned} \rho(z, \tau_1 + \tau) &= e^{i\varphi I_z}\, \rho(z, \tau_1)\, e^{-i\varphi I_z} \\ &= a + b\, \left[I_x \sin\Delta\varphi + I_y \cos\Delta\varphi \right] \end{aligned} \tag{2.37}$$

---

[8]That is, the sign of the Hamiltonian is reversed, whereas with the Hahn-echo pulse sequences the coherences and their phases are changed.

where the net dephasing angle at time $t = \tau_1 + \tau$ is

$$\Delta\varphi(z, \tau_1 + \tau) = \varphi_1 - \gamma_n G z \tau = \gamma_n G z (\tau_1 - \tau) \tag{2.38}$$

For $\tau = \tau_1$ the density operator becomes independent of z owing to $\Delta\varphi = 0$. The complex transverse magnetization is then

$$m(2\tau_1) = m(0+) = i M_0 \tag{2.39}$$

This is the maximum of the gradient-recalled echo, recovering the whole initial transverse magnetization in the absence of relaxation. The transverse magnetization is aligned along the y axis of the rotating frame for the RF pulse phase assumed in this treatment.

2.4
Multiple Echoes

Apart from Hahn echoes of the type described in Sects. 2.1 and 2.2, "multiple echoes" can appear after the same RF pulse sequences provided that a field gradient is present. Under suitable circumstances, a series of echoes then follows the ordinary Hahn echo in multiples of the pulse spacing [58, 122, 138].

With the formalisms depicted in Sects. 2.1 and 2.2, Hahn echoes were shown to be the consequence of the coherence evolution of static nuclei in stationary field gradients. In these treatments it was tacitly assumed that the Larmor frequency is *linearly* related to the externally applied inhomogeneous field at the position of the nucleus (see Eq. 2.1) at any time:

$$\omega = \gamma_n(B_0 + \boldsymbol{G} \cdot \boldsymbol{r}) \tag{2.40}$$

However, in the presence of field gradients this relation becomes a *nonlinear* function of the gradient after the second RF pulse owing to the influence of the demagnetizing field in the sample.[9]

At the beginning of the spin-echo experiment, the demagnetization field is uniform because the magnetization M is generated by the *external* magnetic field. The *internal* field in the sample is reduced by an amount proportional to the external flux density and depending on the sample shape. If the sample may be approached by a homogeneous ellipsoid, the internal magnetic flux density is

$$B_i = B_0 - \frac{\mu_0}{4\pi} N M \tag{2.41}$$

where N is the demagnetization factor.[10]

---

[9]We are dealing here with two-pulse sequences. Recently three-pulse echo phenomena also related to the demagnetizing field, namely the multiple **nonlinear stimulated echoes (NOSE)** were discovered [12].

[10]The demagnetization factor of a long cylinder with the axis aligned along B_0 vanishes; that of a thin slab to which B_0 is normal takes the value 4π.

The situation changes after the second RF pulse of a sequence is applied in the presence of a field gradient: the initially uniform spatial distribution of the magnetization and, hence, that of the internal magnetic flux density are modified by the coherence evolution in the first pulse interval and the action of the second pulse.

As an example, we treat the same pulse sequence as in Sect. 2.1 (Fig. 2.1). The field gradient G is assumed to be spatially constant and aligned along the z direction, that is $G_z = G$ where $B_{0,z} = B_0$. The density operator of the nuclei, situated at positions with the coordinate z, just after the second RF pulse is then (see Eq. 2.11)

$$\rho(z, \tau_1+) = a + b\left[I_x \cos\beta \sin\varphi_1(z) + I_y \cos\varphi_1(z) + I_z \sin\beta \sin\varphi_1(z)\right] \quad (2.42)$$

where

$$\varphi_1(z) = \gamma_n G z \tau_1 = \Omega_G \tau_1 = k_1 z \quad (2.43)$$

The local, gradient-induced frequency offset from resonance is

$$\Omega_G = \gamma_n G z \quad (2.44)$$

Furthermore, a "wave number" describing the spatial modulation of the phase shift φ_1 along the z axis may be defined as

$$k_1 = \gamma_n G \tau_1 \quad (2.45)$$

The three terms in the brackets of Eq. 2.42 correspond to the three magnetization components. We note that all of these are spatially modulated. The z component forms a kind of "magnetization grid,"

$$\begin{aligned} M_z(z, \tau_1+) &= n\gamma_n\hbar \operatorname{Tr}\{\rho(z, \tau_1+)I_z\} \\ &= M_0 \sin\beta \sin(\Omega_G\tau_1) \\ &= M_0 \sin\beta \sin(k_1 z) \end{aligned} \quad (2.46)$$

with extrema at positions

$$z = \pm\frac{(2j+1)\pi}{2k_1} \quad (j = 0, 1, 2, \ldots) \quad (2.47)$$

At these positions, the z component of the magnetization is alternatingly positive and negative.[11]

As the magnetization is now distributed in the form of a "grid," the demagnetizing field generated by the magnetization will be non-uniform too. In principle one can calculate the demagnetizing flux density for any given spatial distribution of the magnetization by first considering the magnetic flux density of the magnetic dipole

---

[11]Note that the spatial modulation of the z magnetization obviously disappears if the flip angle of the second pulse is $\beta = 180°$, but is maximal for $\beta = 90°$.

moment $M(r')\,d^3r'$ of the volume element d^3r' at the position r'. The corresponding dipole field of this infinitesimal dipole at a position r is

$$dB_\delta(r) = \frac{\mu_0}{4\pi}\left(\frac{3[(r-r')\cdot M(r')](r-r')}{|r-r'|^5} - \frac{M(r')}{|r-r'|^3}\right)d^3r' \qquad (2.48)$$

The total demagnetizing flux density at r is then

$$B_\delta(r) = \frac{\mu_0}{4\pi}\int\left(\frac{3[(r-r')\cdot M(r')](r-r')}{|r-r'|^5} - \frac{M(r')}{|r-r'|^3}\right)d^3r' \qquad (2.49)$$

In the case assumed here, i.e., rotational symmetry of the spatial magnetization distribution around B_0, the demagnetizing field component relevant for coherence evolution at a position z can be shown to be that produced by a thin slab normal to B_0. That is [122]

$$B_\delta(z) = \mu_0 M_z(z) u_z \qquad (2.50)$$

where u_z is the unit vector of the z axis. The total offset of the magnetic flux density at a position z is thus

$$\begin{aligned}\Delta B(z) &= B_\delta(z) + Gz = \mu_0 M_z(z) + Gz \\ &= \mu_0 M_0 \sin\beta\,\sin(\Omega_G\tau_1) + Gz \end{aligned} \qquad (2.51)$$

The spatially modulated spin coherences now evolve in the spatially modulated magnetic field which is stationary in the absence of relaxation and diffusion. The sandwich propagator expression for the interval τ after the second RF pulse is

$$\rho(z,\tau_1+\tau) = e^{i\varphi I_z}\rho(z,\tau_1+)e^{-i\varphi I_z} \qquad (2.52)$$

where

$$\begin{aligned}\varphi = \varphi(z) &= \gamma_n\,[\,B_{\delta,z}(z) + Gz\,]\,\tau \\ &= \gamma_n\,[\,\mu_0 M_0\sin(\Omega_G\tau_1) + Gz\,]\,\tau \\ &= (\Omega_\delta + \Omega_G)\,\tau \end{aligned} \qquad (2.53)$$

The local frequency offset by the demagnetizing field is

$$\Omega_\delta = \gamma_n\mu_0 M_0\sin(\Omega_G\tau_1) \qquad (2.54)$$

It is obviously a nonlinear function of the gradient offset Ω_G.

Setting $\beta = \pi/2$ in Eq. 2.42 for maximal efficiency, we obtain

$$\begin{aligned}\rho(z,\tau_1+\tau) &= a + b\left\{I_x\sin\varphi\cos\varphi_1 + I_y\cos\varphi\cos\varphi_1 + I_z\sin\varphi_1\right\} \\ &= a + b\left\{\frac{1}{2}I_x\left[\sin(\varphi-\varphi_1)+\sin(\varphi+\varphi_1)\right]\right. \\ &\quad\left. + \frac{1}{2}I_y\left[\cos(\varphi-\varphi_1)+\cos(\varphi+\varphi_1)\right] + I_z\sin\varphi_1\right\} \\ &= a + b\left\{\frac{1}{4i}I^+\left[e^{i[\Omega_G(\tau-\tau_1)+\Omega_\delta\tau]} + e^{i[\Omega_G(\tau+\tau_1)+\Omega_\delta\tau]}\right]\right. \\ &\quad\left. + cc + I_z\sin(\Omega_G\tau_1)\right\} \end{aligned} \qquad (2.55)$$

For the signal of the whole sample, we must take the average over all positions. Strong gradients lead to an equipartition of the phase angles $\varphi(z) + \varphi_1(z)$ and $\varphi(z)$, so that the averages of the above terms are zero unless they are independent of the position. The only term which can comply with this condition in principle is the factor $\exp\{i(\Omega_G\tau - \Omega_G\tau_1 + \Omega_\delta\tau)\}$ for which one can expect that the phase shifts vanish under certain circumstances.

This becomes obvious by analyzing this phase factor into a Fourier series:[12]

$$e^{i\Omega_G\tau}\,e^{-i(\Omega_G\tau_1-\Omega_\delta\tau)} = e^{i\Omega_G\tau}\sum_{\ell=-\infty}^{\infty} a_\ell\,e^{-i\ell\,\Omega_G\tau_1} \tag{2.56}$$

where

$$a_\ell = \frac{1}{2\pi}\int_{-\pi}^{\pi} e^{i(\Omega_\delta\tau-\Omega_G\tau_1)}\,e^{i\ell\,\Omega_G\tau_1}\,\mathrm{d}(\Omega_G\tau_1) \tag{2.57}$$

The average of the right-hand side of Eq. 2.56 over all positions, that is, all phase angles $\Omega_G\tau_1$, vanishes unless

$$\Omega_G\tau = \ell\,\Omega_G\tau_1 \qquad \text{or} \qquad \tau_1 = \frac{\tau}{\ell} \tag{2.58}$$

In this case, the phase factor before the sum in Eq. 2.56 cancels the corresponding term in the sum. The average over all positions is then no longer subject to destructive superposition of the coherences, and we can expect multiple echoes after the second RF pulse in intervals of integer multiples of the pulse spacing. Note that the demagnetization offset, Ω_δ, normally varies weakly with the position compared with Ω_G, so that the average of the coefficients a_ℓ remains finite. The **ordinary two-pulse Hahn echo** appears for $\ell = 1$ with the relative amplitude

$$a_1 = \frac{\Omega_G}{2\pi}\int_{-\pi/\Omega_G}^{\pi/\Omega_G} e^{i\Omega_\delta\tau_1}\,\mathrm{d}\tau_1 = \frac{1}{\pi}\frac{\Omega_G}{\Omega_\delta}\sin\left(\pi\frac{\Omega_\delta}{\Omega_G}\right) = \mathrm{sinc}\left(\frac{\Omega_\delta}{\Omega_G}\right) \approx 1 \tag{2.59}$$

where the sinc function is defined by

$$\mathrm{sinc}(x) \equiv \sin(\pi x)/(\pi x) \tag{2.60}$$

The ordinary Hahn echo is scarcely influenced by the demagnetizing field because the demagnetization offset tends to obey $\Omega_\delta \ll \Omega_G$. However, the higher-order echoes are pure refocusing effects owing to the demagnetizing-field. The first "non-Hahn" echo of this sort, that is, the **second-order multiple echo** arising for $\ell = 2$, for instance, has a relative amplitude

$$a_2 = \frac{\Omega_G}{2\pi}\int_{-\pi/\Omega_G}^{\pi/\Omega_G} e^{i(2\Omega_\delta+\Omega_G)\tau_1}\,\mathrm{d}\tau_1$$

---

[12]Recall that Ω_δ is a nonlinear function of Ω_G. This is represented here by the Fourier expansion in harmonics of Ω_G.

$$= \frac{1}{\pi} \frac{\Omega_G}{2\Omega_\delta + \Omega_G} \sin\left(\pi \frac{2\Omega_\delta + \Omega_G}{\Omega_G}\right) = \text{sinc}\left(\frac{2\Omega_\delta + \Omega_G}{\Omega_G}\right) \qquad (2.61)$$

Generally, the amplitude of the echo of order ℓ is

$$a_\ell = \text{sinc}\left(\frac{\ell\,\Omega_\delta + (\ell - 1)\Omega_G}{\Omega_G}\right) \qquad (2.62)$$

The amplitudes of multiple echoes depend solely on how the frequency offset due to the demagnetizing field compares with that determined by the gradient. These phenomena disappear if either there is no finite field gradient, or in the limit of low Curie magnetization, that is, for low external magnetic fields, low gyromagnetic ratio, low spin density, and/or high temperature. As multiple echoes represent refocused coherences, they are subject to attenuation by transverse relaxation and translational diffusion as is the case with all other echoes considered up to now.

CHAPTER 3

Rotary Echoes

There is a great variety of spin-coherence evolution experiments in the rotating frame which are quite analogous to laboratory-frame experiments. The far-reaching equivalence of experiments conducted in the laboratory and in the rotating frame has its origin in the fact that the equations of motion in both reference frames are formally identical (see Chap. 48.7). The basic phenomenon behind Hahn echoes is **"Larmor precession"** about the external magnetic field. The rotating-frame equivalent is **"nutation"** about the effective field. The first rotating-frame experiment which we draw attention to is the generation of rotary echoes [463].

The quantization direction in the rotating frame is given by the effective field instead of that of the external field B_0 (see Sect. 48.8). The role of field inhomogeneities, i.e., the B_0 gradients, in the formation of Hahn echoes is therefore taken over by gradients of the amplitude of the rotating RF field (B_1 gradients). Practically any form of RF coil unavoidably generates RF fields which are, at least in the fringe-field range, sufficiently inhomogeneous for the formation of rotary echoes. Experiments with B_1 gradients are nevertheless preferably carried out with probeheads specifically designed for the production of strong B_1 gradients. Simple probe geometries are surface and "anti-Helmholtz" coils ([3, 416], for instance).

A typical rotary-echo experiment consists of an RF pulse with a B_1-gradient. During this pulse the rotating-frame coherences are dephased. A subsequent pulse of the same length but with reversed phase then rephases the rotating-frame coherences so that an "echo" appears in the sense of an alignment of the total magnetization along the z axis. In terms of NMR signals, any misadjustment of the dephasing and rephasing intervals leads to a detectable FID after the two pulses, whereas the rotary echo per se is not accompanied by any signal. The rotary echo may be called the rotating-frame analogue of the gradient echo described in the preceding section (compare Fig. 2.3).

Several other versions of the rotary-echo principle have been suggested in connection with methods for the measurement of diffusion and flow properties [55, 81, 216]. In particular it was pointed out that the dephasing and rephasing intervals of the rotating-frame coherences can be detached from each other. In the following we treat a typical pulse scheme of this sort which consists of two B_1 gradient pulses.[1]

---

[1] If the laboratory-frame coherences contributing to transverse magnetization are entirely spoiled by transverse relaxation or B_0 inhomogeneities, the second RF pulse acts solely on the

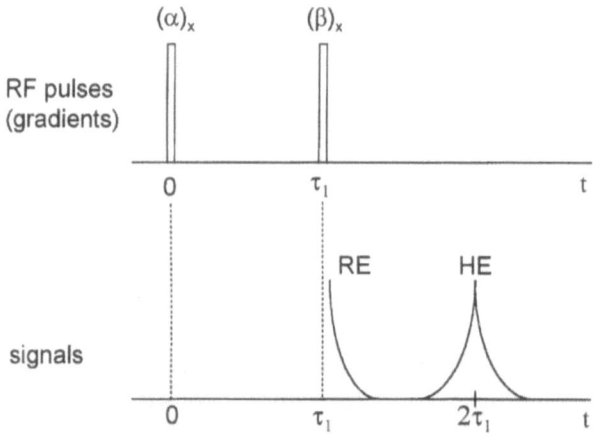

Fig. 3.1. RF pulse scheme for the generation of rotary echo (RE) and Hahn two-pulse echo (HE) signals. The lower traces are to represent envelopes of the expected NMR signals. The pulses are understood to imply B_1 gradients, that is, the flip angles depend on the position of the spins. The RF phases are irrelevant for the rotary echo if $\tau_1 \gg T_2^*$, i.e., if transverse relaxation and B_0 inhomogeneities spoil the coherences excited by the first pulse fast enough.

We consider the RF (B_1 gradient) pulse sequence (Fig. 3.1), that is

$$(\alpha)_x - \tau_1 - (\beta)_x - \text{(acquisition)} \tag{3.1}$$

On the face of it, there is no difference to the pulse scheme of the two-pulse Hahn echo (pulse sequence 2.5). Actually, the signal potentially recorded in the second acquisition interval is a Hahn echo, whereas (laboratory-frame) frame signals immediately after the second RF pulse reflect the formation of a rotary echo. An essential difference, however, is that the flip angles of the pulses, α and β, are now functions of the position due to the B_1 gradient:

$$\begin{aligned} \alpha &= \alpha(r) = \gamma_n B_1(r)\,\tau_\alpha \\ \beta &= \beta(r) = \gamma_n B_1(r)\,\tau_\beta \end{aligned} \tag{3.2}$$

The quantities τ_α and τ_β are the widths of the two pulses. The discussion of the potential influence of spin-lattice relaxation, diffusion and flow in the treatment of the pulse sequence 3.1 is deferred to Part II. The RF phases are only irrelevant for the rotary echo if transverse relaxation and B_0 inhomogeneities spoil the coherences before the second pulse is applied, that is $T_2^* \ll \tau_1 \ll T_1, T_2$. The initial density operator of the "tagged" spin at a position r is

$$\rho(r, 0-) = a + b\,I_z \tag{3.3}$$

---

magnetization left over in the z direction. This appears analogous to the stimulated-echo experiment Fig. 2.2 where the third RF pulse under equivalent field or relaxation conditions merely affects the longitudinal magnetization. The rotary echo produced in this way may therefore be termed "stimulated rotary echo" [131].

The constants a and b are defined in Eqs. 47.26 and 47.27. After the first RF pulse we have

$$\rho(r, 0+) = a + b \left[I_y \sin \alpha + I_z \cos \alpha \right] \tag{3.4}$$

The second term represents coherences for spins at a position r.[2] These evolve in the pulse interval according to

$$\rho(r, \tau_1 -) = a + b \left[I_x \sin \alpha \sin \varphi_1 + I_y \sin \alpha \cos \varphi_1 + I_z \cos \alpha \right] \tag{3.5}$$

where

$$\varphi_1 = \varphi_1(r) = \Omega(r) \tau_1 \tag{3.6}$$

and $\Omega(r)$ is the local frequency offset from the carrier frequency dictating the reference-frame rotation. The second pulse produces

$$
\begin{aligned}
\rho(r, \tau_1 +) = {} & a + b I_x \sin \alpha \sin \varphi_1 \\
& + \frac{b}{2} I_y \left[\sin(\alpha + \beta)(1 + \cos \varphi_1) - \sin(\alpha - \beta)(1 - \cos \varphi_1) \right] \\
& + b I_z \left[\cos(\alpha + \beta) \cos^2(\varphi_1/2) + \cos(\alpha - \beta) \sin^2(\varphi_1/2) \right]
\end{aligned} \tag{3.7}
$$

where we have replaced products of trigonometric functions by functions of the sum or difference of the arguments.

The free evolution of coherences in an interval τ' after the second pulse leads to

$$
\begin{aligned}
\rho(r, \tau_1 + \tau') = {} & \\
a + \frac{b}{2} I_x & \left[\cos(\alpha + \beta) \left\{ \sin \varphi' + \frac{1}{2} \left[\sin(\varphi_1 + \varphi') - \sin(\varphi_1 - \varphi') \right] \right\} \right. \\
& - \sin(\alpha - \beta) \left\{ \sin \varphi' - \frac{1}{2} \left[\sin(\varphi_1 + \varphi') - \sin(\varphi - \varphi') \right] \right\} \\
& \left. + \sin \alpha \left\{ \sin(\varphi_1 - \varphi') + \sin(\varphi_1 + \varphi') \right\} \right] \\
+ \frac{b}{2} I_y & \left[\sin(\alpha + \beta) \left\{ \cos \varphi' + \frac{1}{2} \left[\cos(\varphi_1 - \varphi') + \cos(\varphi_1 + \varphi') \right] \right\} \right. \\
& - \sin(\alpha - \beta) \left\{ \cos \varphi' - \frac{1}{2} \left[\cos(\varphi_1 - \varphi') + \cos(\varphi_1 + \varphi') \right] \right\} \\
& \left. - \sin \alpha \left\{ \cos(\varphi_1 - \varphi') - \cos(\varphi_1 + \varphi') \right\} \right] \\
+ b I_z & \left[\cos(\alpha + \beta) \cos^2(\varphi_1/2) + \cos(\alpha - \beta) \sin^2(\varphi_1/2) \right]
\end{aligned} \tag{3.8}
$$

where

$$\varphi' = \varphi'(r) = \Omega(r) \tau' \tag{3.9}$$

Here we have again replaced products of trigonometric functions by functions of the sum or difference of the arguments.

---

[2]Note that $\langle \rho(r, 0+) \rangle \approx 0$ because of the strong dependence of α on r. That is, no FID signal appears after the first pulse.

The expressions above refer to a certain position r. The signals which can be detected in experiments without spatial resolution reflect the average over all positions within the sample. Owing to the assumptions made in the above treatment, all occurring angles can be position dependent. We will first carry out the average over the phase angles $\varphi_1(r)$ and $\varphi'(r)$ and then over the flip angles $\alpha(r)$ and $\beta(r)$.

3.1
Signal at $\tau' = \tau_1$

A Hahn echo is expected if τ' approaches τ_1, i.e., $\varphi'(r) \approx \varphi_1(r)$. The B_0 inhomogeneities are assumed to be strong enough (or the intervals are set long enough) so that $T_2^* \ll \tau_1, \tau'$ applies, and the phase angles are equally distributed.[3] In this case, all terms vanish which comply to one of the following conditions:

$$
\begin{aligned}
\langle \sin \varphi_1 \rangle_r &= \langle \sin \varphi' \rangle_r &= \langle \cos \varphi_1 \rangle_r &= \langle \cos \varphi_1 \rangle_r \\
&= \langle \sin(\varphi_1 + \varphi') \rangle_r &= \langle \cos(\varphi_1 + \varphi') \rangle_r &= 0
\end{aligned} \tag{3.10}
$$

Equation 3.8 converts then to

$$
\begin{aligned}
\langle \rho(r, \tau_1 + \tau') \rangle_r &= \\
a &+ \frac{b}{4} I_x \left[-\langle \cos(\alpha + \beta) \rangle_r \langle \sin(\varphi_1 - \varphi') \rangle_r + \langle \sin(\alpha - \beta) \rangle_r \langle \sin(\varphi - \varphi') \rangle_r \right. \\
&\left. + 2 \langle \sin \alpha \rangle_r \langle \sin(\varphi - \varphi') \rangle_r \right] \\
&+ \frac{b}{4} I_y \left[\langle \sin(\alpha + \beta) \rangle_r \langle \cos(\varphi_1 - \varphi') \rangle_r + \langle \sin(\alpha - \beta) \rangle_r \langle \cos(\varphi_1 - \varphi') \rangle_r \right. \\
&\left. - 2 \langle \sin \alpha \rangle_r \langle \cos(\varphi_1 - \varphi') \rangle_r \right] \\
&+ \frac{b}{2} I_z \left[\langle \cos(\alpha - \beta) \rangle_r + \langle \cos(\alpha + \beta) \rangle_r \right]
\end{aligned} \tag{3.11}
$$

Inspecting this expression, we recognize that the Hahn echo maximum is reached for $\tau_1 = \tau'$ when the density operator adopts its maximum. Equating $\varphi' = \varphi_1$ leads to

$$
\begin{aligned}
\langle \rho(r, 2\tau_1) \rangle_r &= a + \frac{b}{4} I_y \left[\langle \sin(\alpha + \beta) \rangle_r + \langle \sin(\alpha - \beta) \rangle_r - 2 \langle \sin \alpha \rangle_r \right] \\
&+ \frac{b}{2} I_z \left[\langle \cos(\alpha - \beta) \rangle_r + \langle \cos(\alpha + \beta) \rangle_r \right]
\end{aligned} \tag{3.12}
$$

Hence the Hahn echo amplitude is given by the complex transverse magnetization (Eq. 47.7)

$$
m(2\tau_1) = \frac{i}{4} M_0 \left[\langle \sin(\alpha + \beta) \rangle_r + \langle \sin(\alpha - \beta) \rangle_r - 2 \langle \sin \alpha \rangle_r \right] \tag{3.13}
$$

---

[3]Note that the difference of the phase angles is not necessarily subject to an equipartition, especially if the evolution times approach each other.

The Hahn echo obviously disappears if

$$\alpha(r) = \beta(r) \tag{3.14}$$

and if the flip angles are equally distributed over all possible values, so that $\langle \sin \alpha \rangle_r = 0$.

3.2
Signal at $\tau' = 0$

The proper rotary echo reveals itself only indirectly by the eventual FID following the second RF pulse. Setting $\tau' = 0$ in Eq. 3.8 and executing the average over the phase angles according to Eq. 3.10 as above leads to

$$\langle \rho(r, \tau_1) \rangle_r = a + \frac{b}{4} I_y \left[\langle \sin(\alpha + \beta) \rangle_r - \langle \sin(\alpha - \beta) \rangle_r \right]$$
$$+ \frac{b}{2} I_z \left[\langle \cos(\alpha - \beta) \rangle_r + \langle \cos(\alpha + \beta) \rangle_r \right] \tag{3.15}$$

If the flip angles are equally distributed in the full range, so that[4]

$$\langle \sin(\alpha + \beta) \rangle_r = \langle \cos(\alpha + \beta) \rangle_r = 0 \tag{3.16}$$

this expression turns into

$$\langle \rho(r, \tau_1) \rangle_r = a - \frac{b}{4} I_y \langle \sin(\alpha - \beta) \rangle_r + \frac{b}{2} I_z \langle \cos(\alpha - \beta) \rangle_r \tag{3.17}$$

The proper **rotary echo** is defined for the condition in Eq. 3.14. The complex transverse magnetization vanishes in this case and the total magnetization is aligned along the z direction:

$$M_z(\tau_1) = \frac{1}{2} M_0 \tag{3.18}$$

Thus no detectable signal appears. On the other hand, any deviation from the condition in Eq. 3.14 leads to a complex transverse magnetization given by

$$m(\tau_1) = \frac{i}{4} M_0 \langle \sin(\alpha - \beta) \rangle_r \tag{3.19}$$

This defines the amplitude and the phase of an FID signal acquired immediately after the second RF pulse (see Fig. 3.1).

---

[4]Note that the mean trigonometric functions of the difference of the flip angles, $\langle \sin(\alpha - \beta) \rangle_r$ and $\langle \cos(\alpha - \beta) \rangle_r$, do not vanish if the argument is small enough to prevent equipartition of the argument.

CHAPTER 4

Solid Echoes of Dipolar-Coupled Spins

Hahn echoes arise when the spins are reoriented by RF pulses while the field inhomogeneities are stationary. The recovery of spin coherences is a matter of suitably manipulated spin states. By contrast, the principle of solid echoes due to dipolar (or quadrupolar) interaction is that the spin couplings as the coherence defocusing elements are also affected by the RF pulses. Solid echoes [322, 390] are the consequence of spin-state as well as spin-interaction manipulations. This is the reason why the RF pulse sequences, albeit very similar in many respects, require different pulse phases and flip angles to provide optimal solid-echo amplitudes.

In the following treatments we disregard spin-lattice relaxation and merely take into account effects of the secular dipolar Hamiltonian (see Table 46.5 on page 425). In the time scale in which solid-echo effects can be expected this is certainly a rather uncritical simplification.

4.1
Two-Pulse Dipolar Solid Echoes

Nuclear spins underlying dipolar coupling in solids tend to form very large systems which possibly comprise the whole sample. In this context "solid" means that dipolar interactions are not averaged out by molecular motions in the lifetime of a spin state. A material of this sort can be a rigid solid or just a liquid viscous or anisotropic enough to ensure that there is at least a residual influence of dipolar coupling on the evolution of spin coherences.

Such circumstances appear to be rather complicated for a theoretical consideration. There are, however, situations and approximations enabling one to perform simple straightforward treatments of the coherence evolution. We discuss the following two approaches of the problem.[1]

a) The total spin system is subdivided in groups of a few spins. Dipolar interaction among spins of different groups are neglected, which is justified if the spacings of the spins are much larger than those of interaction partners in a group. In this case the **product operator formalism** is applicable (see Chap. 51). The most elementary problem to be solved on this basis refers to an ensemble of isolated

---

[1]The "intermediate" situation of spin pairs with **weak interpair coupling** was treated in [51]. Under such circumstances the second moment of the interpair contribution to the spectrum can be evaluated.

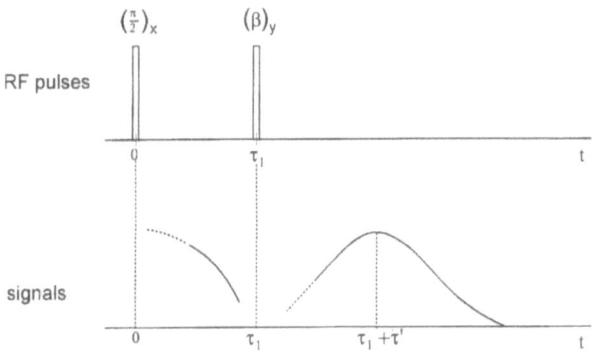

Fig. 4.1. RF pulse sequence for the generation of solid echoes. The lower trace is to represent envelopes of the expected NMR signals.

systems of two like and equivalent spins 1/2. All important experimental features are already revealed at least qualitatively in this case. Although the practical application of the product operator formalism may be restricted to small spin systems not exceeding five members or so in order to avoid expressions of excessive lengths, formalisms of this sort model the behaviour of larger spin systems reasonably well.

b) In the second approach to be dealt with here the size of the spin system is unrestricted. The problem is solved by **series expansions of the sandwich propagator expressions** determining the evolution of spin coherences.

The pulse sequence to be considered is[2]

$$(\pi/2)_x \; - \; \tau_1 \; - \; (\beta)_\delta \; - \; \tau \; - \; \text{(acquisition)} \tag{4.1}$$

A scheme of this sequence with the pulse phases in quadrature is displayed in Fig. 4.1. Resonance offsets by external or internal field inhomogeneities are neglected.

4.1.1
Systems of Two Equivalent Spins 1/2

The treatment in the following is based on the *reduced* density operator as defined in footnote 2 on page 431. The equilibrium value for a two-spin system is (see Sect. 47.2)

$$\sigma(0-) = I_{1z} + I_{2z} \tag{4.2}$$

The RF pulses are assumed to be "hard" enough to neglect all spin interactions as long as the pulses are on. The Hamiltonian effective in the rotating frame is then

---

[2]This is the elementary pulse sequence eligible for the generation of solid echoes. Note that solid echoes can also be refocused repeatedly with the aid of further equidistant and in-phase RF pulses following the second pulse in intervals $2\tau_1$.

$\mathcal{H}'_{rf}$ (Eq. 48.95), so that the first RF pulse produces the single-quantum coherence terms

$$\sigma(0+) = I_{1y} + I_{2y} \tag{4.3}$$

The spin coherences evolve under the action of the bilinear spin operator (see Eq. 51.24):

$$\mathcal{H}_d^{(ev)} = \frac{\mu_0}{4\pi} \frac{3\gamma_n^2 \hbar^2}{2r^3} (1 - 3\cos^2\vartheta) I_{1z} I_{2z} \equiv hc_d I_{1z} I_{2z} \tag{4.4}$$

Bringing the product operator rules (Table 51.3 on page 486) into use leads to the expression

$$\sigma(\tau_1 -) = (I_{1y} + I_{2y})\cos(\pi c_d \tau_1) - (2I_{1x}I_{2z} + 2I_{2x}I_{1z})\sin(\pi c_d \tau_1) \tag{4.5}$$

The phase direction of the second pulse is at first assumed to be the y axis of the rotating frame. The density operator immediately after the second pulse is then

$$\sigma(\tau_1 +)|_{\delta=y} = (I_{1y} + I_{2y})\cos(\pi c_d \tau_1) - \{[2I_{1x}I_{2z} + 2I_{2x}I_{1z}]\cos(2\beta)$$
$$+ [2I_{1x}I_{2x} - 2I_{1z}I_{2z}]\sin(2\beta)\}\sin(\pi c_d \tau_1) \tag{4.6}$$

The last two terms represent multiple-quantum coherences (MQC) and longitudinal order (in this case, dipolar order (DO)) respectively.[3] These phenomena cannot contribute to detectable signals obtained with a two-pulse sequence. For the moment we may therefore skip the consideration of their further evolution. It is relevant in context with pulse sequences consisting of three or more pulses, because then the conversion into detectable single-quantum coherences is feasible (see Sect. 4.2). The DO and MQC terms vanish for the flip angle $\beta = \pi/2$ and become maximal for $\beta = \pi/4$.

After a further evolution interval τ the spin states are characterized by

$$\sigma(\tau_1 + \tau)|_{\delta=y} = (I_{1y} + I_{2y}) \frac{1}{2} \{\cos[\pi c_d(\tau_1 - \tau)] + \cos[\pi c_d(\tau_1 + \tau)]$$
$$- \cos(2\beta)\{\cos[\pi c_d(\tau_1 - \tau)] - \cos[\pi c_d(\tau_1 + \tau)]\}\}$$
$$- (2I_{1x}I_{2z} + 2I_{2x}I_{1z}) \frac{1}{2} \{\sin[\pi c_d(\tau_1 + \tau)] - \sin[\pi c_d(\tau_1 - \tau)]$$
$$+ \cos(2\beta)\{\sin[\pi c_d(\tau_1 - \tau)] + \sin[\pi c_d(\tau_1 + \tau)]\}\}$$
$$+ \text{MQC and DO terms} \tag{4.7}$$

In powders no preferential orientation of the internuclear vectors exists. The polar angle ϑ is correspondingly distributed over all possible values. The internuclear distance r is a further parameter possibly subject to variations from spin system to

---

[3] "Dipolar order" means that the spins are aligned in the local dipole magnetic field produced by their neighbors. The nature of this phenomenon may be visualized by considering spins with different precession frequencies owing to dipolar interaction. The "fast" spins tend to be in the spin-up state, while the "slow" spins adopt the spin-down state or vice versa. The spin states are thus correlated to the local field produced by dipolar coupling.

spin system. The consequence is that the ensemble of spin systems implies a distribution of dipolar coupling constants c_d (see Eq. 4.4). The observable signals are therefore determined by the ensemble average $\langle\ldots\rangle$ with respect to $c_d = c_d(\vartheta, r)$. All sine and cosine terms in Eq. 4.7 must be averaged accordingly.

In the resulting average density operator expression the superpositions of all sine or cosine terms with finite arguments $\pi c_d(\tau_1 + \tau)$ or $\pi c_d(\tau_1 - \tau)$ tend to be destructive. However, if the condition

$$\tau = \tau_1 \tag{4.8}$$

is satisfied the arguments $\pi c_d(\tau_1 - \tau)$ vanish, so that the corresponding terms become independent of the orientation. In particular, this refers to terms implying $\langle\cos[\pi c_d(\tau_1 - \tau)]\rangle$. The density operator then adopts a maximum, which reveals itself as the two-pulse **"solid echo"** of the single-quantum coherences. The echo amplitude is characterized by the first term on the right-hand side of

$$\langle\sigma(2\tau_1)|_{\delta=y}\rangle = \left(I_{1y} + I_{2y}\right) \frac{1}{2}\left[1 - \cos(2\beta)\right] + \text{MQC and DO terms} \tag{4.9}$$

Furthermore, if the flip angle of the second pulse is $\beta = \pi/2$, the amplitude of the initial FID is completely recovered.[4] In this case, the complex transverse rotating-frame magnetization is

$$m(2\tau_1) = m(0+) = i\,M_0 \tag{4.10}$$

The opposite situation arises for $\beta = \pi$. The echo signal then disappears entirely. This is to be compared with the Hahn echo which adopts its maximum amplitude for 180° refocusing pulses (see Eq. 2.17). The solid echo is maximal for pulse phases in quadrature as assumed in the above treatment. In contrast to the two-pulse Hahn echo, the echo signal disappears for in-phase pulses. The reason is that the second RF pulse simultaneously changes the spin states as well as the spin interactions. The consequence of in-phase pulses is that the refocusing tendency, arising from the change of the spin states, is cancelled by the opposite tendency owing to the modified spin interactions, or vice versa. This becomes obvious by considering the reduced density operator after in-phase pulses:

$$
\begin{aligned}
\sigma(\tau_1 + \tau)|_{\delta=x} = {} & \left(I_{1y} + I_{2y}\right)\cos\beta\cos[\pi c_d(\tau_1 + \tau)] \\
& - \left(2I_{1x}I_{2z} + 2I_{2x}I_{1z}\right)\cos\beta\sin[\pi c_d(\tau_1 + \tau)] \\
& - \left(I_{1z} + I_{2z}\right)\sin\beta\cos(\pi c_d\tau_1) \\
& + \text{MQC terms}
\end{aligned}
\tag{4.11}
$$

There is no cosine term with an argument depending on the difference of the free-evolution intervals, $\pi c_d(\tau_1 - \tau)$, so that all trigonometric functions at all finite

---

[4]The completeness in refocusing of dipolar phase shifts is a feature of two-spin systems. Larger systems show echoes with reduced amplitude (see Eq. 4.24).

times are subject to destructive superpositions. For $\beta = \pi/2$, the corresponding transverse magnetization even vanishes entirely.[5]

4.1.2
Approximate Treatment of Multi-Spin 1/2 Systems

Solid echoes appear in multi-spin systems under the same experimental conditions as described above, but the coherences are only partially refocused.[6] These incoherent losses can be calculated using the product operator formalism if the spin system consists of a few spins only and if the mutual arrangement of the coupling partners follows a simple geometry. If the number of coupled nuclei exceeds three or four the treatment becomes more and more intractable. Under such circumstances approximate treatments using series expansions to a certain order are more promising [320, 391, 448, 449].

In the light of the above treatment, the optimum coherence refocusing effect is expected for a pulse sequence consisting of two 90° pulses phase-shifted by 90°. We therefore consider a sequence of the type

$$(\pi/2)_x \; - \; \tau_1 \; - \; (\pi/2)_y \; - \; \tau \; - \; \text{(acquisition)} \tag{4.12}$$

The total spin components are defined by sums over all spins in the system:

$$I_\nu = \sum_k I_{k\nu} \quad \text{where} \quad \nu = x, y, z \tag{4.13}$$

The respective reduced density operators before and just after the first pulse are

$$\sigma(0-) = I_z \tag{4.14}$$

and

$$\sigma(0+) = I_y \tag{4.15}$$

After the first RF pulse, the spin coherences evolve [301] according to the secular dipolar Hamiltonian (see Table 46.5 on page 425)

$$\mathcal{H}_d^{(0)} = \frac{\mu_0}{4\pi} \gamma_n^2 \hbar^2 \frac{1}{2} \sum_{j<k} \frac{1 - 3\cos^2 \vartheta_{jk}}{r_{jk}^3} \, [\, 3 I_{jz} I_{kz} - I_j \cdot I_k \,] \tag{4.16}$$

which now refers to all spins in the sample. The segmented treatment of the time dependence of the density operator involves the following steps which are self-explanatory:

$$\sigma(\tau_1-) = \exp\left\{ -\frac{i}{\hbar} \mathcal{H}_d^{(0)} \tau_1 \right\} \sigma(0+) \exp\left\{ \frac{i}{\hbar} \mathcal{H}_d^{(0)} \tau_1 \right\} \tag{4.17}$$

---

[5]This statement is valid provided that inhomogeneities of the external magnetic field are negligible. If the spin coherences are additionally dephased by sufficiently strong field gradients, the relative phase of the second RF pulse may become irrelevant (compare Chap. 17).

[6]An analogous statement can be made for Hartmann/Hahn cross-polarization (see

$$\sigma(\tau_1+) = \exp\left\{i\frac{\pi}{2}I_y\right\}\sigma(\tau_1-)\exp\left\{-i\frac{\pi}{2}I_y\right\} \tag{4.18}$$

$$\sigma(\tau_1+\tau) = \exp\left\{-\frac{i}{\hbar}\mathcal{H}_d^{(0)}\tau\right\}\sigma(\tau_1+)\exp\left\{\frac{i}{\hbar}\mathcal{H}_d^{(0)}\tau\right\} \tag{4.19}$$

Thus in the acquisition interval the complex transverse rotating-frame magnetization is[7]

$$
\begin{aligned}
m(\tau_1+\tau) &= n\gamma_n\hbar b\,\mathrm{Tr}\{(I_x+iI_y)\sigma(\tau_1+\tau)\} \\
&= n\gamma_n\hbar b\,\mathrm{Tr}\left\{\exp\left(\frac{i}{\hbar}\mathcal{H}_d^{(0)}\tau\right)[I_x+iI_y]\exp\left(-\frac{i}{\hbar}\mathcal{H}_d^{(0)}\tau\right)\right. \\
&\quad \exp\left(i\frac{\pi}{2}I_y\right)\exp\left(-\frac{i}{\hbar}\mathcal{H}_d^{(0)}\tau_1\right)\sigma(\tau_1+) \\
&\quad \left.\exp\left(\frac{i}{\hbar}\mathcal{H}_d^{(0)}\tau_1\right)\exp\left(-i\frac{\pi}{2}I_y\right)\right\}
\end{aligned} \tag{4.20}
$$

where we have used the property $\mathrm{Tr}\{PQ\} = \mathrm{Tr}\{QP\}$. The sandwich operator terms in the parentheses can be expanded in a Taylor series about $t = 0$, leading to (compare Eqs. 48.44 and 48.45)

$$e^{-iQt}Pe^{iQt} = P + it[P,Q] - \frac{t^2}{2!}[Q,[Q,P]] + \frac{it^3}{3!}[Q,[Q,[Q,P]]] - \dots \tag{4.21}$$

Using this expansion twice and rewriting Eq. 4.20 in the form of MacLaurin series as far as possible, leads to

$$
\begin{aligned}
m(\tau_1+\tau) =& \\
iM_0&\left(1 - \frac{(\tau_1-\tau)^2}{2!}\frac{\mathrm{Tr}\{[\mathcal{H}_d^{(0)},I_y]^2\}}{\mathrm{Tr}\{I_y^2\}} + \frac{(\tau_1-\tau)^4}{4!}\frac{\mathrm{Tr}\{[\mathcal{H}_d^{(0)},[\mathcal{H}_d^{(0)},I_y]]^2\}}{\mathrm{Tr}\{I_y^2\}} + \dots\right) \\
&+\left\{-\frac{6\tau_1^2\tau^2}{4!}\frac{\mathrm{Tr}\{[\mathcal{H}_d^{(0)}+\mathcal{H}_d^{(0)},[\mathcal{H}_d^{(0)},I_y]][\mathcal{H}_d^{(0)},[\mathcal{H}_d^{(0)},I_y]]\}}{\mathrm{Tr}\{I_y^2\}} + \dots\right\}
\end{aligned} \tag{4.22}
$$

The coefficients can be expressed by the line moments

$$M_\nu = \int\limits_{-\infty}^{+\infty} g(\Delta\omega)(\Delta\omega)^\nu\,\mathrm{d}(\Delta\omega) \tag{4.23}$$

---

[7]Note that the dipolar Hamiltonian in Eq. 4.16, refers to all spins in the sample. Therefore no extra average over a distribution of internuclear vector orientations is needed. There is only one (multi) spin system. The distribution of internuclear vector orientations is already taken into account by applying the total dipolar Hamiltonian. By contrast, in the treatment of an ensemble of two-spin systems considered in the previous section, the dipolar Hamiltonian referred to a single spin system out of an ensemble so that a corresponding average was essential.

where $g(\Delta\omega)$ is the normalized lineshape, and $\Delta\omega = \omega - \omega_0$ is the deviation from the resonance angular frequency. Thus we obtain

$$m(\tau_1 + \tau) = iM_0 \left(1 - \frac{(\tau_1 - \tau)^2}{2!}M_2 + \frac{(\tau_1 - \tau)^4}{4!}M_4 + \dots\right)$$
$$+ \left\{-\frac{6\tau_1^2\tau^2}{4!}\frac{\mathrm{Tr}\{[\mathcal{H}_d^{(0)} + \mathcal{H}_d^{(0)}, [\mathcal{H}_d^{(0)}, I_y]][\mathcal{H}_d^{(0)}, [\mathcal{H}_d^{(0)}, I_y]]\}}{\mathrm{Tr}\{I_y^2\}} + \dots\right\} \quad (4.24)$$

where

$$M_2 = \frac{\mathrm{Tr}\{[\mathcal{H}_d^{(0)}, I_y]^2\}}{\mathrm{Tr}\{I_y^2\}} \quad (4.25)$$

$$M_4 = \frac{\mathrm{Tr}\{[\mathcal{H}_d^{(0)}, [\mathcal{H}_d^{(0)}, I_y]]^2\}}{\mathrm{Tr}\{I_y^2\}} \quad (4.26)$$

In this way, the problem is transferred to the calculation of commutators. According to the rules given in Chap. 44, we have

$$[I_j \cdot I_k, I_y] = 0 \quad (4.27)$$

so that the scalar product terms of the secular dipolar Hamiltonian do not contribute. The other commutator terms can be conveniently calculated with the aid of Eqs. 42.72 which are valid for spins 1/2. For example, it turns out that

$$[I_{jz}I_{kz}, I_{jy} + I_{ky}] = -i(I_{jx}I_{kz} + I_{jz}I_{kx})$$

$$[I_{jz}I_{kz}, I_{jy} + I_{ky}]^2 = -\frac{1}{8} - \frac{1}{2}I_{jy}I_{ky}$$

$$[I_{jz}I_{kz}, [I_{jz}I_{kz}, I_{jy} + I_{ky}]] = \frac{1}{4}I_{jy} + \frac{1}{4}I_{ky}$$

$$[I_{jz}I_{kz}, [I_{jz}I_{kz}, I_{jy} + I_{ky}]]^2 = \frac{1}{32} + \frac{1}{16}I_{jy}I_{ky} + \frac{1}{16}I_{ky}I_{jy}$$

Relevant traces are (see Chap. 43)

$$\mathrm{Tr}\{I_y^2\} = 2^{N-2}N$$

$$\mathrm{Tr}\{[I_{jz}I_{kz}, I_{jy} + I_{ky}]^2\} = -2^{N-3}N$$

$$\mathrm{Tr}\{[I_{jz}I_{kz}, [I_{jz}I_{kz}, I_{jy} + I_{ky}]]^2\} = 2^{N-5}N$$

According to Eq. 4.24 the signal is expected to be maximal for $\tau_1 = \tau$ as already concluded from the two-spin treatment (see Eq. 4.8). The refocusing of the coherences tends, nevertheless, to be incomplete owing to the last term which summarizes the deviation from the MacLaurin series. While this term vanishes for two-spin 1/2 systems, it represents the tendency towards incoherent behaviour with increasing size of the spin system.

Nevertheless solid echoes are pronounced phenomena. Figure 4.2 shows a signal corresponding to the above treatment (echo in the middle after the second RF pulse). The dependences of the signal amplitude, on the pulse flip angles and phases, has been generally treated in [448]. Solid echoes in the presence of two different spin species are described in [450].

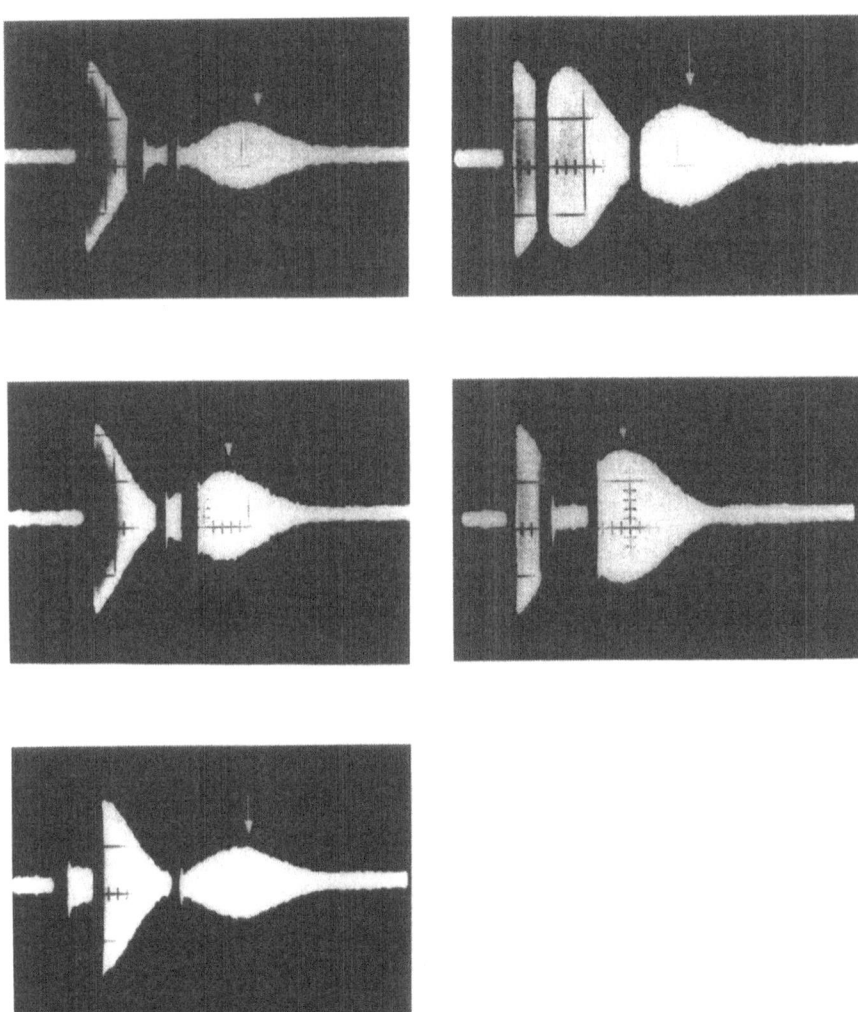

Fig. 4.2. Oscilloscope traces of the five solid echoes produced by the three-pulse sequence 4.28. The arrows indicate the echo maxima. The three RF pulses reveal themselves by black signal breaks. (Reproduced by permission from [448])

4.2
Three-Pulse Dipolar Solid Echoes

As with Hahn echoes, the addition of a third RF pulse leads to a variety of different solid echoes. The treatment suggests five signals of this sort [448], all of which can be distinguished experimentally as shown in Fig. 4.2. The general pulse sequence is

$$(\alpha)_{\varphi_1} \ - \ \tau_1 \ - \ (\beta)_{\varphi_2} \ - \ \tau_2 \ - \ (\gamma)_{\varphi_3} \ - \ \text{(acquisition)} \tag{4.28}$$

The formation of the echoes depends on the flip angles and the pulse phases. In order to simplify the treatment, we specify these parameters according to

$$(\pi/2)_x \ - \ \tau_1 \ - \ (\beta)_y \ - \ \tau_2 \ - \ (\gamma)_y \ - \ \text{(acquisition)} \tag{4.29}$$

Furthermore, we restrict ourselves to systems of two equivalent spins 1/2. The reduced density operator at the beginning of the τ_2 interval is then readily obtained from Eq. 4.6 as

$$\sigma(\tau_1+) = \sigma_{sqc}(\tau_1+) + \sigma_{mqc}(\tau_1+) + \sigma_{do}(\tau_1+) \tag{4.30}$$

where

$$\sigma_{sqc}(\tau_1+) = (I_{1y} + I_{2y}) \cos(\pi c_d \tau_1)$$
$$-[2I_{1x}I_{2z} + 2I_{2x}I_{1z}] \cos(2\beta) \sin(\pi c_d \tau_1) \tag{4.31}$$

is the contribution of **in-phase** and **antiphase single-quantum coherences**,

$$\sigma_{mqc}(\tau_1+) = 2I_{1x}I_{2x} \sin(2\beta) \sin(\pi c_d \tau_1) \tag{4.32}$$

is the **multiple-quantum coherence** term,[8] and

$$\sigma_{do}(\tau_1+) = -2I_{1z}I_{2z} \sin(2\beta) \sin(\pi c_d \tau_1) \tag{4.33}$$

represents **dipolar order**.

As we are dealing with spin systems comprising only two coupling partners, there are no "passive" spins, multiple-quantum coherences do not evolve due to dipolar coupling, and neither does the dipolar order term. Provided that there is no relaxation or molecular reorientation, we have at the end of the τ_2 interval

$$\sigma_{mqc}(\tau_1 + \tau_2-) = \sigma_{mqc}(\tau_1+) \tag{4.34}$$
$$\sigma_{do}(\tau_1 + \tau_2-) = \sigma_{do}(\tau_1+) \tag{4.35}$$

---

[8]This implies zero- and double-quantum coherences (see Chap. 51). Note that double-quantum coherences (two quanta $\hbar\omega$) are to be distinguished from **"overtone coherences"** (one quantum $\hbar 2\omega$) [47]. The latter transitions contribute in a radiationless way to spin-lattice relaxation (see Chap. 12). They can also be generated by cw or pulsed excitation using "brute" force overtone RF or by cross-polarization techniques [105, 400, 436, 485]. The condition is that the spin systems under consideration are strongly coupled so that stationary superpositions of the eigenstates of the uncoupled system exist.

The single-quantum coherences evolve according to

$$
\begin{aligned}
\sigma_{sqc}(\tau_1 + \tau_2-) =\ & (I_{1y} + I_{2y})\frac{1}{2}\{\cos[\pi c_d(\tau_1 - \tau_2)] + \cos[\pi c_d(\tau_1 + \tau_2)] \\
& - \cos(2\beta)\{\cos[\pi c_d(\tau_1 - \tau_2)] - \cos[\pi c_d(\tau_1 + \tau_2)]\}\} \\
& -(2I_{1x}I_{2z} + 2I_{2x}I_{1z})\frac{1}{2}\{\sin[\pi c_d(\tau_1 + \tau_2)] - \sin[\pi c_d(\tau_1 - \tau_2)] \\
& + \cos(2\beta)\{\sin[\pi c_d(\tau_1 - \tau_2)] + \sin[\pi c_d(\tau_1 + \tau_2)]\}\}
\end{aligned} \tag{4.36}
$$

The third pulse modifies these density operators to

$$
\begin{aligned}
\sigma_{(mqc)}(\tau_1 + \tau_2+) =\ & \{2I_{1x}I_{2x}\cos^2\gamma + 2I_{1z}I_{2z}\sin^2\gamma \\
& + \frac{1}{2}(2I_{1x}I_{2z} + 2I_{1z}I_{2x})\sin(2\gamma)\}\sin(2\beta)\sin(\pi c_d\tau_1)
\end{aligned} \tag{4.37}
$$

$$
\begin{aligned}
\sigma_{(do)}(\tau_1 + \tau_2+) =\ & -\{2I_{1z}I_{2z}\cos^2\gamma + 2I_{1x}I_{2x}\sin^2\gamma \\
& - \frac{1}{2}(2I_{1x}I_{2z} + 2I_{1z}I_{2x})\sin(2\gamma)\}\sin(2\beta)\sin(\pi c_d\tau_1)
\end{aligned} \tag{4.38}
$$

$$
\begin{aligned}
\sigma_{(sqc)}(\tau_1 + \tau_2+) =\ & (I_{1y} + I_{2y})\frac{1}{2}\{\cos[\pi c_d(\tau_1 - \tau_2)] + \cos[\pi c_d(\tau_1 + \tau_2)] \\
& - \cos(2\beta)\{\cos[\pi c_d(\tau_1 - \tau_2)] - \cos[\pi c_d(\tau_1 + \tau_2)]\}\} \\
& -\frac{1}{2}\{(2I_{1x}I_{2z} + 2I_{2x}I_{1z})\cos(2\gamma) + (2I_{1z}I_{2z} - 2I_{1x}I_{2x})\sin(2\gamma)\} \\
& \{\sin[\pi c_d(\tau_1 + \tau_2)] - \sin[\pi c_d(\tau_1 - \tau_2)] \\
& + \cos(2\beta)\{\sin[\pi c_d(\tau_1 - \tau_2)] + \sin[\pi c_d(\tau_1 + \tau_2)]\}\}
\end{aligned} \tag{4.39}
$$

where $\sigma_{(mqc)}$, $\sigma_{(do)}$, and $\sigma_{(sqc)}$ are the density operators derived from the expressions for multiple-quantum coherences, dipolar order and single-quantum coherences of the preceding interval.

The further evolution under the influence of dipolar interaction in a subsequent interval τ leads to terms with factors $\cos(\pi c_d\tau)$ and $\sin(\pi c_d\tau)$. These, combined with the trigonometric functions of the above expressions, suggest single-quantum coherence echo maxima at the times indicated in Fig. 4.2.

Echoes which can be traced back to dipolar-order states and multiple-quantum coherences in the second interval are of particular interest. Up to now we have tacitly assumed that there are no resonance offsets (e.g., by field inhomogeneities or chemical shifts) and that relaxation is negligible. Under such conditions $\sigma_{(mqc)}$ and $\sigma_{(do)}$ have evolved in a similar way, and share and share alike. Forming the sum of these two contributions,

$$
\sigma_{(do)}(\tau_1 + \tau_2+) + \sigma_{(mqc)}(\tau_1 + \tau_2+) = (2I_{1x}I_{2z} + 2I_{1z}I_{2x})\sin(2\gamma)\sin(2\beta)\sin(\pi c_d\tau_1) \tag{4.40}
$$

readily leads to

$$\sigma_{(do)}(\tau_1 + \tau_2 + \tau) + \sigma_{(mqc)}(\tau_1 + \tau_2 + \tau) = \frac{1}{2}(2I_{1x}I_{2z} + 2I_{1z}I_{2x})(\sin[\pi c_d(\tau_1 + \tau)]$$
$$+ \sin[\pi c_d(\tau_1 - \tau)]) \sin(2\beta) \sin(2\gamma)$$
$$+ \frac{1}{2}(I_{1y} + I_{2y})(\cos[\pi c_d(\tau_1 - \tau)]$$
$$- \cos[\pi c_d(\tau_1 + \tau)]) \sin(2\beta) \sin(2\gamma)$$

$$(4.41)$$

This expression must be ensemble-averaged over all dipolar coupling constants $c_d = c_d(\vartheta, r)$ as a function of the internuclear distance vector r. The trigonometric terms depending on $\pi c_d(\tau_1 \pm \tau)$ are destructively superimposed unless the argument is zero owing to an adequate value of the time τ. That is $\tau = \tau_1$ in particular. If this condition is satisfied a **dipolar-order/multiple-quantum coherence transfer echo** appears, characterized by the average reduced density operator

$$\langle \sigma_{(do)}(2\tau_1 + \tau_2) + \sigma_{(mqc)}(2\tau_1 + \tau_2)\rangle = \frac{1}{2}(I_{1y} + I_{2y}) \sin(2\beta) \sin(2\gamma) \qquad (4.42)$$

The corresponding complex transverse magnetization is

$$m(2\tau_1 + \tau_2) = \frac{i}{2} M_0 \sin(2\beta) \sin(2\gamma) \qquad (4.43)$$

The flip angles for maximum signal are $\beta = \gamma = \pi/4$ so that

$$m(2\tau_1 + \tau_2) = \frac{i}{2} M_0 \qquad (4.44)$$

Under real conditions, the multiple-quantum coherences in the τ_2 interval tend to be defocused by resonance offsets, or will be relaxed if $\tau_2 \gg T_2$. The eventually detectable single-quantum coherences originating from these multiple-quantum coherences may be refocused as a coherence transfer echo if all resonance offsets remain constant. Refocusing a single-quantum coherence, however, takes twice as long as defocusing a double-quantum coherence (compare Chap. 7). Therefore, $\langle \sigma_{(mqc)}(2\tau_1 + \tau_2)\rangle = 0$, and the echo at the time $2\tau_1 + \tau_2$ will be determined solely by $\langle \sigma_{(do)}(2\tau_1 + \tau_2)\rangle$. The complex transverse magnetization of this **dipolar-order transfer** or **Jeener/Broekaert echo** [208] is, for the optimum flip angles $\beta = \gamma = \pi/4$,

$$m(2\tau_1 + \tau_2) = \frac{i}{4} M_0 \qquad (4.45)$$

CHAPTER 5

Solid Echoes of $I = 1$ Quadrupole Nuclei

Quadrupole nuclei with spin $I = 1$ respond to pulse sequences analogously to pairs of dipolar coupled, equivalent nuclei $I = 1/2$. The reason is that with the two-spin 1/2 system only triplet states (total spin $F = 1$) play a role. Therefore, the same sort of three-level system is active and the eigenfunctions are of an equivalent nature [455]. Nevertheless it may be instructive to treat quadrupolar solid echo pulse sequences directly using the operator formalism given in Chapter 52 for the case of axially symmetric electric field gradients.

5.1
Two-Pulse Quadrupolar Solid Echo

We consider the quadrupolar-echo pulse sequence [110, 462]

$$(\pi/2)_x \ - \ \tau_1 \ - \ (\pi/2)_y \ - \ \tau \ - \ \text{(acquisition)} \tag{5.1}$$

i.e., the pulse scheme shown in Fig. 4.1 in its optimum version for $\beta = \pi/2$.
   The equilibrium reduced density operator is

$$\sigma(0-) = I_z \tag{5.2}$$

The first 90° RF pulse produces

$$\sigma(0+) = I_y \tag{5.3}$$

The corresponding coherences evolve in the first pulse interval under the influence of the high-field secular quadrupole interaction Hamiltonian for axially symmetric electric field gradients (see Table 46.4 on page 424)

$$\mathcal{H}_q^{(0)} = \hbar\omega_q \left(I_z^2 - \frac{1}{3}I^2 \right) = \hbar\omega_q \left(I_z^2 - \frac{2}{3} \right) \tag{5.4}$$

where

$$\omega_q = \frac{3e^2qQ}{8I(2I-1)\hbar}(3\cos^2\vartheta - 1) = \frac{3e^2qQ}{8\hbar}(3\cos^2\vartheta - 1) \tag{5.5}$$

Spin-lattice relaxation, i.e., the nonsecular terms of the quadrupole Hamiltonian, are neglected here. The sandwich propagator expression for the first free-evolution interval is

$$\sigma(\tau_1-) = e^{-(i/\hbar)\mathcal{H}_q^{(0)}\tau_1} \ I_y \ e^{(i/\hbar)\mathcal{H}_q^{(0)}\tau_1} \tag{5.6}$$

The evaluation is based on Eq. 52.19. The commutator of the basis operators needed for this purpose is (see Table 52.1 on page 496)

$$[I_y, I_z^2 - 2/3] = i(I_z I_x + I_x I_z) \tag{5.7}$$

That is, the quantities determining Eq. 52.19 are $a = 1$, $\mathcal{O}_l = I_z I_x + I_x I_z$, and $\varphi = \omega_q \tau_1$. The reduced density operator becomes

$$\sigma(\tau_1 -) = I_y \cos(\omega_q \tau_1) - (I_z I_x + I_x I_z) \sin(\omega_q \tau_1) \tag{5.8}$$

The reduced density operator just after the second pulse, $(\pi/2)_y$, is derived as

$$\sigma(\tau_1 +) = e^{i(\pi/2)I_y} \, \sigma(\tau_1 -) \, e^{-i(\pi/2)I_y} \tag{5.9}$$

In this case, the relevant commutator is (Table 52.1 on page 496)

$$[I_y, I_z I_x + I_x I_z] = -2i(I_z^2 - I_x^2) \tag{5.10}$$

implying $a = 2$, $\mathcal{O}_l = I_z^2 - I_x^2$, and $\varphi = \pi/2$. The evaluation of Eq. 52.20 leads to

$$\sigma(\tau_1 +) = I_y \cos(\omega_q \tau_1) - (I_z I_x + I_x I_z) \sin(\omega_q \tau_1) \tag{5.11}$$

The evaluation of the sandwich propagator expression of the next interval,

$$\sigma(\tau_1 + \tau) = e^{-(i/\hbar)\mathcal{H}_q^{(0)}\tau} \, \sigma(\tau_1 -) \, e^{(i/\hbar)\mathcal{H}_q^{(0)}\tau} \tag{5.12}$$

requires the commutators

$$[I_z^2 - 2/3, I_y] = -i(I_z I_x + I_x I_z) \tag{5.13}$$

(i.e., $a = 1$, $\mathcal{O}_l = I_z I_x + I_x I_z$, $\varphi = \omega_q \tau$) and

$$[I_z^2 - 2/3, I_z I_x + I_x I_z] = iI_y \tag{5.14}$$

(i.e., $a = -1$, $\mathcal{O}_l = I_y$, $\varphi = \omega_q \tau$). The result is

$$\sigma(\tau_1 + \tau) = I_y \cos[\omega_q(\tau_1 - \tau)] - (I_z I_x + I_x I_z) \sin[\omega_q(\tau_1 - \tau)] \tag{5.15}$$

For the derivation of the complex transverse magnetization we need the average over all ω_q values occurring in the sample. One is facing such distributions for powder geometries in particular. In this case, all possible polar angles ϑ of the external magnetic field in the principal axes system of the field gradient tensor must be considered. The average reduced density operator is

$$\langle \sigma(\tau_1 + \tau) \rangle_{\omega_q} = I_y \, \langle \cos[\omega_q(\tau_1 - \tau)] \rangle_{\omega_q} - (I_z I_x + I_x I_z) \, \langle \sin[\omega_q(\tau_1 - \tau)] \rangle_{\omega_q} \tag{5.16}$$

For finite arguments, the average trigonometric functions tend to be zero owing to the ω_q distribution. However, the cosine terms becomes independent of the quadrupole coupling constant if $\tau = \tau_1$. Then the average density operator takes the value

$$\langle \sigma(2\tau_1) \rangle_{\omega_q} = I_y \tag{5.17}$$

This is the condition under which the **quadrupolar solid echo** appears. It coincides with that for dipolar solid echoes (Eq. 4.8). In the absence of relaxation, the complex transverse rotating-frame magnetization is completely recovered:

$$m(2\tau_1) = iM_0 \tag{5.18}$$

In the pulse sequence 5.1 the RF pulse phases were chosen in quadrature which is the optimum adjustment for the solid echo. By contrast, in-phase pulses produce no directly detectable signals after the second pulse. For example, the pulse sequence

$$(\pi/2)_x \;-\; \tau_1 \;-\; (\pi/2)_x \;-\; \tau \;-\; \text{(acquisition)} \tag{5.19}$$

leads to a reduced density operator immediately after the second pulse $(\pi/2)_x$

$$\sigma(\tau_1+) = -I_z + (I_x I_y + I_y I_x) \tag{5.20}$$

The first term corresponds to z magnetization, the second to double-quantum coherences (see Eq. 52.9). Thus, no detectable single-quantum coherences are involved in this case.

5.2
Three-Pulse Quadrupolar Solid Echoes

Extending the pulse sequence by a third RF pulse,

$$(\pi/2)_x \;-\; \tau_1 \;-\; (\beta)_y \;-\; \tau_2 \;-\; (\gamma)_y \;-\; \tau \;-\; \text{(acquisition)} \tag{5.21}$$

gives rise to a number of further echoes [48] in analogy to the dipolar case. A version which is of particular interest is the **Jeener/Broekaert pulse sequence** [208, 466]

$$(\pi/2)_x \;-\; \tau_1 \;-\; (\pi/4)_y \;-\; \tau_2 \;-\; (\pi/4)_y \;-\; \tau \;-\; \text{(acquisition)} \tag{5.22}$$

From Eq. 5.8 we derive the reduced density operator immediately after the second pulse, $(\pi/4)_y$, as

$$\sigma(\tau_1+) = I_y \cos(\omega_q \tau_1) - (I_z^2 - I_x^2) \sin(\omega_q \tau_1) \tag{5.23}$$

The right-hand side of this expression represents a combination of **single-quantum coherences** which are expressed by the first term,

$$\sigma_{sqc}(\tau_1+) = I_y \cos(\omega_q \tau_1) \tag{5.24}$$

and a **spin-alignment** term, i.e., superimposed **quadrupolar order** and **double-quantum coherences,**[1] which is represented by

$$\sigma_{sa} = -(I_z^2 - I_x^2)\sin(\omega_q \tau_1) \tag{5.26}$$

In terms of the basis operators given in Chap. 52 the operator part of the latter can be written in the form

$$(I_z^2 - I_x^2) = \frac{1}{2}(3\mathcal{O}_4 - \mathcal{O}_8) \tag{5.27}$$

The single-quantum coherences are dephased under the action of the quadrupolar Hamiltonian $\mathcal{H}_q = \hbar\omega_q \mathcal{O}_4$, whereas the spin-alignment contribution (second term on the right-hand side of Eq. 5.23) is not affected owing to the vanishing commutators (see Table 52.1 on page 496)

$$[\mathcal{O}_4, \mathcal{O}_4] = 0 \qquad \text{and} \qquad [\mathcal{O}_4, \mathcal{O}_8] \overset{(I=1)}{=} 0 \tag{5.28}$$

At the end of the second pulse interval the reduced density operator is thus

$$\sigma(\tau_1 + \tau_2 -) = [I_y \cos(\omega_q \tau_2) - (I_z I_x + I_x I_z)\sin(\omega_q \tau_2)]\cos(\omega_q \tau_1) - (I_z^2 - I_x^2)\sin(\omega_q \tau_1) \tag{5.29}$$

The third pulse, $(\pi/4)_y$, transfers this into

$$\sigma(\tau_1 + \tau_2 +) = [I_y \cos(\omega_q \tau_2) - (I_z^2 - I_x^2)\sin(\omega_q \tau_2)]\cos(\omega_q \tau_1) + (I_z I_x + I_x I_z)\sin(\omega_q \tau_1) \tag{5.30}$$

After a further evolution period under the action of the quadrupolar Hamiltonian we finally get

$$\sigma(\tau_1 + \tau_2 + \tau) = \sigma_{(sqc)}(\tau_1, \tau_2, \tau) + \sigma_{(sa)}(\tau_1, \tau) \tag{5.31}$$

where the contribution

$$\begin{aligned}\sigma_{(sqc)}(\tau_1, \tau_2, \tau) &= [\{I_y \cos(\omega_q \tau) + (I_z I_x + I_x I_z)\sin(\omega_q \tau)\}\cos(\omega_q \tau_2) \\ &\quad - (I_z^2 - I_x^2)\sin(\omega_q \tau_2)]\cos(\omega_q \tau_1)\end{aligned} \tag{5.32}$$

originates from the single-quantum coherences in the τ_2 interval, and

$$\begin{aligned}\sigma_{(sa)}(\tau_1, \tau) &= \{(I_z I_x + I_x I_z)\cos(\omega_q \tau) - I_y \sin(\omega_q \tau)\}\sin(\omega_q \tau_1) \\ &= \frac{1}{2}(I_z I_x + I_x I_z)\{\sin[\omega_q(\tau_1 + \tau)] + \sin[\omega_q(\tau_1 - \tau)]\} \\ &\quad - \frac{1}{2}I_y\{\cos[\omega_q(\tau_1 - \tau)] - \cos[\omega_q(\tau_1 + \tau)]\}\end{aligned} \tag{5.33}$$

---

[1]The matrix representation of this operator term is

$$I_z^2 - I_x^2 = \frac{1}{2}\begin{pmatrix} 1 & 0 & -1 \\ 0 & -2 & 0 \\ -1 & 0 & 1 \end{pmatrix} \tag{5.25}$$

The off-diagonal elements refer to double-quantum coherences (compare footnote 8 on page 34).

emerges from the spin-alignment term which - in the absence of relaxation and molecular reorientations - is independent of the duration of that interval.

The NMR signal is determined by the average over all $\omega_q = \omega_q(\vartheta, q)$ values occurring in the sample. This average, in particular, refers to the polar angle ϑ which the principal axis of the field gradient tensor forms with the z axis. In samples with powder geometry the average of the products of trigonometric functions in the above density operators tend to cancel unless the arguments are such that no dependence on ω_q remains.

Consider the reduced density operator (Eq. 5.33) of the spin alignment term. The complex transverse magnetization arising from this term is

$$m(\tau_1 + \tau_2 + \tau) = \frac{i}{2}M_0\{\langle\cos[\omega_q(\tau_1 - \tau)]\rangle_{\omega_q} - \langle\cos[\omega_q(\tau_1 + \tau)]\rangle_{\omega_q}\} \qquad (5.34)$$

The first cosine term reaches its maximum for $\tau = \tau_1$ as a **spin-alignment echo**, whereas the second tends to be averaged out owing to its ω_q dependence. In the absence of relaxation or molecular reorientations in the time scale of τ_2 the spin-alignment echo thus recovers half of the signal that was present immediately after the first 90° pulse.

In real systems, double-quantum coherences tend to relax much faster than quadrupolar order so that the **Jeener/Broekaert or quadrupolar-order transfer echo** is determined solely by the quadrupolar-order part of the density operator in the τ_2 interval. The complex transverse magnetization at the echo maximum is then

$$m(2\tau_1 + \tau_2) = \frac{i}{4}M_0 \qquad (5.35)$$

Thus the analogy to the dipolar-order transfer echo appears to be perfect. In this treatment we have neither considered relaxation (see Sect. 12.2.2) nor "exchange" processes (see Chap. 23). The latter are understood in a generalized sense as processes changing the quadrupolar coupling constant in the course of the pulse sequence. This in particular comprises fluctuations of the field-gradient tensor by molecular reorientations.

CHAPTER 6

Dipolar and Quadrupolar Magic Echoes

6.1
Principle

In the last two chapters it became obvious that dipolar and quadrupolar echoes are formally closely interrelated. Equivalent phenomena occur with both interactions. As a further example of a solid echo we now consider homonuclear[1] magic echoes which arise under dipolar [2, 401, 434, 435] as well as quadrupolar [243] interactions. For the sake of brevity we will discuss both coupling types at the same time.

The idea of magic-echo pulse sequences is to produce coherence evolution conditions in subsequent intervals so that the sign of the interaction Hamiltonian is reversed. That is, the dephasing effect of free coherence evolution under the action of dipolar or quadrupolar couplings with "positive" Hamiltonian is compensated in another interval by evolution governed by interaction Hamiltonian with "negative" sign.[2]

The nonsecular parts of the spin interaction Hamiltonians are again neglected, i.e., spin-lattice relaxation is not considered. The coherence evolution is suitably treated in the "tilted rotating frame," where the z axis is aligned along the direction of the effective magnetic field (see Sects. 48.9 and 48.10).

The **secular homonuclear dipolar Hamiltonian** in the laboratory frame is given by (Table 46.5 on page 426)

$$\mathcal{H}_d^{(0)} = \sum_{k<\ell} a_{k\ell} \left[3I_{kz}I_{\ell z} - (\boldsymbol{I}_k \cdot \boldsymbol{I}_\ell) \right] \tag{6.1}$$

---

[1] A heteronuclear variant was suggested in [113].

[2] Another sort of "magic" reversal of the coherence evolution under the action of dipolar coupling is the **polarization echo** [382, 532]. This phenomenon even refers to (immaterial) spin diffusion mediated by flip-flop (zero-quantum) spin transitions. Spin diffusion is known to dissipate the polarization among all spins of a solid whenever the spatial distribution of the spin polarization is perturbed inhomogeneously. The spin displacement probability density has the same form as that of ordinary Fickian diffusion. Nevertheless, spin diffusion can be refocussed with the aid of a cyclic polarization transfer experiment where isolated ^{13}C spins label certain positions within the spin system. The principle of this quantum dynamical phenomenon is the same as with the magic echo: The sign of the effective coupling Hamiltonian is reversed so that the spin system evolves also in reversed direction. The collective nature of the evolution of the spin system was demonstrated by the observation of quantum interference phenomena [383] in a certain analogy to single-photon interference experiments (see ref. [406], for instance).

where

$$a_{k\ell} = \frac{\mu_0}{4\pi} \frac{\gamma^2 \hbar^2}{r_{k\ell}^3} \frac{1}{2} \left(1 - 3 \cos^2 \vartheta_{k\ell} \right) \tag{6.2}$$

The interacting spins, I_k and I_ℓ, are labeled by subscripts k and ℓ. The orientation of the internuclear vector, $r_{k\ell}$, is defined by the polar angle $\vartheta_{k\ell}$ relative to B_0 (i.e., the z axis).

The **secular quadrupolar Hamiltonian** in the laboratory ("Zeeman") frame has the form of (Table 46.4 on page 425)

$$\mathcal{H}_q^{(0)} = \frac{e^2 qQ}{4I(2I-1)} \frac{3}{2} \left(3 \cos^2 \vartheta - 1 \right) \left[3I_z^2 - I^2 \right] \tag{6.3}$$

for axially symmetric molecules. The quadrupole nucleus is characterized by the spin quantum number $I \geq 1$ and the quadrupole moment Q (i.e., the expectation value of the largest principal-axes value of the quadrupole moment tensor for the magnetic quantum number $m = I$). The electric field gradient at the position of the nucleus is specified by the quantity q, i.e., the largest principal-axes value of the electric field gradient tensor divided by the (positive) elementary charge e, and the polar angle ϑ of this particular principal axis relative to B_0.

The unitary transformation to the tilted rotating frame (Chap. 48), which applies in the presence of RF irradiation with the carrier angular frequency ω_c, is

$$\mathcal{H}_{i,TR} = TR \, \mathcal{H}_i^{(0)} \, R^{-1} T^{-1} \qquad \text{where} \quad (i = d, q) \tag{6.4}$$

This transformation makes use of the rotation operator,

$$R \equiv \exp \left\{ -i\omega_c t \sum_j I_{jz} \right\} \tag{6.5}$$

and the tilt operator,

$$T \equiv \exp \left\{ -i\Theta \sum_j I_{jy} \right\} \tag{6.6}$$

where Θ is the polar angle[3] of the effective field in the rotating frame relative to the z axis of the laboratory frame. In the case of dipolar interaction, the sums in the exponents refer to all spin members of the system, whereas quadrupolar interaction refers, of course, to single spins.

The result of the transformation Eq. 6.4 is for the dipolar Hamiltonian

$$\mathcal{H}_{d,TR} = \mathcal{H}_{d,TR}^{(0)} + P_d \, \sin^2 \Theta + Q_d \, \sin \Theta \cos \Theta \tag{6.7}$$

where

$$P_d = \frac{3}{4} \sum_{k<\ell} [I_k^+ I_\ell^+ + I_k^- I_\ell^-] \tag{6.8}$$

---

[3]This angle is not to be confused with the polar angles ϑ and $\vartheta_{k\ell}$ defined before.

$$Q_d = \frac{3}{2} \sum_{k < \ell} [I_{k z}(I_\ell^+ + I_\ell^-) + I_{\ell z}(I_k^+ + I_k^-)] \tag{6.9}$$

$$\mathcal{H}_{d,TR}^{(0)} = \frac{1}{2}(3\cos^2 \Theta - 1)\mathcal{H}_d^{(0)} \tag{6.10}$$

Equation 6.10 represents the **secular dipolar Hamiltonian in the tilted rotating frame.**

Analogously we obtain for the quadrupolar Hamiltonian

$$\mathcal{H}_{q,TR} = \mathcal{H}_{q,TR}^{(0)} + P_q \sin^2 \Theta + Q_q \sin \Theta \cos \Theta \tag{6.11}$$

where

$$P_q = \frac{3}{4}\left[(I^+)^2 + (I^-)^2 \right] \tag{6.12}$$

$$Q_q = \frac{3}{2}\left[I_z(I^+ + I^-) + (I^+ + I^-)I_z \right] \tag{6.13}$$

$$\mathcal{H}_{q,TR}^{(0)} = \frac{1}{2}(3\cos^2 \Theta - 1)\,\mathcal{H}_q^{(0)} \tag{6.14}$$

The latter equation expresses the **tilted rotating-frame secular quadrupolar Hamiltonian.**

The common feature of the Hamiltonians in Eqs. 6.10 and 6.14, is that they depend on the tilt angle in the form of a Legendre polynomial of second order,

$$P_2(\cos \Theta) = \frac{1}{2}(3\cos^2 \Theta - 1) \tag{6.15}$$

The appropriate variation of the tilt angle Θ in the course of a pulse sequence changes the sign of the spin interaction Hamiltonians and, hence, the time direction of coherence evolution. In particular, in the absence of RF irradiation[4] we have $\Theta = 0$ so that $P_2 = 1$, whereas resonant irradiation causes $\Theta = \pi/2$ so that $P_2 = -1/2$. The negative sign in the latter case is interpreted as a **"time reversal effect."** The general principle of magic-echo formation is to match evolution intervals with and without RF irradiation in a ratio 2:1. There is a manifold of operational pulse sequences permitting experiments of this sort [185, 479]. In the following we concentrate on the "magic-sandwich" variant.

6.2
The "Magic Sandwich" Pulse Sequence

Typical RF pulse schemes for magic-echo experiments are shown in Figs. 6.1a, b [479]. The pulse sequence reads

$$(\pi/2)_x - \tau_1 - (\pi/2)_y - (\alpha)_x - (\alpha)_{-x} - (\pi/2)_{-y} - \tau_2 - \text{(magic echo)} \tag{6.16}$$

---

[4]Strictly speaking, in the case of a resonantly rotating frame and no RF irradiation, the effective field vanishes and a direction cannot be defined properly. In this case, one may imagine an infinitesimal resonance offset of the rotation angular frequency leaving a corresponding field component along the z direction.

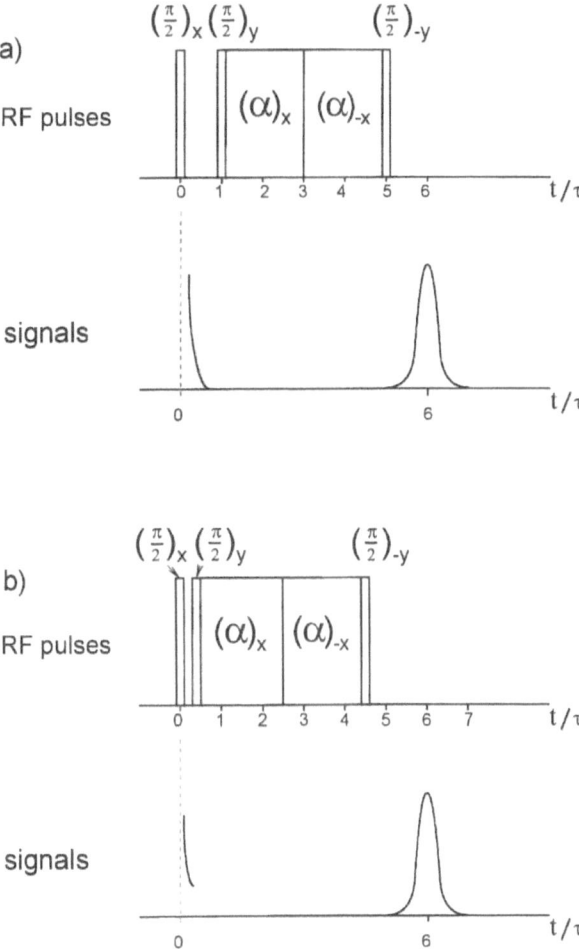

Fig. 6.1. Typical magic-echo pulse sequences: a) with symmetric; b) with asymmetric free-evolution intervals. In both cases, the total free-evolution period is a third of the total echo time. The lower traces are to represent the signal envelopes. The phase difference between the initial excitation pulse and the magic sandwich is uncritical. Other relative phases merely lead to phase-shifted signals.

A $(\pi/2)_x$ RF pulse with a phase direction arbitrarily chosen along the x axis of the rotating frame excites coherences evolving during an interval τ_1 prior to the next RF pulse. Time reversal is provided by the "magic sandwich" pulse applied in the interval between the times $t = \tau_1$ and $t = \tau_1 + 2\tau_\alpha$. The magic sandwich is composed of RF pulses characterized by $(\pi/2)_y$, $(\alpha)_x$, $(\alpha)_{-x}$, and $(\pi/2)_{-y}$. The nominal flip angle α of the "burst pulses", $(\alpha)_x$ and $(\alpha)_{-x}$, corresponds to a pulse length τ_α in each case.

The magic sandwich is followed by another free-evolution interval of length τ_2 after which the magic echo appears, where τ_2 can be shorter than, equal to, or longer than τ_1. The echo maximum representing the completely refocused coherences is reached after the echo time

$$t = \tau_1 + 2\tau_\alpha + \tau_2 = 6\tau \tag{6.17}$$

where we have introduced the scaling unit

$$\tau = \frac{1}{2}(\tau_1 + \tau_2) = \frac{\tau_\alpha}{2} \tag{6.18}$$

The total free-evolution period is $\tau_1 + \tau_2 = 2\tau$, whereas the overall width of the magic sandwich is $2(\tau_1 + \tau_2) = 4\tau$.

The first sandwich pulse, $(\pi/2)_y$, aligns the spins along the effective field produced by the $(\alpha)_{\pm x}$ "burst" pulses. That is, the tilt angle Θ changes from $0°$ to $90°$, and the value of $P_2(\cos\Theta)$ from $+1$ to $-\frac{1}{2}$. The sign reversal indicates that from now on the evolution due to spin interactions takes place in the opposite direction. The $(\pi/2)_{-y}$ pulse terminating the sandwich changes the tilt angle Θ from $90°$ back to $0°$ so that the value $+1$ of $P_2(\cos\Theta)$ is valid again. The burst-pulse width is chosen to be $\tau_\alpha = \tau_1 + \tau_2 = 2\tau$, i.e., equal to the total period for evolution under the tilt angle $\Theta = 0$. The dephasing effect of the dipolar or quadrupolar interaction is then counterbalanced by the reverse evolution. The magic echo appears after a time 6τ.

The phase reversal of the burst pulses, $(\alpha)_{\pm x}$, serves the compensation of the rotating-frame phase shifts caused by B_1 inhomogeneities. In other words, a rotary echo is formed in this way (see Chap. 3). Typical examples of magic-echo experiments are reported in [23] for protons and in [243] for deuterons. Magic-echo schemes turned out to be particularly useful for magnetic-resonance imaging of solid objects (see Sect. 35.2).

6.3
Dipolar-Coupled Two-Spin 1/2 Systems

The purpose of this section is to demonstrate the essence of magic-echo formation in the simplest case, that is, for dipolar coupled two-spin 1/2 systems consisting of equivalent spins I_1 and I_2. The treatment refers to the pulse sequence 6.16 (Fig. 6.1). The rotating-frame dipolar Hamiltonian effective for coherence evolution is (see Eq. 51.24)

$$\mathcal{H}_d^{(ev)} = hc_d I_{1z} I_{2z} \tag{6.19}$$

where

$$c_d = \frac{\mu_0}{8\pi^2} \frac{3}{2} \frac{\gamma_n^2 \hbar}{r^3}(1 - 3\cos^2\vartheta) \tag{6.20}$$

The quantities r and ϑ are the length and the polar angle of the internuclear distance vector.

The Hamiltonian effective in the rotating frame tilted by the angle Θ is modified according to (see Eq. 6.10)

$$\mathcal{H}_{d,TR}^{(ev)} = \frac{1}{2}(3\cos^2\Theta - 1)\mathcal{H}_d^{(ev)} = hc_d' I_{1z} I_{2z} \tag{6.21}$$

From this, the dipolar coupling constant effective in the tilted rotating frame is inferred as

$$c'_d = \frac{\mu_0}{8\pi^2} \frac{3}{4} \frac{\gamma_n^2 \hbar}{r^3} (1 - 3\cos^2 \vartheta)(3\cos^2 \Theta - 1) \tag{6.22}$$

The initial reduced density operator,

$$\sigma(0-) = I_{1z} + I_{2z} \tag{6.23}$$

is converted by the first RF pulse, $(\pi/2)_x$, into

$$\sigma(0+) = I_{1y} + I_{2y} \tag{6.24}$$

The operators on the right-hand side correspond to single-quantum coherences which evolve in the first pulse interval according to (see Table 51.3 on page 487)

$$\sigma(\tau_1-) = (I_{1y} + I_{2y})\cos(\pi c_d \tau_1) - (2I_{1x}I_{2z} + 2I_{2x}I_{1z})\sin(\pi c_d \tau_1) \tag{6.25}$$

The first pulse of the magic sandwich, $(\pi/2)_y$, transforms this into

$$\sigma(\tau_1+) = (I_{1y} + I_{2y})\cos(\pi c_d \tau_1) + (2I_{1x}I_{2z} + 2I_{2x}I_{1z})\sin(\pi c_d \tau_1) \tag{6.26}$$

Up to now the influence of dipolar coupling has been neglected during the RF pulses which are assumed to be sufficiently "hard." During the burst pulses this is no longer justified. The 90° phase shift of the first burst pulse results in a situation in which the spins evolve in the effective field under the action of dipolar interaction.

The next steps, consequently, are to transform the density operator to the tilted rotating frame and to treat the coherence evolution under the action of the secular part of the transformed dipolar Hamiltonian. The unitary transformation leading to the new reference frame is

$$\sigma_{TR}(\tau_1+) = e^{i\Theta(I_{1y}+I_{2y})} \sigma(\tau_1+) e^{-i\Theta(I_{1y}+I_{2y})} \tag{6.27}$$

In the tilted rotating frame, the z axis is aligned along the effective field. Under resonant RF irradiation, the tilt angle is $\Theta = \pi/2$. Evaluating the above sandwich operator expression for this tilt angle[5] gives

$$\sigma_{TR}(\tau_1+) = (I_{1y} + I_{2y})\cos(\pi c_d \tau_1) - (2I_{1x}I_{2z} + 2I_{2x}I_{1z})\sin(\pi c_d \tau_1) \tag{6.28}$$

The expression on the right-hand side seems to be identical with that of Eq. 6.25. However, note that the operator components refer to different coordinate systems.

Each of the burst pulses is of length τ_α. At the end of the first burst pulse, the density operator has evolved to

$$\begin{aligned} \sigma_{TR}(\tau_1 + \tau_\alpha-) = \ & (I_{1y} + I_{2y})\cos[\pi(c_d\tau_1 + c'_d\tau_\alpha)] \\ & - (2I_{1x}I_{2z} + 2I_{2x}I_{1z})\sin[\pi(c_d\tau_1 + c'_d\tau_\alpha)] \end{aligned} \tag{6.29}$$

---

[5]For a treatment of arbitrary tilt angles see [117].

The π phase shift of the second burst pulse is taken into account by a transformation to another tilted rotating frame according to

$$\sigma_{TR}(\tau_1 + \tau_\alpha +) = e^{-i\pi(I_{1y}+I_{2y})} \, \sigma_{TR}(\tau_1 + \tau_\alpha -) \, e^{i\pi(I_{1y}+I_{2y})} \tag{6.30}$$

Under the idealized conditions assumed here, Eq. 6.29 is not affected by this transformation apart from the fact that the spin-operator components are now defined in the new frame:

$$\sigma_{TR}(\tau_1 + \tau_\alpha +) = (I_{1y} + I_{2y}) \cos[\pi(c_d\tau_1 + c'_d\tau_\alpha)] \\ - (2I_{1x}I_{2z} + 2I_{2x}I_{1z}) \sin[\pi(c_d\tau_1 + c'_d\tau_\alpha)] \tag{6.31}$$

However, the phase shift between the two burst pulses becomes crucial if the RF field amplitude B_1 were distributed inhomogeneously. In this case, a rotary echo (see Chap. 3) must be formed at the end of the second burst pulse in order to reach the full recovery of the signal.

The evolution in the second burst interval results in

$$\sigma_{TR}(\tau_1 + 2\tau_\alpha -) = (I_{1y} + I_{2y}) \cos[\pi(c_d\tau_1 + c'_d 2\tau_\alpha)] \\ - (2I_{1x}I_{2z} + 2I_{2x}I_{1z}) \sin[\pi(c_d\tau_1 + c'_d 2\tau_\alpha)] \tag{6.32}$$

The transformation back to the rotating frame, and the effect of the terminal $(\pi/2)_{-y}$ pulse of the magic sandwich, are formally expressed by

$$\sigma(\tau_1 + 2\tau_\alpha +) =$$

$$e^{-i(\pi/2)(I_{1y}+I_{2y})} \underbrace{e^{-i(\pi/2)(I_{1y}+I_{2y})} \, \sigma_{TR}(\tau_1 + 2\tau_\alpha -) \, e^{i(\pi/2)(I_{1y}+I_{2y})}}_{\sigma(\tau_1 + 2\tau_\alpha -)} e^{i(\pi/2)(I_{1y}+I_{2y})}$$

$$\tag{6.33}$$

The combined application of both intercalated unitary transformations leaves the density operator formally unchanged, so that

$$\sigma(\tau_1 + 2\tau_\alpha +) = (I_{1y} + I_{2y}) \cos[\pi(c_d\tau_1 + c'_d 2\tau_\alpha)] \\ - (2I_{1x}I_{2z} + 2I_{2x}I_{1z}) \sin[\pi(c_d\tau_1 + c'_d 2\tau_\alpha)] \tag{6.34}$$

Note that the spin-operator components on the right-hand side are defined by a reference frame other than that relevant for Eq. 6.32.

Accounting for the evolution in the subsequent τ_2 interval, the reduced density operator is found to be

$$\sigma(\tau_1 + 2\tau_\alpha + \tau_2) = (I_{1y} + I_{2y}) \cos[\pi(c_d\tau_1 + c_d\tau_2 + 2c'_d\tau_\alpha)] \\ + (2I_{1x}I_{2z} + 2I_{2x}I_{1z}) \sin[\pi(c_d\tau_1 + c_d\tau_2 + 2c'_d\tau_\alpha)] \tag{6.35}$$

The first operator term on the right-hand side corresponds to in-phase coherences, whereas the second term represents antiphase coherences.

In samples with powder geometry, the dipolar-coupling constants c_d and c'_d, Eqs. 6.20 and 6.22, vary within a certain range according to the orientational distribution of the internuclear vector. We therefore form the powder average $\langle \sigma(\tau_1 + 2\tau_\alpha + \tau_2) \rangle$ of the density operator given in Eq. 6.35. For finite arguments, the cosine and sine factors are destructively reduced in the average or even canceled. However, if the time intervals are chosen in such a way that the argument becomes zero, the cosine term no longer depends on the dipolar coupling constant. That is, the in-phase operator term reaches a maximum, whereas the antiphase term vanishes in the average.

Thus, the condition for the **magic-echo maximum** is

$$\langle \cos[\pi(c_d\tau_1 + c_d\tau_2 + 2c'_d\tau_\alpha)] \rangle = 1 \tag{6.36}$$

or

$$c_d(\tau_1 + \tau_2) = -2c'_d\tau_\alpha \tag{6.37}$$

For the tilt angle $\Theta = \pi/2$ during the burst pulses, the dipolar coupling constants are related as $c'_d = -c_d/2$, so that the magic-echo condition reads

$$\tau_1 + \tau_2 = \tau_\alpha \tag{6.38}$$

At the echo maximum, the average reduced density operator becomes

$$\langle \sigma(3\tau_\alpha) \rangle = I_{1y} + I_{2y} \tag{6.39}$$

so that the complex transverse magnetization is recovered according to

$$m(3\tau_\alpha) = iM_0 \tag{6.40}$$

Obviously the total initial magnetization is restored provided that spin-lattice relaxation is negligible. The condition at Eq. 6.37 also elucidates that it does not matter whether the two free-evolution intervals are symmetrical or not.

6.4
Mixed Echoes

The **magic-echo** pulse sequences shown in Fig. 6.1 selectively recover spin coherences which have been defocused by bilinear homonuclear spin interactions, i.e., dipolar or quadrupolar couplings. On the other hand, phase shifts caused by linear spin interactions in combination with field inhomogeneities and chemical shifts, and by bilinear heteronuclear spin couplings, are not compensated. The reason is that the net effect of the magic sandwich vanishes in these cases.

With **Hahn two-pulse echoes** (see Fig. 2.1) the situation is just reversed. This pulse sequence fails to refocus dephasings by bilinear homonuclear spin interactions as substantiated in Sect. 4.1, whereas those caused by linear spin interactions and bilinear heteronuclear spin couplings are compensated for.

However, a slight modification of the symmetric magic-sandwich pulse technique (Fig. 6.1a) permits one to generate echoes in both respects at one time. The

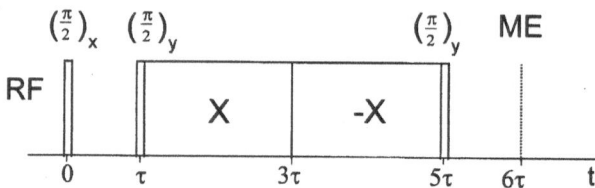

Fig. 6.2. RF pulse sequence for the production of mixed echoes (ME) by refocusing phase shifts due to homonuclear dipolar or quadrupolar couplings as well as those due to heteronuclear dipolar spin interactions and any sort of field inhomogeneities. The phases of the burst pulses are indicated by the letters x and $-x$.

formation of such **"mixed echoes"** recovering coherences dephased due to linear as well as bilinear spin interactions is made possible by the sequence [117, 331]

$$(\pi/2)_x \; - \; \tau \; - \; (\pi/2)_y \; - \; (\alpha)_x \; - \; (\alpha)_{-x} \; - \; (\pi/2)_y \; - \; \tau \; - \; (\text{mixed echo}) \quad (6.41)$$

where the total length of the burst pulses is 4τ (see Fig. 6.2). The only difference to the magic-echo pulse sequence (Fig. 6.1a) is that the phase of the last RF pulse of the magic sandwich is now opposite.

The effect of the modification can be perceived by considering the situation arising when a $(180°)_y$ pulse is inserted into the middle of the magic sandwich. This pulse would act just as in the two-pulse Hahn echo sequence. If the RF amplitude B_1 of the burst pulses is much larger than any local-field offsets ΔB by inhomogeneities or secular heteronuclear interactions, the $180°$ pulse can equivalently be attached back-to-back to the last $(90°)_{-y}$ pulse. That is, in the actual mixed-echo pulse sequence the composite back-to-back pulse, $(90°)_{-y}(180°)_y$, can be replaced by the pulse $(90°)_y$ with the same result. Examining the above treatment makes it obvious that this modification does not affect refocusing of the phase shifts by homonuclear bilinear spin interactions, while Hahn echo formation is activated by the $(180°)_y$ pulse. Mixed echoes turn out to be particularly useful in context with magnetic-resonance imaging schemes for semirigid or heterogeneous materials (see Sect. 35.2).

CHAPTER 7

Coherence Transfer of *J*-Coupled Spins

The influence of indirect spin-spin interaction (briefly *J* coupling) on the evolution of spin coherences becomes appreciable if the much stronger dipolar or quadrupolar interactions are largely averaged out by molecular motions. This is the realm of high-resolution NMR which applies to low-viscous isotropic liquids in particular. Similar conditions can also be produced in solids by experimental averaging procedures such as magic-angle spinning or multi-pulse line narrowing [336]. As outlined before, echoes are understood as coherence dephasing/rephasing phenomena. Therefore any interaction and any measure influencing the coherence evolution and the coherence pathways affect the formation of spin echoes.

In this sense, indirect spin-spin interaction is the origin of a large variety of phenomena. Generally, an RF pulse applied to coupled spin systems gives rise to the conversion of the coherence order whereas free evolution in pulse intervals conserves it (compare Chap. 51). The consequence is that coupled spins pursue different "coherence pathways" in the course of a pulse sequence. In particular, **coherence-transfer echoes** may appear which are indicative for certain coherence pathways.

In the treatment of three-pulse solid echoes arising under the evolution of dipolar or quadrupolar couplings in solids, we have already encountered transfer processes between different orders of multiple-quantum coherences and between dipolar or quadrupolar order on the one hand, and spin coherences on the other. For example, it was shown that single-quantum coherence echoes in the detection interval can be traced back to multiple-quantum coherences or dipolar/quadrupolar order in previous intervals. Analogous effects arise with *J*-coupled spin systems. Generally, whenever three or more RF pulses are applied to a spin system with more than two eigenstates one must reckon with coherence or order transfer echoes [52, 139, 191].

Treatments of coherence pathways in the course of a pulse sequence must refer to the size and type of the spin system. All important phenomena can nevertheless already be demonstrated and examined in the simplest situation, a two-spin 1/2 system AX of weakly coupled, i.e., inequivalent, nuclei A and X. We therefore restrict ourselves to the treatment of this case. Larger spin systems will be considered later in context with volume-selective spectroscopy, for instance (see Chap. 37). All analytical calculations will be performed using the product operator formalism outlined in Chap. 51.

7.1
Two RF Pulses

We reconsider the RF pulse sequence displayed in Fig. 2.1,

$$(\pi/2)_x - \tau_1 - (\beta)_y - \tau_2 - \text{(acquisition)} \tag{7.1}$$

by accounting now for J coupling effects of weakly coupled two-spin 1/2 systems consisting of spins I_1 and I_2. The condition for weak coupling is

$$2\pi|J| \ll |\Omega_1 - \Omega_2| \tag{7.2}$$

where J is the coupling constant. Ω_1 and Ω_2 are the (angular) frequency offsets of the two spins from the RF carrier. The offsets may originate from chemical shifts, field inhomogeneities or deliberately applied field gradients. The subscripts refer to the chemical shift, i.e., the offset contribution which is specific to the spin species.

The RF pulses are assumed to be non-selective ("hard") according to the condition

$$|\gamma_n|B_1 \gg |\Omega_{1,2}| \tag{7.3}$$

so that during the pulses the effective Hamiltonian is determined by the RF contribution alone. As a further simplification, the consideration of relaxation is deferred until later (Part II).

At the beginning of the pulse sequence, the spin systems are regarded as being in thermal equilibrium, so that the initial reduced density operator is

$$\sigma(0-) = I_{1z} + I_{2z} \tag{7.4}$$

The first RF pulse, $(\pi/2)_x$, changes the density operator to

$$\sigma(0+) = I_{1y} + I_{2y} \tag{7.5}$$

corresponding to superimposed single-quantum coherences. These evolve according to the rules summarized in Table 51.3 on page 487.

At the end of the τ_1 interval, the density operator is of the form

$$\begin{aligned}
\sigma(\tau_1-) = \;& [\, I_{1y}\cos(\Omega_1\tau_1) + I_{1x}\sin(\Omega_1\tau_1) \\
& + I_{2y}\cos(\Omega_2\tau_1) + I_{2x}\sin(\Omega_2\tau_1)\,]\cos(\pi J\tau_1) \\
& - [\, 2I_{1x}I_{2z}\cos(\Omega_1\tau_1) - 2I_{1y}I_{2z}\sin(\Omega_1\tau_1) \\
& + 2I_{2x}I_{1z}\cos(\Omega_2\tau_1) - 2I_{2y}I_{1z}\sin(\Omega_2\tau_1)\,]\sin(\pi J\tau_1)
\end{aligned} \tag{7.6}$$

The second pulse is characterized by a tip angle β and a B_1 field direction along the y axis of the rotating frame. This pulse produces five fundamentally different terms,[1]

$$\sigma(\tau_1+) = \sigma_{0qc} + \sigma_{1qc} + \sigma_{2qc} + \sigma_{so} + \sigma_{lm} \tag{7.7}$$

---

[1]The rules for the interpretation of the spin-operator terms are summarized in Table 51.1 on page 482.

The contributions σ_{0qc}, σ_{1qc}, σ_{2qc} refer to **zero-**, **single-**, and **double-quantum co-herences**, respectively. σ_{so} represents **longitudinal scalar order** in complete analogy to the dipolar- or quadrupolar-order phenomena discussed before in context with solids. The reduced density operator of ordinary **longitudinal magnetization** is σ_{lm}.

All signals are generated by transverse magnetization, i.e., by single-quantum coherences. In context with two-pulse sequences, it therefore suffices to concentrate on the term σ_{1qc}. It consists of all sorts of in-phase and antiphase contributions (see Table 51.1 on page 482):

$$
\begin{aligned}
\sigma_{1qc}(\tau_1+) = {} & [\, I_{1x}\sin(\Omega_1\tau_1) + I_{2x}\sin(\Omega_2\tau_1)\,]\cos\beta\cos(\pi J\tau_1) \\
& + [\, I_{1y}\cos(\Omega_1\tau_1) + I_{2y}\cos(\Omega_2\tau_1)\,]\cos(\pi J\tau_1) \\
& - 2I_{1x}I_{2z}\,[\,\cos^2\beta\cos(\Omega_1\tau_1) - \sin^2\beta\cos(\Omega_2\tau_1)\,]\sin(\pi J\tau_1) \\
& - 2I_{2x}I_{1z}\,[\,\cos^2\beta\cos(\Omega_2\tau_1) - \sin^2\beta\cos(\Omega_1\tau_1)\,]\sin(\pi J\tau_1) \\
& + [\, 2I_{1y}I_{2z}\sin(\Omega_1\tau_1) + 2I_{2y}I_{1z}\sin(\Omega_2\tau_1)\,]\cos\beta\sin(\pi J\tau_1) \quad (7.8)
\end{aligned}
$$

Free evolution under the influence of J coupling and frequency offsets in an interval τ_2 following the second RF pulse leads to

$$
\begin{aligned}
\sigma_{1qc}(\tau_1 + \tau_2) = {} & I_{1x}\{\,[\,a_1\cos(\Omega_1\tau_2) + a_3\sin(\Omega_1\tau_2)\,]\cos(\pi J\tau_2) \\
& - [\,a_5\sin(\Omega_1\tau_2) + a_7\cos(\Omega_1\tau_2)\,]\sin(\pi J\tau_2)\,\} \\[4pt]
& + I_{2x}\{\,[\,a_2\cos(\Omega_2\tau_2) + a_4\sin(\Omega_2\tau_2)\,]\cos(\pi J\tau_2) \\
& - [\,a_6\sin(\Omega_2\tau_2) + a_8\cos(\Omega_2\tau_2)\,]\sin(\pi J\tau_2)\,\} \\[4pt]
& - I_{1y}\{\,[\,a_1\sin(\Omega_1\tau_2) - a_3\cos(\Omega_1\tau_2)\,]\cos(\pi J\tau_2) \\
& + [\,a_5\cos(\Omega_1\tau_2) - a_7\sin(\Omega_1\tau_2)\,]\sin(\pi J\tau_2)\,\} \\[4pt]
& - I_{2y}\{\,[\,a_2\sin(\Omega_2\tau_2) - a_4\cos(\Omega_2\tau_2)\,]\cos(\pi J\tau_2) \\
& + [\,a_6\cos(\Omega_2\tau_2) - a_8\sin(\Omega_2\tau_2)\,]\sin(\pi J\tau_2)\,\} \\[4pt]
& + 2I_{1x}I_{2z}\{\,[\,a_1\sin(\Omega_1\tau_2) - a_3\cos(\Omega_1\tau_2)\,]\sin(\pi J\tau_2) \\
& - [\,a_5\cos(\Omega_1\tau_2) - a_7\sin(\Omega_1\tau_2)\,]\cos(\pi J\tau_2)\,\} \\[4pt]
& + 2I_{2x}I_{1z}\{\,[\,a_2\sin(\Omega_2\tau_2) - a_4\cos(\Omega_2\tau_2)\,]\sin(\pi J\tau_2) \\
& - [\,a_6\cos(\Omega_2\tau_2) - a_8\sin(\Omega_2\tau_2)\,]\cos(\pi J\tau_2)\,\} \\[4pt]
& + 2I_{1y}I_{2z}\{\,[\,a_1\cos(\Omega_1\tau_2) + a_3\sin(\Omega_1\tau_2)\,]\sin(\pi J\tau_2) \\
& + [\,a_5\sin(\Omega_1\tau_2) + a_7\cos(\Omega_1\tau_2)\,]\cos(\pi J\tau_2)\,\}
\end{aligned}
$$

$$+2I_{2y}I_{1z} \{ [a_2 \cos(\Omega_2\tau_2) + a_4 \sin(\Omega_2\tau_2)] \sin(\pi J\tau_2)$$
$$+[a_6 \sin(\Omega_2\tau_2) + a_8 \cos(\Omega_2\tau_2)] \cos(\pi J\tau_2) \} \qquad (7.9)$$

where

$$a_1 = \sin(\Omega_1\tau_1) \cos(\pi J\tau_1) \cos\beta$$
$$a_2 = \sin(\Omega_2\tau_1) \cos(\pi J\tau_1) \cos\beta$$
$$a_3 = \cos(\Omega_1\tau_1) \cos(\pi J\tau_1)$$
$$a_4 = \cos(\Omega_2\tau_1) \cos(\pi J\tau_1)$$
$$a_5 = [\cos(\Omega_1\tau_1) \cos^2\beta - \cos(\Omega_2\tau_1) \sin^2\beta] \sin(\pi J\tau_1)$$
$$a_6 = [\cos(\Omega_2\tau_1) \cos^2\beta - \cos(\Omega_1\tau_1) \sin^2\beta] \sin(\pi J\tau_1)$$
$$a_7 = \sin(\Omega_1\tau_1) \sin(\pi J\tau_1) \cos\beta$$
$$a_8 = \sin(\Omega_2\tau_1) \sin(\pi J\tau_1) \cos\beta$$

These formulae, albeit simple and symmetric in structure, look a bit unwieldy. However, as will be seen in the following discussions, this in principle is the general basis of a wealth of phenomena common in NMR.

7.1.1
Correlated Two-Dimensional Spectroscopy

Equation 7.9 represents all single-quantum coherences appearing after the second pulse. This expression simplifies considerably if the tip angle of the second pulse is chosen to be $\beta = \pi/2$:

$$\begin{aligned}
\sigma_{1qc}(\tau_1 + \tau_2) = \; & I_{1x} \{ \tilde{a}_3 \sin(\Omega_1\tau_2) \cos(\pi J\tau_2) - \tilde{a}_5 \sin(\Omega_1\tau_2) \sin(\pi J\tau_2) \} \\
& + I_{2x} \{ \tilde{a}_4 \sin(\Omega_2\tau_2) \cos(\pi J\tau_2) - \tilde{a}_6 \sin(\Omega_2\tau_2) \sin(\pi J\tau_2) \} \\
& - I_{1y} \{ -\tilde{a}_3 \cos(\Omega_1\tau_2) \cos(\pi J\tau_2) + \tilde{a}_5 \cos(\Omega_1\tau_2) \sin(\pi J\tau_2) \} \\
& - I_{2y} \{ -\tilde{a}_4 \cos(\Omega_2\tau_2) \cos(\pi J\tau_2) + \tilde{a}_6 \cos(\Omega_2\tau_2) \sin(\pi J\tau_2) \} \\
& + 2I_{1x}I_{2z} \{ -\tilde{a}_3 \cos(\Omega_1\tau_2) \sin(\pi J\tau_2) - \tilde{a}_5 \cos(\Omega_1\tau_2) \cos(\pi J\tau_2) \} \\
& + 2I_{2x}I_{1z} \{ -\tilde{a}_4 \cos(\Omega_2\tau_2) \sin(\pi J\tau_2) - \tilde{a}_6 \cos(\Omega_2\tau_2) \cos(\pi J\tau_2) \} \\
& + 2I_{1y}I_{2z} \{ \tilde{a}_3 \sin(\Omega_1\tau_2) \sin(\pi J\tau_2) + \tilde{a}_5 \sin(\Omega_1\tau_2) \cos(\pi J\tau_2) \} \\
& + 2I_{2y}I_{1z} \{ \tilde{a}_4 \sin(\Omega_2\tau_2) \sin(\pi J\tau_2) + \tilde{a}_6 \sin(\Omega_2\tau_2) \cos(\pi J\tau_2) \}
\end{aligned}$$

$$(7.10)$$

where

$$\tilde{a}_3 = \cos(\Omega_1\tau_1) \cos(\pi J\tau_1)$$
$$\tilde{a}_4 = \cos(\Omega_2\tau_1) \cos(\pi J\tau_1)$$
$$\tilde{a}_5 = -\cos(\Omega_2\tau_1) \sin(\pi J\tau_1)$$
$$\tilde{a}_6 = -\cos(\Omega_1\tau_1) \sin(\pi J\tau_1)$$

There are terms referring to in-phase or antiphase coherences of the x or y components of each of the two spin species. As summarized in Table 51.1 on page 482, **in-phase coherence** means that the isochromats corresponding to the two doublet lines are in-phase, whereas they are antiphase for **antiphase coherences**. In the latter case, the operator terms are products of a transverse spin-operator component (of the "active" coupling partner) and a longitudinal component (of a "passive" counterpart).

The "coherence pathways" resulting in Eq. 7.10 are of the type

> *single-quantum coherence with* Ω_i → *single-quantum coherence with* Ω_j

where $i, j = 1, 2$. The indices i and j can be different or equal depending on whether the initial single-quantum coherence is converted by the second RF pulse into another single-quantum coherence or not. Evidently, all possible sorts of pathways occur. They are disclosed by products $\cos(\Omega_i \tau_1) \cos(\Omega_j \tau_2)$, $\sin(\Omega_i \tau_1) \sin(\Omega_j \tau_2)$, $\cos(\Omega_i \tau_1) \sin(\Omega_j \tau_2)$, and $\sin(\Omega_i \tau_1) \cos(\Omega_j \tau_2)$ in the coefficient terms of the spin operators.

In particular, the terms with prefactors $\bar{a}_5$ and $\bar{a}_6$ stand for **"coherence transfer"** from single-quantum coherences evolving with the offset frequency Ω_1 in the τ_1 interval to single-quantum coherences precessing with the frequency Ω_2 in the τ_2 domain, and vice versa. The terms indicating transfer between different coherences are characterized by products $\cos(\Omega_{1,2}\tau_1) \sin(\Omega_{2,1}\tau_2)$ and $\cos(\Omega_{1,2}\tau_1) \cos(\Omega_{2,1}\tau_2)$. Such factors reveal **"mixing"** of the coherences in different evolution intervals by the second RF pulse.

This is the basis of the Jeener experiment or - as it is usually called - **homonuclear "correlated spectroscopy"** (COSY). Two-dimensional spectra obtained in this way considerably facilitate the spectroscopic analysis of unknown compounds, and, therefore, are an immensely important tool for the chemist [14, 139].

The complex transverse rotating-frame magnetization corresponding to Eq. 7.10 is

$$m(\tau_1, \tau_2) = \frac{n}{2}\gamma_n \hbar b \ \text{Tr}\left\{ \sigma_{1qc}(\tau_1 + \tau_2)\left(I_{1x} + I_{2x} + iI_{1y} + iI_{2y}\right) \right\} \qquad (7.11)$$

where $n/2$ is the number density of two-spin systems, and b is given by Eq. 47.27. The free-induction signal,

$$S(\tau_1, \tau_2) \propto m(\tau_1, \tau_2) \qquad (7.12)$$

refers to two time domains scaled by $t_1 \equiv \tau_1$ and $t_2 \equiv \tau_2$. The time domain t_1 is probed by incrementing τ_1 in a series of independent transients. The two-dimensional (2D) Fourier transform,

$$\tilde{S}(\omega_1, \omega_2) = \int dt_1 \int dt_2 \ S(t_1, t_2) \ e^{-i\omega_1 t_1} e^{-i\omega_2 t_2} \qquad (7.13)$$

produces a two-dimensional spectrum which may be rendered in the form of stacked plots or contour plots of the magnitude.

Typical COSY plots of an AX spin system are shown in Fig. 7.1. The **diagonal peaks** represent coherences which kept evolving in the second interval with the same

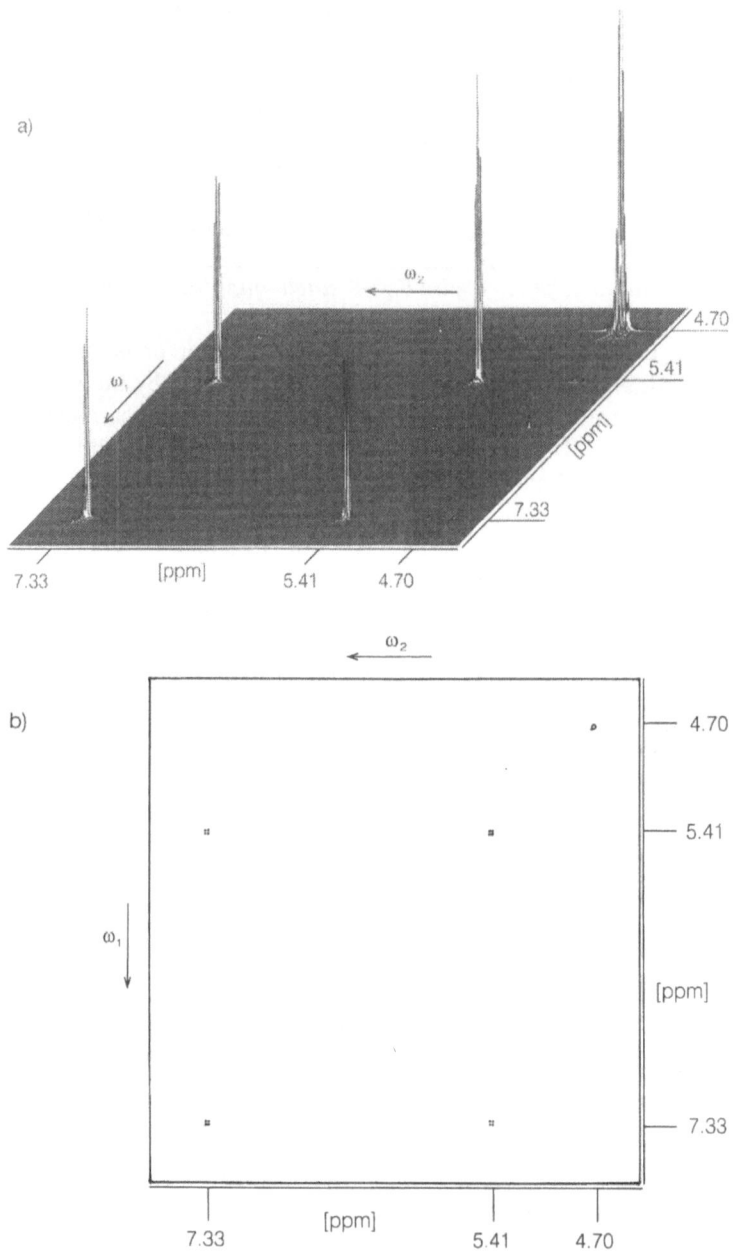

Fig. 7.1. Correlated 300 MHz proton spectra (COSY) of an AX spin system (1.3 mol/l uracil $D_2O/NaOD$ solution): **a)** projection; **b)** contour plot of the magnitude of the two-dimensional Fourier transform. The lines at 4.41 ppm and 7.33 ppm are assigned to the CH groups of the ring compound adjacent to CO/COD and ND groups, respectively. Uncoupled spins produce no cross peaks as demonstrated by the 25-fold more intense HDO line at 4.7 ppm. (courtesy of K.-H. Spohn)

offset frequencies as in the first evolution interval. The **cross peaks**, by contrast, indicate the conversion of coherences as a consequence of the mixing process by the second RF pulse. It is needless to say that uncoupled spins cannot lead to cross peaks.

7.1.2
Echo Formation

Apart from chemical-shift offsets, there may be superimposed offsets due to inho-mogeneities of the external magnetic field. Angular-frequency offsets of this sort, $\Omega_g(r)$, can be assumed to act stationarily and commonly to all spins of a spin sys-tem within a molecule. Even the strongest field gradients occurring under practical conditions do not vary the magnetic field on a molecular length scale in a spectro-scopically perceptible way. Thus, the inhomogeneity offsets are merely a function of the position of the spin system in the spatial field distribution. In the case of an AX spin system, the total offsets are

$$\Omega_1 = \Omega_{cs1} + \Omega_g(r) \tag{7.14}$$
$$\Omega_2 = \Omega_{cs2} + \Omega_g(r) \tag{7.15}$$

where Ω_{cs1} and Ω_{cs2} are the contributions from the chemical shifts.

NMR signals that can be expected under such circumstances indicate that the spatial average of Eq. 7.9 is finite. The complex transverse rotating-frame magneti-zation inducing the signal is

$$m(\tau_1 + \tau_2) = \frac{n}{2}\gamma_n\hbar b \ \mathrm{Tr}\left\{ \langle\sigma_{1qc}(\tau_1 + \tau_2)\rangle_{\Omega_g} (I_{1x} + I_{2x} + iI_{1y} + iI_{2y}) \right\} \tag{7.16}$$

Closer inspection of Eq. 7.9 shows that products of sine and cosine functions of $\Omega_1\tau_1$, $\Omega_1\tau_2$, $\Omega_2\tau_1$, or $\Omega_2\tau_2$ occur. These can be substituted by the trigonometric relations for angle sums and differences, so that the average terms are of the type

$$\langle\sin[\Omega_{1,2}(\tau_1 \pm \tau_2)]\rangle_{\Omega_g} \quad \text{and} \quad \langle\cos[\Omega_{1,2}(\tau_1 \pm \tau_2)]\rangle_{\Omega_g}$$

As with all other echoes discussed so far, trigonometric function terms with finite arguments tend to be destructively averaged out if the argument is distributed uniformly over the full angle range. The only terms relevant for the echo formation are cosine functions of arguments vanishing irrespective of the actual gradient offset. Non-trivially this is fulfilled for

$$\langle\cos[\Omega_{1,2}(\tau_1 - \tau_2)]\rangle_{\Omega_g} \quad \text{if} \quad \tau_2 = \tau_1$$

Therefore we evaluate Eq. 7.9 for the time $t = 2\tau_1$. Let us first assume a tip angle $\beta = \pi/2$ for the second RF pulse. The (unaveraged) reduced density operator then takes the form

$$\sigma_{1qc}(2\tau_1)_{\beta=\pi/2} = I_{1x}\left[\frac{1}{2}\sin(2\Omega_1\tau_1)\cos^2(\pi J\tau_1) + \sin(\Omega_1\tau_1)\cos(\Omega_2\tau_1)\sin^2(\pi J\tau_1)\right]$$

$$+I_{2x}\left[\frac{1}{2}\sin(2\Omega_2\tau_1)\cos^2(\pi J\tau_1) + \sin(\Omega_2\tau_1)\cos(\Omega_1\tau_1)\sin^2(\pi J\tau_1)\right]$$

$$+I_{1y}\left[\cos^2(\Omega_1\tau_1)\cos^2(\pi J\tau_1) + \cos(\Omega_1\tau_1)\cos(\Omega_2\tau_1)\sin^2(\pi J\tau_1)\right]$$
$$+I_{2y}\left[\cos^2(\Omega_2\tau_1)\cos^2(\pi J\tau_1) + \cos(\Omega_1\tau_1)\cos(\Omega_2\tau_1)\sin^2(\pi J\tau_1)\right]$$

$$-\frac{1}{2}\left\{2I_{1x}I_{2z}\left[\cos^2(\Omega_1\tau_1) - \cos(\Omega_1\tau_1)\cos(\Omega_2\tau_1)\right]\right.$$
$$+2I_{2x}I_{1z}\left[\cos^2(\Omega_2\tau_1) - \cos(\Omega_1\tau_1)\cos(\Omega_2\tau_1)\right]$$

$$-2I_{1y}I_{2z}\left[\frac{1}{2}\sin(2\Omega_1\tau_1) - \sin(\Omega_1\tau_1)\cos(\Omega_2\tau_1)\right]$$
$$\left.-2I_{2y}I_{1z}\left[\frac{1}{2}\sin(2\Omega_2\tau_1) - \sin(\Omega_2\tau_1)\cos(\Omega_1\tau_1)\right]\right\}\sin(2\pi J\tau_1)$$

$$(7.17)$$

Forming the average over the gradient-induced phase shifts assumed to be equally distributed in the full range, $0 \le (\Omega_g\tau_1) \le 2\pi$, leads to average reduced density operator determining the **echo amplitude**,

$$\langle\sigma_{1qc}(2\tau_1)_{\beta=\pi/2}\rangle_{\Omega_g} = (I_{1x} - I_{2x})\frac{1}{4}\sin[(\Omega_{cs1} - \Omega_{cs2})\tau_1]\,[1 - \cos(2\pi J\tau_1)]$$

$$+ (I_{1y} + I_{2y})\frac{1}{2}\left\{1 - \sin^2\frac{(\Omega_{cs1} - \Omega_{cs2})\tau_1}{2}\,[1 - \cos(2\pi J\tau_1)]\right\}$$

$$- (2I_{1x}I_{2z} + 2I_{2x}I_{1z})\frac{1}{2}\sin^2\frac{(\Omega_{cs1} - \Omega_{cs2})\tau_1}{2}\,\sin(2\pi J\tau_1)$$

$$- (2I_{1y}I_{2z} - 2I_{2y}I_{1z})\frac{1}{4}\sin[(\Omega_{cs1} - \Omega_{cs2})\tau_1]\,\sin(2\pi J\tau_1) \quad (7.18)$$

This result shows that the echo amplitude is modulated with the difference of the chemical-shift frequency offsets, and with spin-spin coupling constant. In the limit of equivalent nuclei, that is $\Omega_{cs1} = \Omega_{cs2}$, the result for the Hahn spin echo (compare Eq. 2.17) is recovered:

$$\langle\sigma_{1qc}(2\tau_1)_{\beta=\pi/2}\rangle_{\Omega_g} = \frac{1}{2}(I_{1y} + I_{2y}) \quad (7.19)$$

The reduced density operator at $t = 2\tau_1$ becomes particularly simple if the flip angle of the second pulse is chosen as $\beta = \pi$. We deduce from Eq. 7.9

$$\sigma_{1qc}(2\tau_1)_{\beta=\pi} = (I_{1y} + I_{2y})\cos(2\pi J\tau_1) - (2I_{1x}I_{2z} + 2I_{2x}I_{1z})\sin(2\pi J\tau_1) \quad (7.20)$$

This expression for the density operator at the middle of the echo is valid irrespective of the presence of field gradients. It is therefore not affected by the average over the gradient-induced offsets. Without field inhomogeneities, however, no distinct time domain signal can be expected, to which an echo phenomenon could be assigned in the sense of rephased coherences. In the presence of field gradients, all coherences before and after the echo middle tend to be destructively suppressed.

Equation 7.20 also shows, that chemical shift offsets in any case are compensated when the echo center is reached. The maximum of a 180°-pulse echo is not modulated by chemical shifts in contrast to the 90°-pulse echo represented by Eq. 7.18.

The density operator of the 180°-pulse echo, Eq. 7.20, is modulated by spin-spin coupling. In this respect, the 180° pulse obviously does not affect the evolution of the spin coherences. The explanation is that we are dealing with the homonuclear case and with non-selective pulses. It is not only the coherence phase of a spin which is "flipped" by the 180° pulse. Simultaneously the spin state of the coupling partner is changed as well. The spins therefore keep precessing away from the initial phase.

Generally, coherences are refocused in all cases when the frequency offsets are not altered by the second RF pulse. Offsets of this sort may be due to field gradients, chemical shifts, or heteronuclear spin-spin coupling. The same conclusion as for heteronuclear spin systems applies for RF pulses selectively acting on a spin-coupling partner in the homonuclear case.

7.1.3
Spin-Echo Correlated 2D Spectroscopy

Ordinary correlated two-dimensional spectroscopy, i.e., the COSY method described in Sect. 7.1.1, is based on data obtained by two-dimensional Fourier transformation of the FID beginning directly after the second RF pulse. The flip angle of this pulse is preferably adjusted to 90°. In the corresponding plots, the cross-peak coordinates are directly given by the resonance frequencies of the coupling partners.

Alternatively the signal acquisition can be delayed until the second half of the echo, so that the time domains of interest are represented on the one hand by the total echo time, $t_1 = 2\tau_1$, and on the other by the acquisition time, t_2, beginning in the center of the echo. The time domain t_1 is probed by incrementing τ_1 in subsequent transients.

Equation 7.18 reveals that the first time domain reflects the difference of the resonance frequencies of the coupled spins rather than the resonance frequencies themselves. Evolution in the second time domain (not represented by Eq. 7.18) refers to the resonance frequencies in full. Thus, in two-dimensional spectra evaluated on this basis, spectral data for "reduced" chemical shifts are juxtaposed to data referring to the full chemical shifts.

The optimal flip angle of the second pulse is 90° as assumed in the derivation of Eq. 7.18. In the experimental implementation, the only difference to COSY is the delayed data acquisition. The technique has got its own acronym, **SECSY**, standing for "**spin-echo correlation spectroscopy**."

7.1.4
Homonuclear *J* Resolved 2D Spectroscopy

The closer inspection of Eq. 7.20 suggests a further two-dimensional spectroscopy experiment. Increasing the flip angle of the second pulse to 180° enables one to perform **"homonuclear *J* resolved 2D spectroscopy"** [139]. Apart from the tip angle of the second pulse, the same pulse scheme as with the SECSY experiment is applied. In particular the time domains t_1 and t_2 are defined identically.

Equation 7.20 already implies the principle of this technique. The first time domain is probed by "phase encoding" the signal by means of increments of the interval $t_1 = 2\tau_1$ in subsequent transients. The other time domain, t_2, is defined by the second half of the spin echo which is acquired as the signal. This part of the signal is not represented by Eq. 7.20 but can be readily derived from Eq. 7.9 after having formed the spatial average.

From Eq. 7.20 we infer that the first frequency domain solely reflects *J* splittings, whereas the full spectral information is recovered in the second dimension, i.e., the different interactions can readily be separated in a two-dimensional representation. Figure 7.2 shows an application of homonuclear *J* resolved 2D spectroscopy to an AX spin system for comparison with the COSY spectra (Fig. 7.1) of the same sample.

7.2
Three RF Pulses

Analogous to the treatments of three-pulse Hahn echoes and three-pulse solid echoes, we now turn to the pulse sequence

$$(\pi/2)_x \;-\; \tau_1 \;-\; (\beta)_y \;-\; \tau_2 \;-\; (\gamma)_y \;-\; \tau_3 \;-\; \text{(acquisition)} \qquad (7.21)$$

to be discussed in connection with coherence evolution under the influence of *J* couplings. Evidently it is formally identical with the three-pulse Hahn echo sequence shown in Fig. 2.2.[2] However, the application to coupled spin systems gives rise to phenomena which are of a fundamentally different nature. Judged by the coherence pathways the spin systems take in the course of the pulse sequence, there is no real parallel in the family of Hahn three-pulse spin echoes.

The product operator treatment for a two-spin 1/2 system AX until the beginning of the τ_2 interval was outlined above and resulted in Eq. 7.7. Let us continue by specifying and examining the various contributions to the density operator in the course of further evolution.[3]

---

[2]The terminology and symbols for the time intervals normally used in context with three-pulse two-dimensional spectroscopy are partially different from those common with ordinary spin-echo experiments. The 2D nomenclature corresponds to that used in the three-pulse schemes Fig. 7.3, for instance.

[3]A treatment of a similar three-pulse sequence serving volume-selective spectral editing of an A_3X spin system can be found in Sect. 37.1.2.

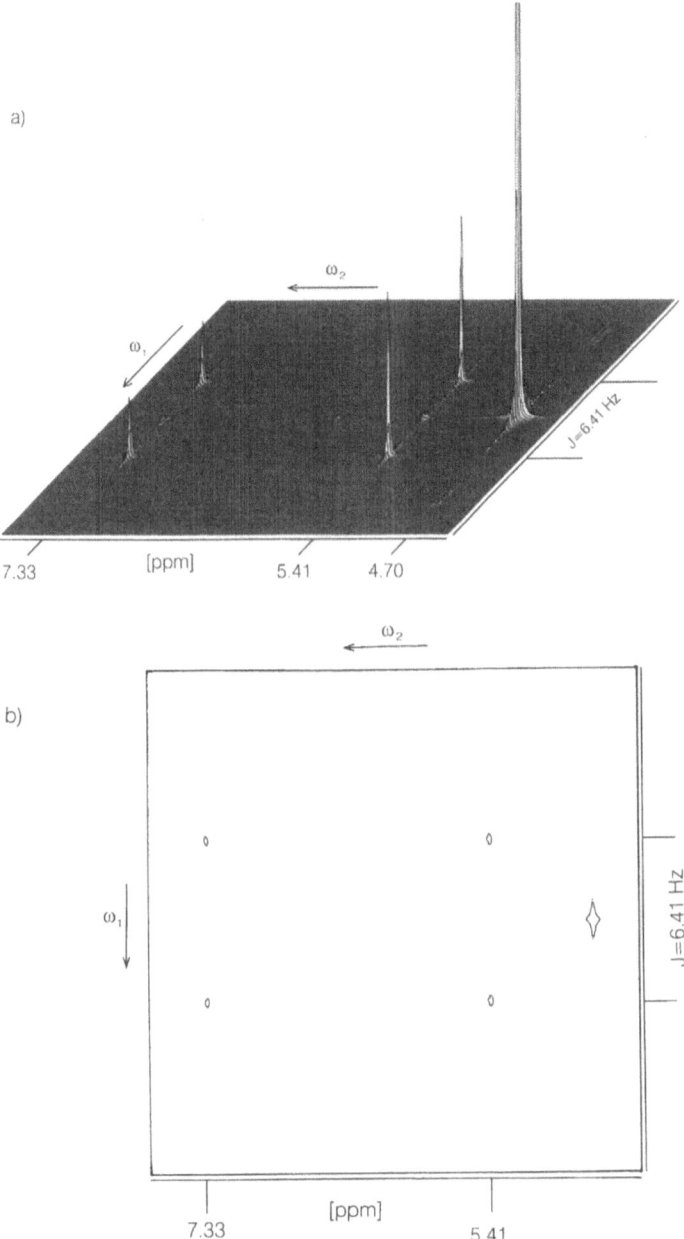

Fig. 7.2. Homonuclear (300 MHz proton) *J* resolved spectra of an AX spin system (same as in Fig. 7.1; 1.3 mol/l uracil D$_2$O/NaOD solution): **a)** projection; **b)** contour plot of the magnitude of the two-dimensional Fourier transform. The two doublets at 4.41 ppm and 7.33 ppm are assigned to the CH groups of the ring compound adjacent to CO/COD and ND groups, respectively (compare the schematic one-dimensional AX spectrum shown in Fig. 50.3). The coupling constant is 6.41 Hz. The intense HDO line appears at 4.7 ppm on the ω_2 scale. (courtesy of K.-H. Spohn)

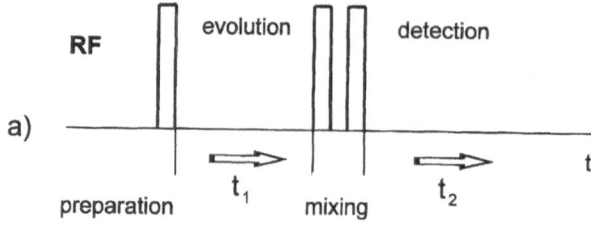

a)

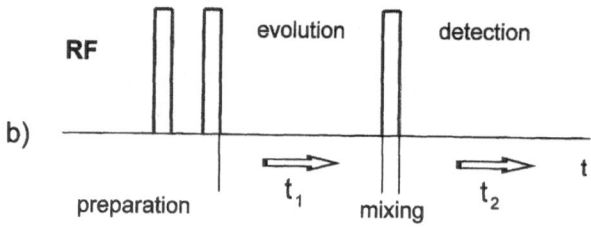

b)

Fig. 7.3. Typical pulse sequences for two-dimensional spectroscopy experiments based on three RF pulses. The evolution and detection intervals are identified with the two time domains, t_1 and t_2, respectively, of the two-dimensional Fourier transformation: **a)** scheme appropriate for DQF-COSY experiments, for instance. In this case, the mixing interval is a small fixed delay between the RF pulses converting single-quantum coherences into double-quantum coherences, and these back to single-quantum coherences. An application is shown in Fig. 7.4; **b)** sequence for multiple-quantum spectroscopy, i.e., the first two RF pulses with a fixed spacing are used for the preparation of multiple-quantum coherences. The second pulse interval acts as evolution interval to be incremented in subsequent transients. The mixing pulse converts the multiple-quantum coherences to detectable single-quantum coherences. A spectrum recorded with the aid of this pulse sequence is shown in Fig. 7.5.

7.2.1
Longitudinal-Magnetization Transfer Echo

The echo species, which is by nature closest to Hahn's stimulated echo, is the longitudinal-magnetization transfer echo. It also develops from single-quantum coherences originating in longitudinal magnetization in the τ_2 interval. However, dissimilar to the stimulated echo, the longitudinal-magnetization transfer echo is modulated by chemical shifts and spin-spin coupling as delineated in the following.

The longitudinal magnetization at the beginning of the τ_2 interval is determined by the reduced density operator

$$\sigma_{lm}(\tau_1+) = [I_{1z}\sin(\Omega_1\tau_1) + I_{2z}\sin(\Omega_2\tau_1)]\cos(\pi J\tau_1)\sin\beta \qquad (7.22)$$

In the absence of relaxation, this expression persists until the end of the interval. The third RF pulse gives rise to single-quantum coherences so that

$$\sigma_{lm}(\tau_1+\tau_2+) = [(I_{1z}\cos\gamma - I_{1x}\sin\gamma)\sin(\Omega_1\tau_1)$$

$$+(I_{2z}\cos\gamma - I_{2x}\sin\gamma)\,\sin(\Omega_2\tau_1)\,]\,\cos(\pi J\tau_1)\sin\beta \quad (7.23)$$

The further free evolution of the single-quantum coherences results in[4]

$$
\begin{aligned}
\sigma_{(lm)}(\tau_1+\tau_2+\tau_3) = &\;\{\,[\,-I_{1x}\cos(\Omega_1\tau_3)+I_{1y}\sin(\Omega_1\tau_3)\,]\,f_1 \\
&+[\,-I_{2x}\cos(\Omega_2\tau_3)+I_{2y}\sin(\Omega_2\tau_3)\,]\,f_2\,\}\,\cos(\pi J\tau_3) \\
&-\{\,[2I_{1x}I_{2z}\sin(\Omega_1\tau_3)+2I_{1y}I_{2z}\cos(\Omega_1\tau_3)]\,f_1 \\
&+[\,2I_{2x}I_{1z}\sin(\Omega_2\tau_3)+2I_{2y}I_{1z}\cos(\Omega_2\tau_3)\,]\,f_2\,\}\,\sin(\pi J\tau_3)
\end{aligned}
$$

$$(7.24)$$

where

$$
\begin{aligned}
f_1 &= \sin\beta\sin\gamma\cos(\pi J\tau_1)\sin(\Omega_1\tau_1) \\
f_2 &= \sin\beta\sin\gamma\cos(\pi J\tau_1)\sin(\Omega_2\tau_1)
\end{aligned}
$$

The residual longitudinal-magnetization terms, which do not contribute to the signal, have been discarded.

The longitudinal-magnetization transfer echo emerges from the pathway in the course of the three pulse intervals,

> *single-quantum coherences → longitudinal magnetization*
> *→ single-quantum coherences*

The contribution to transverse magnetization is

$$m_{(lm)}(\tau_1+\tau_2+\tau_3) = \frac{n}{2}\gamma_n\hbar b\,\mathrm{Tr}\,\{\,\langle\sigma_{(lm)}(\tau_1+\tau_2+\tau_3)\rangle_{\Omega_g}\,(I_{1x}+I_{2x}+iI_{1y}+iI_{2y})\,\} \quad (7.25)$$

The average over the gradient offsets Ω_g refers solely to the intervals τ_1 and τ_3 because this particular pathway does not imply any coherences in the τ_2 interval. The terms to be averaged are sine and cosine functions of $\Omega_1\tau_1$, $\Omega_1\tau_3$, $\Omega_2\tau_1$, and $\Omega_2\tau_3$. The products can be substituted by the known trigonometric relations for angle sums and differences resulting in averages of the type

$$\langle\sin[\Omega_{1,2}(\tau_1\pm\tau_3)]\rangle_{\Omega_g} \quad\text{and}\quad \langle\cos[\Omega_{1,2}(\tau_1\pm\tau_3)]\rangle_{\Omega_g} \quad (7.26)$$

The recovery of coherences in the form of an echo is indicated by vanishing arguments of the cosine functions. The maximum of the **longitudinal-magnetization transfer echo** is therefore reached for

$$\tau_3 = \tau_1 \quad (7.27)$$

The optimum flip angles are $\beta = \gamma = \pi/2$.

The closer inspection of Eq. 7.24 proves that there are no mixed product terms with respect to the offsets Ω_1 and Ω_2, i.e., there is no "cross" transfer between the single-quantum coherences in the first and in the third interval[5]. Coherences of the

---

[4]The subscript in parentheses indicates that we are now dealing with converted coherences. In the present case, $\sigma_{(lm)}$ means single-quantum coherences which can be traced back to longitudinal magnetization in the previous interval.

[5]This also implies that no "exchange" takes place (compare Sect. 23.2).

first interval are merely stored as longitudinal magnetization in the intermediate interval and then recalled as coherences of the same sort as before. In this sense, the components converted to longitudinal-magnetization in the second interval are complementary to those of the other pathways which eventually lead to cross-peaks of the COSY type.

7.2.2
Scalar-Order Transfer Echo

The scalar-order contribution to the reduced density operator after the second pulse is

$$\sigma_{so}(\tau_1+) = -2I_{1z}I_{2z}\,\frac{1}{2}\,[\,\cos(\Omega_1\tau_1) + \cos(\Omega_2\tau_1)\,]\,\sin(2\beta)\sin(\pi J\tau_1) \qquad (7.28)$$

Since only the z components of the spin operators are involved, and since relaxation effects are neglected, the density operator remains unchanged until the end of the τ_2 interval. The third pulse converts the scalar-order spin states partly into zero-quantum, single-quantum, and double-quantum coherences. Only the single-quantum coherences are of interest for detectable signals. Dropping all terms other than those representing single-quantum coherences gives[6]

$$\sigma_{(so)}(\tau_1 + \tau_2+) = \frac{1}{2}(I_{1x}I_{2z} + I_{2x}I_{1z})\,[\,\cos(\Omega_1\tau_1) + \cos(\Omega_2\tau_1)\,]$$

$$\sin(2\beta)\sin(2\gamma)\sin(\pi J\tau_1) \qquad (7.29)$$

These antiphase single-quantum coherences evolve in the τ_3 interval according to

$$\begin{aligned}
\sigma_{(s.o.)}(\tau_1 + \tau_2 + \tau_3) = d\,\{\,&[\,I_{1x}\sin(\Omega_1\tau_3) + I_{1y}\cos(\Omega_1\tau_3) + I_{2x}\sin(\Omega_2\tau_3) \\
&+ I_{2y}\cos(\Omega_2\tau_3)\,]\,\sin(\pi J\tau_3) \\
&+ [\,2I_{1x}I_{2z}\cos(\Omega_1\tau_3) - 2I_{1y}I_{2z}\sin(\Omega_1\tau_3) \\
&+ 2I_{2x}I_{1z}\cos(\Omega_2\tau_3) \\
&- 2I_{2y}I_{1z}\cos(\Omega_2\tau_3)\,]\,\cos(\pi J\tau_3)\,\} \qquad (7.30)
\end{aligned}$$

where

$$d = \frac{1}{4}[\cos(\Omega_1\tau_1) + \cos(\Omega_2\tau_1)]\sin(\pi J\tau_1)\sin(2\beta)\sin(2\gamma)$$

The scalar-order transfer echo arises after the coherence pathway in the three pulse intervals

> *single-quantum coherences $\rightarrow$ longitudinal scalar order $\rightarrow$
> single-quantum coherences*

The contribution to transverse magnetization is

$$m_{(so)}(\tau_1 + \tau_2 + \tau_3) = \frac{n}{2}\gamma_n\hbar b\,\mathrm{Tr}\,\{\,\langle\sigma_{(so)}(\tau_1 + \tau_2 + \tau_3)\rangle_{\Omega_g}\,(I_{1x} + I_{2x} + iI_{1y} + iI_{2y})\,\} \qquad (7.31)$$

---

[6]See footnote 4 on page 63.

The average accounts for the distribution of the local gradient offsets Ω_g (see Eqs. 7.14 and 7.15).

The examination of Eq. 7.30 with respect to gradient evolution terms reveals products of sine and cosine functions of $\Omega_1\tau_1$, $\Omega_1\tau_3$, $\Omega_2\tau_1$, or $\Omega_2\tau_3$. Their replacement by the known trigonometric relations for angle sums and differences leads to average terms of the form

$$\langle \sin[\Omega_{1,2}(\tau_1 \pm \tau_3)] \rangle_{\Omega_g} \quad \text{and} \quad \langle \cos[\Omega_{1,2}(\tau_1 \pm \tau_3)] \rangle_{\Omega_g} \tag{7.32}$$

An echo is formed when the argument of the cosine functions vanishes so that the maximum value is adopted. The condition for the formation of a **scalar-order transfer echo** is consequently

$$\tau_3 = \tau_1 \tag{7.33}$$

Note that this echo is superimposed to the zero-quantum coherence transfer and longitudinal-magnetization transfer echoes, and arises at a time coinciding with that of the Hahn stimulated echo of uncoupled-spin coherences. By contrast to these phenomena, the optimum flip angles for the scalar-order transfer echo are $\beta = \gamma = \pi/4$ whereas it disappears for $\beta = \pi/2$ or $\gamma = \pi/2$ (see Eq. 7.30). A general discussion follows in Sect. 7.4.

7.2.3
Zero-Quantum Coherence-Transfer Echo

At the beginning of the τ_2 interval, the density operator contribution for zero-quantum coherence generated by the second RF pulse is[7]

$$
\begin{aligned}
\sigma_{0qc}(\tau_1+) &= \frac{1}{2}\{ (2I_{1x}I_{2x} + 2I_{1y}I_{2y})\,[\cos(\Omega_1\tau_1) + \cos(\Omega_2\tau_1)]\,\cos\beta \\
&\quad -(2I_{1y}I_{2x} - 2I_{1x}I_{2y})\,[\sin(\Omega_1\tau_1) - \sin(\Omega_2\tau_1)]\}\,\sin\beta\sin(\pi J\tau_1) \\
&= \frac{1}{2}(I_1^+I_2^- + I_1^-I_2^+)\,b_1 - \frac{1}{2i}(I_1^+I_2^- - I_1^-I_2^+)\,b_2
\end{aligned}
\tag{7.34}
$$

where

$$
\begin{aligned}
b_1 &= \frac{1}{2}[\cos(\Omega_1\tau_1) + \cos(\Omega_2\tau_1)]\,\sin(2\beta)\sin(\pi J\tau_1) \\
b_2 &= [\sin(\Omega_1\tau_1) - \sin(\Omega_2\tau_1)]\,\sin\beta\sin(\pi J\tau_1)
\end{aligned}
$$

Free evolution of the AX zero-quantum coherence during τ_2 results in

$$
\begin{aligned}
\sigma_{0qc}(\tau_1 + \tau_2-) =\ & \frac{1}{2}(2I_{1x}I_{2x} + 2I_{1y}I_{2y})\{ b_1 \cos[(\Omega_1 - \Omega_2)\tau_2] - b_2 \sin[(\Omega_1 - \Omega_2)\tau_2]\} \\
& -\frac{1}{2}(2I_{1y}I_{2x} - 2I_{1x}I_{2y})\{ b_1 \sin[(\Omega_1 - \Omega_2)\tau_2] + b_2 \cos[(\Omega_1 - \Omega_2)\tau_2]\}
\end{aligned}
\tag{7.35}
$$

---

[7]The operator terms can equivalently be represented by the single-transition expressions given by Eqs. 42.76.

In this expression, a feature shows up which is characteristic for zero-quantum coherences. The evolution is determined by the **frequency-offset difference**. Therefore all contributions which are unspecific for the two spin species cancel. Zero-quantum coherences evolve independently of any influence of field inhomogeneities or field gradients. The only relevant contribution is the difference of the chemical-shift offsets.[8] According to Eqs. 7.14 and 7.15, that is

$$\Omega_1 - \Omega_2 = \Omega_{cs1} - \Omega_{cs2} \tag{7.36}$$

The zero-quantum coherence of the τ_2 interval is converted by the third RF pulse to a superposition of longitudinal scalar order and zero-, single-, and double-quantum coherences. For signals to be acquired in the τ_3 interval, only the single-quantum coherences existing in this interval are relevant, so that all other density operator contributions can be dropped.

Immediately after the third pulse, the single-quantum coherences originating in the zero-quantum coherence of the τ_2 interval are antiphase. The corresponding reduced density operator is[9]

$$\sigma_{(0qc)}(\tau_1 + \tau_2+) = (2I_{1x}I_{2z} + 2I_{2x}I_{1z})\,b_3 - (2I_{1y}I_{2z} - 2I_{2y}I_{1z})\,b_4 \tag{7.37}$$

where

$$
\begin{aligned}
b_3 &= \frac{1}{4}\{[\cos(\Omega_1\tau_1) + \cos(\Omega_2\tau_1)]\cos[(\Omega_{cs1} - \Omega_{cs2})\tau_2]\cos\beta \\
&\quad - [\sin(\Omega_1\tau_1) - \sin(\Omega_2\tau_1)]\sin[(\Omega_{cs1} - \Omega_{cs2})\tau_2]\}\sin\beta\sin(2\gamma)\sin(\pi J\tau_1) \\
b_4 &= \frac{1}{2}\{[\cos(\Omega_1\tau_1) + \cos(\Omega_2\tau_1)]\sin[(\Omega_{cs1} - \Omega_{cs2})\tau_2]\cos\beta \\
&\quad + [\sin(\Omega_1\tau_1) - \sin(\Omega_2\tau_1)]\cos[(\Omega_{cs1} - \Omega_{cs2})\tau_2]\}\sin\beta\sin\gamma\sin(\pi J\tau_1)
\end{aligned}
$$

These single-quantum coherences evolve in the τ_3 interval according to

$$
\begin{aligned}
\sigma_{(0qc)}(\tau_1 + \tau_2 + \tau_3) = \{ &I_{1x}[\,b_5\sin(\Omega_1\tau_3) + b_6\cos(\Omega_1\tau_3)\,] \\
+ &I_{2x}[\,b_5\sin(\Omega_2\tau_3) - b_6\cos(\Omega_2\tau_3)\,] \\
+ &I_{1y}[\,b_5\cos(\Omega_1\tau_3) - b_6\sin(\Omega_1\tau_3)\,] \\
+ &I_{2y}[\,b_5\cos(\Omega_2\tau_3) + b_6\sin(\Omega_2\tau_3)\,]\}\sin(\pi J\tau_3) \\
+ \{ &2I_{1x}I_{2z}[\,b_5\cos(\Omega_1\tau_3) - b_6\sin(\Omega_1\tau_3)\,] \\
+ &2I_{2x}I_{1z}[\,b_5\cos(\Omega_2\tau_3) + b_6\sin(\Omega_2\tau_3)\,] \\
+ &2I_{1y}I_{2z}[\,b_5\sin(\Omega_1\tau_3) - b_6\cos(\Omega_1\tau_3)\,] \\
+ &2I_{2y}I_{1z}[\,b_5\sin(\Omega_2\tau_3) + b_6\cos(\Omega_2\tau_3)\,]\}\cos(\pi J\tau_3)
\end{aligned} \tag{7.38}
$$

where

$$b_5 = \frac{1}{4}\sin\beta\sin(2\gamma)\sin(\pi J\tau_1)\,\{[\cos(\Omega_1\tau_1) + \cos(\Omega_2\tau_1)]\cos[(\Omega_{cs1} - \Omega_{cs2})\tau_2]\cos\beta$$

---

[8]This is the reason why zero-quantum coherences cannot be spoiled by external field gradients.

[9]Compare footnote 4 on page 63.

$$-[\sin(\Omega_1\tau_1) - \sin(\Omega_2\tau_1)]\sin[(\Omega_{cs1} - \Omega_{cs2})\tau_2]\}$$

$$b_6 = \frac{1}{2}\sin\beta\sin\gamma\sin(\pi J\tau_1)\left\{[\cos(\Omega_1\tau_1) + \cos(\Omega_2\tau_1)]\sin[(\Omega_{cs1} - \Omega_{cs2})\tau_2]\cos\beta\right.$$

$$\left. +[\sin(\Omega_1\tau_1) - \sin(\Omega_2\tau_1)]\cos[(\Omega_{cs1} - \Omega_{cs2})\tau_2]\right\}$$

The transverse magnetization is determined by the density operator at Eq. 7.38, averaged over all field-gradient offsets, Ω_g (see Eqs. 7.14 and 7.15):

$$m_{(0qc)}(\tau_1 + \tau_2 + \tau_3) = \frac{n}{2}\gamma_n\hbar b\ \mathrm{Tr}\left\{\langle\sigma_{(0qc)}(\tau_1 + \tau_2 + \tau_3)\rangle_{\Omega_g}(I_{1x} + I_{2x} + iI_{1y} + iI_{2y})\right\}$$

$$(7.39)$$

where $n/2$ is the number density of two-spin systems. Note that the terms of Eq. 7.38, which actually depend on the field-gradient offset Ω_g, stand for single-quantum coherences in the intervals τ_1 and τ_3, whereas there is no such dependence of the τ_2 dependent factors.

The spatial average of the density operator at Eq. 7.38 exclusively refers to products of sine and cosine functions with arguments $\Omega_1\tau_1$, $\Omega_1\tau_3$, $\Omega_2\tau_1$, or $\Omega_2\tau_3$. The replacement by the known trigonometric relations for angle sums and differences results in prefactors of the form

$$\langle\sin[\Omega_{1,2}(\tau_1 \pm \tau_3)]\rangle_{\Omega_g} \qquad \text{and} \qquad \langle\cos[\Omega_{1,2}(\tau_1 \pm \tau_3)]\rangle_{\Omega_g}$$

In general, these trigonometric functions tend to cancel by destructive superposition of contributions with different local frequency offsets. As already stated several times before in context with the other echo phenomena, the only terms contributing to the formation of an echo are cosine functions with an argument vanishing independently of the position. Non-trivially this is fulfilled for

$$\tau_3 = \tau_1 \qquad (7.40)$$

The pathway

> *single-quantum coherences → zero-quantum coherence →*
> *single-quantum coherences*

thus leads to a coherence-transfer echo at a time when uncoupled spins also show an echo, namely Hahn's stimulated echo. This must be taken into account when interpreting echo signals fulfilling the condition at Eq. 7.40. Note, however, that Hahn's stimulated echo and the zero-quantum coherence transfer echo are intrinsically subject to different relaxation mechanisms in the τ_2 interval. Zero-quantum coherences are attenuated by transverse relaxation whereas longitudinal magnetization approaches its equilibrium value by spin-lattice relaxation (see Chaps. 12 and 13).

7.2.4
Single-Quantum Coherence-Transfer Echoes

The single-quantum coherences of the τ_2 interval are expressed by Eq. 7.9. In principle, the third RF pulse conveys them into scalar order, longitudinal magnetization, apart from coherences of all allowed orders. This transfer is very much like the conversion of the single-quantum coherences by the second RF pulse (see Eq. 7.7). Without application of further RF pulses, only single-quantum contributions are pertinent to the signals detected after a delay τ_3.

All kinds of coherence-transfer echoes arise at times coinciding with those of the primary and secondary echoes of the three-pulse Hahn signals (Sect. 2.2). The midposition times of the echoes are $t = 2\tau_1 + 2\tau_2$, $t = \tau_1 + 2\tau_2$, and $t = 2\tau_2$. However, unlike Hahn three-pulse echoes, these single-quantum coherence transfer echoes exhibit a complicated modulation pattern originating from J coupling and chemical shift offsets.

The coherence pathway in the sequel of the three pulse intervals is

> *single-quantum coherences* → *single-quantum coherences* →
> *single-quantum coherences*

Its treatment, unfortunately, results in lengthy expressions which do not suggest phenomena which, by nature, are different from those already discussed in context with the two-pulse single-quantum coherence transfer experiments. It appears that a three-pulse coherence pathway consisting of single-quantum coherences from the beginning of the pulse sequence until the acquisition interval is of little practical significance. Further examination is therefore skipped.

7.2.5
Double-Quantum Coherence-Transfer Echo

The double-quantum coherences[10] created by the second RF pulse at the beginning of the τ_2 interval are described by[11]

$$\sigma_{2qc}(\tau_1+) = \frac{1}{4}(2I_{1x}I_{2x} - 2I_{1y}I_{2y})\left[\cos(\Omega_1\tau_1) + \cos(\Omega_2\tau_1)\right]\sin(2\beta)\sin(\pi J\tau_1)$$

$$-\frac{1}{2}(2I_{1x}I_{2y} + 2I_{1y}I_{2x})\left[\sin(\Omega_1\tau_1) + \sin(\Omega_2\tau_1)\right]\sin\beta\sin(\pi J\tau_1)$$

$$= \frac{1}{2}(I_1^+I_2^+ + I_1^-I_2^-)c_1 - \frac{1}{2i}(I_1^+I_2^+ - I_1^-I_2^-)c_2 \qquad (7.41)$$

where

$$c_1 = \frac{1}{2}\left[\cos(\Omega_1\tau_1) + \cos(\Omega_2\tau_1)\right]\sin(2\beta)\sin(\pi J\tau_1)\ (= b_1)$$

$$c_2 = \left[\sin(\Omega_1\tau_1) + \sin(\Omega_2\tau_1)\right]\sin\beta\sin(\pi J\tau_1)$$

The double-quantum coherence evolves in the τ_2 interval according to

$$\sigma_{2qc}(\tau_1 + \tau_2-) =$$

---

[10]Compare footnote 8 on page 34.

[11]The operator terms standing for double-quantum coherences can also be expressed by the single-transition operators given by Eqs. 42.77.

$$\frac{1}{2} (2I_{1x}I_{2x} - 2I_{1y}I_{2y}) \{ c_1 \cos[(\Omega_1 + \Omega_2)\tau_2] - c_2 \sin[(\Omega_1 + \Omega_2)\tau_2] \}$$

$$- \frac{1}{2} (2I_{1x}I_{2y} + 2I_{1y}I_{2x}) \{ c_1 \sin[(\Omega_1 + \Omega_2)\tau_2] + c_2 \cos[(\Omega_1 + \Omega_2)\tau_2] \} \quad (7.42)$$

The characteristic feature of double-quantum coherences reveals itself by the dependence on the **frequency-offset sum** $\Omega_1 + \Omega_2$ which governs the evolution in the τ_2 interval. In contrast to zero-quantum coherences, now there is a strong influence of field-gradient offsets Ω_g: it is just doubled relative to single-quantum coherences. From Eqs. 7.14 and 7.15 it follows that

$$\Omega_1 + \Omega_2 = \Omega_{cs1} + \Omega_{cs2} + 2\Omega_g \quad (7.43)$$

The third RF pulse converts the double-quantum coherence to longitudinal scalar order and zero-, single-, and double-quantum coherences. As in the case of the zero-quantum coherence transfer echo, we restrict ourselves to terms leading to detectable signals in the τ_3 interval. Therefore only single-quantum coherences originating from the double-quantum coherence in the τ_2 interval are considered. All other density operator terms are dropped in the following.

Immediately after the third pulse, the single-quantum coherences, into which part of the double-quantum coherence of the τ_2 interval is transferred, are of the antiphase type. The corresponding reduced density operator is

$$\sigma_{(2qc)}(\tau_1 + \tau_2+) = (2I_{1x}I_{2z} + 2I_{2x}I_{1z}) c_3 - (2I_{1y}I_{2z} + 2I_{2y}I_{1z}) c_4 \quad (7.44)$$

where we have used a notation analogous to that defined in footnote 4 on page 63. The operator coefficients are

$$c_3 = \frac{1}{4} \{ [\cos(\Omega_1\tau_1) + \cos(\Omega_2\tau_1)] \cos[(\Omega_1 + \Omega_2)\tau_2] \cos\beta$$
$$- [\sin(\Omega_1\tau_1) + \sin(\Omega_2\tau_1)] \sin[(\Omega_1 + \Omega_2)\tau_2] \} \sin\beta \sin(2\gamma) \sin(\pi J\tau_1)$$

$$c_4 = \frac{1}{2} \{ [\cos(\Omega_1\tau_1) + \cos(\Omega_2\tau_1)] \sin[(\Omega_1 + \Omega_2)\tau_2] \cos\beta$$
$$+ [\sin(\Omega_1\tau_1) + \sin(\Omega_2\tau_1)] \cos[(\Omega_1 + \Omega_2)\tau_2] \} \sin\beta \sin\gamma \sin(\pi J\tau_1)$$

The evolution of these single-quantum coherences in the τ_3 interval results in a reduced density operator of the form

$$\begin{aligned}
\sigma_{(2qc)}(\tau_1 + \tau_2 + \tau_3) = {} & \{ I_{1x} [c_5 \sin(\Omega_1\tau_3) + c_6 \cos(\Omega_1\tau_3)] \\
& + I_{2x} [c_5 \sin(\Omega_2\tau_3) - c_6 \cos(\Omega_2\tau_3)] \\
& + I_{1y} [c_5 \cos(\Omega_1\tau_3) - c_6 \sin(\Omega_1\tau_3)] \\
& - I_{2y} [c_5 \cos(\Omega_2\tau_3) + c_6 \sin(\Omega_2\tau_3)] \} \sin(\pi J\tau_3) \\
& + \{ 2I_{1x}I_{2z} [c_5 \cos(\Omega_1\tau_3) - c_6 \sin(\Omega_1\tau_3)] \\
& + 2I_{2x}I_{1z} [c_5 \cos(\Omega_2\tau_3) + c_6 \sin(\Omega_2\tau_3)] \\
& + 2I_{1y}I_{2z} [c_5 \sin(\Omega_1\tau_3) - c_6 \cos(\Omega_1\tau_3)] \\
& - 2I_{2y}I_{1z} [c_5 \sin(\Omega_2\tau_3) + c_6 \cos(\Omega_2\tau_3)] \} \cos(\pi J\tau_3) \quad (7.45)
\end{aligned}$$

where

$$c_5 = \frac{1}{4} \sin \beta \sin(2\gamma) \sin(\pi J \tau_1) \left\{ [\cos(\Omega_1 \tau_1) + \cos(\Omega_2 \tau_1)] \cos[(\Omega_1 + \Omega_2)\tau_2] \cos \beta \right.$$
$$\left. - [\sin(\Omega_1 \tau_1) + \sin(\Omega_2 \tau_1)] \sin[(\Omega_1 + \Omega_2)\tau_2] \right\}$$

$$c_6 = \frac{1}{2} \sin \beta \sin\gamma \sin(\pi J \tau_1) \left\{ [\cos(\Omega_1 \tau_1) + \cos(\Omega_2 \tau_1)] \sin[(\Omega_1 + \Omega_2)\tau_2] \cos \beta \right.$$
$$\left. + [\sin(\Omega_1 \tau_1) + \sin(\Omega_2 \tau_1)] \cos[(\Omega_1 + \Omega_2)\tau_2] \right\}$$

The pathway in the sequel of the three pulse intervals which finally led to Eq. 7.45 is

> *single-quantum coherences* → *double-quantum coherence* →
> *single-quantum coherences*

The transverse magnetization eventually achieved is given by

$$m_{(2qc)}(\tau_1 + \tau_2 + \tau_3) = \frac{n}{2} \gamma_n \hbar b \operatorname{Tr} \left\{ \langle \sigma_{(2qc)}(\tau_1 + \tau_2 + \tau_3) \rangle_{\Omega_g} (I_{1x} + I_{2x} + iI_{1y} + iI_{2y}) \right\}$$
$$(7.46)$$

where the average refers to the gradient-induced frequency offsets Ω_g entering via Eqs. 7.14 and 7.15. In contrast to the zero-quantum transfer echo, the evolution under the influence of gradients now plays a crucial role in all pulse intervals including τ_2. The inspection of Eq. 7.45 discloses products of sine and cosine functions with arguments $\Omega_1 \tau_1$, $\Omega_1 \tau_3$, $\Omega_2 \tau_1$, $\Omega_2 \tau_3$, or $(\Omega_1 + \Omega_2)\tau_2$. As in the previous echo treatments, these functions may be replaced by the trigonometric relations for angle sums and differences so that the averages to be considered are

$$\left\langle \sin[\, \Omega_{1,2}\tau_1 + (\Omega_1 + \Omega_2)\tau_2 \pm \Omega_{1,2}\tau_3 \,] \right\rangle_{\Omega_g},$$
$$\left\langle \cos[\, \Omega_{1,2}\tau_1 + (\Omega_1 + \Omega_2)\tau_2 \pm \Omega_{1,2}\tau_3 \,] \right\rangle_{\Omega_g},$$
$$\left\langle \sin[\, (\Omega_1 + \Omega_2)\tau_2 + \Omega_{1,2}\tau_3 - \Omega_{1,2}\tau_1 \,] \right\rangle_{\Omega_g},$$
$$\left\langle \cos[\, (\Omega_1 + \Omega_2)\tau_2 + \Omega_{1,2}\tau_3 - \Omega_{1,2}\tau_1 \,] \right\rangle_{\Omega_g},$$
$$\left\langle \sin[\, \Omega_{1,2}\tau_3 + \Omega_{1,2}\tau_1 - (\Omega_1 + \Omega_2)\tau_2 + \Omega_{1,2}\tau_3 \,] \right\rangle_{\Omega_g},$$
$$\left\langle \cos[\, \Omega_{1,2}\tau_3 + \Omega_{1,2}\tau_1 - (\Omega_1 + \Omega_2)\tau_2 + \Omega_{1,2}\tau_3 \,] \right\rangle_{\Omega_g}$$

In case of finite arguments the average trigonometric functions tend to vanish owing to destructive superposition of contributions from different positions in the inhomogeneous field. The only functions relevant for the echo formation therefore are the cosine functions for zero arguments. Non-trivially this is fulfilled under the conditions

$$\tau_3 = \tau_1 + 2\tau_2$$
$$\tau_3 = \tau_1 - 2\tau_2 \quad (> 0) \qquad\qquad (7.47)$$
$$\tau_3 = 2\tau_2 - \tau_1 \quad (> 0)$$

The consequence is a series of **double-quantum coherence transfer echoes** appearing in the τ_3 interval. As outlined below in more detail, these echoes can readily be distinguished from echoes of other coherence pathways. Adding a gradient pulse in the τ_2 interval dephases the multiple-quantum coherences in proportion to the coherence order. That is, a double-quantum coherence transfer echo in the τ_3 interval is generated by a gradient pulse twice as long.

7.3
Multiple-Quantum Coherence Based Spectroscopy

The evolution characteristics of multiple-quantum coherences are widely used for spectroscopic purposes.[12] There are two ways in which multiple-quantum coherences can be employed. Firstly, with **"multiple-quantum/single-quantum correlated spectroscopy,** the multiple-quantum coherence interval is taken as one of the evolution intervals representing the time domains of a two-dimensional spectroscopy experiment (Fig. 7.3b). This interval is incremented in a series of transients. In the spectrum, the multiple-quantum coherences are then correlated with single-quantum coherences.

Secondly, **"multiple-quantum filtered spectroscopy"** is a common technique using the same RF-pulse scheme. However, the multiple-quantum coherence interval is now employed as a mixing interval of fixed length, whereas the time domains of a corresponding two-dimensional spectroscopy experiment are solely to probe the single-quantum coherences occurring in the initial and final evolution intervals (Fig. 7.3a). In the corresponding 2D spectrum, these single-quantum coherences are then correlated with one another.

In experiments of the latter sort the τ_1 interval may be supplemented by a 180° pulse in the middle serving the refocusing of the single-quantum coherences of this interval with respect to all offsets except that due to J coupling. A single echo then arises from the conversion of multiple-quantum coherences to single-quantum coherences in the τ_3 interval. In the case of double-quantum coherence transfer, for instance, the echo appears at $\tau_3 = 2\tau_2$.

---

[12] According to the theory outlined above, double-quantum coherences as well as any other order of non-single-quantum coherences should not occur if spin-spin and dipolar couplings do not affect the evolution of spin-1/2 coherences. In the present case, where complete motional averaging is assumed, this is readily demonstrated by setting $J = 0$ in Eq. 7.45. However, long-range intermolecular dipolar interactions fluctuate more slowly than short-range and intramolecular couplings. Motional averaging may therefore be incomplete for long distances r. Multiple-quantam coherences might arise on these grounds [505]. Remember that dipolar interaction varies proportional to r^{-3}, whereas the number of coupling partners increases proportional to r^2. Also, one should keep in mind that radiation damping and the demagnetizing field can cause striking effects which mimic the coherence-transfer features of J coupled spin systems under motional averaging conditions [64, 65]. Radiation damping is the result of the feedback action of the currents induced in the probe RF coil by the precessing magnetization [66]. The demagnetizing field is caused by the magnetization of the sample as described in Sect. 2.4. The line shifts caused by this field are appreciable in high-resolution NMR at the strong flux densities of modern spectrometers. The demagnetizing field follows the magnetization in the course of a pulse sequence, and, hence, a feedback effect on the evolution occurs [59].

7.3.1
Double-Quantum Filtered Correlated Spectroscopy (DQF-COSY)

This spectroscopy technique uses a three-pulse sequence of the type shown in Fig. 7.3a, but the time domain t_1 is defined in a modified way. In this context it is not identical with the evolution interval τ_1. Rather, the first pulse interval is subdivided in a delay τ_0 kept fixed in the whole experiment, and a period which is incremented in subsequent transients as usual in 2D spectroscopy.

The transfer to double-quantum coherence in the second interval is maximal if the factor $\sin(\pi J \tau_1)$ in Eq. 7.45 takes the value 1. That is,

$$\tau_1 = (2j+1)/(2J) \qquad\qquad (j = 0, 1, 2, \ldots) \qquad\qquad (7.48)$$

One therefore chooses the fixed delay τ_0 in such a manner that the total evolution time is varied in the vicinity of the optimum value for the desired coherence transfer.

If the interval τ_2 is very short (typically only a few microseconds) so that practically no evolution of the multiple-quantum coherences takes place, the signal acquisition may already begin at the beginning of the τ_3 interval. Thus, the third pulse directly plays the role of a "read" pulse.

Rewriting Eq. 7.45 for $\tau_2 \approx 0$, $\tau_3 \approx 0$, and $\beta = \gamma = \pi/2$ gives

$$\sigma_{(2qc)}(\tau_1 + 0 + 0)_{\beta=\gamma=\pi/2} = -\frac{1}{2}\left[\sin(\Omega_1\tau_1) + \sin(\Omega_2\tau_1)\right]$$
$$\sin(\pi J\tau_1)\left(2I_{1y}I_{2z} + 2I_{2y}I_{1z}\right) \qquad (7.49)$$

This expression represents antiphase single-quantum coherences originating from the double-quantum coherence in the τ_2 interval. Single-quantum coherences resulting from other pathways, particularly from the zero-quantum coherence in the τ_2 interval, can be suppressed by phase cycles of the RF pulses in subsequent transients[139]. Correspondingly, other "orders" of multiple-quantum filters can be generated by suitable phase cycles.

Multiple-quantum filtered 2D spectra permit the convenient distinction of coupled spin-systems from each other, and, in particular, from uncoupled spins. The principle of double-quantum filtered correlated spectroscopy is demonstrated in Fig. 7.4 for a simple AX spin system before the background of uncoupled spins of the solvent.

This differentiation of coupled spins is of particular interest in context with ^{13}C spectroscopy. Employing double-quantum filtering, the so-called "incredible natural abundance double-quantum transfer experiment" better known under the acronym **INADEQUATE** permits the elucidation of coupled ^{13}C/^{13}C pairs for the derivation of the carbon connectivity in a compound [25, 86, 139].

In Chap. 37 a "double-quantum filter volume-selective spectral editing" (DQF-VOSING) experiment is treated for an A_3X spin system. With this method the same coherence pathways as outlined above are employed for the discrimination of signals of coupled spins in the compound of interest from superimposed (and even dominating) signals of uncoupled spins.

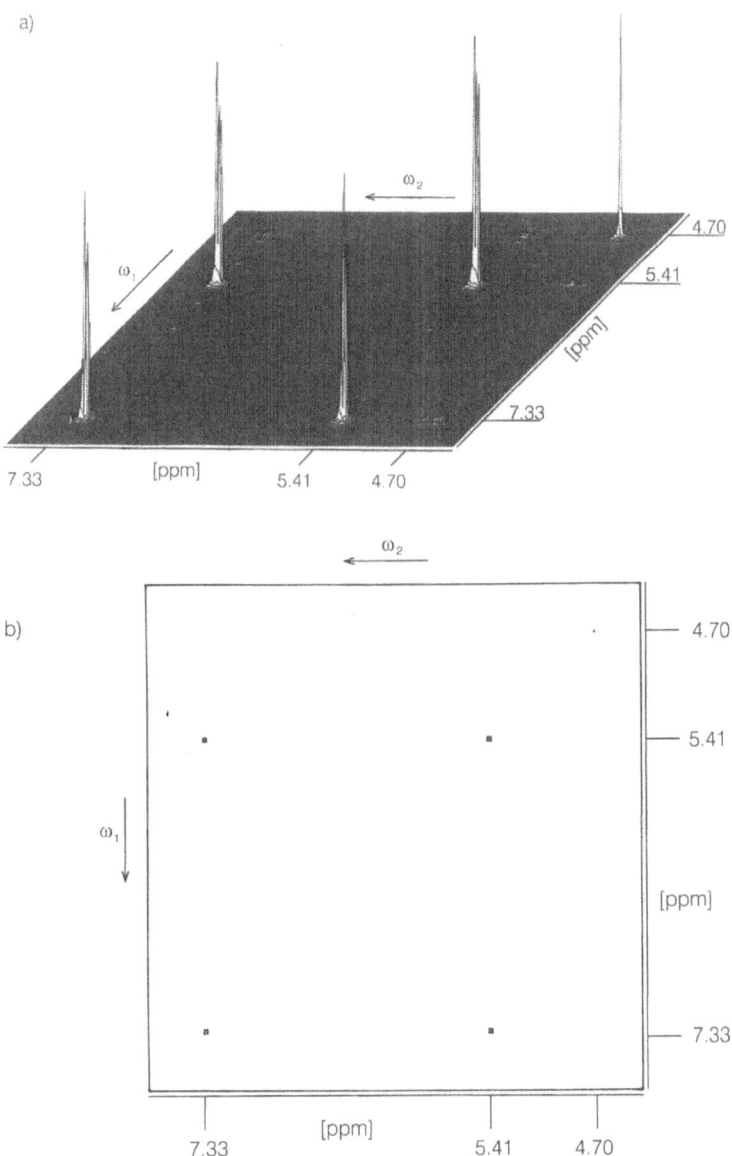

Fig. 7.4. Double-quantum filtered 300 MHz proton spectra of an AX spin system (same as used for the previous 2D spectra; 1.3 mol/l uracil D_2O/NaOD solution) recorded with the aid of pulse sequence at Fig. 7.3a: **a)** projection; **b)** contour plot of the magnitude of the (complex) two-dimensional Fourier transform. ω_1 and ω_2 are the frequency-domain variables conjugate to the time-domain variables τ_1 and τ_3, respectively (following the usual convention, these variables are denoted in Fig. 7.3a by t_1 and t_2, respectively). The lines at 4.41 ppm and 7.33 ppm are assigned to the CH groups of the ring compound adjacent to CO/COD and ND groups, respectively. Uncoupled spins produce no cross peaks and are strongly suppressed. The HDO line at 4.7 ppm is 28 times less intense than in the COSY spectrum (Fig. 7.1) of the same sample. (courtesy of K.-H. Spohn)

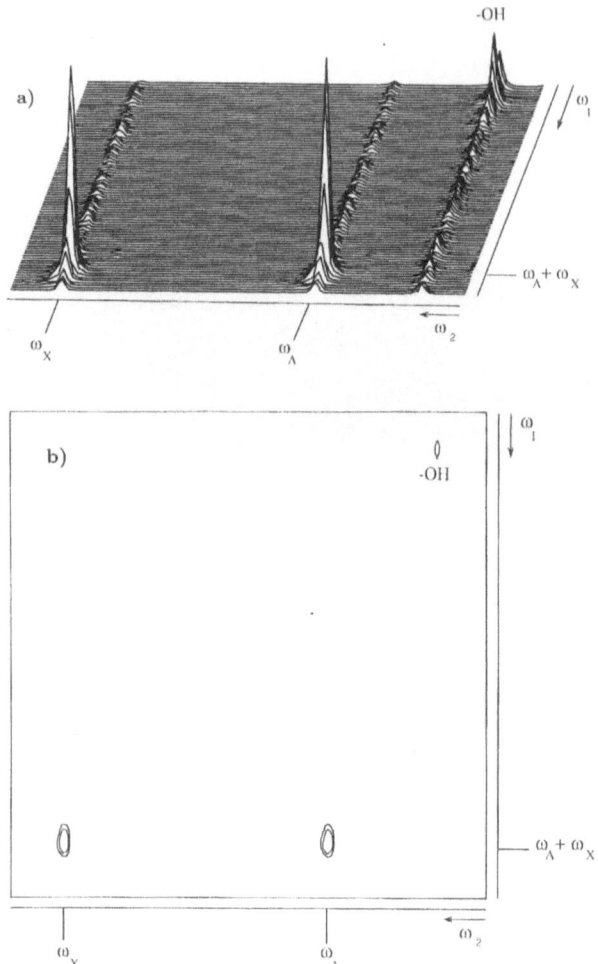

Fig. 7.5. Two-dimensional 300 MHz double-quantum proton spectra of an AX spin system (same as used for the previous 2D spectra; 1.3 mol/l uracil D_2O/NaOD solution) recorded with the aid of pulse sequence at Fig. 7.3b; **a)** projection; **b)** contour plot of the magnitude of the (complex) two-dimensional Fourier transform. (courtesy of R.-O. Seitter)

7.3.2
Double-Quantum/Single-Quantum Correlated Spectroscopy

This method employs the second principle mentioned above. As shown in Fig. 7.3b, the second pulse interval of a three-pulse sequence is considered as the evolution interval defining the first time domain. The objective is to correlate double-quantum coherences in this interval with single-quantum coherences in the detection interval. In the case of AX spin systems, the condition for the first pulse interval τ_1 for optimal transfer to double-quantum coherences is given in Eq. 7.48. Inserting this value into Eq. 7.45 gives the reduced density operator determining the signals to be represented in the form of a 2D plot. Figure 7.5 shows corresponding evaluations of an experiment using the same AX example as in the demonstrations of the other 2D spectroscopy techniques. Evidently the resonance frequencies of the individual lines are correlated with the sum frequencies, whereas the (spurious) HDO peak remains on the diagonal.

7.4
Discrimination of Coherence-Transfer Echoes

Echoes emerging from scalar-order transfer, zero-quantum coherence transfer, and longitudinal magnetization transfer jointly satisfy the maximum condition $\tau_3 = \tau_1$. Moreover, this is the schedule of Hahn's stimulated echo of uncoupled spins. The coincidence is due to the common absence of any coherence evolution in field inhomogeneities during the τ_2 interval. The phenomena may nevertheless reveal themselves in a distinct manner owing to the different dependences on the flip angles, on the pulse phases, and on the evolution under the influence of chemical shifts or spin-spin couplings.

Signals of different provenance can often be discriminated with the aid of homospoil gradient pulses applied in the pulse intervals, and/or of phase cycles of the RF pulses in subsequent transients of signal accumulation. In this manner, undesired coherences can largely be suppressed [139].

Figure 7.6 shows a pulse scheme for **"homospoil gradient-assisted spectral editing"** of single-quantum coherences originating from double-quantum coherences in the τ_2 interval. The principle is that single-quantum rephasing of double-quantum coherence phase shifts needs a refocusing interval twice as long and a gradient

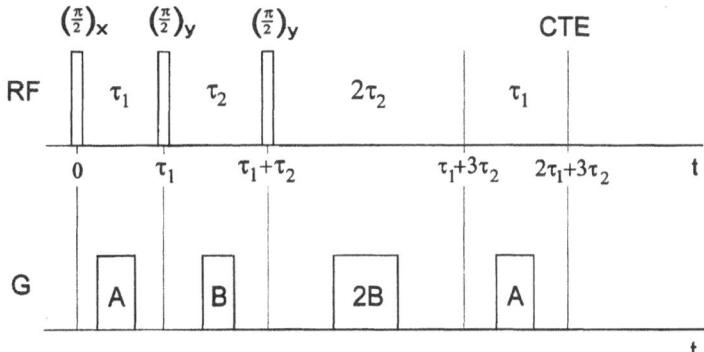

Fig. 7.6. RF and field gradient (G) pulse scheme for "homospoil gradient-assisted spectral editing" of single-quantum coherences that can be traced back to double-quantum coherences in the τ_2 interval. The letters A and B indicate the "areas" under the gradient pulses in the first two RF pulse intervals. In the delay $2\tau_2$ after the third RF pulse the double-quantum phase shifts of the τ_2 interval between the second and the third RF pulse is refocused. In the τ_1 delay preceding the center of the coherence-transfer echo (CTE), the phase shifts adopted in the first τ_1 interval are compensated.

pulse twice as strong. The latter is required for the compensation of the homospoil gradient, the former accounts for background inhomogeneities.

This sequence can also be employed for the sensitive detection of **chemical-shift changes** via the phase resulting from the evolution of double-quantum coherences. The phase shift adopted in the interval τ_2 by the double-quantum coherence of an AX spin system, for instance, is $\varphi_{dqc} = (\Omega_A + \Omega_X)\tau_2$. This is to be compared with the phase shifts single-quantum coherences would take up in the same period, i.e., $\varphi_{sqc} = \Omega_A\tau_2$ or $\Omega_X\tau_2$. If chemical shifts happen to be a function of temperature, the signal phase can correspondingly be calibrated. As an application, this effect was demonstrated for the helix/coil transition range of polypeptides, where - depending on the cooperativity of the transition - the α-CH resonance sensitively varies with the temperature [241].

In Chap. 37 several procedures for **volume-selective editing spectroscopy** are described which take advantage of the selective generation of one of the echo phenomena delineated above. In this context we will come back to the formalisms outlined above. The most prominent spectral-editing principle will be shown to be the fact that uncoupled spins cannot form coherence-transfer echoes. This may be employed for suppressing water signals, for instance, which is crucial in biological applications.

II

Molecular Motion

CHAPTER 8

Survey

Magnetic resonance provides a large variety of techniques suitable for studies of molecular motions in a rather direct and detailed way. The most important phenomena of interest in this respect are nuclear-spin relaxation, self-diffusion, and exchange between sites with different resonance parameters. The methods in use for corresponding studies are based on very different principles. Nevertheless, one may recognize a scheme (Fig. 8.1) common to all three classes of techniques.

An NMR experiment probing molecular motions begins with the preparation of a non-equilibrium state of the sample. This can refer to magnetization vector components deviating from the Curie values, longitudinal order, or coherences encoded by spin interactions or by external field gradients.

Then a certain time interval is allowed for changes of the spin states by molecular fluctuations. As such, reorientations, translational displacements or exchange processes between different sites come into question.

Finally, the coherences still present or generated at the end of the evolution interval are detected. If the pulse sequence starts with the generation and storage of gradient encoded coherences, the detection interval involves the corresponding decoding of coherences.

In this part, magnetic-resonance experiments suitable for probing molecular motions are subdivided into four major classes. We will first deal with spin relaxation. This includes **spin-lattice relaxation** as well as relaxation of spin coherences normally referred to as **transverse relaxation** or - less adequately - spin-spin relaxation. Transverse relaxation is not restricted to single-quantum coherences. In experiments with multiple-quantum coherence evolution intervals, the relaxation of multiple-quantum coherences is of interest too.

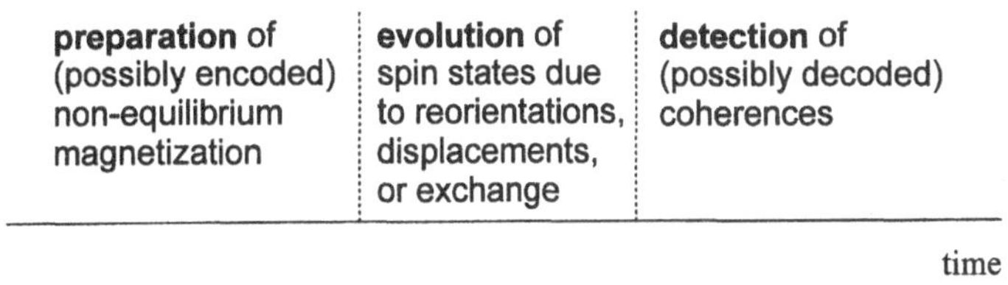

preparation of (possibly encoded) non-equilibrium magnetization	**evolution** of spin states due to reorientations, displacements, or exchange	**detection** of (possibly decoded) coherences

time

Fig. 8.1. Schematic representation of the three intervals which one can distinguish in NMR experiments probing molecular motions.

In systems such as polymers or liquid crystals where molecular motions are anisotropic, residual spin-interactions may occur which are not, or only incompletely, subject to motional averaging. As a method specifically referring to this residual couplings the so-called **dipolar-correlation effect** will be considered.

Translational displacements of molecules by self-diffusion can be probed directly with the aid of **field-gradient NMR diffusometry**. The field gradients may refer to the main magnetic field B_0 or to the RF amplitude B_1. Versions based on pulsed gradients as well as on the steady gradients provided by superconducting magnets in the fringe field will be discussed.

The last chapter of this part is devoted to methods probing exchange phenomena. This category of magnetic-resonance experiments has become particularly popular in the multiple-dimensional form. **Two-dimensional exchange spectroscopy** is now a term comprising different experiments with the common feature that a nucleus faces fluctuating resonance frequencies. The origin of these fluctuations can be of such contrasting natures as chemical exchange between different sites in molecules, isomerization, diffusion between different phases, molecular reorientations and so on.

CHAPTER 9

Categorization of Relaxation Phenomena

9.1
General Remarks

Spin relaxation, according to its very definition, is ubiquitous in magnetic resonance. The prerequisite of the detection of any magnetic-resonance signal is a non-equilibrium state of the spin systems which must be excited before. Principally, such athermal states are subject to relaxation processes. That is, all pulse sequences so far considered in the first part of this book can also be viewed as relaxation experiments although we have neglected any influence of this sort in the formal treatments up to now. Actually some of these pulse sequences were originally developed to the end of relaxation studies. Further experimental schemes suitable for the record of relaxation curves are given in Table 10.1 on 93 and in Chaps. 15 and 17. In view of the vast variety of relaxation-sensitive experiments it appears to be more favorable to classify spin relaxation via the relevant **observable** rather than via the experimental set-ups by which it is probed.

9.1.1
Observables Subject to Relaxation

The treatments of the pulse sequences delineated in the part on spin echoes and coherences already demonstrated the principal classes of spin states and coherences that can arise in NMR experiments. We distinguish

- **longitudinal magnetization**, i.e., the magnetization component along the quantization direction. In laboratory-frame experiments, this direction is given by the external magnetic field B_0. This is replaced by the effective field B_e (see Sect. 48.7) in so-called rotating-frame experiments. The relevant spin energy is defined by the **Zeeman levels**. The relaxation of the populations of these levels toward equilibrium is connected with energy exchange with the mechanical degrees of freedom in the matter (the "lattice"). The relaxation mechanisms of this sort are therefore summarized as **spin-lattice relaxation** with the time constants T_1 and $T_{1\rho}$ in the laboratory and rotating frames, respectively. A synonymous term is **longitudinal relaxation** reminding us of the magnetization component involved. The relevant observable is $\sum_j I_{j,z}$ where the sum in principle concerns all members of the spin system considered.

- **longitudinal (scalar, dipolar, or quadrupolar) order.** This sort of phenomenon arises when bilinear spin interactions apply. The observables are of the type $2I_z S_z$ for scalar or dipolar coupled two-spin-1/2 systems (compare Table 51.1 on page 482), and $I_z^2 - I(I+1)/3$ for quadrupolar coupling (see Eq. 52.11). There is no net magnetization connected with these operators although the z direction is involved again (a more extended discussion can be found in Sect. 51.3, example 3, for instance). That is, the term "observable" should not be taken too literally. However, the evolution of longitudinal order can be probed indirectly by converting the order state into single-quantum coherences (as was repeatedly demonstrated in the first part; see the sections on three-pulse solid echoes and coherence-transfer echoes). The RF pulse managing this transfer may therefore be called "read pulse." Less common variants of transfer to and from longitudinal order are "adiabatic demagnetization in the laboratory or rotating frame" (ADLF or ADRF) and "adiabatic remagnetization in the laboratory or rotating frame" (ARLF or ARRF).[1] The time constants of longitudinal-order relaxation are termed **dipolar-order relaxation time** T_d, **quadrupolar-order relaxation time** T_q, and **scalar-(J)-order relaxation time** T_J, depending on the spin interaction mediating the longitudinal order.
- **zero-quantum coherences in the doubly-rotating frame.** This sort of phenomena arises under Hartmann/Hahn matching conditions bringing two different nuclear species into contact. The establishment of a spin temperature common to both (multi-)spin systems[2] is termed **cross-relaxation** [114].
- **single-quantum coherences.** The "observable" is the rotating-frame magnetization component transverse to the quantization direction, say the component aligned along the x axis. In rotating-frame experiments this axis refers to a tilted rotating coordinate system with the z axis aligned along the effective field direction.[3] The expectation value to be considered is $M_x \propto \langle \sum I_x \rangle$. Under motional averaging conditions, the time constant for the relaxation of this quantitiy is called the **transverse relaxation time** T_2 (in rotating-frame experiments: $T_{2\rho}$).
- **multiple-quantum coherences,** i.e., coherences other than single-quantum coherences. The term "observable" is again not to be taken literally because the signal detection stipulates a transfer to single-quantum coherences. Operators of double-quantum coherences of two-spin-1/2 systems are, written in Cartesian spin operators (see Eqs. 42.77), $(I_x S_x - I_y S_y)$ and $(I_x S_y + I_y S_x)$, or, expressed as irreducible spherical tensor operators $T_{2,\pm 2}$ (see Eqs. 49.22).

---

[1]The adiabatic conduct of experiments will be discussed in more detail in Chaps. 15 and 38.

[2]Small spin systems behave in a more coherent way as pointed out in Chap. 39. See the discussion in Sect. 39.2.

[3]Transverse relaxation in the rotating frame has scarcely been examined so far. This is in contrast to spin-lattice relaxation which is a frequently employed experimental tool (see Sect. 10.3).

9.1.2
Spin Interactions Subject to Fluctuations

Nuclear magnetic relaxation in general stipulates fluctuations of the spin interactions, where temporal fluctuations are a consequence of molecular or lattice dynamics. For instance, rotational diffusion of a molecule gives rise to stochastic modulations of all anisotropic spin couplings within that molecule.

The spin-interaction Hamiltonians relevant for magnetic resonance can be expressed in a common analytical form characterized by (compare Sect. 46.3)

a) a factorization in structure and spin-operator dependent functions
b) an expansion in terms including spin operators which induce different orders of spin transitions.

The general representation is

$$\mathcal{H}_i(t) = f_i \sum_k F^{(k)}(t)\mathcal{O}^{(k)} \tag{9.1}$$

where the factors f_i are interaction-specific constants, the fluctuating structure functions are $F^{(k)}(t)$, and the spin-operator functions are $\mathcal{O}^{(k)}$.

If molecular dynamics merely refer to reorientations, so that the rotation properties are in the foreground, alternative representations based on irreducible spherical tensor operators may be favorable (see Eq. 49.1):

$$\mathcal{H}_i(t) = a_i \sum_{l=0,2} \sum_{m=-l}^{l} (-1)^m A_{l,-m}(t) T_{l,m} \tag{9.2}$$

where the prefactors a_i again are constants specific for the spin interaction. In this expansion, orientation-dependent and, hence, fluctuating functions $A_{l,-m}(t)$ are combined with irreducible spherical tensor operators $T_{l,m}$ which are composed of spin operators as outlined in Chap. 49. The operators $T_{l,m}$ transform under rotations like spherical harmonics of rank l. The subscript m indicates the **order of multiple-quantum transition** induced by the operator. That is, we have separate terms for different multiple-quantum orders. The functions $A_{l,-m}$ are also expressed in terms of spherical harmonics of rank l as far as possible. In the following, we will alternately use the spin-interaction representations at Eqs. 9.1 and 9.2 depending on which version is more adapted to the problem or more convenient.

Dipolar coupling (Table 46.5 on page 426) provides a very efficient relaxation mechanism, although **quadrupole coupling** (Table 46.4 on page 425) normally dominates if present. Electron paramagnetic particles enhance nuclear-spin relaxation by dipolar and **scalar interactions** (Table 46.3 on page 422). They may therefore be employed as "relaxation agents." If dipolar or quadrupolar couplings are comparatively weak or absent, **chemical-shift anisotropy** (Table 46.2 on page 421) often plays a crucial role at high magnetic fields. Finally, in gases and low-viscous liquids, the **spin-rotation interaction** contributes to relaxation.

The spin-operator terms specify the transitions on which relaxation is mainly based. Terms without transition-inducing operators merely mediate transition-less contributions to transverse relaxation. The theory delineated below is restricted to the so-called **"weak-collision case"** stipulating that the perturbation by the fluctuating spin interactions is much smaller than the stationary part of the total spin Hamiltonian. The opposite limit, the **"strong-collision case"** plays a role in dipolar or quadrupolar-order relaxation [4, 158, 523] This sort of solid-state relaxation mechanism will not be pursued any further here.

9.1.3
The Autocorrelation and the Intensity Functions

The stochastic nature of the fluctuating spin interactions is manifested by autocorrelation functions of the structural terms $F^{(k)}(t)$ or $A_{l,-m}(t)$. One can show that, irrespective of the interaction and observable types, the relaxation rates can be expressed in terms of the **reduced autocorrelation function**

$$G(\tau) = \frac{\left\langle F^{(k)}(0)F^{(-k)}(\tau)\right\rangle}{\left\langle |F^{(k)}|^2\right\rangle} \tag{9.3}$$

on the one hand, and

$$G(\tau) = \frac{\left\langle A_{l,-m}(0)A_{l,m}(\tau)\right\rangle}{\left\langle |A_{l,m}|^2\right\rangle} \tag{9.4}$$

on the other. The brackets indicate the ensemble average. Note that $G(\tau)$ is an even function and is normalized for $\tau = 0$. It is normally independent of the sub- or superscripts k, l or m occurring in the spin-interaction Hamiltonians at Eqs. 9.1 or 9.2.

The corresponding **reduced intensity function** (also called "spectral density") is given as the Fourier transform

$$\mathcal{I}(\omega) = \int\limits_{-\infty}^{+\infty} G(\tau)\, e^{-i\omega\tau}\, d\tau \tag{9.5}$$

which is an even and real function obeying the normalization condition

$$\frac{1}{2\pi} \int\limits_{-\infty}^{\infty} \mathcal{I}(\omega)\, d\omega = G(0) = 1 \tag{9.6}$$

These definitions of the reduced correlation and intensity functions imply that the ensemble of spin systems is ergodic. That is, *all* orientations occurring in the ensemble are to be reached by *all* participating spin systems provided that one waits long enough. However, long waiting times may conflict with the assumptions taken for granted in the perturbation theory presented below in Sect. 11.1. This objection matters in case of very anisotropic motions or jump-like rearrangements leaving quasi-stationary **residual correlations** with correlation times longer than the actual relaxation times (for an example, see Sect. 14.2). In this case, the correlation function may be reduced further to the part which is actually relevant for relaxation. Equation 9.3, for instance, is then replaced by

$$G(\tau) = \frac{\left\langle F^{(k)}(0)F^{(-k)}(\tau)\right\rangle - |\left\langle F^{(k)}\right\rangle|^2}{\left\langle |F^{(k)}|^2\right\rangle - |\left\langle F^{(k)}\right\rangle|^2} \tag{9.7}$$

This form ensures that $G(0) = 1$ and $G(\infty) = 0$ where "∞" means a time in the order of the relaxation times. The averages now refer to subensembles which are ergodic on the time scale of spin relaxation. That is, only reorientations taking place on that scale are taken into account. The consequence is that Eq. 9.7 depends on the orientation of the anisotropically reorienting spin system. Hence the local relaxation rates tend to do so as well.

In the case of spin-lattice relaxation in the fast spin-diffusion limit (compare Sect. 23.1.3), merely *average* relaxation rates matter, so that a powder average can be taken already on the stage of the reduced correlation function at Eq. 9.7. On the other hand, if the total spin ensemble is ergodic on the relaxation-time scale, we have $\left\langle F^{(k)}\right\rangle = 0$, and Eq. 9.7 coincides with Eq. 9.3.

9.2
Limits and Definitions for Spin-Lattice Relaxation

Spin-lattice (or longitudinal) relaxation refers to the thermal equilibration of the magnetization components along the quantization field after a non-equilibrium state was initially established. This may refer to the net longitudinal magnetization or to longitudinal order of the spin populations. The quantization field is B_0 in the case of **laboratory-frame relaxation**, the effective field B_e in **rotating-frame variants**, and the local field B_{loc} in **dipolar or quadrupolar-order experiments**.

As the name says, either spin energy is dissipated net in the "lattice" or lattice energy is transferred net to the spin system in a random manner while the magnetization approaches its equilibrium value given by the Curie equation (Eq. 47.31). The lattice comprises all mechanical degrees of freedom in the form of an effectively unlimited heat bath assumed to remain permanently in thermal equilibrium.

The spin transitions induced under weak-collision conditions are accompanied by transitions of the quantum-mechanical system of the lattice in the opposite direction. That is, lattice-transition operators should be implied in principle. However, with the exception of relaxation by **spin-phonon coupling**[4] (e.g., [371, 372])

---

[4]For example, spin-phonon interaction is mediated by dipolar coupling between neighbouring nuclei. The distance dependence of dipolar coupling is proportional to r^{-3} (Table 46.5 on page 426). Expanding this factor about the equilibrium distance r_0 gives $r^{-3} = r_0^{-3} + (d(r^{-3})/dr)_{r_0}\Delta r + \ldots$. The distance variation by phonon modulation is $\Delta r = \sin(r_0 k)q \approx r_0 kq$, where k is the phonon wavenumber, and q is the displacement of a lattice point by the phonon. In the "normal coordinate representation", lattice vibrations are expressed by linear combinations of harmonic oscillator solutions characterized by the eigenfrequency ω and the mass m. The raising and lowering operators of the harmonic-oscillator theory are termed a and a^+. With the aid of the relation $q = \sqrt{\hbar/(2m\omega)}(a+a^+)$, the distance variation may thus be related to these lattice operators in the form $\Delta r \approx r_0 k\sqrt{\hbar/(2m\omega)}(a + a^+)$.

or by low-temperature **rotational tunneling processes** of methyl or ammonium groups, for instance, (e.g., [6, 7]), the quantum mechanics of the lattice need not be taken explicitly into account. The factors $A_{l,-m}$ in Eq. 9.2 will be interpreted as sole functions of the spatial coordinates as such, and not as lattice-operator terms.

Relaxation theories referring to the thermal but not to the quantum-mechanical properties of the lattice are termed **semi-classical**. We restrict ourselves to this sort of description which is adequate for most cases of practical relevance.

9.3
Limits and Definitions for Transverse Relaxation

Transverse relaxation refers to the attenuation of coherences which are entirely absent when thermal equilibrium is reached. In the case of **single-quantum coherences**, the transverse-relaxation process can be directly visualized as the decay of the magnetization transverse to the quantization field. By contrast, **multiple-quantum coherences** must be probed indirectly by transfer to single-quantum coherences after a certain relaxation interval.

It is needless to say that the coherence losses considered here are not due to inhomogeneities of the magnetic field. Their influence can be reversibly refocused with the aid of ordinary spin echoes (see Chaps. 2 and 7).

The proper origin of transverse relaxation are spin interactions as in the case of spin-lattice relaxation. One distinguishes "secular" (slowly varying) from "nonsecular" (rapidly varying) spin interactions. The latter induce spin transitions connected with energy exchange with the lattice, and, hence, mediate the energy transfer between the spin system and the lattice. This is the background of spin-lattice relaxation already addressed above.

The same mechanism also contributes to transverse relaxation, of course. Incoherent spin transitions induced by random perturbations simultaneously destroy any correlation among spin-precession phases. Transformed to the frequency domain, one can speak of lifetime line-broadening in the sense of Heisenberg's uncertainty relation.

However, the secular spin-interaction terms cause additional and often more efficient coherence loss mechanisms. The prominent feature of secular spin interactions is that they either are not effective in inducing spin transitions, or the transitions they induce are spin-energy conserving. That is, spin-energy transfer to the lattice is not included. Despite the absence of any energy dissipation process, transverse relaxation by secular interactions to a certain degree is of an irreversible nature as expected according to thermodynamic principles.

The usual definition of transverse relaxation is closely connected with experimental measuring procedures such as the record of FIDs or, in the presence of inhomogeneities, 180° pulse Hahn echoes. However, the coherence attenuations observed under such circumstances can partly have a reversible character. The mere change of the RF-pulse sequence to more elaborate forms of spin manipulation such as **solid-echo** or **magic-echo** procedures (see Chaps. 4 and 6), for instance, can partly recover coherences *without renewed excitation of virginal coherences starting from thermal equilibrium.*[5]

The factors important for irreversible coherence losses mediated by secular spin interactions are molecular motions causing fluctuations of the spin interactions during the measuring process, and the size of the spin system defined by the number of spins which are linked by interactions. Let us discuss the latter first.

9.3.1
Irreversibility and Spin-System Size

In the absence of molecular motion, an ensemble of isolated small spin systems coherently excited by an RF pulse evolves under secular spin interactions in a largely reversible way. This was demonstrated in Chaps. 4 and 6. Provided that suitable pulse sequences are employed, coherence evolution in an ensemble of dipolar-coupled two-spin 1/2 systems or of spin 1 nuclei can even be entirely reversible if non-secular terms are absent.

However, if the spin systems are larger as they tend to be in solids where a whole network of dipolar interactions may link more or less all spins in the sample, the coherence evolution is getting more complicated. It would then be impossible to invent an RF-pulse sequence for spin manipulation that brings all spins in an ensemble of spin system to the same precession phase at one time. This is a matter of the large number of the spins involved. We are dealing here with a many-particle problem which intrinsically implies a chaotic element. Hence, the coherence losses are quasi-irreversible. Note that this diminution in the coherence-evolution reversibility under secular interactions begins already with spin systems larger than two (also compare the discussion in Sect. 39.2).

9.3.2
Irreversibility and Molecular Motion

Molecular motion is the second reason for the irreversible attenuation of coherences. The refocusing of coherences by solid-echo techniques (Chaps. 4 and 6) requires that the local fields arising from secular spin interactions are stationary on the time scale of the experiment. However, molecular motions cause fluctuations of the spin interactions and, hence, of the local fields. That is, coherence dephasing of even very small systems such as two-spin 1/2 systems or isolated spin 1 nuclei adopts an irreversible character irrespective of any non-secular interactions.

Treatments where such phenomena are relevant will be presented below in context with the Anderson/Weiss theory, the dipolar-correlation effect, and 2D exchange spectroscopy, for instance. It must be stressed that translational diffusion in the presence of magnetic-field gradients also falls into this category of irre-

---

[5]By contrast, **multiple-pulse trains** and **magic-angle spinning** tend to *average* secular spin interactions via the Hamiltonian terms depending on spin and spatial coordinates, respectively. In this case, secular-interaction induced transverse relaxation is slowed down from the very beginning just as is the case with motional averaging in non-viscous liquids.

versible spin-echo coherence attenuation phenomena with all line-broadening consequences. Irreversible coherence losses of this sort may be due to translational diffusion in magnetic-susceptibility induced field inhomogeneities as they occur in porous materials and biological tissue [54], for instance, or intentionally applied field gradients as with field-gradient NMR diffusometry (Chap. 18).

9.3.3
Classification of Transverse Relaxation

For the interpretation of transverse-relaxation experiments, it may be helpful to keep the following distinctions in mind.

- Transverse relaxation of small or large spin systems in the **complete motional-averaging limit**,

$$(\Delta\omega)_{rl}\, \tau_c \ll 1 \tag{9.8}$$

is governed by secular spin interactions for

$$\omega_0 \tau_c \gg 1 \tag{9.9}$$

and by secular as well as non-secular spin interactions for

$$\omega_0 \tau_c \ll 1 \tag{9.10}$$

(**extreme-narrowing limit**[6]). The correlation time of the fluctuating interactions is τ_c, the quantity $(\Delta\omega)_{rl}$ is the linewidth in the absence of molecular motions, and ω_0 is the resonance frequency. The fluctuation rate of the spin interactions exceeds the transverse-relaxation rate in any case. A theory adequate for this situation is the **Bloch/Wangsness/Redfield (BWR) theory** (Chap. 13).

- Transverse relaxation of large spin systems in the **rigid-lattice limit**,

$$(\Delta\omega)_{rl}\, \tau_c \gg 1 \tag{9.11}$$

is governed by secular spin interactions which now tend to be quasi-stationary on the time scale of transverse relaxation. This situation is considered as one of the typical limiting cases of the **Anderson/Weiss (AW) formalism** as outlined below (Sect. 13.2).

- Transverse relaxation of small or large spin systems **halfway between the motional-averaging and rigid-lattice limits**, that is,

$$(\Delta\omega)_{rl}\, \tau_c \approx 1 \tag{9.12}$$

In this case, transverse relaxation is disposed to be governed by secular spin interactions. However, the fluctuations are still not fast enough to fulfill the prerequisites of the BWR theory. Therefore, the AW formalism will be considered again.

---

[6]This term reminds of the relaxation-related linewidth.

- Transverse relaxation of large spin systems for **partially complete/partially incomplete motional averaging**. That is, the motions are strongly anisotropic implying (a) components obeying

$$(\Delta\omega)_{rl} \ \tau_c^{(a)} \ll 1 \tag{9.13}$$

and (b) components complying with

$$(\Delta\omega)_{rl} \ \tau_c^{(b)} \gg 1 \tag{9.14}$$

where $\tau_c^{(a)}$ and $\tau_c^{(a)}$ are the respective correlation times. Transverse relaxation is then governed by secular as well as non-secular interacations as concerns components (a), and by secular interactions alone as regards components (b). This situation of multi-componential motions typically arises with polymers, liquid crystals, and adsorbate molecules on surfaces, for instance. Since (b) is biased to govern the transverse-relaxation rate, the applicability of the BWR theory is still restricted. One therefore often refers to the AW formalism again, or - in context with polymers - to more sophisticated variants emerging from the same anticipated principles [61, 97, 287].

CHAPTER 10

Spin-Relaxation Functions

10.1
The Homonuclear Bloch Equations

The spin-relaxation functions are the time dependences of the observables one measures in relaxation experiments. A list of relaxation functions for typical measuring procedures is given in Table 10.1 on page 93.

The equations of motion yielding **exponential relaxation functions** for homonuclear magnetization components M in a magnetic field B are known as the Bloch equations [40]. Originally, they have been established phenomenologically. However, as will be shown below, they can also be derived in an equivalent form for many important situations directly from the basic quantum-mechanical equations of motion (see Chaps. 12 and 13).

As concerns the time dependence of the z component of the magnetization, minor, though scarcely detectable, deviations from the exponentiality may occur in multi-spin systems.[1] On the other hand, one faces substantial deviations from the exponential relaxation functions inherent to Bloch's equations for the transverse magnetization components if secular spin interactions are not completely averaged out in the NMR time scale which is defined by the reciprocal rigid-lattice linewidth.

In the frame of these limitations, the **laboratory-frame homonuclear Bloch equation** in vectorial form reads

$$\frac{\mathrm{d}M}{\mathrm{d}t} = \underbrace{\gamma_n M \times B}_{\text{precession}} \underbrace{- \frac{M_x}{T_2} u_x - \frac{M_y}{T_2} u_y - \frac{M_z - M_0}{T_1} u_z}_{\text{relaxation}}$$

$$= \gamma_n M \times B - \frac{1}{T_2} \begin{pmatrix} M_x \\ M_y \\ 0 \end{pmatrix} - \frac{1}{T_1} \begin{pmatrix} 0 \\ 0 \\ M_z - M_0 \end{pmatrix} \tag{10.1}$$

The spin-lattice relaxation time T_1 refers to the z component of the magnetization which relaxes towards its equilibrium value M_0, the Curie magnetization (Eq. 47.31). The magnetization transverse to the quantization magnetic field B relaxes with the time constant T_2 towards its equilibrium value 0. The relaxation times are functions

---

[1]Note that this statement refers to relaxation curves of a single well-defined type of spin system. This is not to be confused with heterogeneity effects in multi-component or multi-phase samples which are prone to pronounced non-exponentialities in the absence of fast exchange (see Chap. 23).

of the magnetic field applied to the sample. They are also functions of parameters describing the spin interactions, the molecular structure and the dynamics relevant for the spin system under consideration. This is the reason why these time constants can be so informative for problems of molecular dynamics.

Including the temporary irradiation of RF fields applied perpendicular to the main magnetic field B_0, the total magnetic field is given by

$$B(t) = \begin{pmatrix} 0 \\ 0 \\ B_0 \end{pmatrix} + \begin{pmatrix} B_x(t) \\ B_y(t) \\ 0 \end{pmatrix} = \begin{pmatrix} B_1(t)\cos(\omega_c t) \\ B_1(t)\sin(\omega_c t) \\ B_0 \end{pmatrix} + \begin{pmatrix} B_1(t)\cos(\omega_c t) \\ -B_1(t)\sin(\omega_c t) \\ B_0 \end{pmatrix}$$

$$(10.2)$$

The magnetic-flux vector $2B_1$ of the RF field is assumed to oscillate along the positive and negative x axis of the laboratory frame. It is assumed to be initially aligned along the positive x axis. The linearly polarized RF field may be analyzed into two counterrotating circular polarized RF fields as manifested on the right-hand side of Eq. 10.2. The first and second terms represent components rotating counterclockwise and clockwise, respectively, in the x, y plane as seen when looking down from the positive z axis.

It is more convenient to consider the equations of motion in a reference frame rotating with an angular frequency equal to the carrier frequency ω_c about the z axis in the same sense as the Larmor precession. The corresponding transformation (see Sect. 48.7) and the neglect of any non-secular terms leads to the **rotating-frame Bloch equations**. A matrix representation is

$$\begin{pmatrix} \frac{dM'_x}{dt} \\ \frac{dM'_y}{dt} \\ \frac{dM'_z}{dt} \end{pmatrix} = \begin{pmatrix} -\frac{1}{T_2} & -\Omega & 0 \\ \Omega & -\frac{1}{T_2} & \omega_1 \\ 0 & -\omega_1 & -\frac{1}{T_1} \end{pmatrix} \begin{pmatrix} M'_x \\ M'_y \\ M'_z \end{pmatrix} + \begin{pmatrix} 0 \\ 0 \\ \frac{M_0}{T_1} \end{pmatrix} \qquad (10.3)$$

where $\omega_1 = \gamma_n B_1$. The offset of the carrier frequency ω_c from the resonance frequency $\omega_0 = \gamma_n B_0$ is $\Omega = \omega_c - \omega_0$. The primes indicate that the quantities refer to the rotating frame. The rotating-frame direction of the RF field is assumed along the x' axis.

General solutions of the Bloch equations have been published in Refs. [313, 353, 483]. These are important for the derivation of RF pulse responses in the presence of field gradients, for instance. In the following we consider the much simpler situations in standard relaxation experiments which may be classified in two categories.

10.2
Solutions for Laboratory-Frame Experiments

In the free-evolution intervals of RF pulse sequences we have $\omega_1 = 0$. The Bloch equation (Eq. 10.3) in the resonantly rotating frame, that is, $\Omega = 0$, then reads

$$
\begin{pmatrix} \frac{dM'_x}{dt} \\ \frac{dM'_y}{dt} \\ \frac{dM'_z}{dt} \end{pmatrix} = - \begin{pmatrix} \frac{1}{T_2} & 0 & 0 \\ 0 & \frac{1}{T_2} & 0 \\ 0 & 0 & \frac{1}{T_1} \end{pmatrix} \begin{pmatrix} M'_x \\ M'_y \\ M'_z - M_0 \end{pmatrix} \tag{10.4}
$$

or

$$
\frac{dM'}{dt} = -R \cdot (M' - M_0) \tag{10.5}
$$

where R is the relaxation rate matrix. The standard solutions for the most common pulse sequences are listed in Table 10.1 on page 93.

10.3
Solutions for Rotating-Frame Experiments

The RF pulse scheme usually employed for the production of spin-locked magnetization is shown in Fig. 10.1. During the first RF pulse, the reference frame is assumed to rotate with the resonant carrier frequency, that is $\Omega = 0$. The RF amplitude, B_1, is assumed to be constant, so that ω_1 is constant too. The phase direction of the RF field is arbitrarily chosen along the x' axis of the rotating frame. Equation 10.3 may then be rewritten in the component form

$$
\begin{aligned}
\frac{\partial M'_x}{\partial t} &= -\frac{M'_x}{T_2} \\
\frac{\partial M'_y}{\partial t} &= -\frac{M'_y}{T_1} + \omega_1 M'_z \\
\frac{\partial M'_z}{\partial t} &= -\frac{M'_z - M_0}{T_1} - \omega_1 M'_x
\end{aligned} \tag{10.6}
$$

For the initial condition $M'(0-) = M_0 u'_z$, the solution of the first of these equations evidently is $M'_x(t) = 0$. That is, the third equation is reduced to

$$
\frac{\partial M'_z}{\partial t} = -\frac{M'_z - M_0}{T_1}. \tag{10.7}
$$

This remaining system of differential equations accounts for the situation during the $(\pi/2)_x$ RF pulse of the scheme displayed in Fig. 10.1.

The spin lock pulse, $(SL)_y$, is first assumed to be **resonant**. The magnetization is aligned along the magnetic field effective in the rotating frame, which then is $B_e = B_1 u'_y$ (see Eq. 48.98). This is the quantization field effective during the spin-lock period t_{sl}. The spin-locked magnetization is consequently subject to spin-lattice

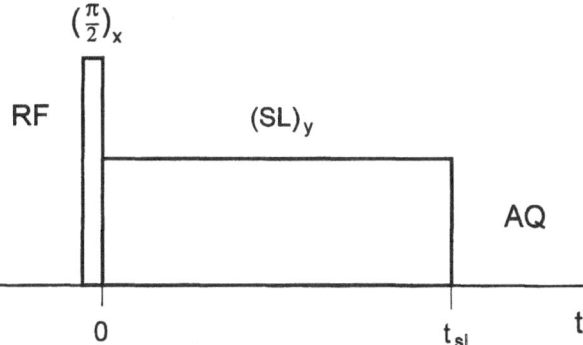

Fig. 10.1. RF pulse scheme for resonant rotating-frame relaxation experiments. The signals are acquired (AQ) after a 90° phase-shifted spin-lock (SL) pulse of variable length t_{sl}.

relaxation (instead of transverse relaxation in the free-evolution case!) with the time constant $T_{1\rho}$ (instead of T_2!), where the subscript ρ indicates the rotating frame.

Under such circumstances, the magnetization is aligned along the effective field so that there is no reason for precession. Under resonant spin-locking conditions, the z' component of the rotating-frame magnetization is zero anyway. That is, we can omit the precession term $\omega_1 M_z'$. The equation of motion of the spin-locked magnetization can thus be rewritten in the form

$$\frac{\partial M_y'}{\partial t} = -\frac{M_y'}{T_{1\rho}} \tag{10.8}$$

Generalizing the spin-lock pulse to the **off-resonant** case, we must take into account that the effective field is tilted by an angle $\Theta \neq 90°$. That is, the flip angle of the initial excitation pulse (which may be assumed to be still resonant) is no longer 90°. Rather it is adjusted to match the tilt angle, so that the total magnetization becomes locked along the effective field. There is still no processing transverse component, but the z' component of the rotating-frame magnetization does not vanish anymore. The correspondingly extended system of equations of motion is

$$\frac{\partial M_y'}{\partial t} = -\frac{M_y'}{T_{1\rho}} \tag{10.9}$$

$$\frac{\partial M_z'}{\partial t} = -\frac{M_z' - M_0}{T_1} \tag{10.10}$$

The magnetization components, M_y' and M_z', form the spin-locked magnetization vector aligned along the effective field defined in Eq. 48.98.

The transformation to a new rotating frame tilted by the angle Θ against the z' axis brings the direction of the effective field B_e into coincidence with that of the z'' axis. This transformation to the tilted rotating frame (indicated by double primes)

Table 10.1. Pulse sequences for typical relaxation experiments and homonuclear relaxation functions.

pulse sequence $[(\varphi = 0) \rightsquigarrow x; (\varphi = \pi/2) \rightsquigarrow y; n = 1, 2, \ldots]$	solutions of Bloch's equations	conditions
$\alpha - \tau - (\beta)_\varphi - AQ$ general two-pulse sequence	$M(0-) = M_0 u_z$ $M_z(\tau-) = M_0 [1 - (1 - \cos\alpha) e^{-\tau/T_1}]$ $m(\tau+) = M_z(\tau-) e^{i\varphi} \sin\beta$	$\tau \gg T_2^*$
$[\frac{\pi}{2}]_n - \tau - (\frac{\pi}{2})_\varphi - AQ$ saturation/recovery (Fig. 2.1)	$M(0-) = M(0-) u_z$ (arb.) $M_z(\tau-) = M_0 (1 - e^{-\tau/T_1})$ $m(\tau+) = M_z(\tau-) e^{i\varphi}$	$\tau \gg T_2^*$
$\pi - \tau - (\beta)_\varphi - AQ$ inversion/recovery	$M(0-) = M_0 u_z$ $M_z(\tau-) = M_0 (1 - 2e^{-\tau/T_1})$ $m(\tau+) = M_z(\tau-) e^{i\varphi} \sin\beta$	
$[\frac{\pi}{2} - \tau]_n - (\frac{\pi}{2})_\varphi - AQ$ progressive saturation	$M(0-) \rightsquigarrow$ arbitrary $M_z(n\tau-) = M_0 (1 - e^{-\tau/T_1})$ $m(n\tau+) = M_z(n\tau-) e^{i\varphi}$	$\tau \gg T_2^*$ $n\tau \gg T_1$
(field cycle) $- (\frac{\pi}{2})_\varphi - AQ$ field-cycling relaxometry (Fig. 15.1)	$M(0-) = M_0(B_p) u_z$ $M_z(\tau-) = M_0(B_r) + [M_0(B_p) - M_0(B_r)] e^{-\tau/T_1(B_r)}$ $m(\tau+) = M_z(\tau-) e^{i\varphi}$	$B_r \gg B_{loc}$

(continued next page)

Table 10.1 (cont.)

pulse sequence $[(\varphi = 0) \rightsquigarrow x; (\varphi = \pi/2) \rightsquigarrow y; n = 1, 2, \ldots]$	solutions of Bloch's equations	conditions
$(\frac{\pi}{2})_x - (SL)_y - AQ$ resonant rotating-frame relaxometry (Fig. 10.1)	$\begin{aligned} M(0-) &= M_0 \boldsymbol{u}_z \\ M'_y(t_{sl}) &= M_0 e^{-t_{sl}/T_{1\rho}} \\ m(t_{sl}+) &= iM'_y(t_{sl}) \end{aligned}$	$B_1 \gg B_{loc}$ $\omega_c = \omega_0$
$\frac{\pi}{2} - \tau_1 - \frac{\pi}{2} - \tau_2 - (\frac{\pi}{2})_\varphi - \tau_1 - AQ$ stimulated echo (Fig. 2.2)	$\begin{aligned} M(0-) &= M_0 \boldsymbol{u}_z \\ m(2\tau_1 + \tau_2) &= \tfrac{1}{2} M_0 e^{-\tau_2/T_1} e^{-2\tau_1/T_2} e^{i\varphi} \end{aligned}$	$\tau_1, \tau_2 \gg T_2^*$ $\tau_1 \ll J^{-1}$ mot. av. $\gamma_n^2 G^2 \tau_2^3 \ll D^{-1}$
$(\frac{\pi}{2})_x - \tau - (\beta)_y - \tau - AQ$ Hahn echo (Fig. 2.1)	$\begin{aligned} M(0-) &= M_0 \boldsymbol{u}_z \\ m(2\tau) &= iM_0 e^{-2\tau/T_2} \sin^2 \tfrac{\beta}{2} \end{aligned}$	$\tau \ll J^{-1}$ mot. av. $\gamma_n^2 G^2 \tau^3 \ll D^{-1}$
$(\frac{\pi}{2})_x - \tau - [(\pi)_y - \tau - AQ - \tau -]_n$ Carr/Purcell/Meiboom/Gill (CPMG)	$\begin{aligned} M(0-) &= M_0 \boldsymbol{u}_z \\ m(2n\tau) &= iM_0 e^{-2n\tau/T_2} \end{aligned}$	$2n\tau \ll J^{-1}$ mot. av. $\gamma_n^2 G^2 n\tau^3 \ll D^{-1}$

is performed using the equations

$$M'_y = M''_z \sin \Theta; \qquad M'_z = M''_z \cos \Theta \qquad (10.11)$$

The spin-locked magnetization is designated by M''_z. Combining Eqs. 10.9 - 10.11 gives

$$\frac{\partial M''_z}{\partial t} = -\frac{M''_z}{T^{(e)}_{1\rho}} + \frac{M_0 \cos \Theta}{T_1} \qquad (10.12)$$

where

$$\frac{1}{T^{(e)}_{1\rho}} = \frac{\cos^2 \Theta}{T_1} + \frac{\sin^2 \Theta}{T_{1\rho}} \qquad (10.13)$$

With the initial condition $M''_z(0+) = M_0$, the solution of Eq. 10.12 is

$$M''_z(t) = (M_0 - M_{0e})\, e^{-t_{sl}/T^{(e)}_{1\rho}} + M_{0e} \qquad (10.14)$$

where

$$M_{0e} = M_0 \frac{T^{(e)}_{1\rho}}{T_1} \cos \Theta \qquad (10.15)$$

This general result implies the case for resonant spin locking, of course. That is, $\Theta = 90°$ and initial $90°$ pulse excitation (see Table 10.1 on page 93):

$$M''_z(t) = M_0\, e^{-t_{sl}/T_{1\rho}} \qquad (10.16)$$

CHAPTER 11

Perturbation Theory of Spin Relaxation

The theory we will outline in the following is based on the Bloch/Wangsness/Redfield (BWR) density-operator perturbation-theoretical approach of the weak-collision case [1, 41, 42, 397, 503]. This formalism is superior to theories based on the standard time-dependent perturbation theory [5, 43, 461] because of its universality.[1] Versions of the BWR theory suitable for more complex spin systems, as they usually are the subject of NOESY experiments (see Sect. 23.2), for instance, have been published in Refs. [478, 500, 514].

11.1
Iterative Approximation

The Hamiltonian is assumed to be composed of a stationary part, $\mathcal{H}_0$, and a small time-dependent perturbation $\mathcal{H}_i(t)$,

$$\mathcal{H} = \mathcal{H}_0 + \mathcal{H}_i(t) \tag{11.1}$$

The operator $\mathcal{H}_0$ primarily refers to the Zeeman energy, but also includes any other time-independent contribution due to chemical or susceptibility shifts etc. The perturbative part, $\mathcal{H}_i(t)$, is due to spin interactions fluctuating stochastically so that the ensemble average vanishes, $\overline{\mathcal{H}_i(t)} = 0$. Any stationary contribution of the spin interactions may be allocated to $\mathcal{H}_0$.

The general equation of motion for the density operator of the spin system is given in the form of the Liouville/von Neumann equation

$$\frac{d\rho}{dt} = -\frac{i}{\hbar} \left[\mathcal{H}_0 + \mathcal{H}_i, \rho \right] = -\frac{i}{\hbar} \left[\mathcal{H}_0, \rho \right] - \frac{i}{\hbar} \left[\mathcal{H}_i, \rho \right] \tag{11.2}$$

In order to get rid of $\mathcal{H}_0$, we transform this equation into the **interaction representation**, i.e., the representation in the resonantly rotating frame (see Sect. 48.7.2). Quantities referring to this representation will be marked by primes. The unitary-transformation relations for the operators in Eq. 11.2 are

$$\rho = \mathcal{U}\rho'\mathcal{U}^{-1}; \qquad \mathcal{H}_0 = \mathcal{U}\mathcal{H}_0'\mathcal{U}^{-1} \equiv \mathcal{H}_0'; \qquad \mathcal{H}_i = \mathcal{U}\mathcal{H}_i'\mathcal{U}^{-1} \tag{11.3}$$

---

[1] Recently it was shown that the same master equation as for the BWR theory can be derived without any perturbation-theoretical approach [143]. However, this exact formalism is more demanding and somewhat unwieldy.

where

$$\mathcal{U} \equiv e^{-(i/\hbar)\mathcal{H}_0 t} \tag{11.4}$$

In order to replace the left-hand side of Eq. 11.2 we need the time derivative of the density operator transformation relation. Deriving the first equation in Eq. 11.3 gives

$$\frac{d\rho}{dt} = -\frac{i}{\hbar}\mathcal{U}\left[\mathcal{H}_0', \rho'\right]\mathcal{U}^{-1} + \mathcal{U}\frac{d\rho'}{dt}\mathcal{U}^{-1} \tag{11.5}$$

The right-hand side of Eq. 11.2 may be rewritten

$$-\frac{i}{\hbar}\left[\mathcal{H}_0, \rho\right] - \frac{i}{\hbar}\left[\mathcal{H}_i, \rho\right] = -\frac{i}{\hbar}\mathcal{U}\left[\mathcal{H}_0', \rho'\right]\mathcal{U}^{-1} - \frac{i}{\hbar}\mathcal{U}\left[\mathcal{H}_i', \rho'\right]\mathcal{U}^{-1} \tag{11.6}$$

Equating eqns 11.5 and 11.6, and multiplying by $\mathcal{U}^{-1}$ from the left and by $\mathcal{U}$ from the right gives the Liouville/von Neumann equation in the interaction (or resonantly rotating) frame,

$$\frac{d\rho'}{dt} = -\frac{i}{\hbar}\left[\mathcal{H}_i', \rho'\right] \tag{11.7}$$

The integral thereof is

$$\rho'(t) = \rho'(0) - \frac{i}{\hbar}\int_0^t \left[\mathcal{H}_i'(\tilde{t}), \rho'(\tilde{t})\right] d\tilde{t} \tag{11.8}$$

An approximate solution of the Liouville/von Neumann equation is found with the aid of the Picard-iteration variant of time-dependent perturbation theory. We start with the zeroth approximation

$$\rho'(t) \approx \rho'(0) \tag{11.9}$$

Replacing $\rho'(t)$ in Eq. 11.8 by the zeroth approximation gives the first-order approximation

$$\rho'(t) \approx \rho'(0) - \frac{i}{\hbar}\int_0^t \left[\mathcal{H}_i'(t_1), \rho'(0)\right] dt_1 \tag{11.10}$$

Inserting this in turn into Eq. 11.8 leads to the second approximation

$$\rho'(t) \approx \rho'(0) - \frac{i}{\hbar}\int_0^t \left[\mathcal{H}_i'(t_2), \rho'(0)\right] dt_2 + \left(\frac{i}{\hbar}\right)^2 \int_0^t dt_2 \int_0^{t_2} dt_1 \left[\mathcal{H}_i'(t_2), \left[\mathcal{H}_i'(t_1), \rho'(0)\right]\right] \tag{11.11}$$

and so on. However, it turns out that higher-order approximations are not necessary, and that the second-order treatment even yields a result coinciding with a formalism not based on perturbation theory [143].

The time derivative of Eq. 11.11 is

$$\frac{d\rho'}{dt} = -\frac{i}{\hbar}\left[\mathcal{H}_i', \rho'(0)\right] - \frac{1}{\hbar^2}\int_0^t \left[\mathcal{H}_i'(t), \left[\mathcal{H}_i'(t_1), \rho'(0)\right]\right] dt_1 \tag{11.12}$$

With the substitution $t_1 = t - \tau$, Eq. 11.12 can be rewritten as

$$\frac{d\rho'}{dt} = -\frac{i}{\hbar} \left[\mathcal{H}'_i, \rho'(0) \right] - \frac{1}{\hbar^2} \int_0^t \left[\mathcal{H}'_i(t), \left[\mathcal{H}'_i(t - \tau), \rho'(0) \right] \right] \, d\tau \tag{11.13}$$

The time dependence of ρ' expected on this basis is the consequence of a certain time dependence $\mathcal{H}'_i(t)$. However, the spin systems in the ensemble under consideration are subject to statistically varying functions $\mathcal{H}'_i(t)$. That is, all quantities measured in real experiments are averages over all these different time dependences.[2]

The spin interactions are assumed to fluctuate stochastically so that $\overline{\mathcal{H}'_i(t)} = 0$. Furthermore, the validity of this treatment is restricted to times $t \gg \tau_c$, where τ_c is the correlation time of the spin interactions. The spin-interaction Hamiltonian at time t, $\mathcal{H}'_i(t)$, and the density operator at time 0, $\rho'(0)$, are therefore uncorrelated and can be averaged independently. That is,[3]

$$\overline{\left[\mathcal{H}'_i(t), \rho'(0) \right]} = \left[\overline{\mathcal{H}'_i(t)}, \overline{\rho'(0)} \right] = 0 \qquad \text{for } t \gg \tau_c \tag{11.14}$$

For the same reason, the ensemble average of the integrand vanishes for $t \gg \tau_c$. That is, the upper integration limit can be equated with ∞. The ensemble average of Eq. 11.13 thus obeys

$$\frac{d\overline{\rho'(t)}}{dt} = -\frac{1}{\hbar^2} \int_0^\infty \overline{\left[\mathcal{H}'_i(t), \left[\mathcal{H}'_i(t - \tau), \rho'(0) \right] \right]} \, d\tau \tag{11.15}$$

The iterative approximation procedure starts with the zeroth order $\rho'(t) \approx \rho'(0)$. As a consequence, this approach for the alteration rate of the density operator is only good *in the vicinity of time o*. Let us denote the range of acceptable accuracy by the symbol Δt. The density operator varies greatly on the time scale of the relaxation times T_1 and T_2, so that the requirement that the density operator is still close to its initial value stipulates $\Delta t \ll T_1, T_2$. On the other hand, the derivation of Eq. 11.15 demands $\Delta t \gg \tau_c$. That is, Eq. 11.15 holds good only in the range $\tau_c \ll \Delta t \ll T_1, T_2$ around $t = 0$.

In order to make Eq. 11.15 relevant for an interval of the same width but shifted to an arbitrary time later, we must replace $\rho'(0)$ in the integrand by $\rho'(t)$.[4] We thus obtain the alteration rate of ρ' at the time t.

This step is allowed because the time dependences of the spin-interaction Hamiltonians $\mathcal{H}'_i(t)$ and $\mathcal{H}'_i(t - \tau)$ in the integrand do not matter in this context, as we

---

[2]This is not to be confused with the ensemble average inherent in the density operator which concerns the distribution of state vectors in the ensemble (i.e., different compositions of eigenstates and different phases). Here we are dealing with an ensemble average over "different time dependences" of the ensemble average over "different state vectors," so to speak.

[3]By contrast, the corresponding average of the Liouville/von Neumann equation (Eq. 11.7) does not vanish because the density operator and the Hamiltonian refer to the *same* time in that case.

[4]In a sense, we insert the zeroth approximation in the reverse direction.

will see shortly. The fluctuating terms enter merely via their correlations after an interval τ. That is, their effect is independent of the absolute time. Thus, at the present stage of the formalism, no dependence on the absolute time is left other than that of the density operator, which can therefore be considered at any time of interest in relaxation experiments.

Introducing the stationary equilibrium density operator, $\rho'(\infty) = \rho'_0$, and omitting the bar over the pure density operator terms from now on, we find for the instantaneous deviation from equilibrium

$$\frac{d(\rho'(t) - \rho'_0)}{dt} = \frac{d\rho'(t)}{dt} = -\frac{1}{\hbar^2} \int_0^\infty \overline{\left[\mathcal{H}'_i(t), \left[\mathcal{H}'_i(t - \tau), (\rho'(t) - \rho'_0)\right]\right]} \, d\tau \quad (11.16)$$

11.2
The Master Equation

Equation 11.16 is the basis for the equation of motion for the expectation value

$$\langle \mathcal{Q} \rangle' = \text{Tr} \left\{ \rho'(t) \mathcal{Q} \right\} \quad (11.17)$$

of an observable $\mathcal{Q}$. As such, the spin operator components will be considered, for instance (see Sect. 9.1.1). Combining this with Eq. 11.16 yields

$$\frac{d\langle \mathcal{Q} \rangle'}{dt} = \text{Tr} \left\{ \frac{d\rho'(t)}{dt} \mathcal{Q} \right\} = -\frac{1}{\hbar^2} \int_0^\infty \text{Tr} \left\{ \overline{\left[\mathcal{H}'_i(t), \left[\mathcal{H}'_i(t - \tau), (\rho'(t) - \rho'_0)\right]\right]} \mathcal{Q} \right\} \, d\tau$$

$$(11.18)$$

The double commutator in the integrand can be written in extended form as an expression consisting of four terms which act as prefactors of the operator $\mathcal{Q}$. The operators may be rearranged with the aid of the rule given in Eq. 43.3 in such a way that $(\rho'(t) - \rho'_0)$ is throughout at the first position. The result is the **master equation**

$$\frac{d\langle \mathcal{Q} \rangle'}{dt} = -\frac{1}{\hbar^2} \text{Tr} \left\{ (\rho'(t) - \rho'_0) \int_0^\infty \overline{\left[\mathcal{H}'_i(t - \tau), \left[\mathcal{H}'_i(t), \mathcal{Q}\right]\right]} \, d\tau \right\}$$

$$= -\frac{1}{\hbar^2} \text{Tr} \left\{ (\rho'(t) - \rho'_0) \mathcal{P} \right\}$$

$$(11.19)$$

where the "fluctuation operator" is

$$\mathcal{P} = \int_0^\infty \overline{\left[\mathcal{H}'_i(t - \tau), \left[\mathcal{H}'_i(t), \mathcal{Q}\right]\right]} \, d\tau \quad (11.20)$$

Equation 11.19 is of vast generality. This becomes obvious by visualizing that it is suitable for treatments with any **spin-quantum number** $\geq \frac{1}{2}$, any **spin-interaction Hamiltonian** $\mathcal{H}_i$, such as those for

- dipolar interaction,
- quadrupolar interaction,
- scalar interaction,
- chemical-shift anisotropy, or
- spin-rotation interaction,

and with any **observable**[5] Q, such as those representing

- the longitudinal magnetization,
- the transverse magnetization,
- the longitudinal order, or
- multiple-quantum coherences.

It is even appropriate for situations when the quantization direction deviates from that of the external magnetic field. Such circumstances arise in the rotating frame under RF irradiation, for instance (Sect. 10.3). The interaction frame is then a coordinate system rotating with the effective resonance frequency about the z axis which is aligned along the effective field. That is, the operators are subjected to unitary transformations leading to the desired reference frame.

---

[5]This also includes "observables" such as longitudinal order or multiple-quantum coherences that can only be indirectly detected by RF signals (see Sect. 9.1.1).

CHAPTER 12

Spin-Lattice Relaxation

12.1
Laboratory-Frame Spin-Lattice Relaxation by Dipolar Coupling

Consider an ensemble of systems consisting of two arbitrary spins I and S. The interaction mechanism is assumed to be dipolar interaction. Spin-lattice relaxation refers to the z components of the magnetizations. The stationary part of the Hamiltonian is given as the Zeeman expression

$$\mathcal{H}_0 = -\hbar\omega_I I_z - \hbar\omega_S S_z \tag{12.1}$$

where $\omega_I = \gamma_I B_0$ and $\omega_S = \gamma_S B_0$. No assumption is yet made concerning the gyromagnetic ratios of the spin-bearing particles, γ_I and γ_S. The spin-interaction Hamiltonian $\mathcal{H}_i$ is identified with that for dipolar coupling $\mathcal{H}_d$.

This Hamiltonian is expanded into terms consisting of the fluctuating functions F^k and the spin-operator expressions $\mathcal{O}^{(k)}$ (compare Table 46.5 on page 426)

$$\mathcal{H}_d = f_d \sum_{k=-2}^{2} F^{(k)} \mathcal{O}^{(k)} \tag{12.2}$$

where

$$f_d = \frac{\mu_0}{4\pi} \gamma_I \gamma_S \hbar^2 \tag{12.3}$$

and

$$
\begin{aligned}
F^{(0)} &= F^{(0)*} &= r^{-3}(1 - 3\cos^2\vartheta) \\
F^{(1)} &= F^{(-1)*} &= r^{-3}(\sin\vartheta\cos\vartheta\, e^{-i\varphi}) \\
F^{(2)} &= F^{(-2)*} &= r^{-3}(\sin^2\vartheta\, e^{-2i\varphi}) \\
\mathcal{O}^{(0)} &= \mathcal{O}^{(0)\dagger} &= I_z S_z - \tfrac{1}{4}(I^+ S^- + I^- S^+) \\
\mathcal{O}^{(1)} &= \mathcal{O}^{(-1)\dagger} &= -\tfrac{3}{2}(I^+ S_z + I_z S^+) \\
\mathcal{O}^{(2)} &= \mathcal{O}^{(-2)\dagger} &= -\tfrac{3}{4}I^+ S^+
\end{aligned}
\tag{12.4}
$$

The spin-operator terms are transformed to the interaction representation by

$$\mathcal{O}^{(k)'} = \mathcal{U}^{-1}\mathcal{O}^{(k)}\mathcal{U} = e^{(i/\hbar)\mathcal{H}_0 t}\mathcal{O}^{(k)} e^{-(i/\hbar)\mathcal{H}_0 t} \tag{12.5}$$

That is,

$$\mathcal{O}^{(0)'} = I_z S_z - \frac{1}{4}I^+ S^- e^{-i(\omega_I - \omega_S)t} - \frac{1}{4}I^- S^+ e^{i(\omega_I - \omega_S)t}$$

$$\mathcal{O}^{(1)'} = -\frac{3}{2}I^+S_z e^{-i\omega_I t} - \frac{3}{2}I_z S^+ e^{-i\omega_S t} \tag{12.6}$$

$$\mathcal{O}^{(2)'} = -\frac{3}{4}I^+S^+ e^{-i(\omega_I+\omega_S)t}$$

Labelling the multiple-quantum order with the superscript k, and the spin transition with the subscript l the dipolar Hamiltonian in the interaction frame can be rewritten as

$$\mathcal{H}_d' = f_d \sum_{k,l} F^{(k)} \mathcal{O}_l^{(k)} e^{-i\omega_l^{(k)}t} \tag{12.7}$$

where

$$
\begin{array}{ll}
\mathcal{O}_0^{(0)} = \mathcal{O}_0^{(0)\dagger} = I_z S_z & \omega_0^{(0)} = 0 \\[4pt]
\mathcal{O}_1^{(0)} = \mathcal{O}_2^{(0)\dagger} = -\frac{1}{4}I^+S^- & \omega_1^{(0)} = \omega_I - \omega_S \\[4pt]
\mathcal{O}_2^{(0)} = \mathcal{O}_1^{(0)\dagger} = -\frac{1}{4}I^-S^+ & \omega_2^{(0)} = \omega_S - \omega_I \\[4pt]
\mathcal{O}_1^{(1)} = -\frac{3}{2}I^+S_z & \omega_1^{(1)} = \omega_I \\[4pt]
\mathcal{O}_2^{(1)} = -\frac{3}{2}I_z S^+ & \omega_2^{(1)} = \omega_S \\[4pt]
\mathcal{O}_1^{(-1)} = \mathcal{O}_1^{(1)\dagger} = -\frac{3}{2}I^-S_z & \omega_1^{(-1)} = -\omega_I \\[4pt]
\mathcal{O}_2^{(-1)} = \mathcal{O}_2^{(1)\dagger} = -\frac{3}{2}I_z S^- & \omega_2^{(-1)} = -\omega_S \\[4pt]
\mathcal{O}_1^{(2)} = -\frac{3}{4}I^+S^+ & \omega_1^{(2)} = \omega_I + \omega_S \\[4pt]
\mathcal{O}_1^{(-2)} = \mathcal{O}_1^{(2)\dagger} = -\frac{3}{4}I^-S^- & \omega_1^{(-2)} = -(\omega_I + \omega_S)
\end{array}
\tag{12.8}
$$

Note that $\omega_l^{(k)} = -\omega_l^{(-k)}$ for $k \neq 0$, and $\omega_1^{(0)} = -\omega_2^{(0)}$.

The operators determining the longitudinal magnetizations are $\mathcal{Q} = I_z$ and $\mathcal{Q} = S_z$. The respective master equations are

$$\frac{d\langle I_z\rangle'}{dt} = \frac{d\langle I_z\rangle}{dt} = -\frac{1}{\hbar^2}\mathrm{Tr}\left\{(\rho'(t) - \rho_0')\mathcal{P}_z^I\right\} \tag{12.9}$$

$$\frac{d\langle S_z\rangle'}{dt} = \frac{d\langle S_z\rangle}{dt} = -\frac{1}{\hbar^2}\mathrm{Tr}\left\{(\rho'(t) - \rho_0')\mathcal{P}_z^S\right\} \tag{12.10}$$

where

$$\mathcal{P}_z^I = f_d^2 \int_0^\infty \overline{\left[\sum_{k,l} F^{(k)}(t-\tau)\mathcal{O}_l^{(k)} e^{-i\omega_l^{(k)}(t-\tau)}, \left[\sum_{k',l'} F^{(k')}(t)\mathcal{O}_{l'}^{(k')} e^{-i\omega_{l'}^{(k')}t}, I_z\right]\right]}\, d\tau \tag{12.11}$$

$$\mathcal{P}_z^S = f_d^2 \int_0^\infty \overline{\left[\sum_{k,l} F^{(k)}(t-\tau)\mathcal{O}_l^{(k)} e^{-i\omega_l^{(k)}(t-\tau)}, \left[\sum_{k',l'} F^{(k')}(t)\mathcal{O}_{l'}^{(k')} e^{-i\omega_{l'}^{(k')}t}, S_z\right]\right]}\, d\tau \tag{12.12}$$

Separating the spin operator commutators and the stochastic functions $F^{(k)}$, and swapping the integration and the sum operations gives

$$P_z^I = f_d^2 \sum_{k,l} \sum_{k',l'} \int_0^\infty \overline{F^{(k)}(t-\tau)F^{(-k')*}(t)} e^{i\omega_l^{(k)}\tau} \, d\tau \; e^{-i(\omega_l^{(k)}+\omega_{l'}^{(k')})t} \left[\mathcal{O}_l^{(k)}, \left[\mathcal{O}_{l'}^{(k')}, I_z\right]\right]$$

(12.13)

$$P_z^S = f_d^2 \sum_{k,l} \sum_{k',l'} \int_0^\infty \overline{F^{(k)}(t-\tau)F^{(-k')*}(t)} e^{i\omega_l^{(k)}\tau} \, d\tau \; e^{-i(\omega_l^{(k)}+\omega_{l'}^{(k')})t} \left[\mathcal{O}_l^{(k)}, \left[\mathcal{O}_{l'}^{(k')}, S_z\right]\right]$$

(12.14)

The integrals imply the correlation functions

$$G_{k,-k'}(t,\tau) \equiv \overline{F^{(k)}(t-\tau)F^{(-k')*}(t)}$$

(12.15)

We recall now that stochastic fluctuations are subject to the properties of **stationarity** and **time-reversal invariance**. That is, the correlation functions are independent of the absolute time and of the sign of the interval τ. Furthermore, we may assume that the ensemble of spin systems complies with the **isotropy** condition.[1] In this case, there is no correlation among the polar coordinates r, φ, ϑ within the ensemble of spin systems. That is, the stochastic functions $F^{(k)}$ of different order k are uncorrelated. Using the Kronecker symbol, we summarize these attributes by

$$G_{k,-k'}(t,\tau) = \delta_{k,-k'} G_k(|\tau|)$$

(12.16)

where $G_k(|\tau|)$ is the **autocorrelation function** of order k. The integrals in Eqs. 12.13 and 12.14 can then be rewritten

$$\int_0^\infty G_k(|\tau|)e^{i\omega_l^{(k)}\tau} \, d\tau = \frac{1}{2} \int_{-\infty}^\infty G_k(|\tau|)e^{i\omega_l^{(k)}\tau} \, d\tau + i \int_0^\infty G_k(|\tau|)\sin(\omega_l^{(k)}\tau) \, d\tau$$

(12.17)

The second term on the right-hand side is purely imaginary, and hence cannot contribute to the (non-oscillatory!) relaxation part of the master equation. It will therefore be dropped in the following.[2] The left integral on the right-hand side is the Fourier transform[3] of the autocorrelation function of order k. In the nomenclature of the Wiener/Khinchin theorem [193], we denote this integral as the **intensity**

---

[1]This holds true for samples with powder geometry. Deviations are only expected with ordered systems permitting only very few orientations of the interdipole vectors.

[2]Besides, the modulus of this term is small compared with that of the first term, because, in the period in which the correlation function retains non-vanishing values, i.e., $\tau \stackrel{<}{\sim} \tau_c$, the sine function either remains minor if $\omega_l^{(k)}\tau_c \stackrel{<}{\sim} 1$, or oscillates if $\omega_l^{(k)}\tau_c \stackrel{>}{\sim} \pi$.

[3]Because of the even character of the autocorrelation function, the sign of τ may be reversed in order to conform to the usual convention of the Fourier transformation (see Sect. 42.3).

function or spectral density of order k

$$J^{(k)}(\omega_l^{(k)}) = \int\limits_{-\infty}^{\infty} G_k(\tau)e^{-i\omega_l^{(k)}\tau}\,d\tau \tag{12.18}$$

In terms of intensity functions, the fluctuation operators at Eqs. 12.13 and 12.14 read

$$\mathcal{P}_z^I = f_d^2 \sum_{k,l} \sum_{l'} \frac{1}{2} J^{(k)}(\omega_l^{(k)})e^{-i(\omega_l^{(k)}+\omega_{l'}^{(-k)})t}\left[\mathcal{O}_l^{(k)},\left[\mathcal{O}_{l'}^{(-k)},I_z\right]\right] \tag{12.19}$$

$$\mathcal{P}_z^S = f_d^2 \sum_{k,l} \sum_{l'} \frac{1}{2} J^{(k)}(\omega_l^{(k)})e^{-i(\omega_l^{(k)}+\omega_{l'}^{(-k)})t}\left[\mathcal{O}_l^{(k)},\left[\mathcal{O}_{l'}^{(-k)},S_z\right]\right] \tag{12.20}$$

According to the angular frequencies specified in Eqs. 12.8, the exponential functions in Eqs. 12.19 and 12.20 take the value 1, i.e., are secular, for the cases (a) $l = l'$, $k \neq 0$; (b) $l = 1$, $l' = 2$, $k = 0$; (c) $l = 2$, $l' = 1$, $k = 0$. Otherwise, they are non-secular.

Non-secular terms oscillate rapidly in the time scale of relaxation. Their influence therefore cancels in the average. That is, the sums referring to the subscript l', and the exponential functions in Eqs. 12.19 and 12.20 can be discarded.

Inserting the operator expressions from Eqs. 12.8 leads to

$$
\begin{aligned}
\mathcal{P}_z^I &= \frac{1}{2}f_d^2 \sum_{k,l} J^{(k)}(\omega_l^{(k)})\left[\mathcal{O}_l^{(k)},\left[\mathcal{O}_{l'}^{(-k)},I_z\right]\right]\\
&= \frac{1}{16}f_d^2 J^{(0)}(\omega_I - \omega_S)\left[I^+S^-,\left[I^-S^+,I_z\right]\right] + \text{H. c.}\\
&\quad \frac{9}{4}f_d^2 J^{(1)}(\omega_I)\left[I^+S_z,\left[I^-S_z,I_z\right]\right] + \text{H. c.}\\
&\quad \frac{9}{16}f_d^2 J^{(2)}(\omega_I + \omega_S)\left[I^+S^+,\left[I^-S^-,I_z\right]\right] + \text{H. c.}
\end{aligned}\tag{12.21}
$$

$$
\begin{aligned}
\mathcal{P}_z^S &= \frac{1}{2}f_d^2 \sum_{k,l} J^{(k)}(\omega_l^{(k)})\left[\mathcal{O}_l^{(k)},\left[\mathcal{O}_{l'}^{(-k)},S_z\right]\right]\\
&= \frac{1}{16}f_d^2 J^{(0)}(\omega_S - \omega_I)\left[S^+I^-,\left[S^-I^+,S_z\right]\right] + \text{H. c.}\\
&\quad \frac{9}{4}f_d^2 J^{(1)}(\omega_S)\left[S^+I_z,\left[S^-I_z,S_z\right]\right] + \text{H. c.}\\
&\quad \frac{9}{16}f_d^2 J^{(2)}(\omega_I + \omega_S)\left[S^+I^+,\left[S^-I^-,S_z\right]\right] + \text{H. c.}
\end{aligned}\tag{12.22}
$$

where the abbreviation H. c. (Hermitian conjugate) represents the preceding expression with all operators replaced by their adjoint forms, and all complex functions substituted by their conjugate complex counterparts.

The commutators in Eqs. 12.21 and 12.22 can readily be evaluated using the rules given in Sect. 42.6 and Chap. 44. The expressions relevant for Eq. 12.21, for instance, are found to be

$$\begin{aligned}
[I^+S_z, [I^-S_z, I_z]] &= 2I_z S_z^2 \\
[I^+S^-, [I^-S^+, I_z]] &= 2I_z(S_x^2 + S_y^2) - 2S_z(I_x^2 + I_y^2) \\
[S^+I^+, [S^-I^-, I_z]] &= 2I_z(S_x^2 + S_y^2 + S_z) + 2S_z(I_x^2 + I_y^2 + I_z)
\end{aligned} \tag{12.23}$$

The commutators in Eq. 12.22 are obtained by interchanging the letters I and S in the above expressions.

For the evaluation of the master equations at Eqs. 12.9 and 12.10, we now need expectation values of the type

$$\begin{aligned}
\text{Tr}\left\{\rho' I_z S_i^2\right\} &= \langle I_z S_i^2 \rangle' \\
\text{Tr}\left\{\rho_0' I_z S_i^2\right\} &= \langle I_z S_i^2 \rangle_0' \\
\text{Tr}\left\{\rho' S_z I_i^2\right\} &= \langle S_z I_i^2 \rangle' \\
\text{Tr}\left\{\rho_0' S_z I_i^2\right\} &= \langle S_z I_i^2 \rangle_0' \\
\text{Tr}\left\{\rho' I_z S_z\right\} &= \langle I_z S_z \rangle' \\
\text{Tr}\left\{\rho_0' I_z S_z\right\} &= \langle I_z S_z \rangle_0'
\end{aligned} \tag{12.24}$$

where $i = x, y, z$.

In the high-temperature approximation, the equilibrium density operator is given by

$$\rho_0' = \rho_0 = a + b(I_z + S_z) \tag{12.25}$$

The constants a and b are defined in Eqs. 47.26 and 47.27. On this basis, one finds for instance

$$\langle I_z S_z \rangle_0' = \langle I_z S_z \rangle_0 = \sum_{m_I, m_S} \langle m_I, m_S | \rho_0 I_z S_z | m_I, m_S \rangle = 0 \tag{12.26}$$

$$\begin{aligned}
\langle I_z S_x^2 \rangle_0' = \langle I_z S_z \rangle_0 &= \sum_{m_I, m_S} \langle m_I, m_S | \rho_0 I_z S_x^2 | m_I, m_S \rangle \\
&= \sum_{m_I, m_S} \sum_{\tilde{m}_I, \tilde{m}_S} \langle m_I, m_S | \rho_0 I_z | \tilde{m}_I, \tilde{m}_S \rangle \langle \tilde{m}_I, \tilde{m}_S | S_x^2 | m_I, m_S \rangle \\
&= \langle I_z \rangle_0 \, \frac{1}{3} I(I+1)
\end{aligned} \tag{12.27}$$

where we have used the relations

$$\begin{aligned}
\langle \tilde{m}_I, \tilde{m}_S | S_x^2 | m_I, m_S \rangle &= \tfrac{1}{3} I(I+1) \delta_{\tilde{m}_I, m_I} \delta_{\tilde{m}_S, m_S} \\
\sum_{m_I, m_S} \langle m_I, m_S | \rho_0 I_z | m_I, m_S \rangle &= \langle I_z \rangle_0
\end{aligned} \tag{12.28}$$

Remembering that spin-state populations deviate in the high-temperature limit by an order of magnitude of only 1 ppm (protons near room temperature), we infer that the variations of the density operator in the course of relaxation experiments are also extremely small. That is, the above expectation values can be transferred to non-equilibrium situations just by omitting the subscript 0. In the high-temperature limit, populations and density operators in equilibrium and any non-equilibrium state are almost the same in each case. *Almost* but not entirely: The finite *differences* of the respective quantities are what matters in NMR experiments.

Calculating the expectation values of the fluctuation operators at Eqs. 12.21 and 12.22 in this way, and inserting the results into the master equations given by Eqs. 12.9 and 12.10, leads to the system of rate equations

$$\frac{d\langle I_z \rangle}{dt} = -\frac{1}{T_1^{II}} (\langle I_z \rangle - \langle I_z \rangle_0) - \frac{1}{T_1^{IS}} (\langle S_z \rangle - \langle S_z \rangle_0) \tag{12.29}$$

$$\frac{d\langle S_z \rangle}{dt} = -\frac{1}{T_1^{SI}} (\langle I_z \rangle - \langle I_z \rangle_0) - \frac{1}{T_1^{SS}} (\langle S_z \rangle - \langle S_z \rangle_0) \tag{12.30}$$

where

$$\frac{1}{T_1^{II}} = \left(\frac{\mu_0}{4\pi}\right)^2 \gamma_I^2 \gamma_S^2 \hbar^2 S(S+1) \left[\frac{1}{12} J^{(0)}(\omega_I - \omega_S) + \frac{3}{2} J^{(1)}(\omega_I) \right.$$
$$\left. + \frac{3}{4} J^{(2)}(\omega_I + \omega_S) \right] \tag{12.31}$$

$$\frac{1}{T_1^{IS}} = \left(\frac{\mu_0}{4\pi}\right)^2 \gamma_I^2 \gamma_S^2 \hbar^2 I(I+1) \left[-\frac{1}{12} J^{(0)}(\omega_I - \omega_S) + \frac{3}{4} J^{(2)}(\omega_I + \omega_S) \right] \tag{12.32}$$

$$\frac{1}{T_1^{SI}} = \left(\frac{\mu_0}{4\pi}\right)^2 \gamma_I^2 \gamma_S^2 \hbar^2 S(S+1) \left[-\frac{1}{12} J^{(0)}(\omega_I - \omega_S) + \frac{3}{4} J^{(2)}(\omega_I + \omega_S) \right] \tag{12.33}$$

$$\frac{1}{T_1^{SS}} = \left(\frac{\mu_0}{4\pi}\right)^2 \gamma_I^2 \gamma_S^2 \hbar^2 I(I+1) \left[\frac{1}{12} J^{(0)}(\omega_I - \omega_S) + \frac{3}{2} J^{(1)}(\omega_S) \right.$$
$$\left. + \frac{3}{4} J^{(2)}(\omega_I + \omega_S) \right] \tag{12.34}$$

Obviously these rates are determined by spectral densities for zero-, single-, and double-quantum transitions as expected in view of the spin-operator terms of the dipolar Hamiltonian. The expectation values of the spin components are proportional to the respective magnetization components. The system of differential equations at Eqs. 12.29 and 12.30 thus defines the time evolution of the longitudinal spin-*I* and spin-*S* magnetizations under the action of dipolar coupling. The solutions, i.e., the relaxation functions, are linear combinations of exponential functions. However, in the limiting cases most important for practical applications, that is, the like-spin and the *S*-spin-equilibrium limits, the system reduces to single differential equations, and the solutions become monoexponential. This will be elucidated in the following sections.

12.1.1
Reduced Dipolar Correlation and Intensity Functions

Above, we have defined the dipolar autocorrelation and intensity functions of order *k* as the Fourier-transform pair (see Eqs. 12.15 and 12.18)

$$G_k(\tau) = \overline{F^{(k)}(\tau) F^{(-k)}(0)} \tag{12.35}$$

$$J^{(k)}(\omega_l^{(k)}) = \int_{-\infty}^{\infty} G_k(\tau)e^{-i\omega_l^{(k)}\tau}\,d\tau \qquad (12.36)$$

The quantities $F^{(k)}$ are functions of the fluctuating polar laboratory-frame coordinates r, φ, ϑ of the interdipole vector as listed in the set of equations 12.4. The initial values of the autocorrelation functions are equal to the mean squared fluctuations of the stochastic functions, i.e., $G_k(0) = \overline{|F^{(k)}|^2}$. In unoriented systems, the interdipole vectors may be assumed to be randomly oriented within the ensemble of two-spin systems. The mean squared fluctuations of the stochastic functions are then related as

$$\overline{|F^{(0)}|^2} : \overline{|F^{(1)}|^2} : \overline{|F^{(2)}|^2} = \frac{48}{15}\left\langle\frac{1}{r^6}\right\rangle : \frac{8}{15}\left\langle\frac{1}{r^6}\right\rangle : \frac{32}{15}\left\langle\frac{1}{r^6}\right\rangle = 6 : 1 : 4 \qquad (12.37)$$

where the terms in the middle of Eq. 12.37 represent the actual ensemble averages of the stochastic functions for random orientations. It therefore suffices to consider merely the mean squared fluctuation of one representative function $F^{(k)}$, e.g.,

$$F^{(1)} = r^{-3}\sin\vartheta\cos\vartheta\,e^{-i\varphi} \qquad (12.38)$$

The reduced dipolar autocorrelation function (see Eq. 9.3) may consequently be expressed as

$$G(\tau) = \frac{\left\langle F^{(1)}(0)F^{(-1)}(\tau)\right\rangle}{\left\langle|F^{(1)}|^2\right\rangle} \qquad (12.39)$$

where we have replaced the bars by brackets. Note that this function is normalized for $\tau = 0$, and that it is independent of the order k. The conjugate reduced dipolar intensity function (Eq. 9.5) is

$$\mathcal{I}(\omega) = \int_{-\infty}^{+\infty} G(\tau)\,e^{-i\omega\tau}\,d\tau \qquad (12.40)$$

12.1.2
S-Spin-Equilibrium Limit

The spin system is to consist of two unlike spins I and S, i.e., $\gamma_I \neq \gamma_S$. The RF pulses are tuned to the I spins whereas the off-resonant S spins merely take effect as interaction partners and are not subject to RF excitation. Furthermore, the S-spin populations are assumed to stay in their thermal equilibrium irrespective of what happens with the I spins. This situation may arise with resonant nuclei (spin I) which are dipolar coupled to quadrupole nuclei or unpaired electrons (spin S) suffering additional and independent relaxation mechanisms. Examples are the interaction of nuclear quadrupoles with electric field gradients and electron-spin-orbit couplings.

If the S-spin relaxation is much faster than that due to the mutual dipolar coupling of the I and S spins, $\langle S_z \rangle$ virtually never leaves its equilibrium value:

$$\langle S_z \rangle(t) \approx \langle S_z \rangle(0) = \langle S_z \rangle_0 \tag{12.41}$$

Under these conditions, the system of master equations, Eqs. 12.29 and 12.30, reduces to

$$\frac{d\langle I_z \rangle}{dt} = -\frac{1}{T_1} (\langle I_z \rangle - \langle I_z \rangle_0) \tag{12.42}$$

$$\frac{d\langle S_z \rangle}{dt} = 0 \tag{12.43}$$

That is, the **rotating-frame Bloch equation** for the z magnetization of the I spins has been reproduced (see Eq. 10.5) with an exponential relaxation function as a solution. The spin-lattice relaxation rate of the I spins is

$$\frac{1}{T_1} \equiv \frac{1}{T_1^{II}} = \left(\frac{\mu_0}{4\pi} \right)^2 \gamma_I^2 \gamma_S^2 \hbar^2 S(S+1)$$

$$\left[\frac{1}{12} J^{(0)}(\omega_I - \omega_S) + \frac{3}{2} J^{(1)}(\omega_I) + \frac{3}{4} J^{(2)}(\omega_I + \omega_S) \right] \tag{12.44}$$

For a fixed interdipole distance r this rate reads in terms of the reduced intensity function at Eq. 9.5

$$\boxed{\frac{1}{T_1} = \left(\frac{\mu_0}{4\pi} \right)^2 \frac{1}{15\,r^6} \gamma_I^2 \gamma_S^2 \hbar^2 S(S+1)\, [\,\mathcal{I}(\omega_I - \omega_S) + 3\mathcal{I}(\omega_I) + 6\mathcal{I}(\omega_I + \omega_S)\,]}$$

$$\tag{12.45}$$

12.1.3
Like-Spin limit

The second limit of interest refers to like spins, i.e., $I \equiv S$, $\gamma_I \equiv \gamma_S = \gamma_n$, and $\omega_I \equiv \omega_0 = \gamma_n B_0$. The consequences are $\langle I_z \rangle = \langle S_z \rangle$ and $\langle I_z \rangle_0 = \langle S_z \rangle_0$. The system of relaxation equations, Eqs. 12.29 and 12.30, thus merges in

$$\frac{d\langle I_z \rangle}{dt} = -\frac{1}{T_1} (\langle I_z \rangle - \langle I_z \rangle_0) \tag{12.46}$$

with

$$\frac{1}{T_1} = \frac{1}{T_1^{II}} + \frac{1}{T_1^{IS}} = \left(\frac{\mu_0}{4\pi} \right)^2 \frac{3}{2} \gamma_n^4 \hbar^2 I(I+1)\, [\, J^{(1)}(\omega_0) + J^{(2)}(2\omega_0)\,] \tag{12.47}$$

In terms of the reduced intensity function at Eq. 9.5 and assuming a fixed internuclear distance r, this expression can be rewritten as

$$\boxed{\frac{1}{T_1} = \left(\frac{\mu_0}{4\pi} \right)^2 \frac{1}{5\,r^6} \gamma_n^4 \hbar^2 I(I+1)\, [\,\mathcal{I}(\omega_0) + 4\mathcal{I}(2\omega_0)\,]}$$

$$\tag{12.48}$$

This essentially is the famous **Bloembergen/Purcell/Pound (BPP) formula** [43].

For systems consisting of more than two dipolar-coupled spins, the dipolar Hamiltonian implies a sum over all pairs in which a given spin participates. The spin-lattice relaxation rate contributions from each of these pairs can then be summed provided that the pair motions are independent from each other. In this case, the ensemble averages of the pair sums which are then implied in Eqs. 12.13 and 12.14 can be carried out separately for each pair with vanishing cross correlations. This permits one to write

$$\frac{1}{T_1} = \frac{3}{2}\left(\frac{\mu_0}{4\pi}\right)^2 \gamma_n^4 \hbar^2 I(I+1) \sum_j \left[J_j^{(1)}(\omega_0) + J_j^{(2)}(2\omega_0) \right] \tag{12.49}$$

where j runs over all dipoles interacting with the reference particle.

However, Eq. 12.49 and the exponential relaxation curve resulting from Eq. 12.42 of the two-spin treatment can also be taken as very good approximations for cases where cross-correlations are present. For example, Hubbard [201] showed that relaxation features of coherently moving three- or four-spin systems without any internal motions can be described by Eqs. 12.42 and 12.49 with an accuracy better than that with the usual experimental error. In other words, cross-correlation effects can safely be neglected under practical measuring conditions.

The insensitivity of multi-spin relaxation to cross-correlation effects is not entirely unexpected. The spin operators inherent to dipolar coupling (see Eqs. 12.8) can merely induce zero-, single-, and double-quantum transitions, but are inert with respect to any collective evolution of spin states involving more than two spins at one time.

The relaxation curves for like spins are again monoexponential in accordance with the **Bloch equation for the z magnetization** (see Eq. 10.5). The frequency and temperature dependences expected on the basis of Eqs. 12.48 or 12.49 for Lorentzian intensity functions (see Sect. 14.1) are plotted in Figs. 12.1 and 12.2.

12.2
Laboratory-Frame Spin-Lattice Relaxation by Other Interactions

12.2.1
Scalar Coupling

Scalar-coupling relaxation mechanisms of unlike spins ($\gamma_I \neq \gamma_S$) can arise in addition to dipolar-interaction-induced equilibration under two different circumstances. The first situation refers to nuclei coupled to unpaired electrons having a finite residence probability in the resonant nucleus. In this case, the electron/nuclear spin coupling takes the form of the **Fermi contact interaction** (see Sect. 46.3.2). The relevant spin-interaction Hamiltonian is

$$\mathcal{H}_i = \mathcal{H}_s = A\, I \cdot S = \frac{A}{2}\left[(I^+ S^- + I^- S^+) + I_z S_z \right] \tag{12.50}$$

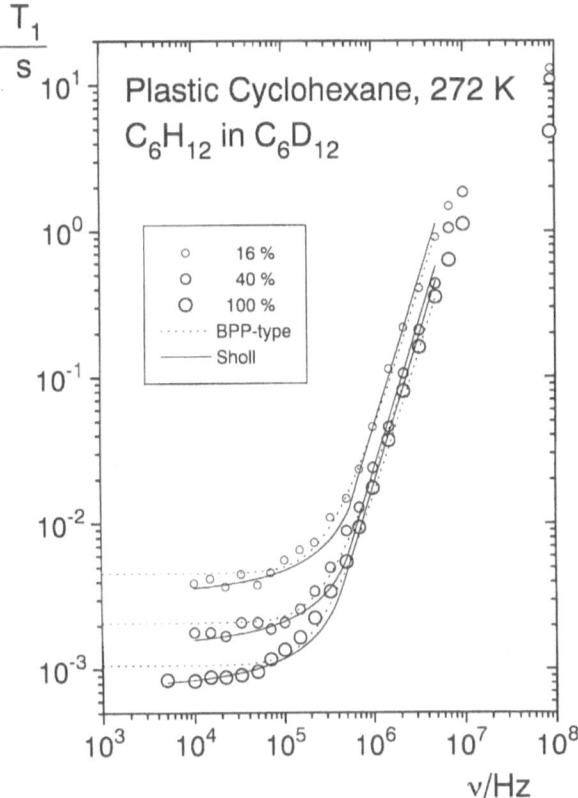

Fig. 12.1. Frequency dependence of the proton spin-lattice relaxation time, T_1, in the plastic phase of cyclohexane with different degrees of dilution with deuterated cyclohexane (given as percentages of the undeuterated solvent). The dashed lines were evaluated from the BPP formula at Eq. 12.48 assuming an exponential correlation function, i.e., a Lorentzian intensity function. Such functions are expected for isotropic continuous rotational diffusion (Sect. 14.1) or for random jumps between two discrete coupling states (Sect. 14.2.1, Eq. 14.21). The solid lines were calculated by taking translational-diffusion modulation of the dipolar coupling into account, i.e., intermolecular interaction (see Eqs. 3.21 and 3.24 in [445]). The influence of this contribution mainly reveals itself by the dilution-dependent relaxation times. The "inflection frequency" where the frequency dependence changes from the low-frequency plateau to the square high-frequency dependence is determined by $\omega_i \tau_c \approx 1$. As a further prediction of the theory, the low-frequency plateaus approached below 10^5 Hz coincide with the low-frequency transverse relaxation times T_2, i.e., $0.3 \times$ the high-frequency values measured at 90 MHz, for instance. (courtesy of S. Stapf)

The coupling constant A is a scalar by contrast to dipolar interaction. A typical example for relaxation phenomena based on this interaction are observed with hydrated paramagnetic ions such as Mn^{2+} [192, 349].

The second case of interest here refers to **indirect spin-spin coupling** of the resonant nucleus to a quadrupole nucleus. One of the interaction partners then suffers additional relaxation mechanisms by quadrupole interaction. In the motional

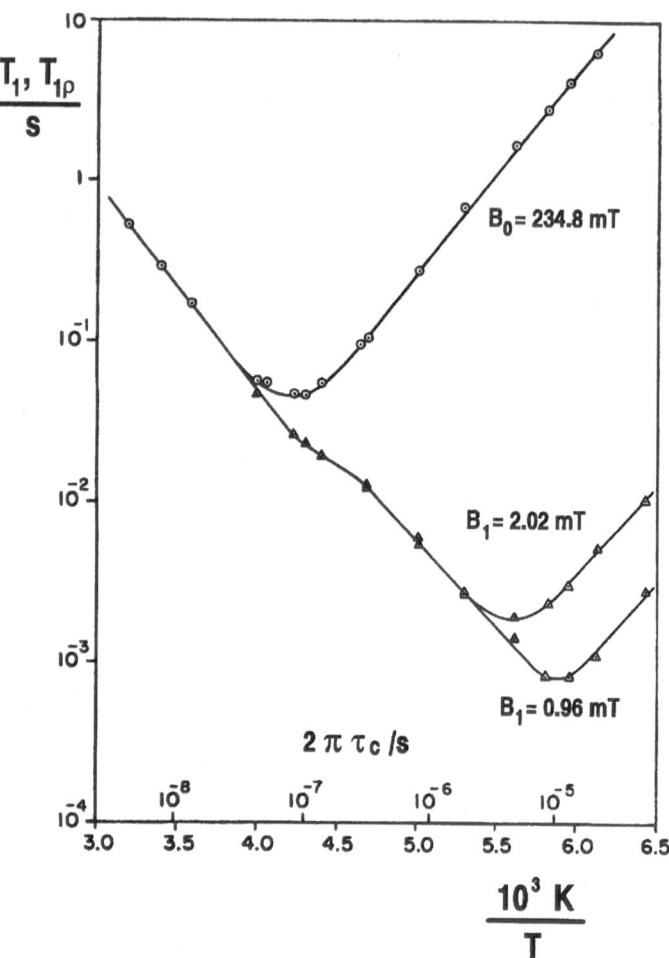

Fig. 12.2. Temperature dependences of the proton spin-lattice relaxation time in the laboratory frame, T_1, and in the rotating frame, $T_{1\rho}$, of crystal water in gypsum (data from [300]). The dipolar-coupling fluctuations arise from 180° rotational flips. The lines were calculated with the aid of Eq. 12.49 (T_1) and Eq. 12.60 ($T_{1\rho}$) in the analogous form for independently fluctuating spin pairs. A monoexponential correlation function, i.e., a Lorentzian intensity function, was assumed corresponding to the two-state jump model treated in Sect. 14.2.1 (see Eq. 14.21). The condition for the minima is $\omega_0 \tau_c \approx 0.62$ and $\omega_1 \tau_c = 1$ for T_1 and $T_{1\rho}$, respectively. (courtesy of S. Stapf)

averaging limit, when only the isotropic part of the coupling tensor matters, the Hamiltonian again has a scalar form analogous to Eq. 12.50 (see Table 46.3 on page 422). A well-known system where this interaction gives rise to significant relaxation effects is water containing ^{17}O as quadrupole nuclei [162, 306, 412].

The scalar-interaction Hamiltonian at Eq. 12.50 is transformed to the interaction representation in complete analogy to Eq. 12.5. The corresponding treatment according to Eq. 11.19 for the observable $Q = I_z$ is practically already implied in the formalism for dipolar coupling because the spin-operator terms of Eq. 12.50 are also part of the dipolar Hamiltonian. Therefore, one expects an additional I spin-lattice relaxation term based on zero-quantum transitions,

$$\left(\frac{1}{T_1}\right)_{scal} \propto A^2 \, S(S+1) \, \mathcal{I}(\omega_S - \omega_I) \tag{12.51}$$

The quantities $\mathcal{I}(\omega) = \mathcal{F}\{G(\tau)\}$ and $G(\tau)$ are the respective reduced intensity and correlation functions for scalar interaction ($G(0) = 1$).

Note that scalar interaction is inherently insensitive to rotational diffusion. Its fluctuations arise either from chemical exchange separating the resonant nucleus from the coupled spin, or from fast spin-lattice relaxation of the interaction partner. The latter case is termed **"scalar relaxation of the second kind."** The condition is that the spin-flip rate caused by additional interaction mechanisms selectively acting on this particle is sufficiently high. For a more detailed discussion, the reader is referred to Abragam's first book [1].

12.2.2
Quadrupolar Coupling ($I = 1$)

Nuclei with an electric-quadrupole moment are subject to "quadrupolar relaxation." Quadrupole interaction, if present, tends to be stronger than dipolar coupling so that it is a very efficient source of relaxation.[4] For example, deuteron relaxation in D_2O as well as in HDO molecules as they occur in mixtures of H_2O and D_2O is dominated by quadrupole coupling. Contributions by homo- or heteronuclear dipolar coupling are negligible in that case as indicated by the independence of the degree of deuteration. Relaxation of quadrupole nuclei is consequently disposed to be a matter of single particles. Even in solids, the effective spin system consists of just one member.

With spins greater than one, substantial deviations from relaxation equations of the Bloch type are found. In the frame of this treatise, we restrict ourselves to the spin-1 case.

Based on the quadrupole-coupling Hamiltonian $\mathcal{H}_i = \mathcal{H}_q$ (Table 46.4 on page 425) and the BWR master equation, Eq. 11.19, one finds the **spin-lattice relaxation rate** of the observable $Q = I_z$ [1, 210, 465]

$$\frac{1}{T_1} = \frac{3}{80}\left(\frac{e^2 qQ}{h}\right)^2 \left(1 + \frac{\eta^2}{3}\right) [\,\mathcal{I}(\omega_0) + 4\mathcal{I}(2\omega_0)\,] \tag{12.52}$$

which is determined by a single- and a double-quantum transition intensity function as expected in view of the quadrupolar-interaction Hamiltonian.

Longitudinal quadrupolar order (Sect. 5.2) may also be regarded to be subject to a sort of spin-lattice relaxation phenomenon. The observable to be inserted in

---

[4]Exceptions are found in cubic- or tetrahedral structure-like molecular environments where the electric-field gradient may even vanish.

the BWR master equation (Eq. 11.19) is $\mathcal{Q} = I_z^2 + \frac{2}{3}$. The resulting **weak-collision quadrupolar-order relaxation rate** is [205]

$$\frac{1}{T_q} = \frac{9}{80}\left(\frac{e^2qQ}{\hbar}\right)^2\left(1 + \frac{\eta^2}{3}\right)\mathcal{I}(\omega_0) \tag{12.53}$$

Note that this rate is governed by the single-quantum transition spectral-density term. In combination with the ordinary spin-lattice relaxation rate at Eq. 12.52, the single-quantum and the double-quantum transition intensity functions can thus be analyzed experimentally [304].

12.2.3
Chemical-Shift Anisotropy

The anisotropy of the chemical-shift tensor produces transverse field components able to induce single-quantum spin transitions. The chemical-shift Hamiltonian (see Table 46.2 on page 421) fluctuates as a consequence of molecular reorientations. For axially symmetric molecules, the spin-lattice relaxation rate following from Eq. 11.19 is

$$\frac{1}{T_1} \propto \gamma_n^2 B_0^2(\sigma_{33} - \sigma_i)^2 \mathcal{I}(\omega_0) \tag{12.54}$$

By contrast to the other spin couplings, the chemical-shift anisotropy (CSA) coupling constant varies proportionally to the external flux density. This is the origin of the factor B_0^2 characteristic for for the CSA mechanism. It becomes important at high fields (above about 6 T), and may even dominate if the competitive spin couplings are weak. This occurs with dipole nuclei on isolated positions, or with quadrupole nuclei in cubic or tetrahedral structures. Nuclear species predestinated for CSA-induced relaxation are ^{13}C and ^{31}P.

12.3
Rotating-Frame Spin-Lattice Relaxation by Dipolar Coupling

Rotating-frame relaxometry is not as widespread as the laboratory-frame techniques. Therefore, we restrict ourselves to the most important case of **dipolar-coupled pairs of like spins**. The treatment follows the same principles as outlined above, but the interaction frame is now the tilted rotating frame with the z axis aligned along the effective field B_e.[5] This is the quantization field we are speaking of, and which now primarily gives the meaning of the attribute "longitudinal."

Experiments of this type (see Sects. 10.3 and 15.2) are usually carried out while the magnetization is "locked" along the (rotating) effective field. Therefore, the angular frequencies of interest now comprise

$$\omega_0 = \gamma_n B_0; \qquad \omega_1 = \gamma_n B_1; \qquad \omega_e = \gamma_n B_e = \sqrt{(\omega_0 - \omega_c)^2 + \omega_1^2} \tag{12.55}$$

---

[5]In [335] an interesting NMR experiment is described that permits the direct observation of the resonance in the effective field acting as the quantization field in the rotating frame.

In the limit $\omega_0 \gg \omega_e$, which represents the actual situation in practically all applications, the effective spin-lattice relaxation rate for dipolar coupling of like spins in the on- or off-resonance (tilted) rotating frame turns out to be [36, 213]

$$\frac{1}{T_{1\rho}^{(e)}} = K \left[C_1 \, \mathcal{I}(\omega_e) + C_2 \, \mathcal{I}(2\omega_e) + C_3 \, \mathcal{I}(\omega_0) + C_4 \, \mathcal{I}(2\omega_0) \right] \tag{12.56}$$

where

$$
\begin{aligned}
K &= \left(\frac{\mu_0}{4\pi} \right)^2 \frac{1}{5 \, r^6} \gamma_n^4 \hbar^2 I(I+1) \\
C_1 &= \frac{3}{2} \sin^2 \Theta \cos^2 \Theta \\
C_2 &= \frac{3}{2} \sin^4 \Theta \\
C_3 &= 1 + \frac{3}{2} \sin^2 \Theta \\
C_4 &= 1 + 3 \cos^2 \Theta
\end{aligned}
\tag{12.57}
$$

Equation 12.56 is a relatively general expression. Considering the fundamental form of Eq. 10.13, it can be analyzed in terms of the spin-lattice relaxation rates in the laboratory frame, i.e., the BPP formula,

$$\frac{1}{T_1} = K \left[\mathcal{I}(\omega_0) + 4\mathcal{I}(2\omega_0) \right] \tag{12.58}$$

and by the spin-lattice relaxation rate of the magnetization spin-locked along the effective field:

$$\frac{1}{T_{1\rho}} = K \left[\frac{3}{2} \mathcal{I}(2\omega_e) + \frac{5}{2} \mathcal{I}(\omega_0) + \mathcal{I}(2\omega_0) \right] \tag{12.59}$$

If the spin-lock pulse is resonant, i.e., $\omega_c = \omega_0$, we have $\Theta = 90°$, $\omega_e = \omega_1$. Equations 10.13 and 12.56 then merge into [300]

$$\frac{1}{T_{1\rho}^{(e)}} = \frac{1}{T_{1\rho}} = \left(\frac{\mu_0}{4\pi} \right)^2 \frac{1}{10 \, r^6} \gamma_n^4 \hbar^2 I(I+1) \left[3\mathcal{I}(2\omega_1) + 5\mathcal{I}(\omega_0) + 2\mathcal{I}(2\omega_0) \right] \tag{12.60}$$

The temperature dependence of $T_{1\rho}$ calculated on the basis of Eq. 12.60 and an Arrhenius law for the correlation time τ_c of an exponential correlation function is plotted in Fig. 12.2.

In the extreme off-resonance case, i.e., $\omega_0 \gg \omega_c$ or $\omega_0 \ll \omega_c$, the tilt angle approaches $\Theta = 0$ or $\Theta = 180°$, respectively. Equations 10.13 and 12.56 then adopt the form of the BPP formula:

$$\frac{1}{T_{1\rho}^{(e)}} = \frac{1}{T_1(\omega_0)} = K \left[\mathcal{I}(\omega_0) + 4\mathcal{I}(2\omega_0) \right] \tag{12.61}$$

The same situation arises if no RF is applied, i.e., $\omega_1 = 0$, of course.

CHAPTER 13

Transverse Relaxation

13.1
Motional-Averaging Limit

The motional-averaging limit anticipates that the spin-interaction fluctuations are fast on the time scale of transverse relaxation. This is a crucial prerequisite of the applicability of the BWR theory as pointed out in Sect. 9.3.3. The time scale of transverse relaxation begins with the coherence attenuation time one expects in the absence of motions, i.e., with the reciprocal rigid-lattice linewidth $(\Delta\omega)_{rl}^{-1}$. Another version of this time scale measure is the root mean squared deviation of the angular-frequency offset due to local fields produced by spin interactions. The motional-averaging condition may thus be expressed as

$$(\Delta\omega)_{rl}\, \tau_c \ll 1 \qquad \text{or} \qquad \gamma_n\sqrt{\langle B_{loc}^2\rangle}\, \tau_c \ll 1 \qquad (13.1)$$

where τ_c is the correlation time of the fluctuating spin interactions.

In the following, the single-quantum coherences under consideration are assumed to be excited by an RF pulse applied in the x rotating-frame direction. The initial coherence phase then corresponds to the y direction, and the observables to be inserted into the BWR master equation, Eq. 11.19, are $Q = I_y$ or $Q = S_y$. The equilibrium expectation values are $\langle I_y\rangle_0 = 0$ and $\langle S_y\rangle_0 = 0$.

13.1.1
Single-Quantum Coherences of Dipolar-Coupled Spin Pairs

In analogy to Eqs. 12.29 and 12.30, the rotating-frame relaxation equations for dipolar coupling between two spins I and S turn out to be

$$\frac{d\langle I_y\rangle}{dt} = -\frac{1}{T_2^{II}}\langle I_y\rangle - \frac{1}{T_2^{IS}}\langle S_y\rangle \qquad (13.2)$$

$$\frac{d\langle S_y\rangle}{dt} = -\frac{1}{T_1^{SI}}\langle I_y\rangle - \frac{1}{T_1^{SS}}\langle S_y\rangle \qquad (13.3)$$

In the S-spin-equilibrium limit, this system reduces to the Bloch-type relaxation equation for the I spins

$$\frac{d\langle I_y\rangle}{dt} = -\frac{1}{T_2}\langle I_y\rangle \qquad (13.4)$$

where the transverse-relaxation rate for single-quantum coherences is

$$
\frac{1}{T_2} = \left(\frac{\mu_0}{4\pi}\right)^2 \gamma_I^2 \gamma_S^2 \hbar^2 S(S+1) \left[\frac{1}{6}J^{(0)}(0) + \frac{1}{24}J^{(0)}(\omega_I - \omega_S)\right.
$$
$$
\left. + \frac{3}{4}J^{(1)}(\omega_I) + \frac{3}{2}J^{(1)}(\omega_S) + \frac{3}{8}J^{(2)}(\omega_I + \omega_S)\right] \tag{13.5}
$$

In terms of the reduced intensity function at Eq. 9.5, and for a fixed interdipole distance r, this rate reads

$$
\frac{1}{T_2} = \left(\frac{\mu_0}{4\pi}\right)^2 \frac{2}{15\,r^6} \gamma_I^2 \gamma_S^2 \hbar^2 S(S+1) \left[\mathcal{I}(0) + \frac{1}{4}\mathcal{I}(\omega_I - \omega_S)\right.
$$
$$
\left. + \frac{3}{4}\mathcal{I}(\omega_I) + \frac{3}{2}\mathcal{I}(\omega_S) + \frac{3}{2}\mathcal{I}(\omega_I + \omega_S)\right] \tag{13.6}
$$

In the like-spin limit, the analytic form of the relaxation equation, Eq. 13.4, holds true again. The corresponding transverse relaxation rate for single-quantum coherences is

$$
\frac{1}{T_2} = \left(\frac{\mu_0}{4\pi}\right)^2 \gamma_n^4 \hbar^2 I(I+1) \left[\frac{3}{8}J^{(0)}(0) + \frac{15}{4}J^{(1)}(\omega_I) + \frac{3}{8}J^{(2)}(\omega_S)\right] \tag{13.7}
$$

Inserting the reduced intensity function at Eq. 9.5, and assuming a fixed interdipole distance r, turns this into

$$
\boxed{\frac{1}{T_2} = \left(\frac{\mu_0}{4\pi}\right)^2 \frac{1}{10r^6} \gamma_n^4 \hbar^2 I(I+1) \left[3\mathcal{I}(0) + 5\mathcal{I}(\omega_0) + 2\mathcal{I}(2\omega_0)\right]} \tag{13.8}
$$

The temperature dependence of T_2 is a function smoothly decaying with increasing reciprocal temperature (compare the low-temperature flank of the $T_{1\rho}$ curve in Fig. 12.2). However, unlike spin-lattice relaxation, no minimum appears.[1] At sufficiently low temperatures, the T_2 data merge into the temperature-independent rigid-lattice limit, where the motional-average formula at Eq. 13.8 is no longer valid.

The frequency dependence of T_2 is characterized by a step centered at the position $\omega_0 \tau_c \approx 1$. The transverse-relaxation rate in the "extreme-narrowing limit" $\omega_0 \tau_c \ll 1$ is (compare Eq. 12.48)

$$
\frac{1}{T_2} = \left(\frac{\mu_0}{4\pi}\right)^2 \frac{1}{r^6} \gamma_n^4 \hbar^2 I(I+1)\,\mathcal{I}(0) = \frac{1}{T_1} \tag{13.9}
$$

whereas the ordinary motional averaging situation, $1/\omega_0 \ll \tau_c \ll 1/(\Delta\omega)_n$, leads to

$$
\frac{1}{T_2} = \left(\frac{\mu_0}{4\pi}\right)^2 \frac{1}{r^6} \gamma_n^4 \hbar^2 I(I+1) \frac{3}{10}\mathcal{I}(0) \tag{13.10}
$$

That is, the step height corresponds to a factor of 10/3.

---

[1]This statement holds true for free evolution. However, in the presence of RF pulse trains, an effective transverse relaxation time T_{2e} applies exhibiting a temperature dependence with a minimum analogous to that of the spin-lattice relaxation times [169, 494].

13.1.2
Single-Quantum Coherences of ($I = 1$) Quadrupole Nuclei

The effect of quadrupole interactions on transverse relaxation can be treated with
the aid of the BWR theory in an analogous manner. Nuclei with spins $I > 1$ show
substantial deviations from Bloch-type relaxation curves contrary to the $I = 1$ case.
Single-quantum coherences of quadrupole nuclei with $I = 1$ relax with a rate

$$\frac{1}{T_2} = \frac{3}{160} \left(\frac{e^2 qQ}{\hbar} \right)^2 \left(1 + \frac{\eta^2}{3} \right) [\, 3\mathcal{I}(0) + 5\mathcal{I}(\omega_0) + 2\mathcal{I}(2\omega_0) \,] \qquad (13.11)$$

13.1.3
Multiple-Quantum Coherences

Multiple-quantum coherences and spin-order states tend to arise in NMR experiments at the same time (see the treatments of the various three-pulse sequences
discussed in Sects. 4.2, 5.2, and 7.2 of the first part of this book). It is clear that
the separate experimental determination of the individual relaxation rates of all
coherences and spin states arising with a given spin system may be a demanding
procedure. Nevertheless, in simple systems a complete analysis is feasible. In the
case of deuterons this was exemplified in [210, 501].

In a sense, multiple-quantum coherences are also of a "transverse" nature albeit they are not directly detectable, of course. The treatment of their relaxation
may again be based on the BWR master equation, Eq. 11.19, which must now be
combined with the respective multiple-quantum transition operator and the spin-
interaction Hamiltonian mediating the relaxation process.

For example, the "observable" for double-quantum coherences of a spin-1 particle subject to quadrupole coupling is $\mathcal{Q} = I_y I_x + I_x I_y$ (see Chap. 52). Taking
quadrupole coupling as the spin interaction dominating the relaxation process, the
relaxation rate of such coherences is found to be

$$\frac{1}{T_{dqc}} = \frac{3}{80} \left(\frac{e^2 qQ}{\hbar} \right)^2 \left(1 + \frac{\eta^2}{3} \right) [\, \mathcal{I}(\omega_0) + 2\mathcal{I}(2\omega_0) \,] \qquad (13.12)$$

13.2
Local-Field Theory

In case the motional-averaging limit at Eq. 13.1 is not or only partially fulfilled, local-
field approaches for the treatment of equilibration of transverse magnetization are
in use instead of the BWR theory. The most elementary theory of this sort is the
Anderson/Weiss (AW) formalism [9].[2] More sophisticated concepts starting from
the same basic assumptions have been developed for polymers [61, 97, 287]. In this

---

[2]The applicability ranges of the two sorts of approaches are overlapping. They are not
entirely complementary in nature. This becomes evident when considering to what extent the
secular and nonsecular terms of dipolar coupling (see Table 46.5 on page 426) are accounted
for.

context, the average orientation of a polymer chain with respect to the external magnetic field may play a crucial role. If the polar angle of the internuclear vector approaches the magic value, the local field vanishes so that the motional-averaging condition at Eq. 13.1 is strongly shifted.

13.2.1
The Anderson/Weiss Ansatz

The local field produced by spin interactions generates the local precession-frequency offset from the resonance frequency,

$$\Omega \equiv \omega - \omega_0 = \gamma_n B_{loc} \tag{13.13}$$

The free-induction decay (FID) resulting as the global effect of such offsets is described by

$$A(t) = \left\langle e^{i\Omega t} \right\rangle = \left\langle e^{i\varphi(t)} \right\rangle \tag{13.14}$$

The brackets indicate an average over the spin ensemble. This is what one expects in the static case. Molecular motion modulates the spin interactions and, hence, the frequency offsets. That is, the precession phase accumulated in an interval $0 \ldots t$ is given as the time integral over the fluctuating frequency offsets:

$$\varphi(t) = \int\limits_0^t \Omega(t') \, dt' \tag{13.15}$$

The ensemble average can be derived with the aid of the distribution function $P(\varphi, t)$ of the phase shifts φ accumulated in time t:

$$A(t) = \left\langle e^{i\varphi(t)} \right\rangle = \int\limits_{-\infty}^{\infty} P(\varphi, t) \, e^{i\varphi} \, d\varphi \tag{13.16}$$

Expanding the exponential function and forming the ensemble average of each expansion term leads to the moment series

$$
\begin{aligned}
A(t) &= \int\limits_{-\infty}^{\infty} P(\varphi, t) \left(1 + i\frac{\varphi}{1!} - \frac{\varphi^2}{2!} - i\frac{\varphi^3}{3!} + \frac{\varphi^4}{4!} \pm \ldots \right) d\varphi \\
&= 1 - \frac{\langle \varphi^2 \rangle}{2!} + \frac{\langle \varphi^4 \rangle}{4!} - \frac{\langle \varphi^6 \rangle}{6!} \pm \ldots
\end{aligned}
\tag{13.17}
$$

The **secular terms** are relevant in both concepts. This holds true for the $I_z S_z$ term as well as for the $(I^+ S^- + I^- S^+)$ contribution which is secular for like spins. The latter term enters the AW treatment via dipolar energy shifts (see Sect. 50.1.1), i.e., local fields. In the frame of the BWR theory, it plays a twofold role: first via local fields, and second via flip-flop spin transitions.

On the other hand, the **non-secular terms** are taken into account only in the BWR formalism. Nevertheless, the AW theory may be considered to be of an equivalent rank as long as the motional averaging is incomplete because the non-secular terms then tend to be negligible anyway. Thus, the AW formalism is applicable when the BWR theory fails, i.e., when the condition $\gamma_n \sqrt{\langle B_{loc}^2 \rangle} \, \tau_c \ll 1$ begins to be violated.

where it is assumed that the phase shift distribution is an even function of the phase shifts, so that all odd moments vanish.

Under normal circumstances, the local field a nucleus experiences is a superposition of contributions of many interaction partners. Therefore, the central-limit theorem applies which expresses that the sum of a sufficiently large number of random contributions is distributed according to a Gauss function irrespective of how the contributions are distributed. If the frequency offsets Ω caused by the local fields are Gauss-distributed, the phase shift accumulated in time t is Gauss-distributed as well, because φ is a linear combination (in the form of an integral) of the Gauss-distributed offsets Ω. Thus

$$P(\varphi, t) = \frac{1}{\sqrt{2\pi \langle \varphi^2 \rangle}} \exp \left\{ -\frac{\varphi^2}{2 \langle \varphi^2 \rangle} \right\} \tag{13.18}$$

The even moments of a Gaussian distribution can be derived from its second moment according to

$$\langle \varphi^{2n} \rangle = 1 \cdot 3 \cdot \cdot 5 \ldots (2n-1) \langle \varphi^2 \rangle^n \qquad (n = 1, 2, 3 \ldots) \tag{13.19}$$

so that the expansion at Eq. 13.17 can be evaluated. This results in

$$A(t) = 1 - \frac{\frac{1}{2} \langle \varphi^2 \rangle}{1!} + \frac{\left(\frac{1}{2} \langle \varphi^2 \rangle \right)^2}{2!} - \frac{\left(\frac{1}{2} \langle \varphi^2 \rangle \right)^3}{3!} \pm \ldots \tag{13.20}$$

or

$$A(t) = \exp \left\{ -\frac{1}{2} \langle \varphi^2 \rangle \right\} \tag{13.21}$$

13.2.2
The Second Moment $\langle \varphi^2 \rangle$

The second moment of the precession-phase distribution after time t can be rewritten in the form

$$\langle \varphi^2 \rangle = \left\langle \left[\int_0^t \Omega(t') \, dt' \right]^2 \right\rangle$$

$$= \left\langle \int_0^t dt' \int_0^t dt'' \, \Omega(t') \Omega(t'') \right\rangle$$

$$= \int_0^t dt' \int_0^t dt'' \, \underbrace{\langle \Omega(t') \Omega(t'') \rangle}_{G_\Omega} \tag{13.22}$$

The **correlation function of the local resonance offsets**, G_Ω, depends on a time difference rather than on the absolute time. This expresses the fact that molecular fluctuations obey the **property of stationarity**. Furthermore, it does not matter which of the two times t' and t'' is first, so that the correlation function must be even:

$$G_\Omega = G_\Omega(|t'' - t'|) \qquad (13.23)$$

The two time variables t' and t'' are now substituted by a new variable pair $\tau = t'' - t'$ and t'. The second moment at Eq. 13.22 then reads

$$
\begin{aligned}
\langle \varphi^2 \rangle &= \int_0^t dt' \int_0^t dt'' G_\Omega(|t'' - t'|) \\
&= \underbrace{\int_0^t d\tau \left[G_\Omega(|\tau|) \int_0^{t-\tau} dt' \right]}_{\text{positive } \tau \text{ values}} + \underbrace{\int_{-t}^0 d\tau \left[G_\Omega(|\tau|) \int_{|\tau|}^t dt' \right]}_{\text{negative } \tau \text{ values}} \\
&= 2 \int_0^t d\tau \left[G_\Omega(|\tau|) \int_0^{t-\tau} dt' \right] \\
&= 2 \int_0^t (t - \tau) G_\Omega(\tau) \, d\tau \qquad (13.24)
\end{aligned}
$$

This result can be perceived with the aid of the scheme in Fig. 13.1 representing the scales of t' and t''. According to the definition in Eq. 13.22, both time scales must be covered by the integration in their full ranges.

Let us first discuss the term in Eq. 13.24 marked for positive τ values. The integrand of the outer integral refers to an interval of a certain length τ, say on the t'' scale (Fig. 13.1a). The variable t' is then varied over all values compatible with the given τ value, i.e., with a fixed value of the correlation function $G_\Omega(|\tau|)$. This is expressed by the inner integral. The outer integral, on the other hand, accounts for all τ values possible on the t'' scale, i.e., $0 \leq \tau \leq t$.

The consideration of negative τ intervals (Fig. 13.1b) and correspondingly modified integration limits leads to the second term in Eq. 13.24. Both terms obviously yield identical results, so that the sum can be replaced by twice the first term.

Inserting the expression for the second moment into Eq. 13.21 gives the **Anderson/Weiss formula** [9, 10, 284] for the free-induction decay envelope of equivalent spins,[3]

$$A(t) = \exp\left\{ -\langle \Omega^2 \rangle \int_0^t (t - \tau) \, G(\tau) \, d\tau \right\} \qquad (13.25)$$

---

[3]The Kubo/Tomita theory [284] leads to this result using the stochastic Liouville/von Neumann equation supplemented by an additional Markov operator term.

a) τ positive:

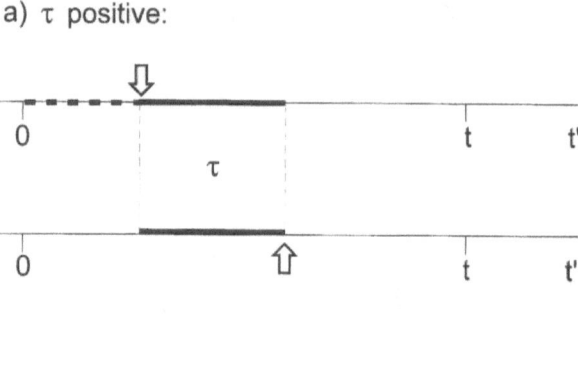

b) τ negative:

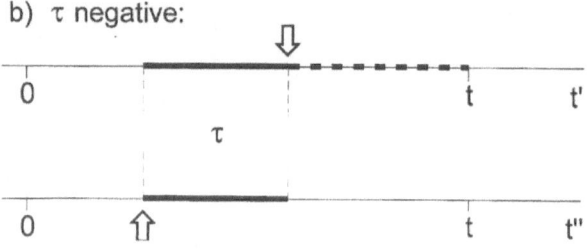

Fig. 13.1. Time scales relevant for Eq. 13.24. The arrows mark the respective times considered for: **a)** positive and; **b)** negative τ intervals. The dashed lines indicate the side on which the t' variation range is to be taken relative to the τ interval.

where

$$G(\tau) \equiv \frac{G_\Omega(\tau)}{\langle \Omega^2 \rangle_{rl}} \qquad (13.26)$$

Here we have introduced the second moment of the resonance line in the absence of molecular motion, $\langle \Omega^2 \rangle_{rl}$, i.e., for stationary local fields. The reduced correlation function $G(\tau)$ is normalized at $\tau = 0$, i.e., $G(0) = 1$ (compare Sect. 9.1.3).

The FID envelope at Eq. 13.25 can also be expressed in terms of the reduced intensity function by replacing $G(\tau)$ by its Fourier transform

$$G(\tau) = \frac{1}{2\pi} \int_{-\infty}^{\infty} \mathcal{I}(\Omega)\, d\Omega \qquad (13.27)$$

and interchanging the integrals. After carrying out the time integration, one finds

$$A(t) = \exp\left\{ -\frac{1}{\pi} \int_{0}^{\infty} \frac{\langle \Omega^2 \rangle_{rl}}{\Omega^2} \mathcal{I}(\Omega)\, [\, 1 - \cos(\Omega t)\,]\, d\Omega \right\} \qquad (13.28)$$

With this representation, it becomes evident that the FID envelope is dominated by fluctuations with rates $\Omega \lesssim \sqrt{\langle \Omega^2 \rangle_{rl}}$ in accordance with the basic motional-averaging rule.

13.2.3
Partial Motional Averaging

The consequences of motional averaging can be exemplified with the aid of an exponential correlation function,

$$G(\tau) = e^{-|\tau|/\tau_c} \tag{13.29}$$

Such a relationship would be expected, for instance, if any change of the local field entails the total loss of correlation. In this case, the correlation time τ_c is identical with the mean lifetime of the local fields.

The **rigid-lattice limit** is defined by the condition $\tau_c \gg t$ where t represents the whole timescale of the FID. In terms of the frequency domain, signal detection is restricted to the range $t \overset{<}{\approx} \Delta\Omega_{rl}^{-1}$ where $\Delta\Omega_{rl}$ is the FWHM value of the rigid-lattice line. The rigid-lattice condition then reads

$$\Delta\Omega_{rl} \tau_c \gg 1 \tag{13.30}$$

That is, the reduced correlation function in Eq. 13.25 may be approximated by $G(\tau) \approx 1$ in the relevant τ range, so that the FID can be described by

$$A(t) \approx \exp\left\{ -\frac{1}{2} \langle \Omega^2 \rangle_{rl} t^2 \right\} \tag{13.31}$$

This Gaussian decay function stipulates a Gaussian lineshape which is often a good approximation for proton lines in rigid materials.

The opposite case is the **motional-averaging limit** already referred to several times. The analogous definition may be written in the form $\tau_c \ll t$, or

$$\Delta\Omega_{rl} \tau_c \ll 1 \tag{13.32}$$

The exponent in Eq. 13.25 can then be approximated by

$$\int_0^t (t - \tau) G(\tau) \, d\tau \approx t \int_0^\infty G(\tau) \, d\tau = t\tau_c \tag{13.33}$$

The integrand can be neglected for $\tau \gg \tau_c$ because $G(\tau)$ has then decayed to imperceptible values. That is, the upper integration limit can be equated with ∞. The τ range in which the integrand contributes to the integral is $\tau \overset{<}{\approx} \tau_c \ll t$. Therefore, τ can be neglected against t. In the motional-averaging limit, Eq. 13.25 thus takes the form of an exponential decay function,

$$A(t) \approx \exp\left\{ - \langle \Omega^2 \rangle_{rl} \tau_c t \right\} \tag{13.34}$$

The Fourier transform yields a Lorentz function for the lineshape,

$$F(\omega) = \frac{2 \langle \Omega^2 \rangle_{rl} \tau_c}{\omega^2 + \left(\langle \Omega^2 \rangle_{rl} \tau_c \right)^2} \tag{13.35}$$

with a FWHM

$$\Delta\Omega = 2 \left\langle \Omega^2 \right\rangle_{rl} \tau_c \tag{13.36}$$

The shorter the correlation time τ_c, the more narrow is the resonance line. It is a matter of course that this expression does not refer to any lifetime broadening by spin transitions as considered in the BPP/BWR theory (Sect. 13.1) as discussed in Sect. 9.3.3.

CHAPTER 14

Examples of Autocorrelation Functions

The principal instrument for the description of the fluctuations of spin interactions following from molecular dynamics is the autocorrelation function, which is preferably used in its reduced form. In the following sections, typical examples of correlation function formalisms will be delineated. There are innumerable accounts of correlation function treatments in the literature to choose from. The purpose of the correlation functions selected here is to represent conceivable situations of a more didactic value.

The motional process already considered at the very beginning of the NMR-relaxation history [1, 43] is **isotropic continuous rotational diffusion**. Therefore, the first example to be considered will be a brief outline of the corresponding formalism. However, one should give due regard to the fact that isotropic continuous rotational diffusion is a rare event in nature. Numerous attempts have been undertaken to develop schemes more adapted to molecular dynamics in condensed matter. This particularly refers to all sorts of anisotropic continuous rotational diffusion models describable by diffusion tensors [465, 519, 520, 521]. Moreover, apart from rotational diffusion, dipolar interaction may also be subject to translational modulation by fluctuations of the interdipole distance. This can become a significant mechanism if intermolecular interactions contribute [445, 484] (see Fig. 12.1 for an experimental example).

Reorientation of molecules or molecular groups may be restricted by microstructural constraints. The degrees of motional freedom are often impeded by potential barriers so that only a limited number of discrete spin-interaction states can be adopted. As an approach of such a situation, **discrete-coupling-state jump models** have been suggested. A corresponding formalism will be presented in Sect. 14.2. An offspring of this model class is defect diffusion in its numerous discrete or continuous variants [230, 232]. Fluctuations of the spin interaction are mediated by the arrival of diffusing structural defects. That is, translational diffusion (of the defects) in potentially restricted geometries forms the rate-limiting factor for the fluctuations of the spin interactions, and, hence, governs the time dependence of the autocorrelation function.

Further on, constraints may arise by the topology of the molecular environment. Adsorbate molecules in the vicinity of surfaces, for instance, are expected to show a strongly modified rotational and translational diffusion behavior compared with the bulk. The third example to be treated below refers to **reorientations mediated by** translational displacements (RMTD) of molecules diffusing along curved surfaces imposing their local orientation on the adsorbate molecules. It is now translational diffusion of the spin-bearing particles themselves which carries them to positions stipulating another molecular orientation. Therefore, losses of the spin-interaction correlation are governed by the effective timescale of translational displacements along the surface as well as by the spatial variation of the surface orientation.

In the so-called "strong adsorption limit" translational diffusion of adsorbate molecules along surfaces was shown to obey **Lévy-walk** statistics [69]. The effective propagator is then a Cauchy distribution rather than an ordinary Gaussian function. As concerns NMR relaxation, a mechanism was suggested by which adsorbate molecules visit the surface in "Lévy dust" form [468]. This mechanism will be detailed in Sect. 14.3.2.

The treatment of (phase spatial) relaxation modes as they typically occur in polymers and liquid crystals requires special considerations. NMR relaxation due to **director fluctuation modes** has been elaborated in [38, 123, 384, 498], for instance. A correlation function of this type is also dealt with in Sect. 17.3.2 in context with the dipolar-correlation effect. NMR relaxation theories of polymers sometimes no longer fit smoothly to the simple autocorrelation/intensity function scheme drawn in Sect. 9.1.3. The NMR relaxation theory based on the **Rouse model of polymer chain fluctuations** was treated by Khazanovich [223] who arrived at a logarithmic intensity function. The same result was reproduced with the aid of a **memory function formalism** [142]. The latter concept is of particular usefulness because it permits the derivation of power-law frequency dependences as so often observed in liquids of entangled polymers [144, 247].

14.1
Isotropic Continuous Rotational Diffusion

Let us consider two dipolar-coupled spins arranged at a fixed distance on a spherically shaped molecule. This molecule is thought to tumble randomly and isotropically in a viscous environment. The rotational nature of this motion suggests the use of the irreducible spherical tensor (IST) variant of the dipolar-coupling Hamiltonian (Table 46.5 on page 426). According to Eqs. 9.2 or 49.1 the autocorrelation function then refers to the fluctuating functions $A_{2,-m}(\vartheta, \varphi) \equiv A_{2,-m}(u)$ characterizing the dependence of dipolar coupling on the orientation of the internuclear axis. This orientation is expressed at time 0 by the unit vector $u_i \equiv u(0)$ with the polar coordinates $r_i = 1$, φ_i, ϑ_i, and at time t by $u_f \equiv u(t)$ with the polar coordinates $r_f = 1$, φ_f, ϑ_f. The reduced autocorrelation function for dipolar coupling at fixed interdipole distance is

$$G(t) = \frac{\langle A_{2,-m}(u_i) A_{2,m}(u_f) \rangle}{\langle |A_{2,-m}|^2 \rangle} = \frac{\langle Y_{2,-m}(u_i) Y_{2,m}(u_f) \rangle}{\langle |Y_{2,m}|^2 \rangle}$$

$$= 4\pi \langle Y_{2,-m}(u_i) Y_{2,m}(u_f) \rangle \tag{14.1}$$

The ensemble average on the right-hand side can be rewritten in terms of of the probability density $P(u_i, 0; u_f, t)$ that the initial orientation is u_i and the final orientation is u_f:

$$
\begin{aligned}
G(t) &= 4\pi \left\langle Y_{2,m}(u_i) Y_{2,-m}(u_f) \right\rangle \\
&= \left\langle \int\int P\left(u_i, 0; u_f, t\right) Y_{2,m}\left(u_i\right) Y_{2,-m}\left(u_f\right) \sin\vartheta_f \, d\vartheta_f \, d\varphi_f \right\rangle_{u_i} \\
&= \int\int \left\langle P\left(u_i, 0; u_f, t\right) Y_{2,m}\left(u_i\right) \right\rangle_{u_i} Y_{2,-m}\left(u_f\right) \sin\vartheta_f \, d\vartheta_f \, d\varphi_f
\end{aligned}
$$

(14.2)

The probability density $P\left(u_i, 0; u_f, t\right)$ may be factorized on the one hand into the a priori probability density $p(u_i)$ for the initial orientation, and, on the other, into the conditional-probability density $P_c(u_i, u_f, t)$ where the final orientation after time t is u_f if the initial orientation was u_i:

$$
P\left(u_i, 0; u_f, t\right) = p(u_i) P_c(u_i, u_f, t)
$$

(14.3)

Under isotropic conditions, the a priori probability for a certain orientation is

$$
p(u_i) \sin\vartheta_i \, d\vartheta_i \, d\varphi_i = \frac{1}{4\pi} \sin\vartheta_i \, d\vartheta_i \, d\varphi_i
$$

(14.4)

For continuous rotational diffusion, the conditional-probability density obeys the rotational variant of the diffusion equation

$$
\frac{\partial P_c(u_i, u_f, t)}{\partial t} = D_r \nabla^2 P_c(u_i, u_f, t)
$$

(14.5)

with the initial condition

$$
P_c(u_i, u_f, 0) = \delta(u_f - u_i)
$$

(14.6)

With regard to the spherical-function basis of the autocorrelation function, it appears favorable to make an ansatz in the form of the spherical-function expansion (see Sect. 42.4)

$$
P_c(u_i, u_f, t) = \sum_{l',m'} c_{l',m'}(t) \, Y_{l',m'}(u_f)
$$

(14.7)

in which the time and the orientation dependences are separated.

Inserting this ansatz into the diffusion equation, Eq. 14.5, applying the eigenvalue equation, Eq. 42.19, for spherical harmonics, multiplying both sides by $Y_{l,m}^*(u_f)$, integrating both sides over the full solid angle with respect to u_f, and taking the orthonormal property of spherical harmonics at Eq. 42.22 into consideration, renders the diffusion equation into the linear differential equation for the time-dependent coefficients,

$$
\frac{dc_{l,m}}{dt} = -D_r \, l(l+1) \, c_{l,m}
$$

(14.8)

Likewise, multiplying both sides of the ansatz at Eq. 14.7 by $Y_{l,m}^*(u_f)$, integrating both sides over the full solid angle with respect to u_f, considering the orthonormal

property of spherical harmonics (Eq. 42.22), and inserting the initial condition for
$P_c(u_i, u_f, 0)$ (Eq. 14.6), gives the initial condition of the coefficients $c_{l,m}(0)$:

$$c_{l,m}(0) = Y_{l,-m}(u_i) \tag{14.9}$$

The solution of Eq. 14.8 is then

$$c_{l,m} = Y_{l,-m}(u_i) \exp\{-D_r\, l(l+1)\, t\} \tag{14.10}$$

Combining this result with the ansatz at Eq. 14.7 yields the conditional probability

$$P_c(u_i, u_f, 0) = \sum_{l,m} Y_{l,-m}(u_i) Y_{l,m}(u_f) \exp\{-D_r\, l(l+1)\, t\} \tag{14.11}$$

which, in turn, can be associated with Eqs. 14.2 and 14.3. Again, utilizing the orthonormal property of spherical harmonics leads to the exponential autocorrelation
function of the second-order spherical harmonics ($l = 2$)

$$G(t) = \frac{\langle Y_{2,m}\,[u(0)]\, Y_{2,-m}\,[u(t)]\rangle}{|Y_{2,m}|^2} = \exp\{-|t|/\tau_c\} \tag{14.12}$$

The time constant $\tau_c = [l(l+1)D_r]^{-1} = (6D_r)^{-1}$ is the rotational correlation time.
The intensity function conjugated to $G(t)$ is

$$\mathcal{I}(\omega) = \mathcal{F}\{G(t)\} = \frac{2\tau_c}{1 + \omega^2 \tau_c^2} \tag{14.13}$$

The laboratory-frame and rotating-frame spin-lattice relaxation times ensuing from
this intensity function are displayed in Fig. 12.2 as a function of τ_c. The frequency
dependence of T_1 is plotted in Fig. 12.1.

14.2
Discrete-Coupling Jump Models

The spin system is supposed to perform random jumps between a well-defined
set of N discrete spin-interaction states. In the course of this process, fluctuations
may arise as a consequence of reorientations or, in the case of dipolar coupling, by
distance variations of the coupled spins. In view of the latter, the autocorrelation
function version at Eq. 9.3 referring to the functions $F^{(k)}(t)$ in the spin-interaction
Hamiltonian at Eq. 9.1 appears to be more opportune than the IST representation.

The discrete coupling states can be considered as a manifestation of a series
of deep wells on a potential surface on which the molecule performs thermally
activated motions. The residence times in the potential wells is assumed to be
much longer than the jump times so that the latter can be ignored.

The analogue of Eq. 14.2 for jumps between the N discrete states is

$$G(t) = \frac{\langle F^{(k)}(0)\, F^{(-k)}(t)\rangle}{\langle\, |F^{(k)}|^2\,\rangle} = \frac{\sum_{i=1}^{N} \sum_{f=1}^{N} p(i)\, P_c(f, i, t)\, F^{(k)}(i)\, F^{(-k)}(f)}{\sum_{i=1}^{N} p(i)\, |F^{(k)}|^2} \tag{14.14}$$

where the numbers i and f label the initial and final interaction states, respectively. The quantity $P_c(f, i, t)$ is the conditional probability that the final state is f if the initial state was i.[1] The a priori probability $p(i)$ and the conditional probability $P_c(f, i, t)$ are normalized, i.e.,

$$\sum_{i=1}^{N} p(i) = 1; \quad \sum_{f=1}^{N} P_c(f, i, t) = 1 \tag{14.15}$$

Expressed in matrix form, Eq. 14.14 reads

$$G(t) = \frac{1}{\langle |F^{(k)}|^2 \rangle} \left(F^{(-k)}(1) \quad \cdots \quad F^{(-k)}(N) \right) \begin{pmatrix} p(1) & \cdots & 0 \\ & \cdot & \\ & \cdot & \\ & \cdot & \\ 0 & \cdots & p(N) \end{pmatrix}$$

$$\begin{pmatrix} P_c(1, 1, t) & \cdots & P_c(1, N, t) \\ \cdot & & \cdot \\ \cdot & & \cdot \\ \cdot & & \cdot \\ P_c(N, 1, t) & \cdots & P_c(N, N, t) \end{pmatrix} \begin{pmatrix} F^{(k)}(1) \\ \cdot \\ \cdot \\ \cdot \\ F^{(k)}(N) \end{pmatrix} \tag{14.16}$$

Analyzing the random jumps between the different interaction states as **Markov chains**, the matrix of the conditional probabilities can be rewritten in the form

$$\begin{pmatrix} P_c(1, 1, t) & \cdots & P_c(1, N, t) \\ \cdot & & \cdot \\ \cdot & & \cdot \\ \cdot & & \cdot \\ P_c(N, 1, t) & \cdots & P_c(N, N, t) \end{pmatrix} =$$

$$\lim_{\Delta t \to 0} \begin{pmatrix} w(1, 1, \Delta t) & \cdots & w(1, N, \Delta t) \\ \cdot & & \cdot \\ \cdot & & \cdot \\ \cdot & & \cdot \\ w(N, 1, \Delta t) & \cdots & w(N, N, \Delta t) \end{pmatrix}^{t/\Delta t} = \lim_{\Delta t \to 0} U\Lambda^{t/\Delta t}U^{-1} \tag{14.17}$$

The matrices U and U^{-1} mediate the transformation to the (diagonal) eigenvalue matrix Λ (see Sect. 48.1.2).

---

[1] Note that we have interchanged the sequence of the labels of the initial and final states, i and f, because in the matrix representation envisaged in the following the first number is to indicate a row whereas the second label specifies a column.

The matrix element $w(j, i, \Delta t)$ is the probability that a jump from state i to state j occurs in an interval Δt. The probability that no jump away from state i takes place is analogously designated by $w(i, i, \Delta t)$. These probabilities are normalized, of course, so that $\sum_{j=1}^{N} w(j, i, \Delta t) = 1$. Denoting the transition rate from state i to state j by $\alpha_{j,i}$, we may write for infinitesimally small intervals Δt

$$w(j, i, \Delta t) = \alpha_{j,i} \Delta t$$
$$w(i, i, \Delta t) = 1 - \sum_{j \neq i} \alpha_{j,i} \Delta t \qquad (14.18)$$

The transition rates are properties intrinsic to the system to be treated.

The matrix of the transition probabilities $w(j, i, \Delta t)$ is raised to the power $t/\Delta t$ by first transforming it into its diagonal eigenvalue matrix. The eigenvalues are obtained as solutions of the characteristic equation formed by equating the secular determinant with zero. Furthermore we recall that a diagonal matrix is readily raised to the power n by raising each diagonal element to the power n (see Sect. 48.1).

The limes $\Delta t \rightarrow 0$ is then directly applied to the eigenvalue powers. The result is a diagonal matrix with elements in exponential function form. Carrying out all matrix products finally leads to the correlation function $G(t)$. On this basis, even relatively complicated random jump models may be tackled. In case the analytic formalism fails, the matrix algebra can be executed numerically [227, 228].

14.2.1
Two-State Jump Model

The prerequisite of the application of the discrete-jump formalism is the detailed knowledge of the structural and energetic features of the interaction states. For a demonstration, let us consider a simple two-state jump model where the jump rates back and forth between the coupling situations "1" and "2" are assumed to be equal, i.e., $\alpha_{1,2} = \alpha_{2,1} = \alpha$. The two states of the interdipole vector connecting two dipolar-coupled spins may be represented by the scheme

$$\boxed{\begin{array}{c} \varphi_1, \vartheta_1 \\ r_1 \end{array}} \quad \begin{array}{c} \alpha \\ \leftarrow \\ \rightarrow \\ \alpha \end{array} \quad \boxed{\begin{array}{c} \varphi_2, \vartheta_2 \\ r_2 \end{array}} \qquad (14.19)$$

Exchange between states of any other spin interaction can be symbolized analogously by replacing the parameters in the boxes by those specifying the interaction.

For this situation the conditional-probability matrix reads

$$\begin{pmatrix} P_c(1,1,t) & P_c(1,2,t) \\ P_c(2,1,t) & P_c(2,2,t) \end{pmatrix} = \lim_{\Delta t \to 0} \begin{pmatrix} 1 - \alpha\Delta t & \alpha\Delta t \\ \alpha\Delta t & 1 - \alpha\Delta t \end{pmatrix}^{t/\Delta t}$$

$$= \frac{1}{2} \begin{pmatrix} 1 & -1 \\ 1 & 1 \end{pmatrix} \begin{pmatrix} 1 & 0 \\ 0 & e^{-2\alpha t} \end{pmatrix} \begin{pmatrix} 1 & 1 \\ -1 & 1 \end{pmatrix} \qquad (14.20)$$

The a priori probabilities are $p(1) = p(2) = 1/2$, so that Eq. 14.16 can readily be evaluated for the two-state jump model:

$$
\begin{aligned}
G(t) &= \frac{|F^{(k)}(1) + F^{(k)}(2)|^2 + |F^{(k)}(1) - F^{(k)}(2)|^2 \, e^{-|t|/\tau_c}}{2\left(|F^{(k)}(1)|^2 + |F^{(k)}|(2)|^2\right)} \\
&= \frac{|\langle F^{(k)} \rangle|^2}{2\langle |F^{(k)}|^2 \rangle} + \frac{|F^{(k)}(1) - F^{(k)}(2)|^2}{2\langle |F^{(k)}|^2 \rangle} \, e^{-|t|/\tau_c}.
\end{aligned}
\tag{14.21}
$$

The correlation time is given by

$$
\tau_c = \frac{1}{2\alpha}
\tag{14.22}
$$

The exponential character of the resulting decay demonstrates that this form of autocorrelation function is not exclusively valid for the isotropic continuous diffusion model considered before. Rather, all processes directly underlying Poisson statistics tend toward correlation functions in the form of exponential functions or of linear combinations thereof.

The intensity function as the Fourier transform of Eq. 14.21 for $\omega > 0$ is found to be

$$
\mathcal{I}(\omega) = \frac{|F^{(k)}(1) - F^{(k)}(2)|^2}{2\langle |F^{(k)}|^2 \rangle} \frac{2\tau_c}{1 + \omega^2 \tau_c^2}
\tag{14.23}
$$

This result is a special case of the more general situation when the jump rates back and forth, now termed α and β, deviate from each other owing to different activation energies $E + \Delta E$ and E, respectively. The jump rates may then be expressed in the form of Arrhenius laws,

$$
\alpha = \alpha_\infty \exp\left\{ -\frac{E + \Delta E}{k_B T} \right\}
\tag{14.24}
$$

$$
\beta = \beta_\infty \exp\left\{ -\frac{E}{k_B T} \right\}
\tag{14.25}
$$

leading to

$$
\mathcal{I}(\omega) = \frac{|F^{(k)}(1) - F^{(k)}(2)|^2}{2\langle |F^{(k)}|^2 \rangle \cosh\left\{ \frac{\Delta E}{k_B T} + \ln \frac{\beta_\infty}{\alpha_\infty} \right\}} \frac{2\tau_c}{1 + \omega^2 \tau_c^2}
\tag{14.26}
$$

The correlation time is now

$$
\tau_c = \frac{1}{\alpha + \beta}
\tag{14.27}
$$

The temperature and frequency dependences of the spin-lattice relaxation time following from this sort of intensity function are plotted in Figs. 12.2 and 12.1, respectively, together with experimental data for appropriate systems.

14.3
Reorientation Mediated by Translational Displacements

The RMTD mechanism is governed by intramolecular interactions of any sort, and is not to be confused with the translational modulation of dipolar coupling which may

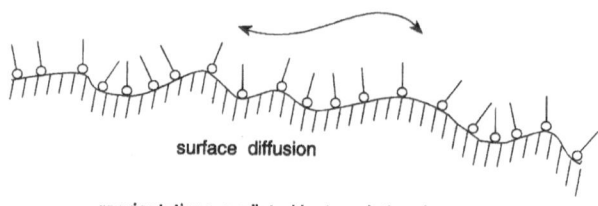

surface diffusion

reorientations mediated by translational
displacements

Fig. 14.1. Schematic representation of the RMTD relaxation mechanism.

arise if intermolecular interactions contribute. We consider molecules diffusing on
a convexly, concavely, or ruggedly shaped surface able to orient adsorbate molecules
perpendicular to the surface. A schematic illustration is shown in Fig. 14.1.

The general expression of the dipolar autocorrelation function for orientational
fluctuations at fixed interdipole distance is

$$G(t) = 4\pi \left\langle Y_{2,m}\left(u_i\right) Y_{2,-m}\left(u_f\right)\right\rangle \tag{14.28}$$

$$= \int \int \left\langle P_c\left(u_i, u_f, t\right) Y_{2,m}\left(u_i\right)\right\rangle_{u_i} Y_{2,-m}\left(u_f\right) \, \sin \vartheta_f \, d\vartheta_f \, d\varphi_f \tag{14.29}$$

The conditional-probability density can be analyzed in a form adapted to the RMTD
process:

$$P_c\left(u_i, u_f, t\right) = \int \Phi\left(u_i, u_f, r\right) \Psi\left(r, t\right) \, d^\zeta r \tag{14.30}$$

The integration refers to the whole topologically ζ dimensional r space. The function
$\Psi\left(r, t\right)$ is the diffusion propagator, i.e., the probability density that a particle is
displaced by a curvilinear distance r in the topological space in a time interval t.
The probability that the surface orientation changes from u_i to u_f in a curvilinear
distance r is symbolized by $\Phi\left(u_i, u_f, r\right)$.

The formal prerequisite for the ansatz at Eq. 14.30 is the stochastic independence
of the orientation at a position r and the displacement leading to this position. This
assumption is readily justified, and the surface propagator $\Psi\left(r, t\right)$ can be considered
to be independent of the orientations of the molecule at the initial as well as at the
final position. The orientations at separated positions are a matter of the surface
structure alone and are merely correlated by the probability $\Phi\left(u_i, u_f, r\right)$.

Using the expansion in terms of spherical harmonics

$$\Phi\left(u_i, u_f, r\right) = \sum_{l',m'} \Gamma_{l',m'}\left(r\right) Y_{l',m'}\left(u_f\right) \tag{14.31}$$

with the initial condition

$$\Phi\left(u_i, u_f, 0\right) = \delta\left(u_i - u_f\right)$$

$$= \sum_{l',m'} Y_{l',-m'}\left(u_i\right) Y_{l',m'}\left(u_f\right) \tag{14.32}$$

we find

$$\Gamma_{l',m'}(0) = Y_{l',-m'}(u_i) \tag{14.33}$$

The coefficients of the expansion can be expressed as

$$\Gamma_{l',m'}(r) = Y_{l',-m'}(u_i)\, g_{l',m'}(r) \tag{14.34}$$

where we have introduced the normalized surface orientation correlation function $g_{l',m'}(r)$ with the value $g_{l',m'}(0) = 1$. Inserting this relation for the coefficients into the expansion at Eq. 14.31 gives

$$\Phi(u_i, u_f, r) = \sum_{l',m'} Y_{l',-m'}(u_i) Y_{l',m'}(u_f)\, g_{l',m'}(r) \tag{14.35}$$

The combination of Eqs. 14.29, 14.30, 14.31, and 14.35 leads to the correlation function

$$G(t) = \int g_{2,m}(r)\, \Psi(r,t)\, d^\zeta r \tag{14.36}$$

where we have made use of the orthonormal properties of spherical harmonics (Eq. 42.22). In the following we omit the indices and use the identity $g(r) \equiv g_{2,m}(r)$.

For ordinary diffusion the propagator $\Psi(r,t)$ is a Gaussian function,

$$\Psi(r,t) = \frac{1}{(4\pi Dt)^{\zeta/2}}\, e^{-r^2/4Dt} \tag{14.37}$$

where D is the diffusion coefficient, and ζ is the dimension of the topological space we are referring to. The integral over the whole topological space is $\int \Psi(r,t)\, d^\zeta r = 1$. The Gaussian function may be expressed by its spatial Fourier transform (see Sect. 42.3)

$$e^{-r^2/4Dt} = \frac{(\pi Dt)^{\zeta/2}}{\pi} \int e^{-Dtk^2}\, e^{i k \cdot r}\, d^\zeta k \tag{14.38}$$

where k is the wave vector. The integral again covers the whole ζ dimensional k space. On this basis, the correlation function may be rewritten as

$$G(t) = \int \tilde{S}(k)\, e^{-Dtk^2}\, d^\zeta k \tag{14.39}$$

where we have introduced the **"orientational structure factor"**

$$\tilde{S}(k) = \frac{1}{2^\zeta \pi} \int g(r)\, e^{i k \cdot r}\, d^\zeta r \tag{14.40}$$

The integral covers the whole topologically ζ dimensional r space accessible to translational diffusion.

In isotropic systems we may equate $g(r) = g(r)$. Under these circumstances a radial orientation-structure factor $S(k)$ can be defined. For example, the hydration shells of surfaces are considered to form a topologically two-dimensional space ($\zeta = 2$) leading to the Hankel transform relation [84]

$$S(k) = \pi k \int_0^\infty r \, g(r) \, J_0(kr) \, dr \tag{14.41}$$

The function $J_0(kr)$ is the Bessel function of zeroth order. The correlation function is consequently

$$G(t) = \int_0^\infty S(k) \, e^{-t/\tau_k} \, dk \tag{14.42}$$

where

$$\tau_k = \frac{1}{Dk^2} \tag{14.43}$$

The corresponding spectral density is given by

$$\mathcal{I}(\omega) = 2 \int_0^\infty G(t) \cos(\omega t) \, dt$$

$$= \int_0^\infty S(k) \frac{2\tau_k}{1 + \omega^2 \tau_k^2} \, dk \tag{14.44}$$

14.3.1
Diffusion on Rugged Surfaces

A typical example where the RMTD formalism can be applied is deuteron low-frequency spin-lattice relaxation of (deuterated) hydration monolayers on globular-protein surfaces [245] as discussed in Sect. 16.3 in more detail. Such surfaces are known to be rugged in the sense of an irregular charge distribution underneath. That is, the electric dipoles of the adsorbed water adopt site-dependent orientations. On the other hand, water adsorbed on protein powders retains a remarkably high translational mobility even if confined as non-frozen liquid between the protein surface and ice [246, 375] . The consequence is that there must be a mechanism reorienting adsorbate molecules while diffusing along the thin, topologically two-dimensional hydration layer.

For the sake of simplicity, an equipartition of wave numbers describing the surface orientation in a certain range is assumed. The upper and lower cut-off magnitude values are designated by k_u and k_l, respectively. The radial orientation-structure factor then reads

$$S(k) = \begin{cases} \frac{1}{k_u - k_l} & \text{if} \quad k_l \le k \le k_u \\ 0 & \text{otherwise} \end{cases} \tag{14.45}$$

From this orientational structure factor the RMTD correlation function is inferred as

$$G_{RMTD}(t) = \int_0^\infty S(k)e^{-Dtk^2} dk$$

$$= \frac{1}{2(k_u - k_l)} \sqrt{\frac{\pi}{Dt}} \left[\mathrm{erf}\left(k_u\sqrt{Dt}\right) - \mathrm{erf}\left(k_l\sqrt{Dt}\right) \right] \quad (14.46)$$

where $\mathrm{erf}(x) = \frac{2}{\sqrt{\pi}} \int_0^x e^{-x'^2} dx'$ is the error function. The quantity D is the average translational water diffusion coefficient in the hydration layer. The lower cut-off wave number is connected with a cut-off correlation time (see Eq. 14.43)

$$\tau = (Dk_l^2)^{-1} \quad (14.47)$$

The intensity function is readily obtained by numerically Fourier transforming Eq. 14.46. Combining the result with the BPP formula, Eq. 12.48, yields the T_1 frequency dispersion plotted in Fig. 14.2 in comparison to experimental field-cycling data for D_2O adsorbed on bovine serum albumin powder.

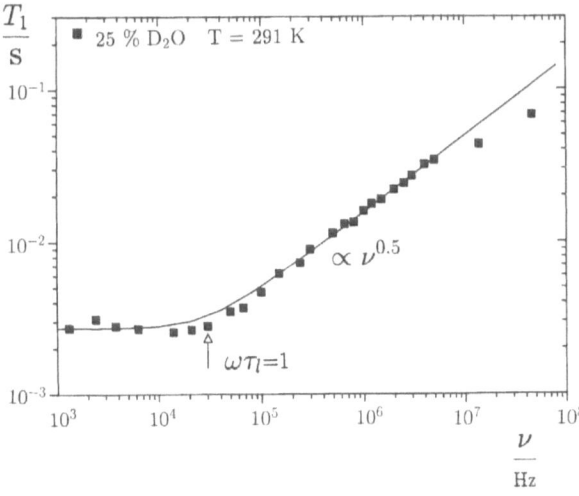

Fig. 14.2. Frequency dependence of the spin-lattice relaxation time T_1 as expected on the basis of the RMTD mechanism for equipartition of wave numbers. The solid line represents Eq. 12.48 in combination with the numerical Fourier transform of Eq. 14.46. The crossover from the low-frequency plateau for $\omega\tau_l \ll 1$ to the square root slope at intermediate frequencies, $\tau_l^{-1} \ll \omega \ll \tau_u^{-1}$, is typical for the wavenumber equipartition assumed. The experimental data points refer to the deuteron spin-lattice relaxation time, T_1, of heavy water adsorbed on bovine serum albumin (BSA) powder at 291 K. The principal measuring technique was field-cycling NMR relaxometry (Chap. 15). The water content of 25 wt% roughly corresponds to a molecular monolayer on the protein surfaces. For a more detailed discussion see Sect. 16.3. (Reprinted by permission from ref. [245])

14.3.2
Lévy-Walk Surface Diffusion

Further applications of the RMTD relaxation mechanism to silica fine particles and porous glasses can be found in [467, 468, 469] and [468]. In the latter reference, a Cauchy distribution was assumed instead of a Gaussian propagator. This probability density function for displacements is typical for Lévy walks in general [257].

The Lévy-walk statistics account for the behavior expected when "strongly adsorbed" molecules perform many desorption/readsorption cycles before escaping to more remote regions. Between desorption and readsorption a more or less extended "excursion" into the free liquid may take place which is subject to ordinary Brownian motion, of course. However, revisiting the surface many times effectively leads to **"bulk mediated surface diffusion"** representing Lévy walks with all known consequences [69]. In terms of NMR relaxation this means that the RMTD mechanism can be relevant even in systems where the adsorbate molecules are not confined to surface layers.

The effective displacements measured in the (in short length-scales) topologically two-dimensional and isotropic surface layer of adsorbate molecules are described by the **Cauchy distribution**:

$$\Psi(r, t) = \frac{1}{2\pi} \frac{ct}{[(ct)^2 + r^2]^{3/2}} \qquad [r \ll (Dt)^{1/2}] \qquad (14.48)$$

where the two-dimensional displacement on the surface is denoted by r, and the bulk diffusion coefficient by D. The quantity $c = D/h$ has the dimension of a velocity. The length h is "adsorption depth." The probability, that a particle is displaced a distance in the range $r \ldots r + dr$ in a time t is then given by $\Psi(r, t)\, 2\pi r\, dr$.

Based on this propagator, we infer from Eq. 14.36 for the time correlation function

$$G(t) = \int_0^\infty g(r)\, \Psi(r, t)\, 2\pi r\, dr$$

$$= ct \int_0^\infty \frac{r\, g(r)}{[(ct)^2 + r^2]^{3/2}}\, dr \qquad (14.49)$$

where $g(r)$ is the surface correlation function. (Particles which are initially adsorbed and are lost to the bulk during t are ignored.)

A simple example for $g(r)$ is a step function

$$g(r) = \begin{cases} 1 & \text{for} \quad r \leq r_c \\ 0 & \text{else} \end{cases} \qquad (14.50)$$

In this case, the time correlation function takes the form

$$G(t) = 1 - \frac{1}{\sqrt{1 + [r_c/(ct)]^2}} \qquad (14.51)$$

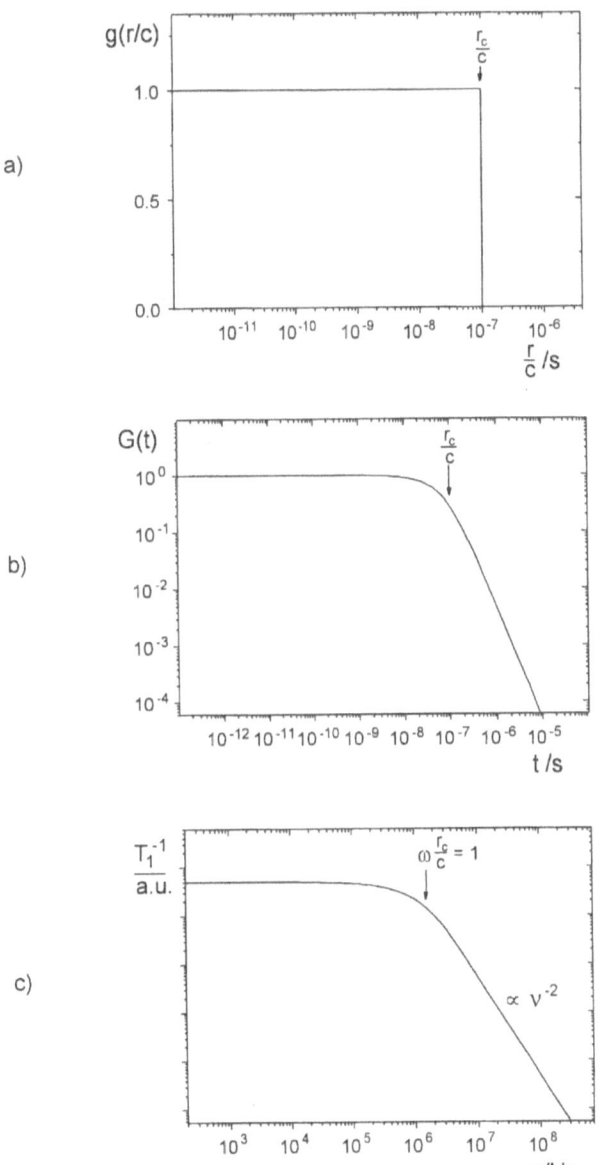

Fig. 14.3. Simple example of the Lévy-walk variant of the RMTD relaxation mechanism for a rectangular surface correlation function (**a**). The corresponding time correlation function and the frequency dispersion of the spin-lattice relaxation rates are plotted in (**b**) and (**c**).

Using Eq. 12.40, the spin-lattice relaxation relaxation rate given by Eq. 12.48 can be evaluated numerically. The result is shown in Fig. 14.3.

CHAPTER 15

Field-Cycling NMR Relaxometry

15.1
Laboratory-Frame Experiments

The main objective of field-cycling experiments [231, 367] is to obtain information on the spectral density of the fluctuating spin interactions in a frequency range as wide as possible.[1] Typical examples are discussed in Chaps. 14 and 16 (see Figs. 12.1 and 14.2, for instance). The origin of the fluctuations is normally molecular dynamics so that these can be characterized in a direct and quantitative way.

Figure 15.1 schematically shows the cycle and the time intervals of a field-cycling relaxometry experiment. The sample is first polarized in the polarization field with the flux density B_p. A non-equilibrium magnetization is produced by switching the external magnetic field to a variable relaxation field B_r. The longitudinal magnetization then relaxes toward its new equilibrium value. The relaxation curve (see Table 10.1 on page 93) is probed one data point after another by incrementing the length of the relaxation interval in a series of transients. For the signal detection, the flux density is switched to a detection field, B_d, chosen as high as technically possible. After having reached a stable field level, the magnetization is recorded with the aid of a 90° read pulse or a spin-echo pulse sequence.

The relaxation field values that can be covered in this way ranges from the local field caused by spin interactions within the sample up to about 2 T. In liquids where local fields are largely averaged out, this typically corresponds to a proton frequency range 10^3 Hz $< \nu < 10^8$ Hz. For deuterons, the range is shifted by a factor of about seven to lower values. The limits are then 10^2 Hz $< \nu < 10^7$ Hz by orders of magnitude.

The accuracy of relaxation time measurements largely depends on how well defined and homogeneously the initial non-equilibrium state can be produced in a sample. With RF pulses, this in particular is a problem of RF-field homogeneity in the probe coil. By contrast, with field-cycling relaxometry, the non-equilibrium state is reached by switching the main magnetic-flux density B_0 which is much

---

[1]There are other variants of field-cycling experiments serving nuclear quadrupole resonance (NQR) spectroscopy [39, 135, 234], or zero-field spectroscopy [513, 529]. A promising biomedical variant of field-cycling experiments is field-cycled proton-electron double-resonance imaging (FC-PEDRI) which was developed to allow the distribution of free radicals in aqueous environments to be imaged via NMR signals [302, 303].

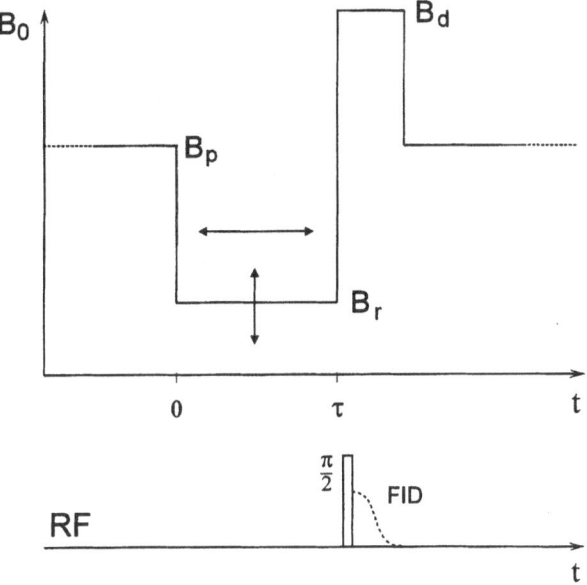

Fig. 15.1. Typical field cycle for spin-lattice relaxometry in a wide frequency range. The polarization field B_p is kept lower than the detection field B_d in view of the Joule's heat produced in the (resistive) magnet. The field levels correspond to the very different intervals these fields need. The B_p interval is chosen in the order of $5T_1(B_p)$ whereas the B_d plateau may in principle be restricted to the FID decay time of the order T_2^*.

more homogeneous than RF fields produced by probe coils. Therefore, field-cycling relaxometry is a rather reliable technique in this respect.[2]

15.1.1
Field-Cycling Magnets

The combination of high magnetic fields, fast switching times, and a reasonable field homogeneity imposes demanding requirements on the magnet-coil design as well as on the magnet power supply. The principal factors determining the magnetic flux density that can be produced in a coil are the number of windings and the current, i.e., the current density. In other words, there is a tendency toward *bulky magnets* in this regard.

On the other hand, fast switching times are favored by *compact coils* so that the volume to be energized and to be deenergized in the course of a field cycle can be

---

[2]The RF pulse homogeneity and the pulse sequence employed for the signal detection are uncritical because the signal is proportional to the transverse magnetization anyway.

kept small.[3] Small resistive high-field magnets stipulate a good cooling efficiency. Actually this is one of the crucial factors limiting the technique.

In order to overcome this difficulty, superconducting magnet coils [431] and liquid-nitrogen cooled resistive coils have been employed. Unfortunately, the advantages of such cryogenic designs are paid for by drawbacks such as the circumstantial handling and the high cryogen consumption. In practice, it turned out to be more favorable to use high-pressure oil-cooled copper magnets optimized for the circulation properties of the cooling medium as well as for the spatial distribution of the current density. Magnet coil geometries optimized for field-cycling purposes are reported in [34, 168, 290, 440].

15.1.2
The Switching Intervals

In order to achieve a good signal sensitivity, the polarization field as well as the detection field should be as high as possible. A second cycle feature of interest is the field-switching time, i.e., the time needed to reach the desired field level with the demanded accuracy and stability. With the relaxation field, an accuracy of 10^{-2} of the adjusted value appears to be reasonable in most applications. The detection field level must be met within a range corresponding to the RF bandwidth of the system.

The latter is also related to the field homogeneity within the sample as a further limiting factor. It turns out that a relative field homogeneity between 10^{-5} and 10^{-4} is sufficient for most applications, so that the switching time for the detection field is defined by the interval needed until the final level is stabilized with a relative accuracy[4] between 10^{-5} and 10^{-4}. Practical switching times defined in this way are of the order 1 ms.

In case the field-switching times are not much shorter than the (low-field) relaxation times, as would be desirable ideally, one faces sensitivity losses, but no systematic experimental errors. This can be substantiated in the following way [499]. During the switching periods, all quantities in Bloch's equation for the z magnetization become time functions (compare Eq. 10.5):

$$\frac{dM_z(t)}{dt} = -\frac{M_z(t) - M_0(t)}{T_1(t)} \tag{15.1}$$

The instantaneous equilibrium magnetization is denoted by $M_0(t)$, the instantaneous spin-lattice relaxation time by $T_1(t)$. This equation is a linear first-order

---

[3]From a practical point of view, "field-cycling" means "field-energy-cycling." Fast switching-up requires a device supplying a high peak power, fast switching-down needs a method for actively transferring the field energy away. Both requirements are best met by switching precharged high-voltage capacitors into the magnet circuit with a polarity matching the desired field change.

[4]Note that phase-sensitive detection as required for signal accumulation stipulates a particularly good stability.

differential equation which may be rewritten in the more common form

$$\frac{dM_z(t)}{dt} + \left(\frac{1}{T_1(t')}\right) M_z(t) = \frac{M_0(t)}{T_1(t)} \tag{15.2}$$

Defining

$$P(t) \equiv \int_{t_0}^{t} \frac{1}{T_1(t')} \, dt' \tag{15.3}$$

the integral for a switching interval $t_0 \leq t \leq t_0 + \Delta t$ can be expressed as

$$M_z(t_0 + \Delta t) = e^{-P(t+\Delta t)} \left[\int_0^{t+\Delta t} e^{P(t'')} \frac{M_0(t'')}{T_1(t'')} \, dt'' + M_z(t_0) \right] \tag{15.4}$$

In a field-cycling experiment, the polarization field B_p and the detection field B_d are default quantities. In the course of a T_1 measurement, the relaxation field B_r also has a predetermined value. The only experimental parameter varied in the field-cycle sequence needed for the determination of a spin-lattice relaxation time in a given relaxation field is thus the relaxation-field interval τ. This, of course, does not affect the temporal variation of the magnetic-flux density during the switching intervals. The time dependence of the magnetic-flux density is an instrumental property which is reproduced in each cycle irrespective of τ provided that all field levels remain constant. The integrals in Eq. 15.4 are therefore constant, and we may write

$$M_z(t_0 + \Delta t) = M_z(t_0) e^{-c_1} + c_2 \tag{15.5}$$

where the switching constants are

$$c_1 = P(t + \Delta t) = \int_{t_0}^{t+\Delta t} \frac{1}{T_1(t')} \, dt' = \text{const} \tag{15.6}$$

$$c_2 = e^{-c_1} \int_{t_0}^{t+\Delta t} e^{P(t'')} \frac{M_0(t'')}{T_1(t'')} \, dt'' = \text{const} \tag{15.7}$$

Switching down from the polarization field B_p to the relaxation field B_r in an interval $(\Delta t)_d$ beginning at $t_0 = 0$ reduces the magnetization from $M_z(0) = M_0^p$ to $M_z[(\Delta t)_d]$ where M_0^p is the equilibrium magnetization in the polarization field. Designating the switching constants at Eqs. 15.6 and 15.7 for this interval by c_1^d and c_2^d, the magnetization at the beginning of the relaxation interval reads

$$M_z[(\Delta t)_d] = M_0^p e^{-c_1^d} + c_2^d \tag{15.8}$$

In this respect, the condition for a relaxation experiment is $M_z[(\Delta t)_d] \gg M_0^r$ where M_0^r is the equilibrium magnetization in the relaxation field.

During the relaxation interval of length τ the magnetization relaxes toward

$$M_z\left[\tau + (\Delta t)_d\right] = \left(M_z\left[(\Delta t)_d\right] - M_0^r\right)\, e^{-\tau/T_1(B_r)} + M_0^r \tag{15.9}$$

where $T_1\,(B_r)$ is the spin-lattice relaxation time in the relaxation field.

The switching-up interval $(\Delta t)_u$ begins at the end of the relaxation interval, i.e., at $t_0 = \tau + (\Delta t)_d$. Denoting the switching constants at Eqs. 15.6 and 15.7 for this interval as c_1^u and c_2^u, the magnetization at the beginning of the detection interval is represented by

$$M_z\left[\tau + (\Delta t)_d + (\Delta t)_u\right] = M_z\left[\tau + (\Delta t)_d\right]\, e^{-c_1^u} + c_2^u \tag{15.10}$$

The magnetization to be detected is

$$M_z\left[\tau + (\Delta t)_d + (\Delta t)_u\right] = \left[\left(M_z\left[(\Delta t)_d\right] - M_0^r\right)\, e^{-\tau/T_1(B_r)} + M_0^r\right]\, e^{-c_1^u} + c_2^u \tag{15.11}$$

The relaxation curve to be evaluated is thus of the form

$$M_z\left[\tau + (\Delta t)_d + (\Delta t)_u\right] - M_z\left[\infty\right] = \left(M_z\left[(\Delta t)_d\right] - M_0^r\right)\, e^{-c_1^u}\, e^{-\tau/T_1(B_r)}$$

$$= \Delta M_z^{\mathrm{eff}}\, e^{-\tau/T_1(B_r)} \tag{15.12}$$

This is an unambiguous function of $T_1(B_r)$, and the switching intervals evidently do not cause any systematic error. However, the accuracy of the evaluation may be reduced owing to the diminution of the effective variation range of the magnetization, $\Delta M_z^{\mathrm{eff}} = \left(M_z\left[(\Delta t)_d\right] - M_0^r\right)\, \exp\{-c_1^u\} = const.$

In view of the experimental accuracy, switching times as short as possible are desirable. On the other hand, there is no upper limit to the field-variation rate with respect to the adiabatic condition (compare Sect. 15.2.1) as long as the field direction remains unchanged. However, at very low external fields the local fields caused by spin interactions begin to dominate the quantization field. The directions of these fields are randomly distributed.

In that case, one has two choices. First, one can vary the field adiabatically so that dipolar or quadrupolar order is produced. This **"adiabatic demagnetization in the laboratory frame"** (ADLF) is the field-cycling counterpart of the Jeener/Broekaert pulse sequence and of the ADRF method. It can be used for measurements of dipolar or quadrupolar-order relaxation times. In the second case, when the local fields are approached non-adiabatically, coherent spin states leading to finite expectation values of the spin components transverse to the local field directions are produced. This permits the so-called **"zero-field NMR or NQR spectroscopy."**[5] That is, the local-field interval is taken as the time domain (coherence evolution interval) of a corresponding FT spectroscopy procedure. This experiment is suitable for studying the evolution of coherences in the local fields independent of any external field direction, so that there is no need to orient the spin systems with the aid of single crystals, for instance.

---

[5] Zero-field NMR or NQR can also be performed by adiabatically switching off the external field and applying "90° pulses" at the beginning and at the end of the zero-field interval. A 90° pulse in the absence of the external field consists of a d.c. field pulse with zero carrier frequency, of course [342, 343].

15.2
Spin-Lock Adiabatic Field-Cycling Imaging Relaxometry

In the rotating-frame analog of laboratory-frame field-cycling experiments the effective field B_e (Eq. 48.98) is cycled instead of B_0. Such techniques can moreover be combined with NMR imaging, so that the whole frequency range accessible is probed at each transient. The method to be delineated in the following is called **spin-lock adiabatic field-cycling imaging** (SLOAFI) [250]. The imaging procedures can be either laboratory-frame (B_0 gradient) or rotating-frame (B_1 gradient) variants. The quantity measured with such techniques is the spin-lattice relaxation time in the (tilted) rotating frame, that is, under off-resonance conditions.

15.2.1
Adiabatic Variation of the Effective Field

The adiabatic characteristic of a process means that no transitions occur: the spin system always resides in eigenstates of the instantaneous Hamiltonian, and all populations remain unchanged.

The transition-inducing spin-vector-operator components are perpendicular to the quantization field, i.e., the effective field B_e. Therefore, the variation of the effective field perpendicular to its instantaneous direction must be considered. The time derivative of B_e may be decomposed into components parallel and perpendicular to B_e:

$$\frac{dB_e}{d\tau} = \frac{dB_{e\parallel}}{d\tau} + \frac{dB_{e\perp}}{d\tau} \tag{15.13}$$

The RF amplitude B_1 is assumed to remain constant, so that the variation refers either to the main field B_0 or to the carrier frequency ω_c. The vector dB_e is consequently aligned along B_0. The decomposition of the effective-field vector then reads

$$\frac{dB_e}{d\tau} = \frac{d(B_0 - \omega_c/\gamma_n)}{d\tau} \cos\Theta \, u_\parallel + \frac{d(B_0 - \omega_c/\gamma_n)}{d\tau} \sin\Theta \, u_\perp \tag{15.14}$$

where $u_\parallel$ and $u_\perp$ are unit vectors parallel and perpendicular to B_e, respectively, in the plane spanned by B_0 and B_e. The tilt angle of B_e is designated by Θ. Since only the second term is relevant for potential spin transitions, we consider from now on

$$\frac{dB_{e\perp}}{d\tau} = \frac{d(B_0 - \omega_c/\gamma_n)}{d\tau} \sin\Theta \, u_\perp \tag{15.15}$$

The next step is to find out how fast $B_0 - \omega_c/\gamma_n$ may be varied without inducing spin transitions. The corresponding condition follows from Heisenberg's energy/time uncertainty relation $\varepsilon\tau \geq \hbar$. Transition energies ε that can be conveyed to or from the system in this period obey the condition $\varepsilon \geq \frac{\hbar}{\tau}$. The minimum transition energy transferable in a period τ is thus $\varepsilon = \frac{\hbar}{\tau}$. The lowest transfer rate is consequently $|d\varepsilon/d\tau| = \hbar/\tau^2$. *Adiabatic conduct* of a process means that the lowest transfer rate is *not* reached at any instant, so that no transition can take place. The

adiabatic condition is therefore

$$\left|\frac{d\varepsilon}{d\tau}\right| \ll \frac{\hbar}{\tau^2} \tag{15.16}$$

In the present case, the energy quantum is $\varepsilon = \hbar\gamma_n B_e$. The right-hand side may be rewritten on this basis using $\tau = \hbar/\varepsilon = 1/(\gamma_n B_e)$. The left-hand side may be reformulated as

$$\left|\frac{d\varepsilon}{d\tau}\right| = \hbar\gamma_n\left|\frac{dB_e}{d\tau}\right| \le \hbar\gamma_n\left|\frac{d(B_0 - \omega_c/\gamma_n)}{d\tau}\right| \tag{15.17}$$

Expressing the flux densities by the corresponding angular frequencies thus leads to the **condition for the adiabatic variation of the angular-frequency offset**,

$$\boxed{\left|\frac{d(\omega_0 - \omega_c)}{d\tau}\right| \ll \frac{\omega_e^2}{\sin\Theta}} \tag{15.18}$$

In the spin-lock field-cycling experiments to be described in the following, the angular-frequency offset $\Omega \equiv \omega_0 - \omega_c$ is varied in a cyclic way. The optimal time dependence of Ω follows from the differential equation

$$\frac{d\Omega}{d\tau} = c\,\frac{\omega_e^2}{\sin\Theta} = c\,\frac{(\Omega^2 + \omega_1^2)^{3/2}}{\omega_1} \tag{15.19}$$

where $c \ll 1$. This is another but entirely equivalent formulation of Eq. 15.18. The solution is readily found by integrating both sides:

$$\int (\Omega^2 + \omega_1^2)^{-3/2}\,d\Omega = c\,\omega_1^{-1}\int d\tau \tag{15.20}$$

For the initial condition $\Omega(0) = 0$ the result is

$$\Omega(\tau) = \frac{c\,\omega_1^2\,\tau}{\sqrt{1 - (c\,\omega_1\,\tau)^2}} \tag{15.21}$$

Figure 15.2 illustrates this function for various c values. The slope is initially flat, i.e., as long as the tilt angle Θ is close to 90°. At $\tau = 1/(\omega_1 c)$, a singularity appears corresponding to $\Theta = 0$.

15.2.2
Spin-Lock Field-Cycling Laboratory-Frame Imaging Relaxometry

Figure 15.3 shows the pulse scheme to be considered. On the RF side, the sequence consists of a 90° excitation pulse followed by a pulse whose phase is shifted by 90° so that the magnetization is spin locked. At the beginning, the spin-lock pulse is resonant. Then a magnetic-field gradient pulse is adiabatically switched on so that the carrier frequency deviates from resonance depending on the position along the

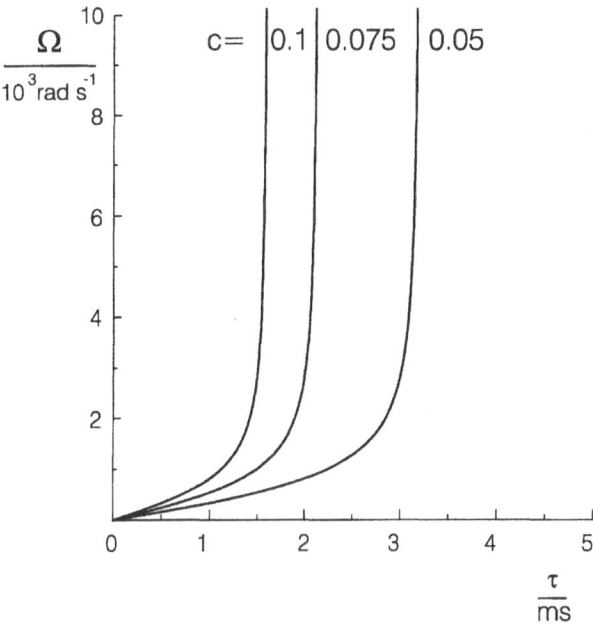

Fig. 15.2. Graphical representation of the time dependence of the angular-frequency offset Ω for adiabatic changes of the effective field (see Eq. 15.21). The curve parameter is the constant c. The value of ω_1 was assumed to be $2\pi \cdot 10^3$ Hz. (Reprinted by permission from ref. [250])

field-gradient direction. Let us assume that the gradient is directed along the x axis of the laboratory frame, so that

$$G = \frac{\partial B_0}{\partial x} \tag{15.22}$$

Depending on the position in the sample, the effective field is tilted toward the z direction at angles other than 90°. The tilt angle thus becomes a function of the position in the sample, $\Theta = \Theta(x)$. The adiabatic characteristic of the gradient pulse (Eq. 15.21) ensures that the local magnetization follows the local effective field $B_e = B_e(x)$ and remains spin-locked at all instants.

According to the applied field gradient, the deviation from resonance varies across the sample. That is, in the interval τ_R, spin-lattice relaxation (see Eqs. 10.13 and 12.59) is determined by effective angular frequencies $\omega_e = \omega_e(x)$ which are spatially distributed according to the field gradient. At the end of the interval τ_R, the field gradient is adiabatically switched off. Adiabatic now means that the field-gradient decay follows the mirrored rise function (Fig. 15.2). In this way, the local magnetization is led back to resonant spin-locking at all positions within the sample. The magnetization profile along the field gradient corresponds to the partial relaxation depending on the duration of the relaxation interval τ_R and on the local effective angular frequency $\omega_e = \omega_e(x)$.

The last part of the procedure serves to render an image of the profile of the partially relaxed magnetization (see Sect. 25.3). The free-induction decay (FID) fol-

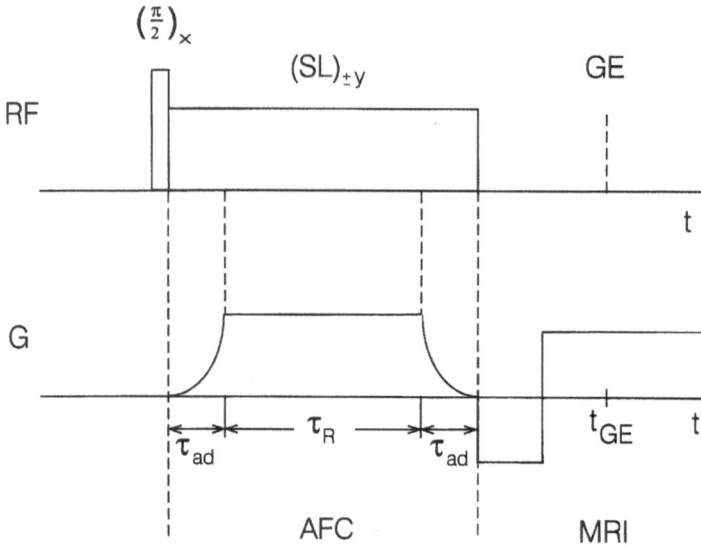

Fig. 15.3. Pulse scheme for the B_0-gradient SLOAFI relaxometry technique. A section in which adiabatic spin-lock field cycling (AFC) takes place is followed by a one-dimensional magnetic-resonance imaging (MRI) interval. The RF part consists of a 90° excitation pulse and a spin-lock pulse (SL) which may be subject to phase cycling as indicated. The B_0 gradient, G, is adiabatically switched on and off during the intervals τ_{ad}. The effective angular-frequency offset follows Eq. 15.21 in the switching-on period, and, in the specular sense, while the gradient is switched off. The field gradient causes a spatial distribution of the effective angular frequency ω_e across the sample along the gradient direction. The spin-locked magnetization consequently relaxes during the interval τ_R at rates depending on the effective angular frequency ω_e, i.e., on the position. The partially relaxed magnetization is imaged by acquiring the gradient echo (GE). (Reprinted by permission from ref. [250])

lowing the spin-lock pulse is refocused and read in the form of a gradient echo. The one-dimensional image obtained after Fourier transformation directly reflects the magnetization profile. Recording a series of such profiles for intervals τ_R incremented in subsequent transients permits one to evaluate the relaxation curves (Eq. 10.14) at different positions of the magnetization profile. As the positions x are unambiguously related to the effective angular frequencies $\omega_e(x)$ at these positions during the interval τ_R, we obtain the desired dependence $T_{1\rho}^{(e)} = T_{1\rho}^{(e)}(\omega_0, \omega_e)$ (see Eq. 10.13 in combination with Eqs. 12.58 and 12.59).

The ideal excitation tip angle is 90°, so that the total magnetization is transverse and can be spin-locked. However, in practical experiments, the B_1 field may be inhomogeneous, and the local excitation may be imperfect. The consequence is a residual z component of the local magnetization immediately after the excitation pulse. These residual z magnetizations may cancel when averaged over the whole sample, but nevertheless exist locally. During the spin-lock pulse and adiabatic field cycling, they precess about the instantaneous effective field B_e. That is, they always remain perpendicular to the spin-locked magnetization component. They are therefore subject to transverse relaxation in the rotating frame and decay with the time constant $T_{2\rho}$ [37]. If $T_{2\rho} \ll T_{1\rho}^{(e)}$, no contribution to the acquired signal is expected, and the spin-lattice relaxation experiment is not affected. Otherwise, spurious signals arising on these grounds can be eliminated by an appropriate phase cycle.

15.2.3
Spin-Lock Field-Cycling Rotating-Frame Imaging

Figure 15.4 represents an alternative variant where the effective field is adiabatically cycled by modulating the carrier frequency of the spectrometer. All RF pulses imply gradients of B_1 which in principle need not be constant nor uniformly directed. Instead of a B_0 gradient, the amplitude of the RF field now is a function of the position, e.g., of the x coordinate of the laboratory frame, so that $B_1 = B_1(x)$ for the

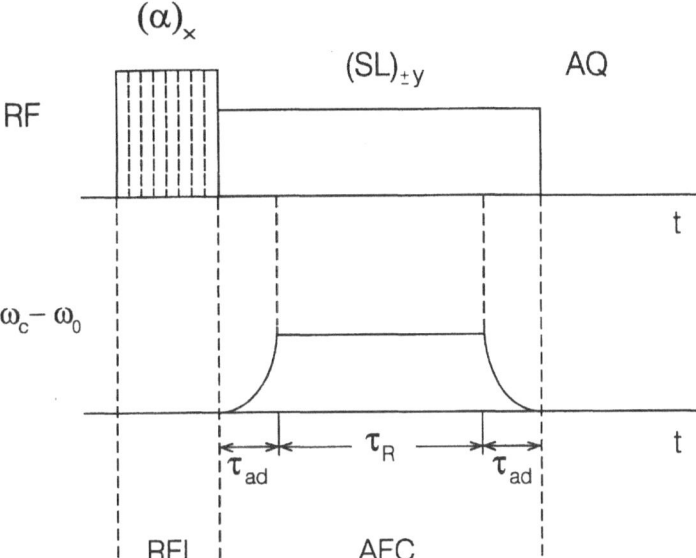

Fig. 15.4. Pulse scheme for the B_1-gradient SLOAFI relaxometry technique. Adiabatic spin-lock field cycling (AFC) is combined with one-dimensional rotating-frame imaging (RFI) (Chap. 34). All pulses refer to RF with amplitudes distributed across the sample according to a B_1 gradient. The RF part consists of an initial excitation pulse and a spin-lock pulse (SL) which is subject to phase cycling as indicated. The angular-frequency offset, $\omega_c - \omega_0$, is adiabatically switched on and off by varying the carrier frequency ω_c during the intervals τ_{ad} so that the effective angular-frequency offset follows Eq. 15.21 directly or in the specular sense while rising and decaying, respectively. The B_1 gradient causes a spatial distribution of the effective angular frequency ω_e across the sample along the gradient direction. The spin-locked magnetization consequently relaxes during the interval τ_R at rates depending on the effective angular frequency ω_e, i.e., on the position. The partially relaxed magnetization is imaged by incrementing the excitation pulse in subsequent scans and acquiring the FIDs after the pulse sequence. (Reprinted by permission from ref. [250])

spin-lock pulse as well as for the initial excitation pulse. Based on this B_1 gradient, the profile of the partially relaxed magnetization can be rendered as an image by combining spin-lock field-cycling with rotating-frame imaging (see Chap. 34).

The length (or, equivalently, the amplitude) of the initial excitation pulse is incremented in subsequent transients. The FID is then acquired as a function of the excitation pulse width and of the relaxation interval τ_R.

After the excitation pulse, local magnetization components exist which are perpendicular to the spin-locking effective field. Note that they also remain perpendicular to the effective field when the effective-field direction is changed, provided that this is done in an adiabatic manner. If $T_{2\rho} \ll T_{1\rho}^{(e)}$, they relax to zero before the signal of interest is detected, and they need not be considered further here. Otherwise, the spin-lock pulse can be phase cycled $\pm y$ in the course of signal accumulation. The initial phase of the FID arising from the spin-locked magnetization is always that of the excitation pulse, whereas the phase of signal contributions from unlocked magnetizations alternates with the phase of the spin-lock pulse.

The main advantage of the B_1-gradient SLOAFI technique is that the FIDs are recorded in the absence of field gradients. That is, the full spectroscopic information is preserved. The hardware prerequisite is that the carrier frequency can be varied during the experiment in a freely programmable way. As a source of B_1 gradients, one can use the fringe field of solenoid RF probe coils as they are in use in standard probeheads. Other probe geometries such as toroid cavity detectors [516] or coaxial resonators [22] are also of interest in this context.

CHAPTER 16

Field-Cycling Relaxometry in Biosystems

Systems of biological origin tend to be *heterogeneous* in the sense that the investigated nuclei occur in different environments with different molecular dynamics and spin interactions. It is then a matter of exchange rates whether spin-lattice relaxation is determined by average intensity functions ("fast exchange" relative to relaxation times) so that monoexponential relaxation decays arise, or whether multiexponential distributions occur (see Chap. 23). In context with water relaxation in biological systems, the two-site/fast-exchange model turned out to be very successful in accounting for experimental findings. Note, however, that the term "exchange" in this context may refer to material transport as well as to (immaterial) spin diffusion [136, 225].

The element usually detected in T_1 frequency dispersion experiments with biological systems is hydrogen via proton or deuteron resonance.[1] Other nuclear species can be probed indirectly via cross relaxation if their relaxation rates are fast enough to act in combination with proton spin diffusion as relaxation sinks. This can be expected in particular with quadrupole nuclei. The cross relaxation then manifests itself as **"quadrupole dips"** at the resonance crossing frequencies [234] as demonstrated in Fig. 16.1 for polyalanine protons cross-relaxed by the amide ^{14}N relaxation sinks.

The T_1 frequency dispersion may further be influenced by electron paramagnetic constituents such as molecular oxygen [229], paramagnetic ions [221, 272], or heme groups [226, 274], for instance. On the other hand, NMR relaxometry studies have been performed with many purely diamagnetic systems of biological origin. Typical examples are dissolved [225, 273], wet [238], or dry [369] proteins, DNA [234], lipid bilayers [232, 410], tissue [238], eye lenses [53, 276], or small animals in vivo [233].

The principal constituents of biological systems are biopolymers and water. In vitro, the water component (including all exchangeable hydrogen atoms of the biopolymers) can be conveniently prepared in deuterated form so that separate studies of the dynamics in the two components can be performed. Figure 16.2 shows the T_1 frequency dispersions of water deuterons on the one hand, and of nonexchangeable biopolymer protons on the other which were measured in the same solution of bovine serum albumin (BSA) in D_2O [370]. The entirely different nature of the fluctuations within these two constituents is obvious.

---

[1] In exceptional cases, ^{31}P relaxation studies may also be feasible [431].

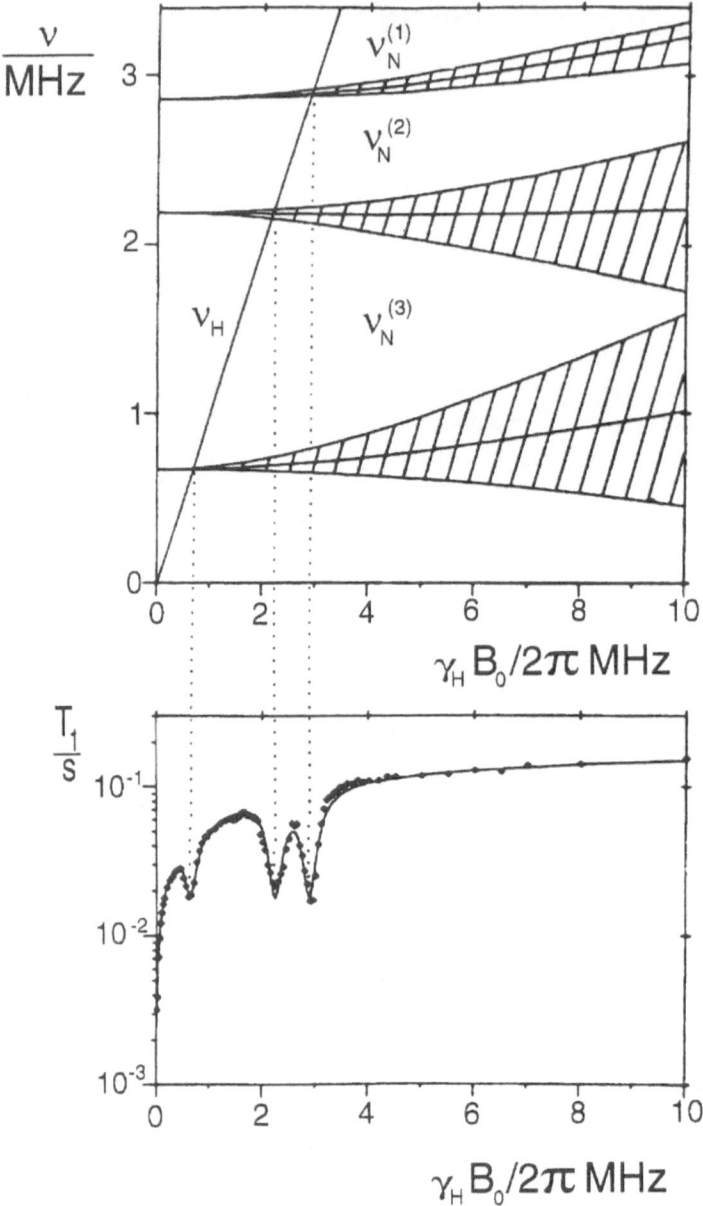

Fig. 16.1. Top: Field dependence of the ^{1}H and the three ^{14}N low-field quadrupole resonance frequencies of amide groups. The hatched areas indicate the range covered in a powdery sample. The solid lines within these areas represent the powder averages. The magnetic flux density B_0 is expressed in units of the proton Larmor frequency. **Bottom:** T_1 frequency dispersion of poly-L-alanine at $-1°$C. The three ^{14}N-^{1}H quadrupole dips at the resonance crossings are obvious. (Reprinted by permission from ref. [234])

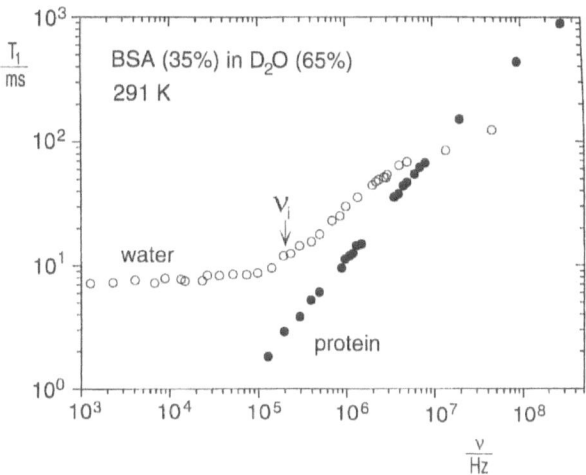

Fig. 16.2. Frequency dependence of the proton (•) and deuteron (○) spin-lattice relaxation times of a D_2O solution of bovine serum albumin (35 wt%) at 291 K. The data in the vicinity of the $^{14}N^1H$ quadrupole dips are omitted. The different T_1 frequency dispersions indicate different fluctuation processes dominant for water (deuterons including exchangeable protein hydrogens) and for the protein molecules (protons). (Reprinted by permission from ref. [370])

16.1
Fluctuations in Proteins

Molecular motions of proteins can comprise main chain ("backbone") fluctuations, side-group motions, and tumbling of the whole macromolecule. The latter can easily be identified by considering the concentration dependence of the reorientation rates (see below). Side-group motions, such as restricted rotational diffusion and ring flips, are a matter of the local structure and chemical composition. Proteins, for instance, therefore tend to be characterized by a broad distribution of the side-group correlation times.

Backbone fluctuations in proteins and polypeptides, on the other hand, appear to be governed by homogeneous modes leading to an apparently universal T_1 dispersion behavior at low frequencies where side-group motions are of minor importance. The frequency dependence in the absence of molecular tumbling was found to obey the power law [234] $T_1 \propto v^{0.74\pm0.06}$ in a range of several orders of magnitude. As an explanation, one-dimensional multiple trapping diffusion of defects locally dilating the structure was suggested [369]. The correlation function attributed to backbone fluctuations is given by the error function expression

$$G_{bb}(t) = a_1 \left[\mathrm{erf}\left(\frac{b^2}{2\langle\xi^2\rangle} \right)^{1/2} - \left(\frac{2\langle\xi^2\rangle}{\pi b^2} \right)^{1/2} \left(1 - \exp\left\{ -\frac{b^2}{2\langle\xi^2\rangle} \right\} \right) \right] + a_2 \quad (16.1)$$

where $a_2 = 1 - a_1$ represents the residual correlation at long times, b is the extension of the defect, and $\langle\xi^2\rangle \propto t^{1/2}$ the (anomalous) mean-square displacement of the defects in a time interval t.

16.2
Fluctuations in Lipid Bilayers

Lipid bilayers form another system whose dynamics are governed by dynamic modes. In experiments with suitably deuterated samples, one can distinguish between the alkyl chain and the headgroup part of the bilayers. In the gel phase, one-dimensional restricted diffusion of defects, defined as chain orientation conserving rotational isomers, permits the description of the T_1 frequency dispersion of the alkyl chains [232]. The corresponding intensity function for defect diffusion is characterized by the limits

$$
I_{dd}(\omega) =
\begin{cases}
\frac{2}{3}\left(\tau_b^{1/2}\tau_d^{1/2} - \tau_b\right) & \text{for} \quad \omega\tau_d \ll 1 \\
\left(\tau_d\tau_b^{1/2} - 2\tau_b\tau_d^{1/2}\right)\left(\tau_d^{1/2} - \tau_b^{1/2}\right)^{-2}\omega^{-1/2} & \text{for} \quad \omega\tau_b \ll 1 \ll \omega\tau_d \\
\tau_d^{1/2}\left(\tau_d^{1/2}\tau_b^{1/2} - \tau_b\right)^{-1}\omega^{-3/2} & \text{for} \quad \omega\tau_b \gg 1
\end{cases}
$$

$$(16.2)$$

where τ_d is the mean diffusion time across the bilayer (from headgroup to headgroup), and τ_b is the mean diffusion time over a length equal to the width of the defect.

In the liquid crystalline phase, collective modes ("director fluctuations") tend to dominate at low frequencies [410] (see also Sect. 17.3.2). The behavior is, however, much more complex because other mechanisms such as rotational isomerism, chain rotations, and diffusion along the curved shape of the bilayer are superimposed.

16.3
Deuteron T_1 Frequency Dispersion of Protein Solutions

The proton signals observed in tissue, diluted aqueous biopolymer solutions, and diluted aqueous lipid bilayer dispersions are normally dominated by water. In such systems, the polar surfaces of the macromolecular constituents are covered by hydration shells. This water phase certainly forms only a minor fraction of the total water. Nevertheless, a strong enhancement of spin-lattice relaxation of water occurs as first discovered in protein solutions [109].

Also at a rather early stage, it was found by NMR spectroscopy that water molecules in the hydration shell are oriented relative to the biopolymer surface [28, 345] and congeal far below the freezing point of bulk water [291]. Nonfreezing water may actually be a useful definition of hydration water, the fraction of which can be determined quantitatively in this way.

Another striking finding is that translational diffusion within the hydration shells is relatively fast. Even in frozen protein solutions not far below 0 °C, where diffusion is restricted to thin (about one to two monolayer thick) films of liquid water on the surface of the macromolecules, the diffusion coefficient is merely reduced by one order of magnitude relative to liquid bulk water [246]. On the other hand, orientation correlation times up to more than six orders of magnitude longer than

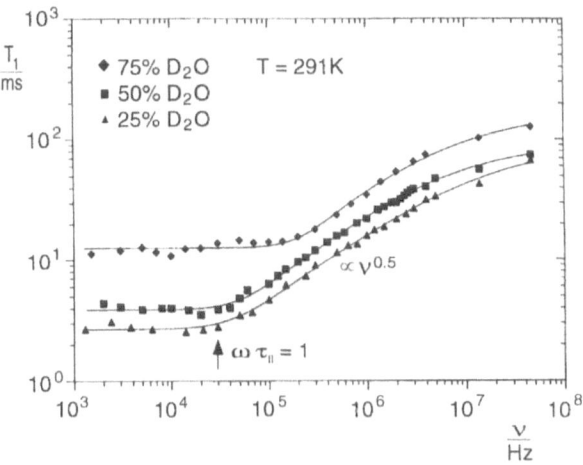

Fig. 16.3. Deuteron T_1 frequency dispersion of D_2O hydrated bovine serum albumin at 291 K. The solid lines were calculated using the RMTD formalism taking into account protein tumbling (Eq. 16.8) and restricted rotational diffusion of the water molecules (high-frequency regime). Note that the T_1 frequency dispersions at a water content of 25 wt% (powder; no free water) and 50 wt% (with free water) are qualitatively the same apart from differences in the absolute values of the relaxation times. In these cases, the low-frequency dispersion is dominated by RMTD. At 75 wt%, the inflection point is shifted to higher frequencies as a consequence of the competitive action of protein tumbling. (Data from ref. [238])

those in bulk water must be concluded from T_1 frequency dispersion experiments revealing inflection frequencies as low as 30 kHz (see Fig. 16.3).

The T_1 frequency dispersion of the water phases in protein/D_2O solutions was selectively measured using deuteron resonance [238]. In the frame of the two-site fast-exchange model (Sect. 23.1.3), the effective water spin-lattice relaxation rate is given by

$$\frac{1}{T_1} \approx \frac{1 - p_h}{T_1^f} + \frac{p_h}{T_1^h} \tag{16.3}$$

where T_1^f and T_1^h are the spin-lattice relaxation times in the "free" (i.e., bulk-like) and "hydration" water phases, respectively, and p_h is the fraction of hydration water.

The description of the T_1^h frequency dispersion of hydration water is principally based on four competitive mechanisms.

- Tumbling of the macromolecule including its hydration shell (correlation time τ_t).
- Restricted rotational diffusion of water molecules about axes perpendicular to the local surface (correlation time τ_r).
- Exchange between free and hydration water (correlation time = residence time in the hydration shell $\tau_\perp$).

- Reorientations mediated by translational displacements (RMTD) along the more or less rugged and curved surface of the macromolecule (see Sect. 14.3.1 and Fig. 14.1). The longest correlation of this mechanism is designated by $\tau_{\parallel}$.

The RMTD correlation function is (see Eq. 14.46)

$$G_{RMTD}(t) = \frac{1}{2(k_u - k_l)} \sqrt{\frac{\pi}{Dt}} \left[\mathrm{erf}\left(k_u\sqrt{Dt}\right) - \mathrm{erf}\left(k_l\sqrt{Dt}\right) \right] \qquad (16.4)$$

where k_u and k_l are the upper and lower cut-off wave numbers of the equipartition assumed for the surface structure. D is the translational water diffusion coefficient effective in the hydration shell. The lower cut-off wave number is connected with a cut-off correlation time

$$\tau = (Dk_l^2)^{-1} \qquad (16.5)$$

revealing itself as the "inflection point" $2\pi\nu_i\tau_{\parallel} \approx 1$ in the absence of molecular tumbling (see Figs. 14.2 and 16.3). This correlation function produces the square root frequency dependence typical for water relaxation in aqueous protein solutions at intermediate frequencies.

Restricted rotational diffusion, exchange with free water, and macromolecular tumbling can be represented by exponential correlation functions. Restricted rotational diffusion is only important for the high-frequency dispersion, whereas the exchange mechanism is normally too slow (relative to the other contributions) to limit the reorientation rate. What remains as low-frequency mechanisms are macromolecular tumbling and RMTD.

On the basis of this scheme, the deuteron T_1 frequency dispersion of D_2O in aqueous protein solutions can be described in full detail [238]. The cut-off correlation time τ (Eq. 16.5) revealing itself via the inflection point at a protein concentration prohibiting macromolecular tumbling (see below) can be combined with the water diffusion coefficients experimentally determined with the NMR field-gradient technique at the corresponding water content [246]. From the wavenumber evaluated in this way, a length scale can be estimated which equals half of the protein circumference as required by RMTD (Figs. 14.2 and 16.3). In other words, the mean diffusion time around half of the circumference of the protein molecule is calculated (on the basis of the experimental diffusion coefficient and the known short and long axes of the protein) to be in the order of 10^{-6} s in accordance with the inflection point measured in the absence of macromolecular tumbling.

The RMTD mechanism accounts very well for the conspicuously weak temperature dependence of spin-lattice relaxation. The long orientation correlation times found in protein/water systems are due to the geometry of the system rather than to high binding energies at certain sites. Transient irrotational binding of a small percentage of the hydration water at certain sites as suggested in several studies [109, 276] would require binding energies twice as high as the apparent activation energies estimated from the experimental T_1 data.

Another question is the origin of the long residence times in the hydration shells apparently needed for the observation of correlation times in the order of up to 10^{-4} s. This, of course, is no problem, in principle, if free water is absent or frozen. However, adding more and more water to a protein sample initially having saturated hydration shells but no free water does not change the qualitative behavior of relaxation (see Figs. 14.2 and 16.3) and diffusion [246]. Apart from the onset of macromolecular tumbling, there is no evidence for a principally new relaxation mechanism in the hydration water owing to the presence of free water. Note, however, that the possibility of surface diffusion is enhanced by Lévy walks as discussed in Sect. 14.3.2.

The relaxation scheme outlined above can also be applied to systems modeling hydrated globular proteins to a certain degree. Water adsorbed on agglomerates of silica fine-particles of a similar diameter shows a T_1 frequency dispersion again reflecting the surface structure (see Sect. 14.3.2).

While the surfaces of silica fine particles are expected to be smooth and chemically homogeneous, those of proteins are extremely heterogeneous with respect to polarity and hydrogen binding ability. On such surfaces water diffusion likely resembles more a hopping process among the preferential binding sites or areas. Translational displacements may then be governed by waiting time and step-length distributions modifying the Lévy walk RMTD mechanism. In any case, exchange processes take part in the sense that molecules (or hydrogens) leave and reenter the hydration shell to and from other environments such as free water or sites within the protein. The surface is probed only selectively by a set of discrete binding sites.

From the experimental point of view, the fast jump/exchange limit reveals itself by monoexponential attenuation curves in diffusion as well as in spin-lattice relaxation experiments [246]. That is, effective diffusion coefficients, D_{eff}, and effective spin-lattice relaxation times, $T_{1,eff}$, are evaluated. Assume a discrete set of microphases (or molecular environments) characterized by statistical weights p_i, diffusion coefficients D_i, and spin-lattice relaxation times $T_{1,i}$. The following fast-exchange formulae are then expected to hold true:

$$D_{eff} = \sum_i p_i D_i \qquad (16.6)$$

$$\frac{1}{T_{1,eff}} = \sum_i \frac{p_i}{T_{1,i}} \qquad (16.7)$$

where $\sum_i p_i = 1$.

The quantity D_{eff} tends to be governed by environments with high molecular mobility, whereas $1/T_{1,eff}$ is dominated by the slowly reorienting molecule fraction, residing relatively long on sites with certain orientations. Thus, high effective diffusion coefficients, long orientation correlation times, and low thermal activation energies do not contradict each other a priori.

16.4
Critical Water Contents

In the low-frequency limit, the T_1 frequency dispersion of water in protein solutions terminates at the "inflection frequency" ν_i in a plateau (see Figs. 14.2, 16.2, and 16.3).

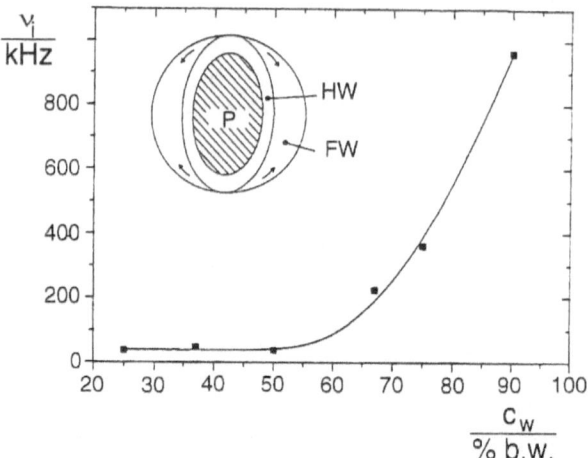

Fig. 16.4. Inflection frequency of the deuteron T_1 dispersion of bovine serum albumin solutions in D_2O at 291 K vs water content. This dependence can be described by the "free-water volume" model of macromolecular tumbling illustrated in the upper left-hand corner of the diagram. A protein molecule (P) including hydration water (HW) can only tumble if sufficient free water (FW) is available. The minimum free-water volume corresponds to the circumscribing sphere. Below a critical water content $c_0 \approx 65$ wt%, macromolecular tumbling is slower than the RMTD process of water on the surface so that the inflection frequency is governed by RMTD. Above c_0, tumbling becomes faster than this limit. The inflection frequency is then determined by the tumbling rate of the macromolecules. (Reproduced by permission from ref. [239])

As an indication of the competitive nature of RMTD and macromolecular tumbling, this quantity depends greatly on the water content c_w (Fig. 16.4.).

Tumbling can only take place if the water content exceeds the saturation concentration c_s defined by the saturation of the hydration shells. Water added in excess to that saturating the hydration shells is identified with "free water." If this is present, there is a certain probability, that a macromolecule is surrounded by enough free water so that it is enabled to perform rotational jumps. The "free-water volume" formalism [238] leads to a macromolecular tumbling correlation time

$$\tau_t = \tau_t^0 \exp\left\{ \gamma^* \left(r - 1\right) \frac{1 - c_w + c_s}{c_w - c_s} \right\} \qquad (c_w > c_s) \qquad (16.8)$$

where $\tau_t^0 = \eta(v_p + v_s)/(k_B T)$ is the Stokes/Einstein expression for tumbling of a particle of volume $v_p + v_s$ (bare protein plus the saturated hydration shell) in a solvent with viscosity η; $0.5 < \gamma^* < 1$ is a numerical constant, and r is the ratio of the volumes of the circumscribing sphere and the hydrated protein molecule itself approximated by an ellipsoid (see Fig. 16.4.). The critical water content, c_0, at which macromolecular tumbling becomes competitive to the correlation time τ corresponding to the lower cut-off wave number k_l of the RMTD process, is then defined by the condition $\tau = \tau_t$. Evaluating this condition leads to

$$c_0 = c_s + \frac{\tilde{\gamma}}{1 + \tilde{\gamma}} \qquad (16.9)$$

where $\bar{\gamma} = (r - 1)\gamma^* / \ln(\tau / \tau_t^0)$. For aqueous bovine serum albumin solutions, one finds $c_s \approx 30$ wt% and $c_0 \approx 65$ wt%, i.e., $\bar{\gamma} \approx 0.5$.

16.5
Proton Relaxation in Tissue

The T_1 frequency dispersion of a number of different tissues reveals significant differences [148, 195, 238, 275]. One is therefore interested in principal schemes enabling one to discuss such findings.

Although water is normally the most abundant compound in tissue, and therefore provides the prevailing contribution to hydrogen signals recorded in relaxation experiments, the behavior observed in tissue is determined by the interaction and cross relaxation with the macromolecular constituents. The description must therefore refer to fluctuations in the water phases as well as in the macromolecules. Assuming fast "exchange" between all components, it is possible to reproduce the experimental T_1 frequency dispersion of tissue by considering the combined action of RMTD in the hydration shells, tumbling and backbone fluctuations of the proteins, restricted rotational diffusion of hydration water molecules, and (with regard to the quadrupole dip frequency region) $^{14}N^1H$ cross relaxation. Figure 16.5 shows two typical data sets [238].

Contrasts in ordinary magnetic resonance imaging are largely dominated by relaxation (see Sect. 26.3). The low-frequency behavior of the T_1 dispersion corresponds to that of the transverse relaxation time which often acts as the dominating

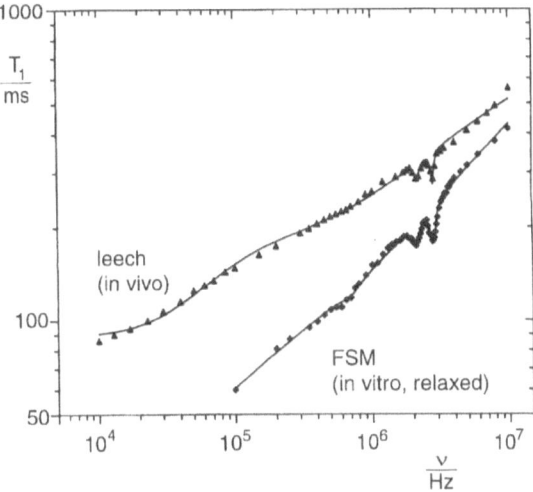

Fig. 16.5. Proton T_1 frequency dispersion of a leech (in vivo) and freshly excised frog sartorius muscle (FSM) in the relaxed state in vitro. The solid lines were calculated on the basis of macromolecular tumbling, RMTD, restricted rotational diffusion of hydration water, protein backbone fluctuations, and terms representing the $^{14}N^1H$ cross-relaxation quadrupole dips. (Reprinted by permission from ref. [238])

contrast parameter. The same relaxation scheme can indeed be used for modelling the T_2 behaviour [237]. Relaxation schemes for tissue may also be helpful for the design of magnetic resonance imaging pulse sequences revealing specific tissue properties. An example is given in Chap. 27 where a magnetic resonance imaging method is delineated which renders $T_{1\rho}$ frequency dispersion related image contrasts.

CHAPTER 17

The Dipolar-Correlation Effect

The dipolar-correlation effect refers to the unaveraged part of dipolar coupling. It is a particularly favorable way of exploiting residual dipolar interactions for studies of molecular dynamics on a time scale typically beginning with 10^{-3} s.

The stimulated echo following a sequence of three RF pulses with flip angles of preferably 90° was originally introduced assuming isolated spins in an inhomogeneous magnetic field (see Sect. 2.2). This situation applies to low-viscous liquids with equivalent spins, for instance, where dipolar interactions are motionally averaged and indirect spin-spin couplings are absent.

By contrast, we are now dealing with situations where motional averaging is incomplete (compare Sect. 9.3.3). This in particular includes the case where the motional narrowing condition, $(\Delta\omega)_{rl}\tau_c \ll 1$, which relates the correlation time of the fluctuation, τ_c, and the rigid-lattice linewidth, $(\Delta\omega)_{rl}$, is fulfilled only for one motional component, whereas it is violated otherwise. Typical examples are liquid crystals or polymers, where motional averaging tends to be restricted to rotational motions about the molecular axis.

In the following, we reconsider the pulse sequence already known from Sect. 2.2.

$$\overbrace{(\pi/2)_x \ldots \tau_1 \ldots (\pi/2)_{-x}}^{\text{1-st} \qquad \text{2-nd}} \ldots \tau_1 \ldots (\text{pr. echo}) \ldots (\tau_2 - \tau_1) \ldots$$

$$\overbrace{(\pi/2)_y \ldots \tau_1 \ldots (\text{stim. echo})}^{\text{3-rd interval}} \qquad (17.1)$$

already known from Sect. 2.2. The only echoes of interest in this context are the **primary echo** and the **stimulated echo** modified by the presence of residual and possibly fluctuating dipolar couplings. All other Hahn-type echoes are not examined. The particular choice of the pulse phases and of the flip angles of 90° prevents the occurrence of a dipolar-order transfer echo (Sect. 4.2) which would otherwise be superimposed on the residual-coupling modified stimulated echo and, hence, complicate the treatment.

The attenuation factors of the modified primary and stimulated-echo amplitudes are designated as $A_{pr}(2\tau_1)$ and $A_{st}(2\tau_1 + \tau_2)$, respectively. Under complete motional-averaging conditions the irreversible echo attenuation factors of homogeneous isotropic systems of equivalent spins would be attenuated solely by independent factors due to spin-lattice relaxation, A_{r1}, transverse relaxation, A_{r2}, and translational self-diffusion, A_{diff}. That is,

$$A_{pr}(2\tau_1) = A_{diff}(2\tau_1) \exp\{-2\tau_1/T_2\}$$
$$\equiv A_{diff}(2\tau_1) A_{r2}(2\tau_1) \tag{17.2}$$
$$A_{st}(2\tau_1 + \tau_2) = A_{diff}(2\tau_1 + \tau_2) A_{r2}(2\tau_1) \exp\{-\tau_2/T_1\}$$
$$\equiv A_{diff}(2\tau_1 + \tau_2) A_{r2}(2\tau_1) A_{r1}(\tau_2) \tag{17.3}$$

The formation of distinct echoes requires magnetic field inhomogeneities as already delineated in Chap. 2. These, however, can be assumed to be small enough to neglect echo attenuation by translational self-diffusion in the systems of interest here. Therefore we may approximate $A_{diff}(2\tau_1) \approx A_{diff}(2\tau_1 + \tau_2) \approx 1$.

Full or partial lack of motional averaging gives rise to further factors, $A_{dc}(2\tau_1)$ and $A_{dc}(2\tau_1 + \tau_2)$, independently affecting the primary and stimulated-echo amplitudes, respectively:

$$A_{pr}(2\tau_1) = A_{r2}(2\tau_1) A_{dc}(2\tau_1) \tag{17.4}$$
$$A_{st}(2\tau_1 + \tau_2) = A_{r2}(2\tau_1) A_{r1}(\tau_2) A_{dc}(2\tau_1 + \tau_2) \tag{17.5}$$

These factors are due to the **dipolar-correlation effect** [251], arising as a consequence of unaveraged dipolar interactions. The factor $A_{r2}(2\tau_1)$ now represents transverse relaxation only as concerns the motionally averaged component of the dipolar interaction, whereas the "dipolar-correlation factors" $A_{dc}(2\tau_1)$ and $A_{dc}(2\tau_1 + 2\tau_2)$ express components which are not subject to motional averaging ("residual coupling").

In Sect. 4.2 we have already demonstrated that the formation of echoes is strongly affected by dipolar interactions. That treatment referred to rigid lattices without molecular motion. We now turn to the case including molecular motions on the timescale of the pulse intervals in such a way that the secular part of dipolar interaction is largely but not completely averaged out. Thus, the nature of the echoes is partly "Hahn" (Chap. 2) and partly "dipolar" (Sect. 4.2).

17.1
Outline of Attenuation Mechanisms and Time Scales

The dipolar-coupling Hamiltonian is given in Table 46.5 on page 426. The quantity crucial for further treatment is the dipolar-coupling constant (in angular-frequency units) of two equivalent spins 1/2,

$$\Omega_d = \frac{3\mu_0 \gamma_n^2 \hbar (1 - 3\cos^2\vartheta)}{8\pi r^3} \tag{17.6}$$

where r and ϑ are polar coordinates of the internuclear vector. The fluctuations of the dipolar interaction in systems such as liquid polymers or liquid crystals can be assumed to consist of (at least) two contributions with rates above and below the motional averaging criterion. We will discuss these contributions separately.

17.1.1
The Motional-Averaging Contribution to Echo Attenuation

The secular dipolar interactions are assumed to be partially averaged by molecular reorientations that are fast but restricted with respect to the solid-angle range of the internuclear vector. The correlation time of this contribution, τ_s, is assumed to obey the motional-averaging condition

$$(\Delta\omega)_{rl}\tau_s \ll 1 \tag{17.7}$$

Quantitatively, this condition is typically $\tau_s \ll 10^{-5}$ s for protons, whereas the experimental time scale defined by the spacings in the pulse sequence 17.1 is much longer:

$$\tau_s \ll \tau_1, \tau_2 \tag{17.8}$$

The residual dipolar coupling constant is defined as the average $\overline{\Omega}_d$. For instance, the addition theorem of spherical harmonics gives, for fast rotation about a fixed axis,

$$\overline{\Omega}_d = \frac{3\mu_0\gamma_n^2\hbar}{16\pi r^3}(1 - 3\cos^2\vartheta')(3\cos^2\beta - 1) \tag{17.9}$$

where ϑ' is the polar angle of the rotation axis, and β is the angle spanned by the internuclear vector and the rotation axis.

Due to this fast but restricted contribution the transverse-magnetization is subject to an attenuation factor which, according to the Anderson/Weiss theory (Sect. 13.2.3), suggests itself as

$$A_s(2\tau_1) = \exp\left\{-(\langle\Omega_d^2\rangle - \langle\overline{\Omega}_d^2\rangle)\tau_s 2\tau_1\right\} \exp\left\{-\frac{1}{2}\langle\overline{\Omega}_d^2\rangle(2\tau_1)^2\right\} \tag{17.10}$$

The first exponential factor on the right-hand side represents the attenuation by the fast reorientation component fulfilling the motional averaging condition. This factor is already implied in A_{r2} according to the ordinary definition of transverse relaxation (see the discussion in footnote 2 on page 118). The second term, which depends on the residual dipolar coupling constant $\overline{\Omega}_d$, is the basis for the dipolar-correlation effect.

17.1.2
The Residual-Coupling Contribution to Echo Modulation

The Gaussian factor in Eq. 17.10 reflects the fraction of secular dipolar coupling not underlying motional averaging in the sense of Eq. 17.7. On the other hand, this residual secular dipolar coupling may be subject to slow reorientations on a timescale characterized by a typical time constant $\tau_l \gg 10^{-5}$ s:

$$(\Delta\omega)_{rl}\tau_l \gg 1 \tag{17.11}$$

Monitoring such fluctuations via the pulse sequence at Eq. 17.1 requires that the experimental time scale defined by the pulse spacings is of the order

$$\tau_1, \tau_2 \approx \tau_l \tag{17.12}$$

where the upper limit is a matter of spin-lattice relaxation.

This is the proper situation relevant for the dipolar-correlation effect. The spin coherences can only be refocused in the form of echoes as far as the residual secular dipolar couplings are correlated in the whole echo formation process. Thus, the dipolar-correlation function in the limit beyond motional averaging is the quantity to be probed by detecting modulations of the primary and stimulated echo amplitudes by the dipolar-correlation factors $A_{dc}(2\tau_1)$ and $A_{dc}(2\tau_1 + \tau_2)$. The dipolar coupling constant effective for the dipolar-correlation effect is $\overline{\Omega}_d$ (Eq. 17.9) (the bar will be omitted in the following for simplicity).

17.2
Density-Operator Formalism for Equivalent Two-Spin 1/2 Systems

The subsequent density-operator treatment refers to a representative pair of two equivalent spins 1/2 with the vector operators I_k and I_l. The local rotating-frame Hamiltonian responsible for the evolution of spin coherences in the course of the pulse sequence 17.1 is composed of an RF term, $\mathcal{H}_{rf}$, a second term representing the local field gradient offset, $\mathcal{H}_g$, and, in particular, the incompletely averaged secular part of dipolar coupling, $\mathcal{H}_d^{(0)}$,

$$\mathcal{H}' = \mathcal{H}_{rf} + \mathcal{H}_g + \mathcal{H}_d^{(0)} \tag{17.13}$$

where

$$\mathcal{H}_g = -\hbar\gamma_n\,(G \cdot r)\,(I_{kz} + I_{lz}) = -\hbar\Omega_g(r)(I_{kz} + I_{lz}) \tag{17.14}$$

and

$$\mathcal{H}_d^{(0)} = \hbar\Omega_d\left(I_{kz}I_{lz} - \frac{1}{3}I_k \cdot I_l\right) \tag{17.15}$$

The secular dipolar Hamiltonian effective for the evolution of spin coherences is given by (see Sect. 51.4 and Eq. 51.24)

$$\mathcal{H}_d^{(ev)} = \hbar\Omega_d I_{kz}I_{lz} \tag{17.16}$$

The pulse sequence is to begin in thermal equilibrium. The corresponding (reduced) density operator is

$$\sigma(0-) = I_{kz} + I_{lz} \tag{17.17}$$

All RF pulses are assumed to be "hard"; i.e., they excite the spin systems irrespective of local field-gradient offsets or of dipolar interaction. The first pulse, $(\pi/2)_x$, produces

$$\sigma(0+) = I_{ky} + I_{ly} \tag{17.18}$$

During the first pulse interval, the corresponding coherences evolve under the influence of the field-gradient offset and the dipolar interaction. The resulting density operator is

$$\sigma(\tau_1-) = \left[(I_{ky} + I_{ly})\cos\varphi_{d1} - 2(I_{kx}I_{lz} + I_{lx}I_{kz})\sin\varphi_{d1}\right]\cos\varphi_{g1}$$
$$+ \left[(I_{kx} + I_{lx})\cos\varphi_{d1} + 2(I_{ky}I_{lz} + I_{ly}I_{kz})\sin\varphi_{d1}\right]\sin\varphi_{g1} \tag{17.19}$$

where

$$\varphi_{g1} = \int_0^{\tau_1} \Omega_g(t') \, dt'$$

$$\varphi_{d1} = \frac{1}{2} \int_0^{\tau_1} \Omega_d(t') \, dt'$$

The second RF pulse, $(\pi/2)_{-x}$, converts this into

$$\sigma(\tau_1+) = \left[(I_{kz} + I_{lz}) \cos \varphi_{d1} + 2(I_{kx}I_{ly} + I_{lx}I_{ky}) \sin \varphi_{d1} \right] \cos \varphi_{g1}$$
$$+ \left[(I_{kx} + I_{lx}) \cos \varphi_{d1} - 2(I_{kz}I_{ly} + I_{lz}I_{ky}) \sin \varphi_{d1} \right] \sin \varphi_{g1} \quad (17.20)$$

The term $2(I_{kx}I_{ly} + I_{lx}I_{ky}) \sin \varphi_{d1}$ represents zero- and double-quantum coherences. It can be ignored in further treatment, because coherences tend to relax fast on the timescale of the interval between the second and third RF pulses in the presence of the residual dipolar couplings assumed here. Anyway, double-quantum coherences would not contribute to the echoes of interest (see Sect. 4.2).

Among the other terms, two groups can be distinguished:

$$\sigma(\tau_1+) = \sigma_{st}(\tau_1+) + \sigma_p(\tau_1+) \quad (17.21)$$

The first contribution,

$$\sigma_{st}(\tau_1+) = (I_{kz} + I_{lz}) \cos \varphi_{d1} \cos \varphi_{g1} \quad (17.22)$$

eventually leads to the modified stimulated echo. Further evolution of the second contribution,

$$\sigma_p(\tau_1+) = \left[(I_{kx} + I_{lx}) \cos \varphi_{d1} - 2(I_{kz}I_{ly} + I_{lz}I_{ky}) \sin \varphi_{d1} \right] \sin \varphi_{g1} \quad (17.23)$$

results in the modified primary echo which reaches its maximum at time $t = 2\tau_1$.

17.2.1
The Modified Primary Echo

The reduced density operator representing the maximum of the modified primary echo is

$$\sigma_p(2\tau_1) = \frac{1}{2} \left\{ (I_{kx} + I_{lx}) \left[\sin(\varphi_{g2} + \varphi_{g1}) - \sin(\varphi_{g2} - \varphi_{g1}) \right] \right.$$
$$\left. + (I_{ky} + I_{ly}) \left[\cos(\varphi_{g2} + \varphi_{g1}) - \cos(\varphi_{g2} - \varphi_{g1}) \right] \right\} \cos(\varphi_{d2} - \varphi_{d1}) \quad (17.24)$$

where

$$\varphi_{g2} = \int_{\tau_1}^{2\tau_1} \Omega_g(t') \, dt', \qquad \varphi_{d2} = \frac{1}{2} \int_{\tau_1}^{2\tau_1} \Omega_d(t') \, dt'$$

In this expression, we have omitted all antiphase-coherence terms because they generate no signals contributing to the echo maximum. Moreover, under low-resolution conditions, i.e., when the field-gradient offsets exceed dipolar splitting by far, the signal contributions of this sort cancel anyway.

The (complex) amplitude of the modified primary echo is

$$A_{pr} = \big\langle \, \mathrm{Tr}\{\, [\, (I_{kx} + I_{lx}) + i(I_{ky} + I_{ly}) \,] \, \sigma_{pr}(2\tau_1) \} \big\rangle \, A_{2r}(2\tau_1) \tag{17.25}$$

where the brackets indicate the ensemble average with respect to the phase shifts $\varphi_{g1}, \varphi_{g2}, \varphi_{d1}$, and φ_{d2}. The factor $A_{2r}(2\tau_1)$ accounts for transverse relaxation in the sense discussed before (Sect. 17.1). That is, it represents the dipolar-interaction component which is subject to motional averaging. The gradient- and dipolar-interaction-induced phase shifts are independent of each other, so that

$$\begin{aligned} A_{pr} \, = \, \frac{1}{2} \big[\, & \langle \sin(\varphi_{g2} + \varphi_{g1}) \rangle - \langle \sin(\varphi_{g2} - \varphi_{g1}) \rangle \\ & + i \langle \cos(\varphi_{g2} + \varphi_{g1}) \rangle - i \langle \cos(\varphi_{g2} - \varphi_{g1}) \rangle \, \big] \\ & \langle \cos(\varphi_{d2} - \varphi_{d1}) \rangle A_{2r}(2\tau_1) \end{aligned} \tag{17.26}$$

Let us now assume that the gradient-induced phase shifts in the pulse intervals are distributed over a broad range. The phase-shift sums in the arguments of the trigonometric functions are then widely distributed too. The superposition of these trigonometric functions is therefore destructive, so that $\langle \sin(\varphi_{g2} + \varphi_{g1}) \rangle \approx \langle \cos(\varphi_{g2} + \varphi_{g1}) \rangle \approx 0$.

On the other hand, $\langle \cos(\varphi_{g2} - \varphi_{g1}) \rangle$ remains finite because the gradient-induced phase shifts tend to cancel each other provided that translational diffusion is negligible. Analogously, the dipolar phase shifts in the terms $\langle \cos(\varphi_{d2} - \varphi_{d1}) \rangle$ are correlated if reorientations by rotational diffusion are restricted in the experimental time scale. This is the basis of the modified primary echo.

Note that this echo implies the refocusing of a dipolar contribution although the phases of the first two RF pulses are not in quadrature so that a solid echo per se should not appear (Sect. 4.1). The reason is that the strong influence of the field gradients spoils any sensitivity to the relative phase of the second RF pulse. On the other hand, in case of perfect homogeneity of the external magnetic field, the gradient-induced phase shifts would be zero, so that the density operator at Eq. 17.24 vanishes identically as expected from Eq. 4.11.

The distribution function of the net phase shifts can safely be assumed to have an even parity, so that $\langle \sin (\varphi_{d2} - \varphi_{d1}) \rangle = 0$. The cosine term can then be replaced by an exponential function leading to the echo amplitude

$$A_{pr} \approx -\frac{i}{2} A_{dc}(2\tau_1) A_{r2}(2\tau_1) \tag{17.27}$$

where the dipolar-coupling factor is

$$A_{dc}(2\tau_1) = \big\langle \exp\{ -i \left(\varphi_{d2} - \varphi_{d1} \right) \} \big\rangle \tag{17.28}$$

In case of static spin systems, i.e., $\varphi_{d1} = \varphi_{d2}$, we have $A_{pr} = -(i/2)A_{r2}(2\tau_1)$ as it must be with two-spin-1/2 systems.

17.2.2
The Modified Stimulated Echo

The reduced density operator term, to which the modified stimulated echo can be traced back in the second pulse interval, depends only on the z components of the spin operators, so that no evolution takes place. That is,

$$\sigma_{st}(\tau_1 + \tau_2-) = (I_{kz} + I_{lz}) \cos\varphi_{d1} \cos\varphi_{g1} \qquad (17.29)$$

The third RF pulse, $(\pi/2)_y$, arbitrarily assumed in phase direction y, converts this into

$$\sigma_{st}(\tau_1 + \tau_2+) = -(I_{kx} + I_{lx}) \cos\varphi_{d1} \cos\varphi_{g1} \qquad (17.30)$$

The maximum of the modified stimulated echo is reached at time $t = 2\tau_1 + \tau_2$. Omitting all antiphase-coherence terms for the same reason as above, the reduced density operator is found to be

$$\sigma_{st}(2\tau_1 + \tau_2) = -\frac{1}{4}(I_{kx} + I_{lx})[\cos(\varphi_{d3} - \varphi_{d1}) + \cos(\varphi_{d3} + \varphi_{d1})]$$
$$[\cos(\varphi_{g3} - \varphi_{g1}) + \cos(\varphi_{g3} + \varphi_{g1})]$$
$$-\frac{1}{4}(I_{ky} + I_{ly})[\cos(\varphi_{d3} - \varphi_{d1}) + \cos(\varphi_{d3} + \varphi_{d1})]$$
$$[\sin(\varphi_{g3} - \varphi_{g1}) + \sin(\varphi_{g3} + \varphi_{g1})] \qquad (17.31)$$

where

$$\varphi_{g3} = \int_{\tau_1+\tau_2}^{2\tau_1+\tau_2} \Omega_g(t')\,dt'$$

$$\varphi_{d3} = \frac{1}{2}\int_{\tau_1+\tau_2}^{2\tau_1+\tau_2} \Omega_d(t')\,dt'$$

The (complex) amplitude of the modified stimulated echo is then

$$A_{st} = \langle \text{Tr}\{[(I_{kx} + I_{lx}) + i(I_{ky} + I_{ly})]\sigma_{st}(2\tau_1 + \tau_2)\}\rangle\, A_{r2}(2\tau_1)A_{r1}(\tau_2) \qquad (17.32)$$

The brackets indicate the ensemble average with respect to the phase shifts φ_{g1}, φ_{g3}, φ_{d1}, and φ_{d3}. The additional attenuation factors $A_{r2}(2\tau_1)$ and $A_{r1}(\tau_2)$ account for transverse relaxation and spin-lattice relaxation, respectively. As defined before, the factor $A_{r2}(2\tau)$ represents the dipolar-interaction component subject to motional averaging. Carrying out the trace leads to

$$A_{st} = -\frac{1}{4}[\langle\cos(\varphi_{d3} - \varphi_{d1})\rangle + \langle\cos(\varphi_{d3} + \varphi_{d1})\rangle]$$
$$[\langle\cos(\varphi_{g3} - \varphi_{g1})\rangle + \langle\cos(\varphi_{g3} + \varphi_{g1})\rangle]$$
$$-\frac{i}{4}[\langle\cos(\varphi_{d3} - \varphi_{d1})\rangle + \langle\cos(\varphi_{d3} + \varphi_{d1})\rangle]$$
$$[\langle\sin(\varphi_{g3} - \varphi_{g1})\rangle + \langle\sin(\varphi_{g3} + \varphi_{g1})\rangle]$$
$$A_{r2}(2\tau_1)A_{r1}(\tau_2) \qquad (17.33)$$

As with the modified primary echo, the gradient-induced phase shifts can again be assumed to be distributed over a wide range. That is, the gradient is assumed to be strong enough so that all coherences are spoiled within the intervals. The superposition of the trigonometric functions of the phase-shift sums then tends to be destructive, i.e., $\langle\sin(\varphi_{g2}+\varphi_{g1})\rangle \approx \langle\cos(\varphi_{g2}+\varphi_{g1})\rangle \approx 0$. Taking this into account leads to the amplitude of the modified stimulated echo

$$A_{st} \approx -\frac{1}{2}A_{dc}(2\tau_1 + \tau_2)A_{r2}(2\tau_1)A_{r1}(\tau_2) \tag{17.34}$$

where the attenuation factor for dipolar coupling is

$$A_{dc}(2\tau_1 + \tau_2) = \frac{1}{2}\left(\langle\exp\left\{-i(\varphi_{d3} - \varphi_{d1})\right\}\rangle + \langle\exp\left\{-i(\varphi_{d3} + \varphi_{d1})\right\}\rangle\right) \tag{17.35}$$

In this expression, the cosine term has again been replaced by an exponential function. Note that the dipolar phase shifts vanish in the case of complete motional averaging on the time scales of the first and last τ_1 intervals, so that $A_{dc}(2\tau_1+\tau_2) = 1$ as expected for ordinary stimulated-echo experiments.

17.3
The Dipolar-Correlation Quotient

In the quotient of the amplitudes of the modified stimulated and the modified primary echo the attenuation factors due to transverse relaxation (motional-averaging component) cancel. The interval τ_2 can be kept constant while varying τ_1. Therefore the factor $A_{r1}(\tau_2)$ is also constant and can be normalized in the limit $\tau_1 \to 0$.

The quotient of the echo amplitudes normalized for $\tau_1 \to 0$ (indicated by the subscript n) is therefore dominated by the dipolar-correlation factors (residual-coupling component). The dipolar-correlation quotient is

$$Q_{dc} \equiv \left.\frac{A_{st}}{A_{pr}}\right|_n = \frac{A_{dc}(2\tau_1 + \tau_2)}{A_{dc}(2\tau_1)}$$

$$= \frac{1}{2}\left(\frac{\langle\exp\{-i(\varphi_{d3} - \varphi_{d1})\}\rangle + \langle\exp\{-i(\varphi_{d3} + \varphi_{d1})\}\rangle}{\langle\exp\{-i(\varphi_{d2} - \varphi_{d1})\}\rangle}\right) \tag{17.36}$$

The polar coordinates $\vartheta_{kl} = \vartheta_{kl}(t)$ and $r_{kl} = r_{kl}(t)$ of the internuclear vector fluctuate according to the molecular dynamics. As a consequence the dipolar coupling constant fluctuates as well. The following consideration includes **ordered** materials such as liquid crystals. In this case the ensemble average of the residual dipolar coupling constant, $\langle\Omega_d\rangle$, does not vanish. the coupling constant may then be analyzed as

$$\Omega_d = \langle\Omega_d\rangle + \delta\Omega_d \tag{17.37}$$

where $\delta\Omega_d$ is the part of the coupling constant fluctuating on the time scale of the experiment. The phase shifts in the pulse intervals can then be written:

$$\varphi_{d1} = \frac{1}{2}\left(\langle\Omega_d\rangle\tau_1 + \int_0^{\tau_1} \delta\Omega_d(t')\,dt'\right) = \langle\varphi_{d1}\rangle + \delta\varphi_{d1}$$

$$\varphi_{d2} = \frac{1}{2}\left(\langle \Omega_d \rangle \tau_1 + \int_{\tau_1}^{2\tau_1} \delta\Omega_d(t')\, dt' \right) = \langle \varphi_{d1} \rangle + \delta\varphi_{d2} \tag{17.38}$$

$$\varphi_{d3} = \frac{1}{2}\left(\langle \Omega_d \rangle \tau_1 + \int_{\tau_1+\tau_2}^{2\tau_1+\tau_2} \delta\Omega_d(t')\, dt' \right) = \langle \varphi_{d1} \rangle + \delta\varphi_{d3}$$

The distribution of the phase shifts $\delta\varphi_{d1}, \delta\varphi_{d2}, \delta\varphi_{d3}$ is an even function so that the averages of sine functions of the phase shifts vanish. The average cosine may therefore be replaced by the average of an exponential function according to Euler's formula. That is,

$$\langle \cos(\delta\varphi_{d2} - \delta\varphi_{d1}) \rangle = \left\langle e^{-i(\delta\varphi_{d2}-\delta\varphi_{d1})} \right\rangle$$

$$\langle \cos(\delta\varphi_{d3} - \delta\varphi_{d1}) \rangle = \left\langle e^{-i(\delta\varphi_{d3}-\delta\varphi_{d1})} \right\rangle \tag{17.39}$$

$$\langle \cos(\varphi_{d3} + \varphi_{d1}) \rangle = \cos\left(2\langle \varphi_{d1} \rangle\right) \left\langle e^{-i(\delta\varphi_{d3}+\delta\varphi_{d1})} \right\rangle$$

Thus

$$Q_{dc} = \frac{\left\langle e^{-i(\delta\varphi_{d3}-\delta\varphi_{d1})} \right\rangle + \cos\left(2\langle \varphi_{d1}\rangle\right) \left\langle e^{-i(\delta\varphi_{d3}+\delta\varphi_{d1})} \right\rangle}{2\left\langle e^{-i(\delta\varphi_{d2}-\delta\varphi_{d1})} \right\rangle} \tag{17.40}$$

Assuming a Gaussian distribution of the phase shifts in accordance with the central-limit theorem, we find (compare Sect. 13.2.1)

$$Q_{dc} = \frac{\exp\left\{-\frac{1}{2}\left\langle (\delta\varphi_{d3} - \delta\varphi_{d1})^2 \right\rangle\right\} + \cos\left(2\langle \varphi_{d1}\rangle\right) \exp\left\{-\frac{1}{2}\left\langle (\delta\varphi_{d3} + \delta\varphi_{d1})^2 \right\rangle\right\}}{2\exp\left\{-\frac{1}{2}\left\langle (\delta\varphi_{d2} - \delta\varphi_{d1})^2 \right\rangle\right\}}$$

$$= \frac{1}{2} e^{\langle \delta\varphi_{d1}\delta\varphi_{d3}\rangle - \langle \delta\varphi_{d1}\delta\varphi_{d2}\rangle} \left(1 + \cos\left(2\langle \varphi_{d1}\rangle\right) e^{-2\langle \delta\varphi_{d1}\delta\varphi_{d3}\rangle} \right) \tag{17.41}$$

or

$$\boxed{Q_{dc} = \frac{1}{2} e^{\langle \delta\varphi_{d1}\delta\varphi_{d3}\rangle - \langle \delta\varphi_{d1}\delta\varphi_{d2}\rangle} \left(1 + \cos\left(\langle \Omega_d\rangle\tau_1\right) e^{-2\langle \delta\varphi_{d1}\delta\varphi_{d3}\rangle} \right)} \tag{17.42}$$

The correlation function of the fluctuating part of the dipolar coupling constant is

$$G_d(\tau) \equiv \langle \delta\Omega_d(t')\delta\Omega_d(t'') \rangle = \langle \delta\Omega_d^2 \rangle\, G(\tau) \tag{17.43}$$

where we have assumed stationarity, i.e., the correlation function depends solely on $\tau = t'' - t'$. The reduced correlation function, $G(\tau)$, is defined by the condition $G(0) = 1$, and obeys $G(\tau) = G(-\tau)$.

The phase shift correlation functions in Eq. 17.42 can be derived in a way analogous to the treatment delineated in Sect. 13.2.2 (see also Fig. 13.1). Let us first consider $\langle \delta\varphi_{d1}\delta\varphi_{d3}\rangle$:

$$\langle \delta\varphi_{d1}\delta\varphi_{d3}\rangle = \frac{1}{4}\left\langle \int_0^{\tau_1} \delta\Omega_d(t')\, dt' \int_{\tau_1+\tau_2}^{2\tau_1+\tau_2} \delta\Omega_d(t'')\, dt'' \right\rangle$$

$$= \frac{1}{4}\langle\delta\Omega_d^2\rangle \int_0^{\tau_1} dt' \int_{\tau_1+\tau_2}^{2\tau_1+\tau_2} dt'' \, G(t'' - t') \tag{17.44}$$

The integration variables can be substituted by $\tau' = t'' - t'$ and $\tilde{t} = t'$. The relevant range of these variables is split into two regions, (a) $\tau_2 \le \tau' \le \tau_1 + \tau_2$, that is, $\tau_1 + \tau_2 - \tau' \le \tilde{t} \le \tau_1$, and (b) $\tau_1 + \tau_2 \le \tau' \le 2\tau_1 + \tau_2$, that is, $0 \le \tilde{t} \le 2\tau_1 + \tau_2 - \tau'$, so that Eq. 17.44 becomes

$$\langle\delta\varphi_{d1}\delta\varphi_{d3}\rangle = \frac{1}{4}\langle\delta\Omega_d^2\rangle \left[\int_{\tau_2}^{\tau_1+\tau_2} d\tau' \, G(\tau') \int_{\tau_1+\tau_2-\tau'}^{\tau_1} d\tilde{t} + \int_{\tau_1+\tau_2}^{2\tau_1+\tau_2} d\tau' \, G(\tau') \int_0^{2\tau_1+\tau_2-\tau'} d\tilde{t} \right]$$

$$= \frac{1}{4}\langle\delta\Omega_d^2\rangle \left[\int_{\tau_2}^{\tau_1+\tau_2} d\tau' \, (\tau' - \tau_2)G(\tau') + \int_{\tau_1+\tau_2}^{2\tau_1+\tau_2} d\tau' \, (2\tau_1 + \tau_2 - \tau')G(\tau') \right] \tag{17.45}$$

Substituting the integration variables in the first and second integrals by $\tau' = -\tau + \tau_1 + \tau_2$ and $\tau' = \tau + \tau_1 + \tau_2$, respectively, leads to

$$\langle\delta\varphi_{d1}\delta\varphi_{d3}\rangle = \frac{1}{4}\langle\delta\Omega_d^2\rangle \left[\int_0^{\tau_1} d\tau \, (\tau_1 - \tau)G(\tau_1 + \tau_2 - \tau) \right.$$

$$\left. + \int_0^{\tau_1} d\tau \, (\tau_1 - \tau)G(\tau_1 + \tau_2 + \tau) \right] \tag{17.46}$$

The phase shift correlation function $\langle\varphi_{d1}\varphi_{d2}\rangle$ is readily obtained by equating $\tau_2 = 0$:

$$\langle\delta\varphi_{d1}\delta\varphi_{d2}\rangle = \frac{1}{4}\langle\delta\Omega_d^2\rangle \left[\int_0^{\tau_1} d\tau \, (\tau_1 - \tau)G(\tau_1 + \tau) + \int_0^{\tau_1} d\tau \, (\tau_1 - \tau)G(\tau_1 - \tau) \right] \tag{17.47}$$

17.3.1
Exponential Correlation Function

The first dipolar correlation function to be considered is a simple exponential function:

$$G(\tau) = e^{-|\tau|/\tau_c} \tag{17.48}$$

The phase-shift correlation functions are then calculated as

$$\langle\delta\varphi_{d1}\delta\varphi_{d2}\rangle = \frac{1}{4}\langle\delta\Omega_d^2\rangle \, \tau_c^2 e^{-\tau_1/\tau_c} \left[e^{-\tau_1/\tau_c} + e^{\tau_1/\tau_c} - 2 \right] \tag{17.49}$$

$$\langle\delta\varphi_{d1}\delta\varphi_{d3}\rangle = \langle\delta\varphi_{d1}\delta\varphi_{d2}\rangle \, e^{-\tau_2/\tau_c} \tag{17.50}$$

so that the dipolar-correlation quotient turns out to be

$$Q_{dc} = \frac{1}{2} \exp \left\{ -\frac{1}{4} \langle \delta\Omega_d^2 \rangle \, C_1 \right\} \left(1 + \cos\left(\langle \Omega_d \rangle \tau_1 \right) \exp \left\{ -\frac{1}{2} \langle \delta\Omega_d^2 \rangle \, C_2 \right\} \right) \quad (17.51)$$

where

$$C_1 = \tau_c^2 \left(e^{-2\tau_1/\tau_c} - 2e^{-\tau_1/\tau_c} + 1 \right) \left(1 - e^{-\tau_2/\tau_c} \right) \quad (17.52)$$

$$C_2 = \tau_c^2 \left(e^{-2\tau_1/\tau_c} - 2e^{-\tau_1/\tau_c} + 1 \right) e^{-\tau_2/\tau_c} \quad (17.53)$$

17.3.2
Correlation Function for Liquid-Crystal Director Fluctuations

In a nematic crystal, for instance, the director is subject to collective director fluctuations characterized by wave numbers q [89, 112, 127, 384]. The normalized correlation function is predicted to be

$$G(\tau) = \frac{1}{q_c} \int_0^{q_c} \exp \left\{ -\left(\frac{K}{\eta} + D \right) q^2 \tau \right\} dq \quad (17.54)$$

where q_c is the upper cutoff wave number, K the elastic constant, η the viscosity, and D is the diffusivity. With the "mode correlation time"

$$\tau_q = \frac{1}{\left(\frac{K}{\eta} + D \right) q^2} \quad (17.55)$$

Eq. 17.54 becomes

$$G(\tau) = \frac{\sqrt{\tau_l}}{2} \int_{\tau_l}^{\infty} \tau_q^{-3/2} e^{-\tau/\tau_q} \, d\tau_q \quad (17.56)$$

where

$$\tau_l = \frac{1}{\left(\frac{K}{\eta} + D \right) q_c^2} \quad (17.57)$$

is the shortest mode correlation time.[1]

Replacing the correlation time τ_c in Eq. 17.51 by τ_q, and Eqs. 17.52 and 17.53 by

$$\tilde{C}_1 = \frac{\sqrt{\tau_l}}{2} \int_{\tau_l}^{\infty} \sqrt{\tau_q} \left(e^{-2\tau_1/\tau_q} - 2e^{-\tau_1/\tau_q} + 1 \right) \left(1 - e^{-\tau_2/\tau_q} \right) d\tau_q \quad (17.58)$$

$$\tilde{C}_2 = \frac{\sqrt{\tau_l}}{2} \int_{\tau_l}^{\infty} \sqrt{\tau_q} \left(e^{-2\tau_1/\tau_q} - 2e^{-\tau_1/\tau_q} + 1 \right) e^{-\tau_2/\tau_q} \, d\tau_q \quad (17.59)$$

---

[1]In this context "shortest" means "shortest beyond the motional-averaging regime" (Eq. 17.7) in contrast to the general director-fluctuation theory. Modes contributing to motional averaging are irrelevant for the dipolar-correlation effect as discussed in Sect. 17.1.

leads to the dipolar-correlation quotient [165]

$$Q_{dc} = \frac{1}{2} \exp\left\{-\frac{1}{4}\langle\delta\Omega_d^2\rangle\tilde{C}_1\right\}\left(1 + \cos\left((\langle\Omega_d\rangle\tau_1)\exp\left\{-\frac{1}{2}\langle\delta\Omega_d^2\rangle\tilde{C}_2\right\}\right)\right) \qquad (17.60)$$

17.3.2.1
Limiting Cases

The coefficients $\tilde{C}_1$ and $\tilde{C}_2$ in Eq. 17.60 may be rewritten as

$$\tilde{C}_1 = \frac{\sqrt{\tau_l}}{2}\int_{\tau_l}^{\infty}\sqrt{\tau_q}\left(1 - e^{-\tau_1/\tau_q}\right)^2\left(1 - e^{-\tau_2/\tau_q}\right)d\tau_q \qquad (17.61)$$

$$\tilde{C}_2 = \frac{\sqrt{\tau_l}}{2}\int_{\tau_l}^{\infty}\sqrt{\tau_q}\left(1 - e^{-\tau_1/\tau_q}\right)^2 e^{-\tau_2/\tau_q}d\tau_q \qquad (17.62)$$

The integration variable τ_q can be substituted by $x = \tau_2/\tau_q$ so that

$$\tilde{C}_1 = \frac{\tau_2^{3/2}\sqrt{\tau_l}}{2}F_1 \qquad (17.63)$$

$$\tilde{C}_2 = \frac{\tau_2^{3/2}\sqrt{\tau_l}}{2}F_2 \qquad (17.64)$$

where

$$F_1 = \int_0^{\tau_2/\tau_l} x^{-5/2}\left(1 - e^{-x\tau_1/\tau_2}\right)^2\left(1 - e^{-x}\right)dx \qquad (17.65)$$

$$F_2 = \int_0^{\tau_2/\tau_l} x^{-5/2}\left(1 - e^{-x\tau_1/\tau_2}\right)^2 e^{-x}dx \qquad (17.66)$$

The integrals F_1 and F_2 may be approximated for diverse limiting cases.

A typical experimental situation is $\tau_1 \ll \tau_2$. Integral F_2 already converges when x approaches a value of the order of 1. Therefore, we may assume $x\tau_1/\tau_2 \ll 1$. Let us now consider the following three cases:

Limit I: $\tau_1 \ll \tau_2 \ll \tau_l$

$$F_1 \approx \left(\frac{\tau_1}{\tau_2}\right)^2\int_0^{\tau_2/\tau_l} x^{1/2}dx = \frac{2}{3}\tau_1^2\tau_2^{-1/2}\tau_l^{-3/2} \qquad (17.67)$$

$$F_2 \approx \left(\frac{\tau_1}{\tau_2}\right)^2\int_0^{\tau_2/\tau_l} x^{-1/2}dx = 2\tau_1^2\tau_2^{-3/2}\tau_l^{-1/2} \qquad (17.68)$$

where we have used the lowest-order approximations

$$e^{-x} \approx 1$$
$$1 - e^{-x} \approx x$$
$$\left(1 - e^{-x\tau_1/\tau_2}\right)^2 \approx \left(\frac{\tau_1}{\tau_2}\right)^2 x^2 \tag{17.69}$$

In this limit the dipolar-correlation quotient becomes

$$Q_{dc} \approx \frac{1}{2} \exp\left\{-\frac{1}{4} \langle \delta\Omega_d^2 \rangle \frac{\tau_2\tau_1^2}{3\tau_l}\right\} \left(1 + \cos\left(\langle\Omega_d\rangle\tau_1\right) \exp\left\{-\frac{1}{2} \langle\delta\Omega_d^2\rangle \tau_1^2\right\}\right) \tag{17.70}$$

Limit II: $\tau_1 \ll \tau_l \ll \tau_2$

$$F_1 \approx \left(\frac{\tau_1}{\tau_2}\right)^2 \int_0^{\tau_2/\tau_l} x^{-1/2}\left(1 - e^{-x}\right)\, dx \approx \left(\frac{\tau_1}{\tau_2}\right)^2 \left(\int_0^{\tau_2/\tau_l} x^{-1/2}\, dx - \int_0^{\infty} x^{-1/2} e^{-x}\, dx\right)$$
$$= \left(\frac{\tau_1}{\tau_2}\right)^2 \left(2\sqrt{\frac{\tau_2}{\tau_l}} - \sqrt{\pi}\right) \tag{17.71}$$

$$F_2 \approx \left(\frac{\tau_1}{\tau_2}\right)^2 \int_0^{\infty} x^{-1/2} e^{-x}\, dx = \sqrt{\pi} \left(\frac{\tau_1}{\tau_2}\right)^2 \tag{17.72}$$

where we have again used the relations 17.69. The upper integration boundary has been replaced by ∞ in all cases where the integrand implies an exponentially decaying factor.

On this basis, the dipolar-correlation factor may be approximated as

$$Q_{dc} \approx \frac{1}{2} \exp\left\{-\frac{1}{4} \langle\delta\Omega_d^2\rangle \tau_1^2 \left(1 - \frac{1}{2}\sqrt{\frac{\pi\tau_l}{\tau_2}}\right)\right\}$$
$$\left(1 + \cos\left(\langle\Omega_d\rangle\tau_1\right) \exp\left\{-\frac{1}{4} \langle\delta\Omega_d^2\rangle \tau_1^2 \sqrt{\frac{\pi\tau_l}{\tau_2}}\right\}\right) \tag{17.73}$$

Limit III: $\tau_l \ll \tau_1 \ll \tau_2$

$$F_1 \approx \int_0^{\infty} x^{-5/2}\left(1 - e^{-x\tau_1/\tau_2}\right)^2 \left(1 - e^{-x}\right)\, dx$$
$$= \int_0^{\infty} x^{-5/2}\left(1 - e^{-x\tau_1/\tau_2}\right)^2 dx - \int_0^{\infty} x^{-5/2}\left(1 - e^{-x\tau_1/\tau_2}\right)^2 e^{-x}\, dx$$
$$\approx \left(\frac{\tau_1}{\tau_2}\right)^{3/2} \int_0^{\infty} x'^{-5/2}\left(1 - e^{-x'}\right)^2 dx' - \left(\frac{\tau_1}{\tau_2}\right)^2 \int_0^{\infty} x^{-1/2} e^{-x}\, dx$$
$$= \frac{8}{3}\sqrt{\pi}\left(\sqrt{2} - 1\right)\left(\frac{\tau_1}{\tau_2}\right)^{3/2} - \sqrt{\pi}\left(\frac{\tau_1}{\tau_2}\right)^2 \tag{17.74}$$

$$F_2 \approx \left(\frac{\tau_1}{\tau_2}\right)^2 \int\limits_0^\infty x^{-1/2} e^{-x} \, dx = \sqrt{\pi} \left(\frac{\tau_1}{\tau_2}\right)^2 \tag{17.75}$$

In the derivation of Eqs. 17.74 and 17.75, the upper integration boundary was set at ∞ because the integrand either implies an exponentially decaying factor or decays proportional to $x^{-5/2}$.

The dipolar-correlation quotient may thus be rewritten as

$$Q_{dc} \approx \frac{1}{2} \exp\left\{-\frac{1}{4} \langle \delta\Omega_d^2 \rangle \, \tau_1^2 \left(\sqrt{\frac{\tau_l}{\tau_1}} - \frac{1}{2}\sqrt{\frac{\pi\tau_l}{\tau_2}}\right)\right\}$$
$$\left(1 + \cos\left(\langle\Omega_d\rangle\tau_1\right) \exp\left\{-\frac{1}{4} \langle\delta\Omega_d^2\rangle \, \tau_1^2 \sqrt{\frac{\pi\tau_l}{\tau_2}}\right\}\right) \tag{17.76}$$

17.4
Applications of the Dipolar-Correlation Effect

As outlined in the introductory section, the dipolar-correlation effect is particularly suitable for investigations of systems with anisotropic molecular motions so that a residual dipolar coupling escapes from motional averaging. Typical applications may be classified into two categories depending on whether the system is macroscopically ordered or not.

In the treatment of the dipolar-correlation effect in ordered media presented above, it was assumed that the dipolar-coupling constant causes much stronger spectral effects than chemical shifts of inequivalent nuclei potentially contributing to the signal. Otherwise, chemical-shift offsets are superimposed and modulate the attenuation curves of the dipolar-correlation quotient in an analogous manner. Phenomena of this sort can be eliminated by inserting 180° pulses in the middle of each τ_1 interval. Single-spin frequency offsets are then refocused whereas the evolution due to dipolar coupling is not affected (see Sect. 4.1). The dipolar-correlation effect can thus be recorded in unconcealed form.

17.4.1
Macroscopic Order

In macroscopically ordered systems the term $\cos(\langle\Omega_d\rangle\tau_1)$ in Eq. 17.42 is expected to have a finite value according to the residual dipolar coupling. This is the typical situation in liquid crystalline phases. The consequence is that the dipolar-correlation quotient becomes an undulated function of the pulse spacing τ_1. In simple cases, i.e., if dipolar coupling is determined by a distinct dipole-dipole distance r, and if the director is aligned preferentially along the external magnetic field, the mean dipolar-coupling constant is directly related to the order parameter according to

$$\langle\Omega_d\rangle = \frac{3\mu_0 \gamma_n^2 \hbar}{8\pi r^3} \langle 1 - 3\cos^2\vartheta\rangle \tag{17.77}$$

Figure 17.1 shows typical dipolar-correlation quotient data of a liquid crystal plotted as a function of the pulse spacing τ_1 with τ_2 as a curve parameter [165].

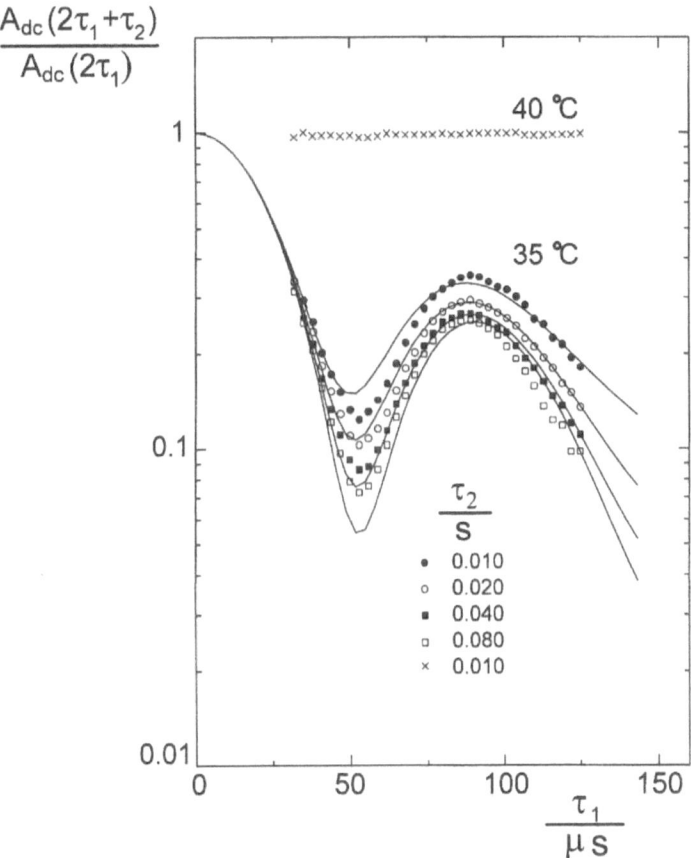

Fig. 17.1. Dipolar-correlation quotient $Q_{dc} = A_{dc}(2\tau_1 + \tau_2)/A_{dc}(2\tau_1)$ of the nematic 4'-n-pentyl-4-cyanobiphenyl (5CB) as a function of the pulse interval τ_1. The data were measured in the nematic and the isotropic phase at 35 and 40 °C, respectively. The curve parameter is τ_2. The solid lines represent a least-squares fit of Eq. 17.73. (Reprinted by permission from ref. [165])

The nematic phase exhibits a pronounced modulation pattern. This effect is entirely absent in the isotropic phase. The modulation directly visualizes the influence of the finite value of $\langle \Omega_d \rangle$.

The shortest mode correlation time accessible was fitted to be 3.9 ms. In reality much faster processes occur, of course. However, these are subject to motional averaging and do not contribute to the dipolar-correlation effect. An impression of the length of the modes that can be probed in this sort of experiment is given by the τ_2 values in Fig. 17.1 which range up to the order of 100 ms.

The time scale of this experiment is to be compared with that of field-cycling NMR relaxometry (Chap. 15) which was reported for the same liquid crystal [271, 524, 531]. At the low-frequency end of the scale, which can be probed by that method, director fluctuations also appear to be the dominating process of molecular dynamics. The dipolar-correlation effect is sensitive to much slower motions. Therefore it favorably supplements field-cycling NMR relaxometry.

The nematic order reveals itself by modulated attenuation curves of the dipolar-correlation quotient. The modulation frequency is determined by the mean dipolar-coupling constant which vanishes in isotropic media. This is not to be confused with motional averaging, because even in an amorphous solid no modulation of this sort would appear. The mean value of the dipolar-coupling constant rather has to be taken as an ensemble average in case the system is not ergodic on the time scale of the experiment.

17.4.2
Short-Range Order and Polymers

The dipolar-correlation quotient function of disordered systems such as polymer melts does not show any undulated pattern provided that chemical-shift offsets are absent. This holds true although motional averaging tends to be incomplete in polymers [251]. The reason is that subensembles with different values of $\langle \Omega_d \rangle$ are averaged in this case. The superposition of the corresponding cosine functions in Eq. 17.42 is destructive, so that $\overline{\cos\left(\langle \Omega_d \rangle \tau_1\right)} \approx 0$.

The undulations may even vanish in locally short-range ordered, but macroscopically disordered, systems. For example, this situation was found in a liquid crystal confined in porous glasses [166], where the (local) director is oriented by the surface rather than by the magnetic field. As a further consequence of the confinement, the distribution of director fluctuation modes is restricted to the pore size in contrast to the bulk. Moreover, according to the boundary conditions, the spectrum of the allowed modes is discrete rather than continuous [534].

Another example is the mesophase of poly(dialkylsiloxanes) such as poly(diethylsiloxane). Dipolar-correlated stimulated-echo studies suggested the existence of exchange fluctuations between a more ordered and a less ordered segment environment. Although macroscopic order is known to be characteristic in such mesophases, no undulations of the dipolar-correlation factor as a function of the pulse spacing τ_1 were observed. This again is a consequence of an order-parameter distribution [167].

CHAPTER 18

Survey of NMR Diffusometry

Self-diffusion or translational displacements of molecules as a consequence of Brownian motions [88] is to be distinguished from **interdiffusion** [104] of molecules of different species which are initially separated. Both phenomena can favorably be investigated by magnetic-resonance methods.

Interdiffusion refers to the intermingling of initially separated components. The evolution toward thermodynamic equilibrium can be studied by temporally resolved magnetic-resonance imaging of the spatial distributions of the molecular species involved. The experimental distinction of the components is feasible on the basis of spectral parameters such as the chemical shift or the gyromagnetic ratio of different atom species (e.g., hydrogen and fluorine). Another strategy is to label compounds by enrichment of rare isotopes that can be identified by NMR (e.g., ^{2}H, ^{13}C). Furthermore, a differentiation is possible based on concentration dependent relaxation rates. Typical applications of **diffusion-profile imaging** have been published in [96, 188, 194, 327, 363, 423]. Homo- or heteronuclear magnetic-resonance imaging methods suitable for this purpose are discussed in Part III.

The following chapters are mainly devoted to **field gradient NMR diffusometry**, i.e., to self-diffusion. The gradients either refer to the main magnetic field, B_0, or to the RF amplitude, B_1. Versions employing pulsed or stationary gradients are in use.

Diffusion experiments of this sort are usually discussed in terms of spin echoes, the amplitude of which is attenuated by translational displacements of the spin-bearing particles in the course of the pulse sequence [70, 217, 316]. That is, coherences are first dephased by field gradients and then rephased in the form of a Hahn or a gradient echo. Displacements in the presence of field gradients prevent the superposition of coherences, entirely constructive at the time when the spin echo is to appear, so that the echo amplitude is irreversibly reduced.

An echo form often employed for field-gradient diffusometry [481] is the **stimulated echo** (Sect. 2.2). In this case, the effective diffusion time is prolonged by an intermittent pulse interval. This pause of free coherence evolution is merely restricted by spin-lattice relaxation which tends to be much slower than transverse relaxation unless the "extreme narrowing limit" is fulfilled (see Eq. 13.9).

In principle, any type of spin echo is suitable for diffusometry. For example, **multiple-quantum coherence transfer echoes** of coupled spin systems (see Sects. 4.2, 5.2 and 7.2) have been suggested for this purpose [528]. Multiple-quantum coherences of order n have the appealing feature of being n times as sensitive to field gradients so that the field gradient has an n-fold efficiency. Since the effective field gradient enters quadratically into the echo-attenuation factor (Table 19.1 on page 181), multiple-quantum diffusometry is n^2 times as sensitive.

Field-gradient NMR diffusometry has another reading which is equivalent to the coherence dephasing/rephasing picture, but is very instructive with respect to methodology [248, 428]. Coherence evolution in a spatially constant field gradient results in a periodic distribution of the magnetization vectors (**"the magnetization grid"**). That is, the magnetization becomes a function of the position. In stimulated-echo experiments, for instance, the first two RF pulses produce a z-magnetization grid. In the course of two-pulse Hahn echo experiments, the magnetization grid evolves in the form of circularly polarized transverse magnetization vectors helically distributed along the gradient direction.[1]

The effect of diffusion on the grid is a levelling-off tendency as long as this inhomogeneous distribution of the magnetization vector exists. This is what one measures in indirect form when recording echo attenuation curves. The direct method is to visualize the grid temporally resolved with the aid of NMR imaging techniques.

18.1
The Diffusion Propagator

Displacements by translational diffusion are characterized by the distribution of the probability density, $p(r, t)$, that a molecule is displaced a distance r (components x, y, z) in an interval t. In other words, the probability that the particle is found in the region $x \cdots x+dx$, $y \cdots y+dy$, $z \cdots z+dz$ is $p(r, t)d^3r$. For ordinary, isotropic diffusion this "propagator"[2] obeys the diffusion equation

$$\frac{\partial p}{\partial t} = D \, \nabla^2 p \tag{18.1}$$

where

$$\nabla^2 = \frac{\partial^2}{\partial x^2} + \frac{\partial^2}{\partial y^2} + \frac{\partial^2}{\partial z^2} \tag{18.2}$$

is the Laplace operator expressed in orthogonal Cartesian coordinates.

In the case of **ordinary isotropic unrestricted diffusion**, the diffusion coefficient (or diffusivity), D, is a scalar constant. Furthermore, in an isotropic, homogeneous medium, only the magnitude of the displacement is relevant. With the initial condition $p(r, 0) = \delta(r)$ and the boundary condition $p(r, t) \rightarrow 0$ for $r \rightarrow \infty$, the solution of Eq. 18.1 is the normalized **Gaussian function**

$$p(r, t) = \frac{1}{(4\pi Dt)^{3/2}} \exp\left\{-\frac{r^2}{4Dt}\right\} \tag{18.3}$$

---

[1]The 90° - τ_1 - 180° Hahn echo two-pulse sequence may be interpreted as a 90° - τ_1 - 90° - τ_2 - 90° three-pulse sequence in the limit $\tau_2 \rightarrow 0$.

[2]Apart from propagator, the probability density for a displacement r may also be called Green's function, i.e., the response to an initial distribution given by a delta function.

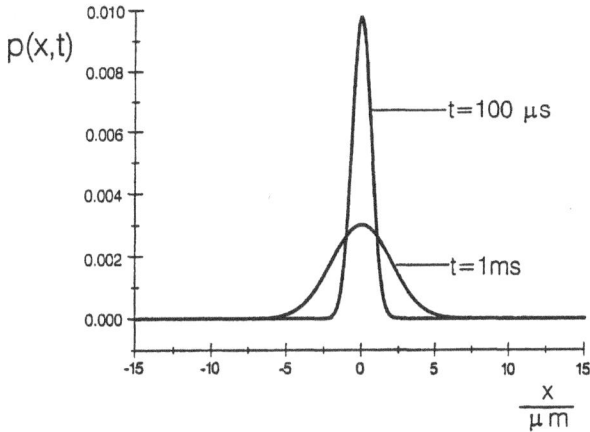

Fig. 18.1. Displacement distribution function (Eq. 18.4) calculated for water self-diffusion at 24 °C, $D = 2.15 \cdot 10^{-9}$ m²/s [344].

where $\zeta = 1, 2, 3$ is the dimensionality of the Euclidean space in which the diffusion process under consideration takes place.[3] The distribution function at Eq. 18.3 depends on the net displacement r of a particle, but not on the initial position. In this sense, we are dealing with a **Markov process**.

In the case of ordinary diffusion in three dimensions, the width of the displacement distribution function is suitably characterized by its second moment, i.e., the mean-squared displacement

$$\langle r^2(t) \rangle = \int\limits_{-\infty}^{+\infty} \int\limits_{-\infty}^{+\infty} \int\limits_{-\infty}^{+\infty} (x^2 + y^2 + z^2)\, p(r, t)\, \mathrm{d}x \mathrm{d}y \mathrm{d}z = 6\,D\,t \tag{18.5}$$

Translational diffusion in one dimension is often modeled as a random walk with constant step length l and a step rate ν. The mean- squared displacement after N steps is then [88]

$$\langle x^2 \rangle = Nl^2 = 2\frac{DN}{\nu} \tag{18.6}$$

so that

$$D = \frac{1}{2}l^2\nu \tag{18.7}$$

---

[3]Note that in NMR diffusometry experiments carried out with unidirectional field gradients, only one displacement component, say x, is probed. The corresponding probability density is then

$$p(x, t) = \frac{1}{(4\pi Dt)^{1/2}} \exp\left\{ -\frac{x^2}{4Dt} \right\} \tag{18.4}$$

where $\int\limits_{-\infty}^{+\infty} p(x, t)\, \mathrm{d}x = 1$. Figure 18.1 shows representative displacement distributions as expected for water self-diffusion.

Deviations from Gaussian distribution functions arise if the translational displacements are restricted by geometrical constraints, or the space in which diffusion takes place is not of an Euclidean nature. An example of the latter case often referred to as **"anomalous diffusion"** is diffusion on random-percolation clusters which are known to be fractal within the correlation length ξ [68, 373]. The time dependence of the mean-squared displacement by anomalous diffusion often obeys a power law of the form

$$\langle r^2 \rangle = \alpha t^{\kappa} \tag{18.8}$$

Diffusion on random-percolation clusters is characterized by exponents $0 < \kappa < 1$ ("subdiffusive behavior"). For ordinary unrestricted diffusion (Eq. 18.5) the exponent is $\kappa = 1$, and the coefficient is $\alpha = 2\zeta D$. The case $\kappa > 1$ is referred to as "superdiffusive behavior." Examples of the latter are certain consequences of Lévy walks [69] already discussed in Sect. 14.3.2, and displacements caused by turbulent flow [403].

The description of diffusion on a fractal object requires a generalization of the diffusion equation leading to a modified distribution function. Corresponding approaches were reported in [172, 256, 341, 408, 409]. The consequence may be the appearance of "stretched Gaussian" distribution functions, where the fractal dimensionality plays a crucial role.

CHAPTER 19

Main-Field Gradient NMR Diffusometry

19.1
The Principle

The magnetic-resonance signal usually recorded with this measuring technique is either a two-pulse Hahn echo or a stimulated echo. The RF pulse sequences (see Chap. 2) are applied in the presence of a strong pulsed or steady main-field gradient

$$G = \frac{\partial B}{\partial z} \tag{19.1}$$

where we have arbitrarily assumed that the gradient is aligned along the z axis of the laboratory frame. Field gradients constant within the sample are particularly favorable. Figure 19.1 shows the time dependences of the field gradients and of the RF pulses which are typically in use.

In Chap. 2, Hahn spin echoes were treated under the assumption that coherence refocusing is entirely reversible. In reality, irreversible phenomena such as relaxation and translational diffusion tend to attenuate the echo amplitude. The complex transverse magnetization at the time t_e, when the echo maximum occurs, is subject to diverse attenuation factors:

$$m(t_e) = m_0 \, A_{diff}(t_e) \, A_{dc}(t_e) \, A_r(t_e) \tag{19.2}$$

The quantity m_0 is the complex transverse magnetization expected in the absence of irreversible processes as assumed in the treatments in Chap. 2. The attenuation factors refer to translational diffusion, the dipolar-correlation effect, and relaxation, respectively. Translational displacements of the spin-bearing particles in the course of the pulse sequence cause residual shifts $\varphi(t_e)$ of the Larmor-precession phase at time t_e, which are not rephased in the echo-formation process. The diffusion attenuation factor is therefore given by the ensemble average[1]

$$A_{diff}(t_e) = \left\langle e^{i\varphi(t_e)} \right\rangle \tag{19.5}$$

---

[1] Note that this attenuation factor is analogous to the **incoherent scattering function** playing a crucial role in quasi-elastic incoherent neutron scattering experiments. Consider a stimulated-echo experiment with a spatially and temporally constant field gradient (Fig. 19.1). The time intervals are supposed to obey $\tau_2 \gg \tau_1$. Formally introducing the "wave vector" $q = \gamma_n \tau_1 G$ suggests the phase shift

$$\varphi(t_e) = q \cdot r(t_e) - q \cdot r(0) \tag{19.3}$$

so that the echo attenuation factor becomes

$$A_{diff}(t_e) = \left\langle e^{-i q \cdot r(0)} e^{i q \cdot r(t_e)} \right\rangle \tag{19.4}$$

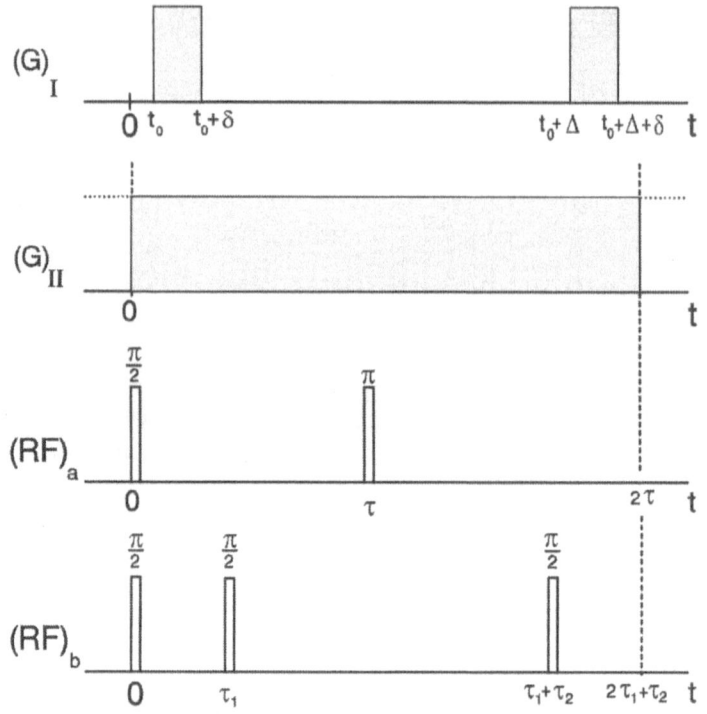

Fig. 19.1. Programs for main-field gradient diffusometry. Pulsed gradients (sequence $(G)_I$) or steady gradients (sequence $(G)_{II}$) in combination with two-pulse echo sequences (pulse train $(RF)_a$) or stimulated echoes (pulse train $(RF)_b$) are in common use. The echoes occur at times $t_e = 2\tau$ or $t_e = 2\tau_1 + \tau_2$. The echo-attenuation factors for all these gradient/RF pulse combinations are summarized in Table 19.1.

which characterizes the degree to which coherences are refocused at time t_e. Clearly, in the absence of diffusion or within the limit of very short diffusion times, the phase shifts of all coherences are zero at t_e, and the attenuation factor is equal to 1. In the opposite limit, the residual phase shifts are widely distributed, and the attenuation factor suppresses the formation of an echo.

The local flux density of the main magnetic field in the presence of a constant gradient G along the z axis of the laboratory frame is given by

$$B(z) = B_0 + G\,z \tag{19.6}$$

This expression has the structure of the incoherent scattering function. The wave vector q is more than of a formal nature. As will be shown below, the pulse sequences employed in field-gradient NMR diffusometry produce wavelike magnetization grids characterized by just this wavenumber. The effect of translational diffusion is then to smear out the grid. This is the real-space variant for explaining NMR diffusometry. Because of the real and static character of the "wave" we prefer in the following to use the symbol k instead of q (see Chap. 20).

Table 19.1. Echo-attenuation factors for ordinary diffusion and mono-exponential relaxation (see Eq. 19.2). The pulse sequences refer to the gradient (I or II) and RF (a or b) programs shown in Fig. 19.1. M_0 is the equilibrium magnetization. The attenuation factors by diffusion are treated in Sects. 19.2 and 19.3.

sequence	t_e	m_0	A_{diff}	$A_r = A_{r1}A_{r2}$
Ia	2τ	M_0	$\exp\left\{-\gamma_n^2 G^2 D\delta^3\left(\frac{\Delta}{\delta}-\frac{1}{3}\right)\right\}$	$\exp\left\{-\frac{2\tau}{T_2}\right\}$
Ib	$2\tau_1+\tau_2$	$\frac{M_0}{2}$	$\exp\left\{-\gamma_n^2 G^2 D\delta^3\left(\frac{\Delta}{\delta}-\frac{1}{3}\right)\right\}$	$\exp\left\{-\frac{2\tau_1}{T_2}-\frac{\tau_2}{T_1}\right\}$
IIa	2τ	M_0	$\exp\left\{-\frac{2}{3}\gamma_n^2 G^2 D\tau^3\right\}$	$\exp\left\{-\frac{2\tau}{T_2}\right\}$
IIb	$2\tau_1+\tau_2$	$\frac{M_0}{2}$	$\exp\left\{-\gamma_n^2 G^2 D\tau_1^3\left(\frac{2}{3}+\frac{\tau_2}{\tau_1}\right)\right\}$	$\exp\left\{-\frac{2\tau_1}{T_2}-\frac{\tau_2}{T_1}\right\}$

A spin-bearing particle diffuses from a position $z(0)$ at time 0 to a position $z(t)$ in time t. The coherence-phase shift at time t is

$$\varphi(t) = \gamma_n \int_0^t [B(t') - B(0)]\, dt' = \gamma_n \int_0^t G(t')[z(t') - z(0)]\, dt' \qquad (19.7)$$

In a homogeneous infinite medium, the displacements $z(t') - z(0)$ are distributed according to a Gaussian function $p = p[z(t')-z(0), t']$ at any instant (see Eq. 18.4).[2] The coherence-phase shifts resulting from the time integral at Eq. 19.7 where $G(t')$ is assumed to be either constant or zero, consequently are also Gauss distributed [128] so that

$$p_\varphi(\varphi, t) = \frac{1}{[2\pi\langle\varphi^2(t)\rangle]^{1/2}} \exp\left\{-\frac{\varphi^2}{2\langle\varphi^2(t)\rangle}\right\} \qquad (19.9)$$

The mean squared phase shift, $\langle\varphi^2(t)\rangle$, has a minimum at time $t = t_e$, when the echo occurs. Then all "static" phase shifts are refocused as a consequence of the echo-formation principle. The remaining echo-attenuation factor is (compare the formalism in Sect. 13.2.1)

$$A_{diff}(t_e) = \int_{-\infty}^{\infty} p_\varphi(\varphi, t_e)e^{i\varphi}\, d\varphi$$

$$= e^{-\frac{1}{2}\langle\varphi^2(t_e)\rangle} \qquad (19.10)$$

---

[2]This is a consequence of the **central limit theorem** [147]. It generally states that sums of the form

$$\zeta_N = \sum_{j=1}^{N} \xi_j \qquad (19.8)$$

which are repeatedly formed of N random and statistically independent variables ξ_j, are distributed according to a Gaussian function when $N \to \infty$. An immensely important characteristic of this theorem is that the random distribution, with which the variables themselves occur, may be arbitrary provided that it is the same for each variable and that its second moment is finite.

The problem is thus reduced to the calculation of the mean squared deviation of the residual phase shift at the time t_e.

In the following, anomalous diffusion and pulsed main-field gradients will be considered first. The cases for ordinary diffusion and stationary gradients will then be derived from the resulting formalism similar to the treatment presented by Kärger et al. in [218].

19.2
Pulsed-Gradient Spin-Echo (PGSE) Diffusometry

Consider two main-field gradient pulses, i.e., the sequence $(G)_I$ in Fig. 19.1, in combination with any spin-echo RF pulse sequel. The gradients are assumed to be aligned along the z axis of the laboratory frame. In order to derive the echo-attenuation factor, the mean squared phase shift at time t_e must be related to the particle displacements in the diverse intervals and pulse delays of the pulse sequence. For this purpose, we first derive the general correlation function of the displacements in two intervals t' and $t' + T$.

19.2.1
The Displacement-Correlation Function

The system in which the diffusion process takes place is assumed to be isotropic so that translational diffusion is also isotropic. The correlation function of displacements r in the three-dimensional space is then related to the correlation function of the displacement components along the z axis (i. e., the field-gradient direction) according to

$$\langle z(t')z(t' + T)\rangle = \frac{1}{3}\langle r(t') \cdot r(t' + T)\rangle \tag{19.11}$$

Let us now consider the identities

$$\langle r(t') \cdot r(t' + T)\rangle \equiv \langle r^2(t')\rangle + \langle r(t') \cdot [r(t' + T) - r(t')]\rangle \tag{19.12}$$

and

$$\begin{aligned}
\langle r^2(t' + T)\rangle &\equiv \langle [r(t') + \{r(t' + T) - r(t')\}]^2\rangle \\
&= \langle r^2(t')\rangle + \langle [r(t' + T) - r(t')]^2\rangle \\
&\quad + 2\langle r(t') \cdot [r(t' + T) - r(t')]\rangle
\end{aligned} \tag{19.13}$$

From the latter equation we deduce

$$\langle r(t') \cdot [r(t' + T) - r(t')]\rangle = \frac{1}{2}\langle r^2(t' + T)\rangle - \frac{1}{2}\langle r^2(t')\rangle - \frac{1}{2}\langle [r(t' + T) - r(t')]^2\rangle \tag{19.14}$$

where the mean squared displacement in the last term may be replaced by

$$\langle [r(t' + T) - r(t')]^2\rangle = \langle r^2|T|\rangle \tag{19.15}$$

Inserting Eq. 19.14 into Eq. 19.12, the displacement-correlation function at Eq. 19.11 may be expressed in the form

$$\langle z(t')z(t'+T)\rangle = \frac{1}{6}\langle r^2(t')\rangle + \frac{1}{6}\langle r^2(t'+T)\rangle - \frac{1}{6}\langle r^2|T|\rangle \tag{19.16}$$

This equation relates the displacement-correlation function with the mean squared displacements in the intervals considered. Inserting Eq. 18.8 leads to the **displacement-correlation function for anomalous diffusion**

$$\langle z(t')z(t'+T)\rangle = \frac{\alpha}{6}\left[(t'+T)^\kappa + t'^\kappa - |T|^\kappa\right] \tag{19.17}$$

19.2.2
The Mean Squared Phase Shift

The mean square of the residual phase shift at the echo maximum can be calculated using Eq. 19.7. The times defining the gradient and RF pulse intervals are given in Fig. 19.1. The echoes occur at times $t_e = 2\tau$ or $t_e = 2\tau_1 + \tau_2$ for two-pulse Hahn or stimulated echoes, respectively. "Static" phase shifts by inhomogeneities and gradients are compensated in the echo-formation process. What remains are phase shifts caused by incomplete refocusing owing to diffusional displacements. The phase shift of a representative ("tagged") spin is the difference between the phase changes adopted in the presence of the two gradient pulses,

$$\varphi(t_e) = \gamma_n G \left[\int_{t_0}^{t_0+\delta} z(t')\,dt' - \int_{t_0+\Delta}^{t_0+\Delta+\delta} z(t')\,dt'\right] \tag{19.18}$$

The minus sign stands for the coherence-phase shift induced by the RF pulses between the gradient pulses (compare the Hahn-echo treatments in Chap. 2). Taking the square of Eq. 19.18 and averaging over the spin ensemble leads to the mean squared phase shift (see the gradient intervals of pulse train $(G)_I$ in Fig. 19.1)

$$\begin{aligned}
\langle \varphi^2(t_e)\rangle &= \gamma_n^2 G^2 \left\langle \left[\int_{t_0}^{t_0+\delta} z(t')\,dt' - \int_{t_0+\Delta}^{t_0+\Delta+\delta} z(t')\,dt'\right]^2\right\rangle \\
&= \gamma_n^2 G^2 \left[\int_{t_0}^{t_0+\delta} dt' \int_{t_0}^{t_0+\delta} dt'' \langle z(t')z(t'')\rangle + \int_{t_0+\Delta}^{t_0+\Delta+\delta} dt' \int_{t_0+\Delta}^{t_0+\Delta+\delta} dt'' \langle z(t')z(t'')\rangle \right. \\
&\qquad\qquad \left. - 2\int_{t_0}^{t_0+\delta} dt' \int_{t_0+\Delta}^{t_0+\Delta+\delta} dt'' \langle z(t')z(t'')\rangle\right]
\end{aligned}$$

$$= \gamma_n^2 G^2 \left[\int_{t_0}^{t_0+\delta} dt' \int_{t_0-t'}^{t_0+\delta-t'} dT \langle z(t')z(t'+T) \rangle \right.$$

$$+ \int_{t_0+\Delta}^{t_0+\Delta+\delta} dt' \int_{t_0+\Delta-t'}^{t_0+\Delta+\delta-t'} dT \langle z(t')z(t'+T) \rangle - 2 \int_{t_0}^{t_0+\delta} dt' \int_{t_0+\Delta-t'}^{t_0+\Delta+\delta-t'} dT \langle z(t')z(t'+T) \rangle \left. \right]$$

$$(19.19)$$

where t'' has been substituted by a new integration variable defined by

$$T = t'' - t' \qquad (19.20)$$

Recalling the stationarity property of equilibrium systems, we may arbitrarily choose $t_0 = 0$ for convenience. That is, the experiment is, effectively, to begin at time 0. Any prehistory is irrelevant. Replacing the integrands by Eq. 19.17 and carrying out the integrations leads to the mean squared phase shift

$$\langle \varphi^2(t_e) \rangle = \frac{2\alpha\gamma_n^2 G^2}{3(\kappa+1)(\kappa+2)} \left[\frac{1}{2}(\Delta+\delta)^{\kappa+2} + \frac{1}{2}(\Delta-\delta)^{\kappa+2} - \Delta^{\kappa+2} - \delta^{\kappa+2} \right]$$

$$(19.21)$$

19.2.3
The Echo-Attenuation Factor for Anomalous Diffusion

With the above result the echo-attenuation factor at Eq. 19.10 reads

$$\boxed{\begin{aligned} A_{diff}(\delta, \Delta) = \\ \exp\left\{ -\frac{\alpha\gamma_n^2 G^2}{3(\kappa+1)(\kappa+2)} \left[\frac{1}{2}(\Delta+\delta)^{\kappa+2} + \frac{1}{2}(\Delta-\delta)^{\kappa+2} - \Delta^{\kappa+2} - \delta^{\kappa+2} \right] \right\} \end{aligned}}$$

$$(19.22)$$

In the limit of short field-gradient pulses, $\delta \ll \Delta$, which is often fulfilled under practical conditions, this reduces in second-order approximation to

$$A_{diff}(\delta, \Delta) \approx \exp\left\{ -\frac{\gamma_n^2 G^2 \delta^2 \alpha}{6} \Delta^\kappa \right\} = \exp\left\{ -\frac{k^2 \alpha}{6} \Delta^\kappa \right\} \qquad (19.23)$$

where we have introduced the "wavenumber"

$$k = \gamma_n G \delta \qquad (19.24)$$

These results have been derived assuming the general mean-squared displacement power law at Eq. 18.8. It should be noted, however, that a crucial assumption in the derivations was that the probability density of the displacements is Gaussian (see

Eq. 18.3). Strictly speaking this contradicts the character of anomalous diffusion, but - with some caution - may be taken as an approximation of non-Gaussian distribution function arising if the displacement steps underlie topological restraints.[3]

19.2.4
The Echo-Attenuation Factor for Ordinary Diffusion

In the case of unrestricted ordinary diffusion, i.e., $\kappa = 1$ and $\alpha = 6D$, Eq. 19.22 adopts the form of the **Stejskal/Tanner expression** [472]

$$A_{diff}(\delta, \Delta) = \exp\left\{ -\gamma_n^2 G^2 D\delta^3 \left(\frac{\Delta}{\delta} - \frac{1}{3} \right) \right\} \tag{19.25}$$

Considering the limit $\delta \ll \Delta$ again, we have

$$A_{diff}(\delta, \Delta) \approx \exp\left\{ -\gamma_n^2 G^2 D\delta^2 \Delta \right\} = \exp\left\{ -k^2 D\Delta \right\} \tag{19.26}$$

where the wavenumber k is given by Eq. 19.24.

19.2.5
Direct Evaluation of the Mean Squared Displacement

In the limit $\delta \ll \Delta$, the echo-attenuation factor may be represented by

$$A_{diff}(k, \Delta) = \exp\left\{ -\frac{1}{6}k^2 \langle r^2(\Delta) \rangle \right\} \tag{19.27}$$

commonly for ordinary and anomalous diffusion (see Eqs. 19.23 and 19.26), where we have inserted Eq. 18.8. Solving this equation results in

$$\langle r^2(\Delta) \rangle = -6 \frac{\ln\left\{ A_{diff}(k, \Delta) \right\}}{k^2} \tag{19.28}$$

Recording the echo-attenuation factor for constant wavenumber k thus permits one to evaluate the time dependence of the mean squared displacement $\langle r^2 \rangle = \langle r^2(\Delta) \rangle$.

19.3
Steady-Gradient Spin-Echo (SGSE) Diffusometry

The PGSE method described and treated in Sect. 19.2 has the advantage that the gradients can be varied without distorting the homogeneous excitation of the sample by the RF pulses. Varying the gradient strength means that all time intervals of the pulse sequences (see Fig. 19.1) can be kept constant, so that the time scale to which the experiment refers is well defined.

---

[3] An explicit treatment for a non-Gaussian probability density of displacements can be found in Sect. 22.2 for polymer diffusion ("reptation").

Nevertheless, steady-gradient spin-echo diffusometry, which was already suggested when spin echoes were discovered and first applied for the determination of dynamic parameters of liquids [83, 186, 308, 518], remains of considerable interest. It was recently revived as a method of exploiting the extremely strong and stable magnetic-field gradients in the fringe fields of ordinary superconducting magnets [240]. Superconducting "anti-Helmholtz" coil magnets especially designed for this purpose produce even stronger field gradients [153]. This **supercon-fringe field steady-gradient spin-echo (SFF-SGSE)** technique does not require any power supply for the generation of the gradients, and is suitable for the investigation of very slow diffusion phenomena.

Compared with the conventional PGSE technique [70, 217, 474] and the MASSEY variant of it [74], the available magnetic field gradients are extremely strong (40 to 200 T/m) and perfectly stable as far as vibrations of the building can be excluded. The encoding efficiency in the dephasing and rephasing intervals is optimal because no time is lost by switching the gradients on and off. Thus relatively low mean-squared displacements can be studied even in viscous systems having short transverse relaxation times.

Due to the limited RF bandwidth, slices rather than the whole samples are excited in the presence of such strong field gradients (see Sect. 25.1). However, the slice widths are easily much greater than the maximal root mean squared displacement of the molecules. Diffusion into or out of the excited regions during the intervals sensitive to displacements can therefore be regarded as negligible, and need not to be considered as an echo-attenuation mechanism.

19.3.1
Echo-Attenuation Factors

The echo-attenuation factors for the SGSE variant are already implied in the PGSE formulae derived in the previous sections. We must merely set $t_0 = 0$, $\delta = \Delta = \tau$ in the case of two-pulse Hahn spin echoes (sequence (RF)$_a$ in Fig. 19.1), or $t_0 = 0$, $\delta = \tau_1$, $\Delta = \tau_2 + \tau_1$ in context with stimulated echoes (sequence (RF)$_b$ in Fig. 19.1). The results are as follows.

Two-pulse Hahn spin echo; anomalous diffusion:

$$A_{diff}(\tau) = \exp\left\{ -\frac{2\alpha\gamma_n^2 G^2}{3(\kappa+1)(\kappa+2)} (2^\kappa - 1)\, \tau^{\kappa+2} \right\} \qquad (19.29)$$

Two-pulse Hahn spin echo; ordinary diffusion:

$$A_{diff}(\tau) = \exp\left\{ -\frac{2}{3}\gamma_n^2 G^2 D\tau^3 \right\} \qquad (19.30)$$

Stimulated echo; anomalous diffusion:

$$A_{diff}(\tau_1, \tau_2) = \exp\left\{ -\frac{\alpha\gamma_n^2 G^2}{3(\kappa+1)(\kappa+2)} \left[\tfrac{1}{2}(2\tau_1 + \tau_2)^{\kappa+2} + \tfrac{1}{2}\tau_2^{\kappa+2} - (\tau_1 + \tau_2)^{\kappa+2} - \tau_1^{\kappa+2} \right] \right\}$$

$$(19.31)$$

In the limit $\tau_2 \gg \tau_1$, the latter formula can be approximated by

$$A_{diff}(\tau_2 \gg \tau_1) \approx \exp\left\{ -\frac{1}{6}k_1^2 \alpha \tau_2^\kappa \right\} = \exp\left\{ -\frac{1}{6}k_1^2 \langle r^2(\tau_2) \rangle \right\} \tag{19.32}$$

where

$$k_1 = \gamma_n G \tau_1 \tag{19.33}$$

Stimulated echo; ordinary diffusion:

$$A_{diff}(\tau_1, \tau_2) = \exp\left\{ -\gamma_n^2 G^2 D \tau_1^3 \left(\frac{2}{3} + \frac{\tau_2}{\tau_1} \right) \right\} \tag{19.34}$$

With the limit $\tau_2 \gg \tau_1$, this expression approaches

$$A_{diff}(\tau_2 \gg \tau_1) \approx \exp\left\{ -k_1^2 D \tau_2 \right\} = \exp\left\{ -\frac{1}{6}k_1^2 \langle r^2(\tau_2) \rangle \right\} \tag{19.35}$$

19.3.2
Relaxation-Compensated Pulse Sequences

The SFF variant of NMR diffusometry excludes the variation of the gradient. Therefore the attenuation factors due to relaxation, $A_r(t_e)$, as well as to diffusion, $A_{diff}(t_e)$, affect the echo amplitude (Eq. 19.2). In order to be able to measure the latter without knowledge of the former, a number of special pulse sequences have been suggested permitting relaxation-compensated evaluations [118, 248]. Examples are shown in Fig. 19.2. The prerequisite of these methods is that there is no residual dipolar (or quadrupolar) coupling. Otherwise the dipolar-correlation effect (or its quadrupolar-coupling counterpart) must be taken into account (Chap. 17).

The pulse trains in Fig. 19.2 consist of three or five pulses. Because of the partial refocusing effect of the $180°$ pulses in pulse train (b), the relevant echo appears an interval 2τ later than with the three-pulse sequence (a). The insensitivity of the diffusion decays to the influence of relaxation is achieved by forming the quotient of the stimulated and primary echo signals and/or by keeping the relaxation sensitive intervals constant.

Three-Pulse SFF Method (Fig. 19.2a)

The primary and the stimulated echoes are acquired in each transient. Half of the transverse magnetization dephased in the first δ interval is refocused by the second $90°$ pulse and forms the primary echo at time $t = 2\delta$. For ordinary diffusion and mono-exponential relaxation curves, the amplitude is (Table 19.1 on page 181)

$$\hat{m}_p(k_1) = \frac{1}{2}M_0 \exp\left\{ -\frac{2}{3}Dk_1^2\delta - \frac{2\delta}{T_2} \right\} \tag{19.36}$$

where

$$k_1 = \gamma_n G\delta = \gamma_n G t_1 \tag{19.37}$$

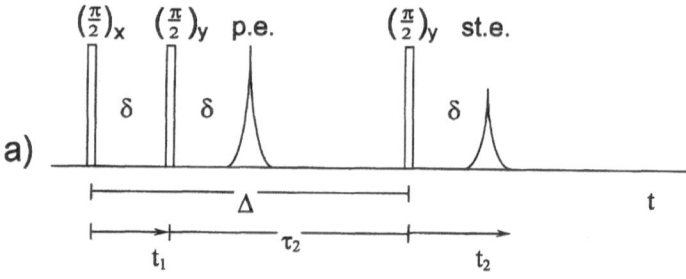

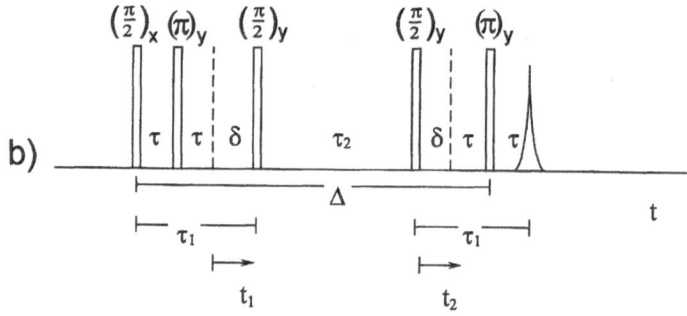

Fig. 19.2. RF pulse sequences for the determination of the echo-attenuation factor due to diffusion, A_{diff}, without knowledge of that for relaxation, A_r. The pulse sequences are applied in the presence of a steady gradient. a) three-pulse Hahn echo sequence – the amplitudes of the primary (p.e.) and the stimulated echoes (st.e.) are recorded and divided by each other; b) five-pulse sequence - the coherences are partially refocused by π pulses in the (constant) τ_1 intervals. The time domains considered in two-dimensional evaluations are designated by t_1 and t_2. Only those spin echoes which are relevant for the technique and which are acquired in the experiments are indicated. The prerequisite of these techniques is that there is no residual dipolar or quadrupolar coupling (compare Chap. 17).

The other half of the dephased transverse magnetization is "stored" as $\pm$ components of the z magnetization and is read out in the form of the stimulated echo after a delay Δ. The free-induction decay envelope obeys

$$m(k_1, k_2) = M_0 \int e^{i(k_1 - k_2)z} \, e^{-Dk_1^2 \, (\Delta - \delta/3)} \, e^{-(\Delta - \delta)/T_1} \, e^{-2\delta/T_2} \, dz \qquad (19.38)$$

where

$$k_2 = \gamma_n G t_2 \qquad (19.39)$$

The integral covers the whole slice width along the z gradient direction. The maximum of the stimulated echo is reached for the condition $k_1 = k_2$.

The quotient of the stimulated-echo FID (Eq. 19.38) and the amplitude of the primary echo (Eq. 19.36) is

$$S(k_1, k_2) = \frac{m(k_1, k_2)}{\hat{m}_p(k_1)}$$

$$= 2 \int e^{-Dk_1^2 \tau_2} \, e^{-\tau_2/T_1} \, e^{i(k_1 - k_2)z} \, dz \qquad (19.40)$$

where $\tau_2 = \Delta - \delta$. In this expression, the transverse relaxation factor has been canceled whereas the spin-lattice relaxation term can be kept constant for fixed τ_2 intervals. In pure k-domain experiments only the maximum of the stimulated echo is of interest, and the signal function is reduced to

$$S(k_1 = k_2) = \int e^{-Dk_1^2 \tau_2} \, e^{-\tau_2/T_1} \, dz \qquad (19.41)$$

Note that the quantity $S(k_1, k_2)$ represents a "two-dimensional k-domain signal."

Five-Pulse SFF Method (Fig. 19.2b)

In the pulse scheme at Fig. 19.2b, the dephased coherences are partially rephased in the first (constant) τ_1 interval using a 180° pulse. The remaining dephasing of the coherences is analogous to a PGSE experiment where the two δ sections of the stationary gradient correspond to the gradient "pulses." The following diffusion and relaxation attenuation factors (Table 19.1 on page 181) and coherence-evolution terms (see Sect. 2.2) must be taken into account:

- losses by transverse relaxation in the two τ_1 intervals, i.e., a factor $\propto \exp\{-2\tau_1/T_2\}$
- losses by longitudinal relaxation in the τ_2 interval, i.e., a factor $\propto \exp\{-\tau_2/T_1\}$
- attenuation by translational diffusion in the four τ intervals during which further dephasing of the coherences is compensated, i.e., a factor $\propto \exp\left\{-\frac{1}{6}D\gamma_n^2 G^2 (\tau_1 - \delta)^3\right\}$
- attenuation by translational diffusion during and between the two δ intervals, i.e., a factor $\propto \exp\left\{-D\gamma_n^2 G^2 \delta^2 \left(\tau_2 + \frac{2}{3}\delta\right)\right\}$
- coherence evolution during the first δ interval in gradient induced frequency offsets (dephasing from the initial y direction of the rotating frame), i.e., a factor $\propto i \exp\{-ik_1 z\}$
- the same in time t_2, i.e., a factor $\propto i \exp\{ik_2 z\}$

Taking all factors together and integrating over the sensitive slice leads to the signal function (divided by $iM_0/2$)

$$S(k_1, k_2) = \int e^{-\frac{1}{6}\gamma_n^2 G^2 D[\tau_1^3 - 3\delta\tau_1^2 + 3\delta^2(\tau_1 + 2\tau_2) + 3\delta^3]} \, e^{-\tau_2/T_1} \, e^{-2\tau_1/T_2} \, e^{i(k_2 - k_1)z} \, dz$$

$$= \int e^{-\frac{1}{6}D\left[\gamma_n^2 G^2 \tau_1^3 - 3\gamma_n G \tau_1^2 k_1 + 3(\tau_1 + 2\tau_2)k_1^2 + \frac{3}{\gamma_n G}k_1^3\right]} \, e^{-\tau_2/T_1} \, e^{-2\tau_1/T_2} \, e^{i(k_2 - k_1)z} \, dz$$

$$(19.42)$$

The condition for the maximum of the stimulated echo is $k_1 = k_2$, i.e., $t_2 = \delta$. The signal is then given by

$$S(k_1 = k_2) = \int e^{-\frac{1}{\delta}D\left[\gamma_n^2 G^2 \tau_1^3 - 3\gamma_n G\tau_1^2 k_1 + 3(\tau_1 + 2\tau_2)k_1^2 + \frac{3}{\gamma_n G}k_1^3\right]} e^{-\tau_2/T_1} e^{-2\tau_1/T_2} \, dz \qquad (19.43)$$

Diffusion experiments are carried out by varying time $t_1 = \delta$, but keeping the intervals τ_1 and τ_2 constant. The intervals τ therefore must be varied complementarily. The relaxation attenuation factors then remain constant, and the decay of the stimulated echo amplitude is determined solely by translational diffusion.

CHAPTER 20

Reciprocal- vs Real-Space Representations

In the above description of NMR diffusometry experiments, a quantity k was introduced which was formally identified with a "wavenumber." As already stated several times, the wavenumber addressed by coherence evolution under the influence of a field gradient G of duration τ_1 is

$$k = \gamma_n G \tau_1 \qquad (20.1)$$

In the following we will show that the wavenumber is not just of a formal nature in this context. It is real in the sense that wavelike magnetization patterns[1] are generated in the course of field-gradient NMR diffusometry experiments. The wavelength corresponds exactly to the wavenumber at Eq. 20.1. The "waves" can be directly visualized with the aid of magnetic-resonance imaging techniques.

This suggests two alternative views of field-gradient NMR diffusometry which are of particular interest if diffusion is affected by geometrical constraints, as is the case in porous media, for instance (compare also [346, 347, 441]).

- First, the method can be regarded as an experiment probing the **reciprocal space** in a certain analogy to optical diffraction [75, 77, 100] (compare footnote 1 on page 179). The idea is to translate formalisms known for light diffraction at geometrical objects into rules for translational diffusion in similar geometrical confinements. If the wavenumber at Eq. 20.1 matches the length scale of the constraints, one expects "diffraction-like" patterns in the echo-attenuation curves.

- The second reading is that the field-gradient NMR diffusometry experiment produces an undulated non-equilibrium magnetization distribution in **real space**. This wave-like pattern is then reexamined at the end of the diffusion interval concerning effects by translational displacements. In the case of geometrical confinements, the wavelike magnetization distribution is modulated by the structure of the material. If the wavelength matches the length scale of the constraints, one expects length-scale related diffusion effects. If desired, the phenomena can be probed with the aid of NMR imaging methods with the same spatial sensitivity as relevant for the reciprocal-space interpretation (see Eq. 20.18 and Chap. 32).

---

[1] An analogous phenomenon has already been discussed in the treatment of multiple echoes (Sect. 2.4).

20.1
The Generalized Reciprocal-Space Formalism

Assuming the gradient aligned along the z axis, the wavenumber k and the displacement z can be considered to form a **pair of conjugated variables**. The relaxation-compensated amplitude of a stimulated echo generated by a three-pulse sequence with the intervals τ_1 and τ_2 (see Fig. 19.2, for instance) can be expressed in the form of the convolution [76]

$$A_{diff}(k, \tau_2) = \overline{|S_0(k)|^2} \, \tilde{p}(k, \tau_2) \otimes \tilde{L}(k) \qquad (\tau_2 \gg \tau_1) \qquad (20.2)$$

Three factors are important.

a) The Fourier transform of the **displacement distribution** is

$$\tilde{p}(k, \tau_2) = \mathcal{F}_z\{p(z, \tau_2)\}, \qquad (20.3)$$

where $p(z, \tau_2)$ is the normalized distribution of displacements z in the interval τ_2.

This formalism is to include the case of systems with geometrical constraints or other heterogeneities. Therefore, a distribution of local diffusion coefficients may occur. We take this into account by considering the *mean* diffusion coefficient $\bar{D}$. As an average it is independent of the displacement, so that $p(z, \tau_2)$ may be represented by the Gaussian function (Eq. 18.3)

$$p(z, \tau_2) = \frac{1}{\sqrt{4\pi \bar{D}\tau_2}} \exp\left\{-z^2/(4\bar{D}\tau_2)\right\} \qquad (20.4)$$

The Fourier transform is (Table 42.3 on page 401)

$$\tilde{p}(k, \tau_2) = \mathcal{F}_z\{p(z, \tau_2)\} = \exp(-k^2 \bar{D}\tau_2) \qquad (\tau_2 \gg \tau_1) \qquad (20.5)$$

The width of this diffusion profile function is characterized by

$$(\Delta k)_d = \left(\bar{D}\tau_2\right)^{-1/2} \qquad (20.6)$$

b) The Fourier transform of the **microstructural correlation function,**

$$\tilde{L}(k) = \mathcal{F}_z\{L(z)\}, \qquad (20.7)$$

characterizing the spatial correlation of the diffusion properties at different positions. This function in particular reflects any structural periodicity with respect to diffusion properties. In the ideal case of a lattice-like structure, it adopts the form of a comb of delta functions. In amorphous systems, on the other hand, the correlation of diffusion properties is expected to vary analogously to the familiar density-density correlation functions of ordered media. The function $\tilde{L}(k)$ thus causes a modulation of the echo decay curves by the microstructural modes of the system. The first peak or shoulder of the modulated echo decay curve is expected at

$$k_s \approx 2\pi b^{-1} \tag{20.8}$$

where b is the correlation length. To reveal itself in diffusion experiments, $\tilde{L}(k)$ must be quasi-stationary, i.e., the mean lifetime of the structural modes must exceed the diffusion time τ_2.

c) The average **form factor**, $\overline{|S_0(k)|^2}$, finally characterizes the shape of the regions in which diffusion takes place homogeneously without any constraint. The form factor for spherical domains, for instance, is [76]

$$\overline{|S_0(k)|^2} = \frac{9\,[ka\cos(ka) - \sin(ka)]^2}{(ka)^6}$$

$$\approx \exp\{-k^2 a^2/5\} \quad \text{for} \quad ka \ll 1 \tag{20.9}$$

where a is the radius. $\overline{|S_0(k)|^2}$ is also assumed to be quasi-stationary relative to the time scale of diffusion experiments.

In the absence of any structural heterogeneity in the sample, the echo amplitude, $A_{diff}(k, \tau_2)$, decays solely according to the diffusion profile function at Eq. 20.5. This is the regular situation one expects in homogeneous liquids. On the other hand, if the liquid is confined to a porous medium, for instance, diffusion is restricted by the microstructure of the pore cavities and capillaries. Callaghan et al. [75, 77] have demonstrated with such systems that the diffusion attenuation curves are modulated in a pronounced way if the pore radius a and the nearest neighbor pore distance b comply with $k \approx 2\pi b^{-1}$ or $k \approx 2\pi a^{-1}$. Vice versa, modulations of this sort in experimental diffusion decay curves indicate microstructural constraints of diffusion.

20.2
The Real-Space Representation

In this notion, a field-gradient diffusometry experiment consists of a preparation interval in which a non-equilibrium magnetization distribution is produced in the form of a "**magnetization grid**" (or grating). That is, the magnetization is spatially modulated along the gradient direction. The consequence of translational displacements is then that the grid tends to be leveled off. This process can directly be observed by imaging the grid after a certain diffusion interval [248, 428].

20.2.1
The Longitudinal-Magnetization Grid

A stimulated-echo pulse sequence (Figs. 19.1 or 19.2a) may be discussed in terms of three subsequent steps (compare Fig. 8.1). Initially, the desired magnetization distribution is prepared with the aid of a hard composite pulse consisting of the first two RF pulses of the sequence. "Hard" means that the RF amplitude is much greater than any gradient induced frequency-offsets. Then an interval follows in which the spatial magnetization distribution is modified by diffusive displacements.[2] The last section serves the detection of the final magnetization distribution.

The nominal flip angle of each moiety of the initial excitation double-pulse is assumed to be 90°. It is clear that the pulse spacing τ_1 must not exceed the mean lifetime of the locally coherent spin states, i.e., the transverse relaxation time T_2.[3] Otherwise it would not be permitted to regard the first two RF pulses as a composite excitation unit (apart from the fact that the whole experiment would not make much sense in that case).

For the sake of simplicity, the composite excitation pulse is treated as a pair of delta function pulses separated by the interval τ_1. The first pulse is applied at time 0. The Fourier transform is then

$$F(\omega) = 1 + \exp(i\Omega\tau_1) \tag{20.10}$$

(compare Table 42.1 on page 399). That is, the sample is excited with a sinusoidal profile.

Let us now assume a constant B_0 gradient in the z direction, for instance, so that the local frequency offset from resonance is $\Omega = \Omega(z)$. The hard double RF pulse obviously leaves the magnetization untouched at all positions where

$$\Omega(z) = \frac{(2j+1)\pi}{\tau_1} \qquad (j = 0, 1, 2, ...) \tag{20.11}$$

This point of view shows that spins located at these positions do not "sense" the double pulse at all. On the other hand, spins at positions corresponding to

$$\Omega(z) = \frac{2j\pi}{\tau_1} \qquad (j = 0, 1, 2, ...) \tag{20.12}$$

are excited the most. That is, the longitudinal magnetization is reversed at these locations.

After the double pulse, the longitudinal magnetization consequently forms a sinusoidal magnetization grid. As a non-equilibrium distribution, this magnetization grating is disposed to become more or less leveled off owing to translational diffusion (apart from relaxation) in the course of the subsequent interval.[4] The third RF

---

[2] As a competitive mechanism, **relaxation** also contributes in a competitive manner, but, as described above, can be compensated for experimentally. Furthermore, **coherent flow** may be relevant if present. This will be considered in more detail in context with the MAGROFI method (Sect. 21).

[3] This does not refer to the dephasing time T_2^* in the field gradient which tends to be much less than the pulse separation, of course.

[4] There is a complete analogy [428] to the optical **"forced Rayleigh scattering"** technique (also called "holographic grating") [21, 137, 297, 387, 456]. With the forced Rayleigh scattering method the diffusing molecules are labeled with photochromic groups. A grid of excited regions with different indices of refraction is produced by crossed laser beams at the beginning of the experiment. The diffusion of molecules into the excited regions (and vice versa) is then measured by diffraction of a probe beam incident at the Bragg angle. In the NMR case, the magnetization is the analog to the concentration of the photochromic molecules.

pulse of the sequence finally "reads" the magnetization grid. This may be performed in an accumulated form of a stimulated echo (as depicted before) or directly as a frequency-encoding magnetic resonance imaging procedure (see below).[5]

The spatial distribution of the magnetization, $M_z(z_0, 0)$, immediately after the double-pulse excitation is now considered as the initial situation of the subsequent diffusion period t. The diffusion propagator is (Eq. 18.4)

$$p(z - z_0, t) = \frac{1}{\sqrt{4\pi Dt}} \exp\left\{-\frac{(z - z_0)^2}{4Dt}\right\} \tag{20.13}$$

where z_0 and z are the gradient-relevant coordinates of a spin at the times 0 and t, respectively. The magnetization profile at time t is composed in the form of the integral over all initial positions, i.e., of the convolution

$$M_z(z, t) = \int_{-\infty}^{\infty} M_z(z_0, 0)\, p(z - z_0, t)\, dz_0 = M_z(z, 0) \otimes p(z, t) \tag{20.14}$$

That is, the spin ensemble defined by the gradient-relevant coordinate z at time t cannot be traced back to a certain initial position. Only probabilistic statements are possible.

The initial magnetization grid may be described by

$$M_z(z_0, 0) = M_0 \cos(kz_0) \tag{20.15}$$

Inserting this and Eq. 20.13 into Eq. 20.14 gives

$$M_z(z, t) = \frac{M_0}{\sqrt{4\pi Dt}} \int_{-\infty}^{\infty} \cos(kz_0) \exp\left\{-\frac{(z - z_0)^2}{4Dt}\right\} dx_0 = M_0 \cos(kz)\, e^{-k^2 Dt} \tag{20.16}$$

Evidently the initially sinusoidal magnetization profile is conserved at later times. The grid amplitude merely decays towards zero independent of the position while the grid wavelength remains the same (see Fig. 20.1).

The reason why this magnetization grid does not become so evident in the usual experimental field-gradient diffusometry procedure is that the signals are usually detected in the form of a stimulated echo in the absence of field gradients. In the presence of read-out field gradients, however, one records a reciprocal-space (k space) signal, the Fourier transform of which directly renders a one-dimensional image of the magnetization grid in real space spread along the z axis (for z gradients) (compare Sect. 25.3). This can be done either on the basis of the free-induction decay directly after the reading pulse or by recording the stimulated echo.

---

[5]Note that the third RF pulse is usually applied after a time much longer than the lifetime of the locally coherent spin states, T_2, so that it does not take part in the primary-excitation scheme described above. It therefore only makes sense to consider the first two RF pulses as a composite excitation pulse rather than the whole sequence.

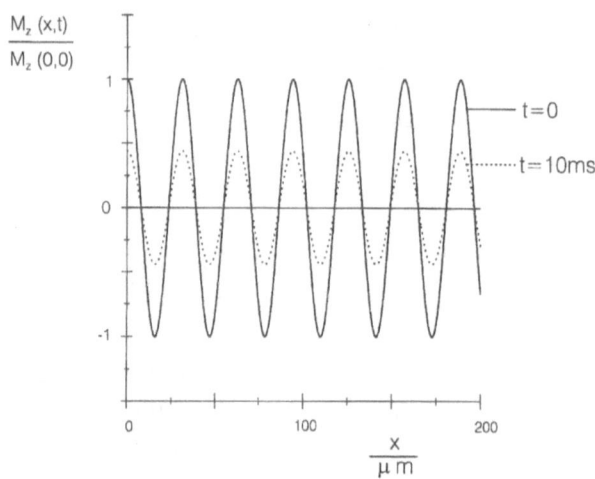

Fig. 20.1. Attenuation of the longitudinal-magnetization grid by self-diffusion. The field gradient is assumed to be aligned along the x axis. The profiles were calculated on the basis of Eq. 20.16 assuming for the diffusivity that of water at room temperature ($D = 2 \cdot 10^{-9}$ m²/s). The wavenumber was set $k = 10^7$ m⁻¹ corresponding to a τ_1 interval of 75 μs for proton resonance and a field gradient of 10 Tm⁻¹.

20.2.2
One-Dimensional Real-Space Evaluation

As a typical example, we consider the signal function $S(k_1, k_2)$ in Eq. 19.40 which was derived in context with the three-pulse SFF technique. The one-dimensional Fourier transform to the z_2 domain conjugated to the k_2 domain is

$$\tilde{S}(k_1, z_2) = \frac{1}{2\pi} \int_{-\infty}^{\infty} S(k_1, k_2) \, e^{ik_2 z_2} \, dk_2$$

$$= 2e^{-Dk_1^2 \tau_2} \, e^{-\tau_2/T_1} \, e^{ik_1 z_2} \tag{20.17}$$

The cosine of the real part of Eq. 20.17 indicates the modulation of the signals along the z_2 direction. The wavelength is

$$d = \frac{2\pi}{k_1} = \frac{2\pi}{\gamma_n G \delta} \tag{20.18}$$

Figure 20.2 shows a series of experimental grid patterns.

20.2.3
Two-Dimensional Real-Space Evaluation

The z_2 domain function $\tilde{S}(k_1, z_2)$ still depends on k_1 according to Eq. 20.17. A second Fourier transform now with respect to k_1 leads to the two-dimensional

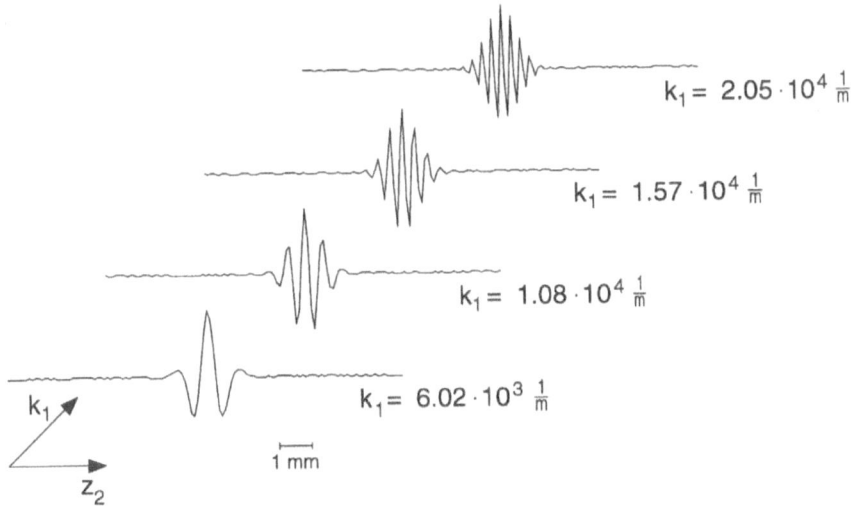

$k_1 = 2.05 \cdot 10^4 \, \frac{1}{m}$

$k_1 = 1.57 \cdot 10^4 \, \frac{1}{m}$

$k_1 = 1.08 \cdot 10^4 \, \frac{1}{m}$

$k_1 = 6.02 \cdot 10^3 \, \frac{1}{m}$

k_1

z_2

1 mm

Fig. 20.2. Magnetization grids in the z_2 domain derived as the numerical k_2 Fourier transform of experimental stimulated-echo data of glycerol. The measurements were carried out with pulse sequence at Fig. 19.2a. The parameter $k_1 = \gamma_n G \delta$ was varied as indicated whereas τ_2 was kept constant. The envelope is due to the finite bandwidth of the RF pulses and corresponds to the profile $g(z_2)$ across the excited slice. (This factor is not included in Eq. 20.17). The plotted signal is proportional to $\tilde{S}(k_1, z_2) g(z_2)$. (Reproduced by permission from ref. [248])

intensity distribution:

$$\tilde{\tilde{S}}(z_1, z_2) = \frac{1}{2\pi} \int_{-\infty}^{\infty} \tilde{S}(k_1, z_2) \, e^{ik_1 z_1} \, dk_1$$

$$= \frac{1}{\pi} e^{-\tau_2/T_1} \sqrt{\frac{2}{D\tau_2}} \, e^{-\frac{(z_1 - z_2)^2}{4D\tau_2}} \qquad (20.19)$$

The maximal intensity is concentrated on the diagonal defined by $z_1 = z_2$. The diagonal intensity function is

$$\tilde{\tilde{S}}(z_1 = z_2) = \frac{1}{\pi} e^{-\tau_2/T_1} \sqrt{\frac{2}{D\tau_2}} \qquad (20.20)$$

The "cross intensity" for $z_1 \neq z_2$ reflects the diffusion of nuclei from the initial position z_1 to the final position z_2. Thus, the two-dimensional real-space treatment of stimulated-echo/field-gradient experiments directly mirrors the probability $P(z_1, 0 | z_2, t)$ that a particle is at a position z_1 at time 0 and at a position z_2 at time t [248].

20.2.4
Geometrical Confinements

In the presence of confinements of the diffusing particles in a pore space such as a percolation cluster (of sufficient extension), for instance, the magnetization grid is modulated itself by the excluded volume. The initial magnetization grid may then be represented by

$$M_z(z_0, 0) = M_0\, \rho(z_0)\, \cos(kz_0) \qquad (20.21)$$

where $\rho(z_0)$ is the probability that the position z_0 is in the pore space, and M_0 is the equilibrium magnetization in the pores. In an interval τ the grid evolves to

$$M_z(z, \tau) = \int_{-\infty}^{\infty} P(z_0, z, \tau) M_z(z_0, 0)\, dz_0 \qquad (20.22)$$

The conditional probability density that a particle is found at z at time τ if it was at z_0 at time 0 is denoted by $P(z_0, z, \tau)$. This function may be analyzed into

$$P(z_0, z, \tau) = \rho(z) p(z - z_0, \tau) \qquad (20.23)$$

where $p(z - z_0, \tau)$ is the diffusion propagator. We now introduce the **pore-space correlation function**

$$G_\rho(z - z_0) = \frac{\langle \rho(0)\rho(z - z_0)\rangle}{\langle \rho^2\rangle} \qquad (20.24)$$

which is assumed to be even and independent from the absolute position.[6] The brackets indicate the average over all positions with the initial and final coordinates z_0 and z. Note that the pore-space correlation function approaches a finite constant value for distances large compared with the correlation length.

The substitution $\zeta = z_0 - z$ leads to the combined formula for the average magnetization at the coordinate z

$$M_z(z, \tau) = M_0 \int_{-\infty}^{\infty} G_\rho(\zeta)\, p(\zeta, \tau)\, \cos[k(z + \zeta)]\, d\zeta \qquad (20.25)$$

The cosine function can be broken down into cosine and sine functions of the individual terms of the sum argument. Only terms with even parity contribute to the integral so that

$$M_z(z, \tau) = M_0 \cos(kz) \int_{-\infty}^{\infty} G_\rho(\zeta)\, p(\zeta, \tau)\, \cos(k\zeta)\, d\zeta \qquad (20.26)$$

This convolution may be rewritten in the form

$$M_z(z, \tau) = M_0 \cos(kz) \int_{-\infty}^{\infty} \tilde{G}_\rho(k')\tilde{p}(k - k', \tau)\, dk' \qquad (20.27)$$

---

[6]This property implies "coarse-grain homogeneity and isotropy" of the system.

where $\tilde{G}_\rho$ and $\tilde{p}$ are the spatial Fourier transforms of pore-space correlation function G_ρ and the propagator p, respectively.

For ordinary diffusion in a homogeneous medium in bulk, we have $G_\rho(\zeta) = 1$, i.e., $\tilde{G}_\rho(k) = 2\pi\delta(k)$, and $\tilde{p}(k,\tau) = \exp\{-k^2 D\tau\}$. Inserting this into Eq. 20.27 readily reproduces the result at Eq. 20.16, i.e.,

$$M_z(z,\tau) = M_0 \cos(kz)\, e^{-k^2 D\tau} \tag{20.28}$$

CHAPTER 21

RF-Field-Gradient NMR Diffusometry

Diffusion can be also studied with the aid of gradients of the RF amplitude, B_1, instead of B_0 gradients considered up to now. Several experimental procedures have been suggested on this basis [55, 81, 131, 216, 249]. In the following, a B_1-gradient method for diffusion and flow measurements will be delineated which is entirely analogous to the above real-space evaluation scheme for B_0-gradient diffusometry. The principle is to prepare a magnetization grid and then, after a certain displacement interval, to image the modified grid [249]. This **magnetization-grid rotating-frame imaging (MAGROFI)** technique is a pure magnetization pattern generation/imaging procedure.

21.1
Magnetization-Grid Rotating-Frame Imaging

Figure 21.1 shows the RF-pulse sequence to be considered in the following. It consists of two pulses which are varied in length or, equivalently, in amplitude. The RF coil geometry is chosen in such a way that a strong gradient of the RF amplitude B_1 is produced. Due to this B_1 gradient, the sample is excited inhomogeneously and the flip angle varies along the gradient direction. The first pulse thus produces a longitudinal-magnetization grid apart from transverse magnetization.

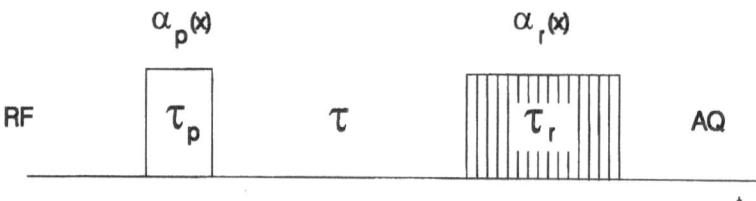

Fig. 21.1. RF-pulse sequence of the magnetization-grid rotating-frame imaging (MAGROFI) technique. The second (reading) pulse images the z-magnetization grid produced by the first (preparation) pulse. The imaging procedure is rotating-frame imaging (Chap. 34). The flip angles of the pulses, α_p and α_r, depend on the position x according to the B_1 gradient assumed in the x direction. AQ, acquisition interval. The reading pulse is incremented in length or, equivalently, in amplitude.

The initially excited transverse magnetization is of no further interest. It decays by transverse relaxation or is spoiled by moderate B_0 inhomogeneities which may deliberately be applied for this particular purpose.[1]

In the interval τ between the RF pulses, displacements by diffusion (or flow) tend to level (or shift) the longitudinal-magnetization grid. Competitively the grid is fading away by spin-lattice relaxation, of course. As several phenomena affect the magnetization grid during τ, one may generally speak of the "evolution interval." The final distribution of the longitudinal magnetization is then imaged with the aid of rotating-frame imaging (Chap. 34) using the same RF coil, i.e., the same B_1 gradients. A pseudo free-induction decay is generated by incrementing the length or amplitude of the reading pulse. The Fourier transform of these nutation-frequency encoded signal data forms a one-dimensional image which directly renders the effects of diffusion, flow, and relaxation on the magnetization grid visible.

The B_1 gradient is assumed to be aligned along the x axis of the laboratory frame. The influence of relaxation, diffusion, and flow during the RF pulses is neglected. That is, the RF pulses are assumed to be sufficiently short.

Assuming a homogeneous sample, the initial magnetization can be set homogeneously equal to the Curie magnetization,

$$M_z(x,0-) = M_0 = \text{const} \tag{21.1}$$

The first RF pulse serves for the preparation of the magnetization grid by producing a local flip angle $\alpha_p(x) = \gamma_n B_1(x)\tau_p$, where τ_p is the preparation pulse length. The z magnetization is thus

$$M_z(x,0+) = M_0 \cos \alpha_p(x) \tag{21.2}$$

The origin of the x axis is arbitrarily defined by any position where the effective flip angle is zero. For the sake of simplicity but without loss of generality, we further assume a linear variation of B_1 with x, i.e., a constant B_1 gradient G_1. The preparation pulse then defines the wavenumber

$$k_p = \gamma_n |G_1| \tau_p \tag{21.3}$$

Equation 21.2 can be rewritten with $\alpha_p(x) = k_p x$ as

$$M_z(x,0+) = M_0 \cos(k_p x) \tag{21.4}$$

The longitudinal magnetization after the interval τ is

$$M_z(x,\tau-) = M_0 - \left[M_0 - \overline{M_z(x,0+)}^{\tau} \right] e^{-\tau/T_1} \tag{21.5}$$

where we have assumed homogeneous and monoexponential spin-lattice relaxation which is not affected by transport of matter. This assumption again serves for ease

---

[1]Note that no spin echo or double-pulse excitation are employed in this case. The B_0 gradients are therefore irrelevant for the diffusion measurement.

of treatment and does not restrict the general applicability of the principle. The quantity $\overline{M_z(x,0+)}^\tau$ is the magnetization formed by those spins which are located at position x at time τ, irrespective of their initial position. We are dealing with partial spin ensembles defined by common positions x at time τ, but more or less extended spatial distributions at time zero. The link is given by the propagator for diffusion and flow during τ.

With the attenuation factor for ordinary diffusion (Eq. 19.26) and the phase shift expected for coherent flow in the interval τ, we obtain

$$\overline{M_z(x,0+)}^\tau = M_0 \, \cos(k_p x + k_p v \tau) \, e^{-D k_p^2 \tau} \tag{21.6}$$

where v is the flow velocity component along the B_1 gradient. A schematic representation of this longitudinal-magnetization grid can be found in Fig. 20.1. Inserting Eq. 21.6 into Eq. 21.5 gives

$$M_z(x,\tau-) = M_0 \left[1 + \left\{ \cos(k_p x + k_p v \tau) e^{-D k_p^2 \tau} - 1 \right\} e^{-\tau/T_1} \right] \tag{21.7}$$

This magnetization is then read out by the second RF pulse with the local flip angle $\alpha_r(x_r) = \gamma_n |G_1| x_r \tau_r$ incremented in a series of cycles. The pulse phase is assumed corresponding to the $-y$ rotating-frame axis. The free-induction signal at time t_{aq} is represented by

$$S(k_p, k_r, \tau, t_{aq}) \propto M_0 e^{i\omega_0 t_{aq} - t_{aq}/T_2}$$
$$\int \left[1 + \left\{ \cos(k_p x + k_p v \tau) e^{-D k_p^2 \tau} - 1 \right\} e^{-\tau/T_1} \right] B_1(x_r) \sin(k_r x_r) \, dx_r \tag{21.8}$$

where $k_r = \gamma_n |G_1| \tau_r$ and $\omega_0 = \gamma_n B_0$. The factor $B_1(x_r)$ takes into account the dependence of the detection sensitivity on the voxel-to-coil distance. Transverse relaxation is assumed to be monoexponential for simplicity again without loss of generality.

The magnetization-grid shape function is then reproduced by the Fourier transform from the k_r to the x_r space:

$$\begin{aligned}
\tilde{S}(k_p, x_r, \tau, t_{aq}) &= \mathcal{F}_{k_r}\{S(k_p, k_r, \tau, t_{aq})\} \\
&\propto M_0 B_1(x_r) e^{i\omega_0 t_{aq} - t_{aq}/T_2} \left\{ \left[1 - e^{-\tau/T_1} \right] \right. \\
&\quad \left. + e^{-\tau/T_1} \cos(k_p x_r + k_p v \tau) e^{-D k_p^2 \tau} \right\}
\end{aligned} \tag{21.9}$$

The curled bracket expression on the right-hand side consists of two terms. The first refers solely to spin-lattice relaxation during the evolution interval. The second term represents the magnetization grid modified by diffusion, flow, and spin-lattice relaxation. All three contributions can be unambiguously distinguished and determined in experimental evaluations.

- The baseline of the magnetization grid is determined by the first term. Evaluating the τ dependence of the baseline shift permits one to determine the **spin-lattice relaxation time** T_1 independent of any influence of flow or diffusion.

- The **diffusion coefficient** D can be evaluated by plotting the amplitude of the grid function (second term) vs k_p^2. For a given diffusion time τ, the amplitude is attenuated exponentially with increasing k_p^2. As only the grid amplitude is evaluated, this procedure is independent of any baseline shift by spin-lattice relaxation or of any flow shift of the grid. Note that neither T_1 nor T_2 losses during τ affect this evaluation.
- Alternatively, the **diffusion coefficient** can be evaluated on the basis of the τ instead of the k_p^2 dependence. In this case, the grid amplitudes can be corrected for attenuation by spin-lattice relaxation by exploiting the baseline shift recorded in the same experiment. Thus one first evaluates T_1 and then the diffusion coefficient.
- The **flow-velocity component** v along the gradient direction can be directly evaluated from the τ dependence of the grid phase. None of the other observables are required for this determination. The resolution of the velocity determination becomes better with increasing k_p values, i.e., with shorter repeat distances of the grid.

It should be noted that all evaluations mentioned above are intrinsically carried out with a certain spatial resolution. As one images the whole grid, the information is available for all positions along the B_1 gradient.

21.1.1
Rapid MAGROFI Diffusometry

The conventional rotating-frame imaging technique can be replaced by a multipulse variant producing a pseudo-FID in a single scan (Sect. 34.3 [314, 340]). In the absence of any frequency offsets, such pulse sequences are suitable for rapid data acquisition. A corresponding pulse scheme is shown in Fig. 21.2.

It may be favorable not to image the actual grid as it is effective during the diffusion interval, but to stretch the grid wavelength just for imaging purposes so that the required spatial resolution is feasible and independent of the actual grid wavelength during the diffusion interval. This can be achieved by applying a pulse compensating the variable part of the preparation pulse before the grid is imaged using the rapid rotating-frame imaging method. The diffusion coefficient is determined from the t_1 dependence of the amplitude of the stretched longitudinal-magnetization grid which is rendered as an image by a sine Fourier transformation of the signal $S = S(t_2)$ "stroboscopically" acquired in the gaps of the rapid rotating-frame imaging pulse train.

21.1.2
Experimental Aspects of MAGROFI Diffusometry

In the above treatment, (locally) constant B_1 gradients have been assumed for simplicity. This is not a stringent condition. Since the magnetization grid is rendered as an image, the local wave number k_p is known. That is, the local transport quantities can be evaluated as long as one refers to the same position within one data set. This

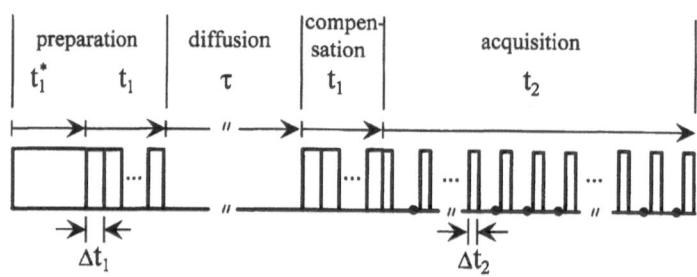

Fig. 21.2. B_1 gradient pulse sequence of the rapid rotating-frame imaging variant of MAGROFI diffusometry. The preparation pulse producing the longitudinal-magnetization grid is incremented in steps Δt_1 starting from a width t_1^*. The subsequent diffusion interval allows for displacements by self-diffusion or flow to be detected in the experiment. A further B_1 gradient pulse compensates the increments of the preparation pulse. The z magnetization grid is then rendered as a one-dimensional image with the aid of the rapid rotating-frame imaging sequence consisting of a train of read B_1 gradient pulses. The dots on the baseline indicate the dwell-time intervals in which the signal data points are acquired "stroboscopically". (Reproduced by permission from ref. [452])

is an important feature because the RF field gradients produced by the usual coil geometries tend to be rather inhomogeneous.

The required gradients of the RF amplitude can be produced by surface coils, anti-Helmholtz coils, or other nonsolenoidal geometries [81, 174, 199, 416, 417, 516]. For example, toroid cavity probes have been successfully employed for MAGROFI experiments [517]. In the case of this geometry, the radial RF field distribution offers an extremely wide k_p range to be covered at one time. In analogy to the supercon fringe field version of B_0 gradient diffusometry (Sect. 19.3), it is also possible to use the fringe RF field of solenoid coils of standard NMR probeheads [452].

For studies of small displacements, large wave numbers k_p are required, i.e., large B_1 gradients. Suitable coil geometries must therefore be optimized for this purpose. The B_1 gradients of surface coils of solenoid probes can be improved by reducing the coil diameter as far as possible. The fringe-field gradient of a three-turn solenoid with a diameter of 5 mm reaches 3.3 T/m.

Figure 21.3 shows a series of experimental magnetization grids recorded with the aid of the MAGROFI pulse sequence (Fig. 21.1) as described above. The influence of the evolution interval τ on the baseline, and the variation of the grid wave number k_p with τ_p, is demonstrated. Longer evolution intervals τ lead to reduced grid amplitudes as a common consequence of diffusion and relaxation (trace **(b)** relative to trace **(a)** in Fig. 21.3). The diffusion effect alone reveals itself by the reduction of the grid amplitude with increasing k_p values (keeping τ constant), as demonstrated by the profiles (Fig. 21.3c) measured in acetone at room temperature. As a consequence of spin-lattice relaxation, the baseline of the magnetization grid is shifted upward with increasing evolution intervals (traces **(b)** and **(c)** relative to trace **(a)** in Fig. 21.3).

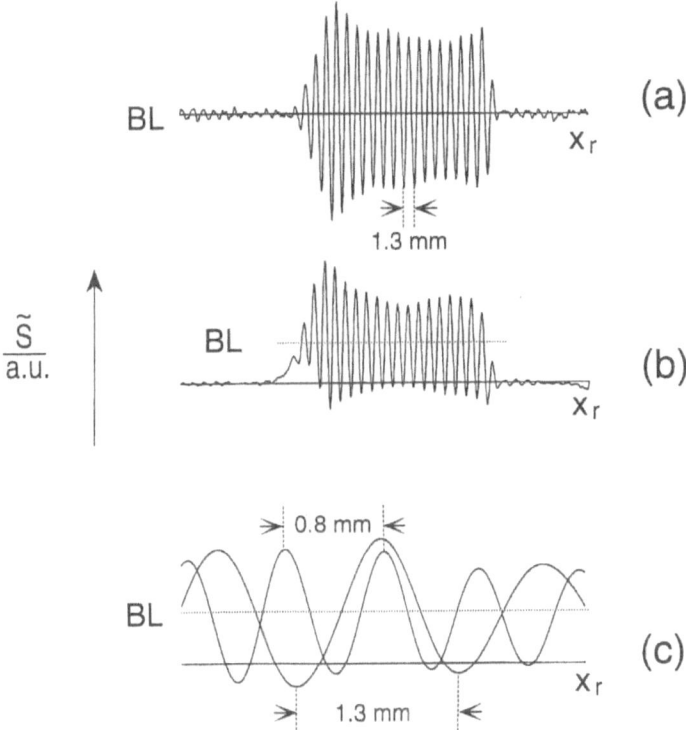

Fig. 21.3. Typical magnetization grids prepared in water or acetone samples by using different parameters of the MAGROFI pulse sequence (Fig. 21.1). The B_1 gradient was produced with the aid of a conventional surface coil with a diameter of 5 cm. The grids are just as measured, i.e., the voxel-to-coil distance dependence of the sensitivity and the nonlinearity of the B_1 decay with the distance have not been compensated. The reading pulse was incremented 256 times in steps of 39 μs. With an average B_1 gradient corresponding to 16 kHz/cm, this results in a pixel resolution of 0.2 mm. The maximum length of the reading pulse was 10 ms. The field of view was 5.1 cm. The displayed grids were recorded with $\tau_p = 3$ ms (wavelength, 1.3 mm) and $\tau_p = 5$ ms (wavelength, 0.8 mm). The evolution interval was: a) $\tau = 14$ ms; b,c) $\tau = 1$ s. BL, baseline. (Reproduced by permission from ref. [249])

Figure 21.4 shows four typical magnetization grids of a water sample investigated with the rapid rotating-frame imaging variant of MAGROFI in the fringe field of a solenoid probe coil [452]. The grid amplitude decays faster close to the coil, where the RF amplitude B_1 and its gradient are higher. As the quantity to be further evaluated, the integral over the half-waves of the grid can be formed (shaded areas in Fig. 21.4). For each half-wave position the dependence of the integral values on the grid wave number k_p can be evaluated by fitting the function $A_1 \exp(-k_p^2 D\tau) + A_2$, where A_1 and A_2 are constant parameters.

Rotating-frame images recorded with B_1 gradients varying in space must be evaluated in the proper way, of course. The usual Fourier transform processing of the pseudo FID assumes constant gradients. With gradients varying with the position,

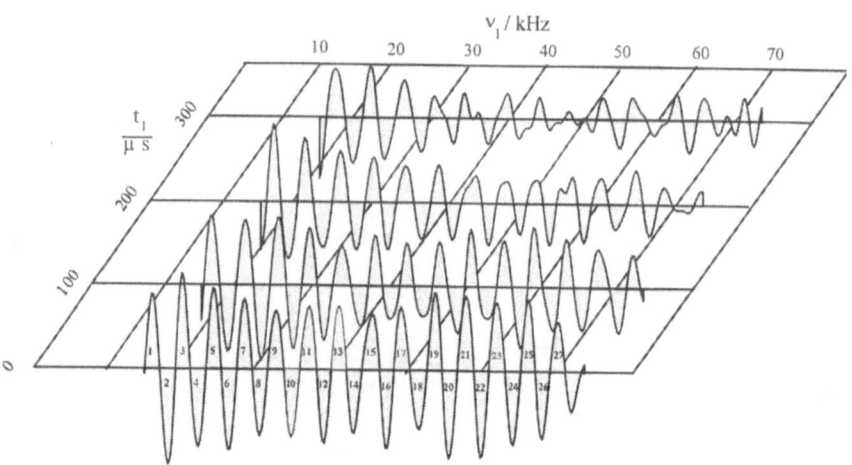

Fig. 21.4. Typical magnetization grids recorded with the rapid rotating-frame imaging variant of MAGROFI diffusometry (Fig. 21.2) in a water sample at room temperature. The B_1 gradients (up to 3.3 T/m) originated from the fringe RF field of a three-turn probe coil with a diameter of 5 mm. The right-hand side of the grids corresponds to the position closest to the coil, where the B_1 gradient is strongest. The parameters of the pulse sequence were $t_1^* = 220$ μs, $\tau = 0.1$ s and $t_2 = 128 \times 4$ μs. (Reproduced by permission from ref. [452])

the coordinate axis of the Fourier transform is correspondingly distorted. It can, however, be corrected easily if the spatial distribution of the RF field is known. The $B_1(r)$ distribution can either be calculated from the coil geometry or measured by conventional B_0 gradient imaging of the excitation distribution induced by the RF coil. The corrected rotating-frame images of the magnetization grid then permits the evaluation of the local k_p values.[2]

The only prerequisite for the compensation of variations of the B_1 gradient within the sample by nonlinear calibration of the imaging axis is that locations of equal gradients are simultaneously positions of equal B_1 values. The reason is that rotating-frame imaging renders the superimposed projections of the magnetization on the local gradient directions globally referred to as the "imaging axis." In order to avoid the influence of different preparation wave numbers at the same position of the imaging axis, it is advisable to use sample shapes adapted to the symmetry of the RF field.

21.2
Comparison of B_0 and B_1 Gradient Methods

The complete analogy of the MAGROFI measuring principle to that of PGSE and SGSE techniques (in the real-space reading) became evident with the above treat-

---

[2]Carrying this option to the extreme should even make it possible to cover at one time the whole k_p range needed for a diffusion measurement.

ment. It is also obvious that magnetization grid experiments based on B_1 and on B_0 gradients in principle could even be "mixed" if there was any need to do so. That is, the preparation could be performed using B_1 gradients and the imaging part could be based on B_0 gradients, and vice versa.

The MAGROFI method provides information on diffusion (or incoherent flow), coherent flow, and spin-lattice relaxation at one time without mutual interference. Hahn, stimulated, or rotary echoes are not exploited in the measuring principle. Transverse relaxation in the free-evolution intervals - a crucial factor limiting NMR B_0 gradient studies - is therefore irrelevant. Apart from signal detection, the method is based solely on the longitudinal component of the magnetization.

As no free evolution of coherences is involved, B_0 gradients or internal gradients due to variations of the magnetic susceptibility in heterogeneous samples are inessential as long as the RF bandwidth of the pulses exceeds the width of the resonance distribution. In this respect, B_1 gradient-techniques are certainly superior to methods employing gradients of the main field. Moreover, alternating-gradient sequences are required as a remedy of susceptibility artifacts in heterogeneous samples [103].

While there is no free coherence evolution involved in the MAGROFI measuring principle, transverse relaxation during the RF pulses must be considered, because this may spoil the preparation of the grid or the rotating-frame imaging procedure if the RF pulses are not kept short enough [131, 340]. This problem is, however, more of a technical nature. With sufficient transmitter power and suitable probeheads it should be possible to restrict the pulses to lengths much shorter than the transverse relaxation time in the presence of RF fields. Likewise the potential influence of translations during the RF pulses [35] can be kept negligible. Thus, spin-lattice relaxation in the diffusion interval remains the only time scale limit inherent to MAGROFI diffusometry.

It is often desirable to combine diffusometry with high-resolution spectroscopy (e.g. [525]). As no B_0 gradient is employed, the MAGROFI method is ideally suited for this purpose. In this case, the FIDs as well as the pseudo FIDs must be recorded and evaluated in full. Diffusion properties can thus be correlated with spectral lines and with the position.

CHAPTER 22

Examples for Anomalous Self-Diffusion

Anomalous self-diffusion[1] is the result of restrictions and of the dimensionality of the space accessible by the diffusing particles. This, of course, is a matter of the length and time scales considered. Displacements far beyond the correlation length characterizing the structural restrictions tend to follow the ordinary Einstein law (Eq. 18.5) with a correspondingly reduced diffusion coefficient. On shorter length scales, topological constraints cause all sorts of deviations. Even the molecular interaction with a plane surface may give rise to strong anomalies as already discussed in Sect. 14.3.2 in context with Lévy walks.

A frequent assumption is that the mean squared displacement follows a power law of the type at Eq. 18.8 which is expected for diffusion on random percolation clusters, for instance. The consequences for field-gradient NMR diffusometry have already been treated under the assumption that the probability density of displacements can be approximated by a Gaussian function (see Eqs. 19.22, 19.23, and 19.31). As a typical application we will consider **adsorbate diffusion in lacunar systems**.

In cases where Gaussian probability densities are not justified, a treatment specifically adapted to the system under consideration is required. An example is **segment diffusion in entangled-polymer liquids**. In this context we will examine the echo-attenuation behavior expected for the so-called reptation/tube model [126].

22.1
Anomalous Diffusion in Lacunar Systems

Random percolation clusters are often considered as model structures for lacunar systems. The question whether self-diffusion on such clusters shows anomalous features or not is a matter of the displacement length scale probed in the experiments. Only below the correlation length of the confining system does diffusion tend to be anomalous [373]. NMR diffusometry is appropriate for such studies, because the length scales intrinsic to this method are often in the order of typical correlation lengths, and moreover can be varied by choosing different field gradients and diffusion times.

---

[1]This is to be distinguished from anomalous interdiffusion which may deviate from Fick's law as a consequence of combined diffusion/reaction processes, for instance. Examples studied by magnetic-resonance imaging are briefly discussed in Sect. 26.3.

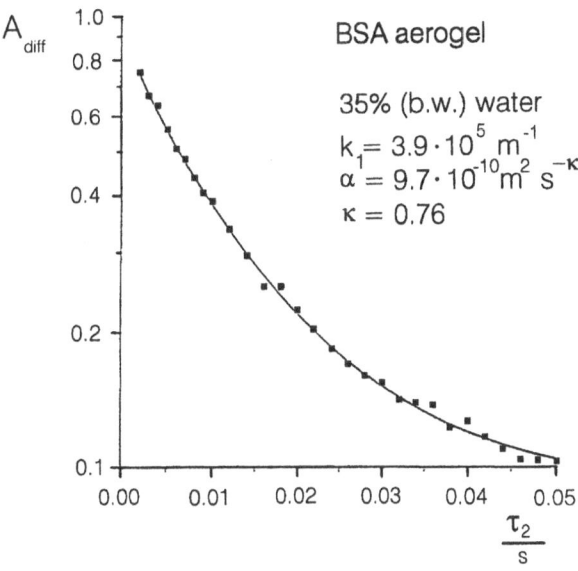

Fig. 22.1. Echo attenuation curve of water adsorbed in an aerogel of BSA prepared by lyophilization of an aqueous solution. The line represents a fit of Eq. 19.32. The field gradient was 9.04 T/m. (Reproduced by permission from ref. [246])

Fine particle agglomerates and aerogels are suitable lacunar systems in which anomalous diffusion shows up in NMR diffusometry. The attenuation curve in Fig. 22.1, for instance, was measured in a hydrated protein aerogel. It clearly indicates anomalous behavior, i.e., a non-exponential decay. This is in contrast to ordinary hydrated protein samples which were not prepared as an aerogel.

22.2
Reptation/Tube Model

22.2.1
The Doi/Edwards Limits

A polymer chain in a matrix of other polymer chains with molecular weights above the critical value M_c is subject to topological constraints often referred to as the "tube." The tube is a replacement scheme, i.e., a model globally representing the interactions with the matrix. The tube model is characterized by a number of specific power laws concerning translational diffusion. Among these, the time dependences of the mean-squared segment displacements are of particular interest.

Consider linear polymer chains consisting of N Kuhn segments of length b. The number of Kuhn segments corresponding to half of the critical molecular weight, M_c, is designated by N_e. The characteristic length scales are defined by the step length of the so-called primitive path, a, and by the radius of gyration

$$R_g = b \left(\frac{N}{6}\right)^{1/2} \tag{22.1}$$

The characteristic time scales are specified by a series of time constants. τ_s is the segment reorientation time representing the orientationally restricted segment motions taking place locally in an Å length scale. The "entanglement time"

$$\tau_e = N_e^2 \tau_s \tag{22.2}$$

is the period during which lateral displacements occur which are not yet affected by the tube constraint effect. These motions are obviously of a semi-local and incoherent nature. The so-called Rouse relaxation time

$$\tau_R = N^2 \tau_s \tag{22.3}$$

is the time after which *coherent* motions of the whole chain become relevant. Finally the time constant

$$\tau_d = \frac{N^3}{N_e} \tau_s \tag{22.4}$$

is defined as the tube disengagement time. It represents the ultimate correlation time because a chain having migrated out of its initial tube tends to have a conformation entirely uncorrelated to the initial one.

With these definitions the following limits for the mean-squared displacement behavior are expected on the basis of the Doi/Edwards model [126, 141].
Limit I ($\tau_s \ll t \ll \tau_e$ or $b^2 < \langle r^2 \rangle < a^2$):

$$\langle r^2 \rangle = b^2 \left(\frac{t}{\tau_s}\right)^{1/2} \tag{22.5}$$

Limit II ($\tau_e \ll t \ll \tau_R$ or $a^2 < \langle r^2 \rangle < \sqrt{6} R_g a$):

$$\langle r^2 \rangle = b^2 N_e^{1/2} \left(\frac{t}{\tau_s}\right)^{1/4} \tag{22.6}$$

Limit III ($\tau_R \ll t \ll \tau_d$ or $\sqrt{6} R_g a < \langle r^2 \rangle < 6 R_g^2$):

$$\langle r^2 \rangle = b^2 \left(\frac{N_e t}{N \tau_s}\right)^{1/2} \tag{22.7}$$

Limit IV ($t \gg \tau_d$ or $6 R_g^2 < \langle r^2 \rangle$):

$$\langle r^2 \rangle = 6 D t \tag{22.8}$$

where D is the center-of-mass self-diffusion coefficient of the chain.

Limit IV corresponds to segment displacements large enough to be identified with those of the center of mass of a chain. The center-of-mass diffusion coefficient was predicted to obey [111]

$$D = \frac{\pi^2 a^2 k_B T}{6 b^2 \zeta N^2} \tag{22.9}$$

where ζ is the friction coefficient of a Kuhn segment, k_B is Boltzmann's constant, and T is the absolute temperature. The chain-length dependence inherent to this expression has been verified experimentally several times [299].

With the excessive field gradients of the fringe field of superconducting magnets, root mean-square displacements down to the 10 nm range are accessible by NMR. This is to be compared with the radius of gyration of polydimethylsiloxane (PDMS) coils, for instance, which exceeds several hundred Å for molecular weights above 100 000. Thus one is already dealing with processes in length scales within the polymer coil to which the above limits of the Doi/Edwards should apply.

22.2.2
Evaluation Formula for PGSE/SGSE Experiments

According to Eq. 19.5, the echo attenuation factor may be written in the form of the ensemble average

$$A_{diff}(k^2, t) = \langle \exp\{i\mathbf{k} \cdot \mathbf{r}(t)\} \rangle_{\mathbf{r}} \tag{22.10}$$

where the "wave vector", $\mathbf{k} = \gamma_n \delta \mathbf{G}$, is used as above. δ is the width of the field-gradient pulses in PGSE experiments, and the length of the coherence evolution intervals in SGSE experiments. $t \gg \delta$ is the effective diffusion time between the gradient pulses, and $\mathbf{G}$ is the gradient of the magnetic flux density during the gradient pulses.

Equation 22.10 is now to be applied to the reptation/tube model [145]. The segment displacement, $\mathbf{r}(t)$, during t may be analyzed into $\mathbf{r}(t) = \mathbf{r}_{cm}(t) + \bar{\mathbf{r}}(t)$, where $\mathbf{r}_{cm}(t)$ is the displacement of the center of gravity of the polymer chain, and $\bar{\mathbf{r}}(t)$ is the segment displacement relative to a reference frame fixed in the center of gravity. In the limit $t \ll \tau_d$, $\bar{\mathbf{r}}(t)$ essentially is the segment displacement in the initial tube.

Center-of-gravity diffusion follows Fick's law, whereas segment displacements in the tube do not. As the two displacement contributions are uncorrelated, we may factorize Eq. 22.10 resulting in

$$A_{diff}(k^2, t) = \langle \exp\{i\mathbf{k} \cdot \bar{\mathbf{r}}(t)\} \rangle_{\bar{\mathbf{r}}} \langle \exp\{i\mathbf{k} \cdot \mathbf{r}_{cm}(t)\} \rangle_{\mathbf{r}_{cm}} = \tilde{A}_{diff}(k^2, t) \exp\{-k^2 D t\} \tag{22.11}$$

where $\tilde{A}_{diff}(k^2, t) = \langle \exp\{i\mathbf{k} \cdot \bar{\mathbf{r}}(t)\} \rangle_{\bar{\mathbf{r}}}$, and D is the center-of-gravity diffusion coefficient. The problem is now to find an expression for the attenuation factor for segment diffusion in the tube.

The segment displacement in the Euclidean space, $\bar{\mathbf{r}}(t)$, is connected with a displacement $s(t)$ measured in curvilinear coordinates along the tube. That is, the end-to-end vector of the curvilinear path of length $s(t)$ is given by $\bar{\mathbf{r}}(t)$. Assuming a Gaussian probability density for the end-to-end vector of a given path length $s(t)$, we find

$$
\tilde{A}_{diff}(k^2, t) = \left\langle \int \left(\frac{2\pi}{3} a|s(t)|\right)^{-3/2} \exp\left\{-\frac{3\bar{r}^2}{2a|s(t)|}\right\} \exp\{ik \cdot \bar{r}\} \, d^3\bar{r} \right\rangle_s
$$

$$
= \left\langle \exp\left\{-\frac{1}{6}k^2 a|s(t)|\right\}\right\rangle_s \tag{22.12}
$$

For $t \ll \tau_d$ the curvilinear segment displacements may be considered as the result of a one-dimensional ordinary diffusion process, i.e., they are distributed according to a Gaussian probability density. Hence

$$
\tilde{A}_{diff}(k^2, t) = \int_{-\infty}^{\infty} (2\pi\langle s^2(t)\rangle)^{-1/2} \exp\left\{-\frac{s^2}{2\langle s^2(t)\rangle}\right\} \exp\left\{-\frac{1}{6}k^2 a|s(t)|\right\} \, ds
$$

$$
= \exp\left\{\frac{k^4 a^2 \langle s^2(t)\rangle}{72}\right\} \text{erfc}\left\{\frac{k^2 a\sqrt{\langle s^2(t)\rangle}}{6\sqrt{2}}\right\} \tag{22.13}
$$

The mean-squared curvilinear segment displacement is given by [126]

$$
\langle s^2(t)\rangle = 2\frac{D_0}{N}t + \frac{2Nb^2}{3\pi^2}\sum_{p=1}^{\infty}\frac{1}{p^2}\left(1 - \exp\left\{-\frac{tp^2}{\tau_R}\right\}\right)
$$

$$
\approx 2\frac{D_0}{N}t + \frac{2b\sqrt{D_0 t}}{\sqrt{3\pi} + 18\frac{\sqrt{D_0 t}}{Nb}} \tag{22.14}
$$

where $D_0 = k_B T/\zeta$ is the monomeric diffusion coefficient. ζ is the friction constant of a Kuhn segment of length b. The polymer chain is assumed to be composed of N Kuhn segments. The approximation implies the correct limits for $t \ll \tau_R$ and $t \gg \tau_R$ (see [126]).

Equation 22.14 is valid in the limit $t \ll \tau_d$. In the opposite limit, i.e., $t \gtrsim \tau_d \gg \tau_r$, the treatment of the mean-squared curvilinear segment displacement can be considered as a one-dimensional restricted-diffusion problem with reflecting boundaries at curvilinear tube coordinates $x_s = 0$ and $x_s = L \approx L_t$ where L_t is the tube length. The "free-diffusion" length, L, depends on the position of the considered segment in the tagged chain. Let $P(x_s, x_{s0}, t)$ be the probability density that the segment is at a position x_s at time t if it was at the position x_{s0} at time 0. The diffusion equation,

$$
\frac{\partial}{\partial t}P(x_s, x_{s0}, t) = \frac{D_0}{N}\frac{\partial^2}{\partial x_s^2}P(x_s, x_{s0}, t) \tag{22.15}
$$

must be solved for the boundary conditions

$$
\frac{\partial}{\partial x_s}P(x_s, x_{s0}, t)|_{x_s=0,L} = 0 \tag{22.16}
$$

The solution is

$$P(x_s, x_{s0}, t) = \frac{1}{L} + \sum_{p=1}^{\infty} \frac{2}{L} \cos\left(\frac{\pi}{L} p x_s\right) \cos\left(\frac{\pi}{L} p x_{s0}\right) \exp\left\{-\left(\frac{\pi}{L} p\right)^2 \frac{D_0}{N} t\right\} \quad (22.17)$$

The mean-squared curvilinear displacement is then

$$\langle s^2(t) \rangle = \int_0^L dx_s \int_0^L \frac{dx_{s0}}{L} (x_s - x_{s0})^2 P(x_s, x_{s0}, t) \quad (22.18)$$

Combining Eqs. 22.17 and 22.18 leads to

$$\langle s^2(t) \rangle = \left(\frac{4L}{\pi^2}\right)^2 \sum_{p \, odd} \frac{1}{p^4} \left(1 - \exp\left\{-\left(\frac{\pi p}{L}\right)^2 \frac{D_0}{N} t\right\}\right) \approx \frac{2\frac{D_0}{N} t}{1 + \frac{12 D_0 t}{NL^2}} \quad (22.19)$$

$$= \begin{cases} 2\frac{D_0}{N} t & \text{if} \quad t \ll \frac{L^2 N}{\pi^2 D_0} \propto \tau_d \\ \frac{L}{6} & \text{if} \quad t \gg \frac{L^2 N}{\pi^2 D_0} \propto \tau_d \end{cases} \quad (22.20)$$

The latter formulae are valid in the limit $t \gg \tau_R$. An expression of the mean-squared curvilinear segment displacement in the whole time range of interest is obtained by combining the second term of Eq. 22.14 and Eq. 22.19, i.e.,

$$\langle s^2(t) \rangle = \frac{2\frac{D_0}{N} t}{1 + \frac{12 D_0 t}{NL^2}} + \frac{2b\sqrt{D_0 t}}{\sqrt{3\pi} + 18\frac{\sqrt{D_0 t}}{Nb}} \quad (22.21)$$

According to Eqs. 22.11 and 22.13, the total spin-echo attenuation factor is

$$A_{diff}(k^2, t) = \exp\left\{\frac{k^4 a^2 \langle s^2(t) \rangle}{72}\right\} \text{erfc}\left\{\frac{k^2 a \sqrt{\langle s^2(t) \rangle}}{6\sqrt{2}}\right\} \exp\{-k^2 Dt\} \quad (22.22)$$

where $\langle s^2(t) \rangle$ is given by Eq. 22.21.

Let us now determine the quantity L. In the limit $t \gg \tau_d$, $k \to 0$, Eq. 22.10 may be approximated by

$$A_{diff}(k^2, t) = \exp\left\{-\frac{1}{6} k^2 \left(2R_g^2 + 6Dt\right)\right\} \quad (22.23)$$

where $R_g = Nb^2/6$ is the radius of gyration. Equation 22.22 may be approximated in the same limit by

$$A_{diff}(k^2, t) = \exp\left\{-\frac{1}{6} k^2 \left(\sqrt{\frac{2}{\pi}} k^2 a \sqrt{\langle s^2(t) \rangle} + 6Dt\right)\right\} \quad (22.24)$$

where we have used the approximation $\text{erfc}(x) \approx 1 - 2x/\sqrt{\pi} \approx \exp\{-2x/\sqrt{\pi}\}$ which is valid for $x \ll 1$.

Equations 22.23 and 22.24 must become identical in the limit $t \gg \tau_d$. That is

$$\sqrt{\frac{2}{\pi}} k^2 a \sqrt{\langle s^2(t) \rangle} = k^2 2R_g^2 = \frac{1}{3} N b^2 k^2 \tag{22.25}$$

Using Eq. 22.20, we find

$$\sqrt{\frac{2}{\pi}} a \sqrt{\frac{L^2}{6}} = \frac{1}{3} N b^2 \tag{22.26}$$

so that

$$L = \sqrt{\frac{2}{\pi} \frac{N b^2}{a}} = \sqrt{\frac{\pi}{3}} L_t \approx L_t \tag{22.27}$$

where $L_t = N b^2/a$ [126].

The final result of the spin-echo attenuation by segment diffusion in the reptation/tube model is

$$A_{diff}(k^2, t) = \exp \left\{ \frac{k^4 a^2 \langle s^2(t) \rangle}{72} \right\} \mathrm{erfc} \left\{ \frac{k^2 a \sqrt{\langle s^2(t) \rangle}}{6\sqrt{2}} \right\} \exp\{-k^2 D t\} \tag{22.28}$$

$$\langle s^2(t) \rangle = \frac{2\frac{D_0}{N} t}{1 + \frac{12 D_0 t}{N L_t^2}} + \frac{2b\sqrt{D_0 t}}{\sqrt{3\pi} + 18\frac{\sqrt{D_0 t}}{Nb}} \tag{22.29}$$

where $D_0 = k_B T/\zeta$, $L_t = N b^2/a$, $D = D_0 N_e/(3N^2)$, $a = b\sqrt{N_e}$, and N_e is the number of Kuhn segments corresponding to the step length a of the primitive path.

In the limit $t \gg \tau_d$, the factor $\exp\{-k^2 D t\}$ dominates. The spin-echo attenuation curve then corresponds to ordinary diffusion of the center of gravity. The opposite limit, $t \ll \tau_d$, is connected with anomalous (segment) diffusion, i.e., the above factor virtually does not vary in this time scale. The spin-echo attenuation is then governed by

$$A_{diff}(k^2, t) = \exp \left\{ \frac{k^4 a^2 \langle s^2(t) \rangle}{72} \right\} \mathrm{erfc} \left\{ \frac{k^2 a \sqrt{\langle s^2(t) \rangle}}{6\sqrt{2}} \right\}$$

$$\approx \begin{cases} 1 - \sqrt{\frac{2}{\pi}} \frac{k^2 a \sqrt{\langle s^2(t) \rangle}}{6} & \text{if } \frac{k^2 a \sqrt{\langle s^2(t) \rangle}}{6\sqrt{2}} \ll 1 \\[2ex] 6\sqrt{\frac{2}{\pi}} \frac{1}{k^2 a \sqrt{\langle s^2(t) \rangle}} & \text{if } \frac{k^2 a \sqrt{\langle s^2(t) \rangle}}{6\sqrt{2}} \gg 1 \end{cases} \tag{22.30}$$

where we have used the approximations $\mathrm{erfc}(x) \approx 1 - 2x/\sqrt{\pi}$ for $x \ll 1$, and $\mathrm{erfc}(x) \approx \exp\{-x^2\}/(x\sqrt{\pi})$ for $x \gg 1$. Equation 22.30 implies the situations expected for limit II, i.e., $\tau_e \ll t \ll \tau_R$, and limit III, i.e., $\tau_R \ll t \ll \tau_d$, of the tube/reptation model. For limit II, we must put $\langle s^2(t) \rangle \approx 2b\sqrt{D_0 t}/\sqrt{3\pi}$, whereas $\langle s^2(t) \rangle \approx 2 D_0 t/N$ in limit III. A discussion of Eq. 22.30 in comparison with experimental data can be found in [149, 421].

The result Eq. 22.30 is to be compared with the attenuation formulae found for anomalous diffusion assuming a Gaussian propagator for Euclidean-space displacements (see Eqs. 19.23 and 19.31). The discrepancy emphasizes that the latter assumption is inadequate for the situation in the tube/reptation model. In this case, the curvilinear displacements inside the tube have a Gaussian character, whereas the Euclidean-space displacements strongly deviate from this behavior.

CHAPTER 23

Exchange

The objective of one- or multiple-dimensional exchange spectroscopy is to probe fluctuations between different resonance frequencies a spin experiences. The exchange processes as well as the fluctuations of the resonance frequency can be of extremely different natures. The resonance frequency may vary for example due to

- sites subject to different chemical shift
- isomers with different chemical shift
- environments of different magnetic susceptibility
- anisotropic spin interactions in combination with different orientations in solid-like materials
- positions in a field gradient

There may be a discrete or continuous variation of spin environments which are connected with different resonance frequencies. The term "exchange" is understood to comprise such different mechanisms as

- chemical exchange of the spin-bearing particle
- structural or isomeric fluctuations of the spin environment so that the resonance frequency becomes a function of time
- diffusion of the considered molecule in environments with varying magnetic susceptibility
- immaterial spin exchange by spin flip/flop processes (spin diffusion or cross relaxation)
- reorientations of molecules so that the secular terms of anisotropic spin interactions are modulated
- translational diffusion in the presence of a field gradient (see Chap. 18)
- time-varying coherent flow (if "resonance frequency" is considered to be equivalent to "phase-variation rate" of velocity phase-encoding by gradient pulses [80])

In this chapter we restrict ourselves to the treatment of a finite set of discrete spin environments between which exchange can occur. We will first consider the effect of exchange on the evolution of spin states and coherences, and then turn to two- (or multiple-) dimensional exchange experiments.

23.1
Equation of Motion for Discrete Spin Environments

The exchange network is assumed to comprise N discrete different "environments" the spins can experience. A spin environment is characterized by the chemical composition, the molecular mobility, the orientation of the spin system, the spin interactions that contribute, and possibly other factors. These environments are specified by the resonance frequencies, $\omega_{0,j}$, and longitudinal and transverse relaxation times, $T_{1,j}$, and $T_{2,j}$. Other relaxation times like those referring to longitudinal order or multiple-quantum coherences can also be brought into play if relevant in the course of the pulse sequence considered. The interaction between spins in different environments is neglected.

For each spin environment, a complex transverse partial magnetization

$$m_j = M_{x,j} + i M_{y,j} \tag{23.1}$$

and a partial z magnetization $M_{z,j}$ with the equilibrium values $M_{0,j}$ can be defined where $j = 1 \ldots N$. On the basis of these magnetizations, N-dimensional magnetization vectors can be formed:

$$
m = \begin{pmatrix} m_1 \\ \cdot \\ \cdot \\ m_j \\ \cdot \\ \cdot \\ m_N \end{pmatrix}, \qquad
M_z = \begin{pmatrix} M_{z,1} \\ \cdot \\ \cdot \\ M_{z,j} \\ \cdot \\ \cdot \\ M_{z,N} \end{pmatrix}, \qquad
M_0 = \begin{pmatrix} M_{0,1} \\ \cdot \\ \cdot \\ M_{0,j} \\ \cdot \\ \cdot \\ M_{0,N} \end{pmatrix} \tag{23.2}
$$

We will designate these vectors as transverse and longitudinal **"magnetization system vectors"** in order to distinguish them from ordinary magnetization vectors. Bloch's equations can then be extended by exchange terms and rewritten as new matrix equations of motion

$$
\boxed{
\begin{aligned}
\frac{d m(t)}{d t} &= \ell \, m(t) \\[2mm]
\frac{d M_z(t)}{d t} &= L \, [M_z(t) - M_0]
\end{aligned}
} \tag{23.3}
$$

called **Hahn/Maxwell/McConnell equations** [187, 309]. The coefficients ℓ and L are referred to as HMM matrices.

Recall that, as concerns transverse relaxation, Bloch's as well as these Hahn/Maxwell/McConnell (HMM) equations stipulate beforehand that the motional-averaging condition is fulfilled. The general solutions of the HMM equations, Eqs. 23.3, are

$$m(t) = \exp\{\ell t\} \, m(0) \tag{23.4}$$
$$M_z(t) = M_0 - \exp\{L t\} \, [M_0 - M_z(0)] \tag{23.5}$$

23.1.1
Interpretation of the HMM Matrices

The matrices ℓ and L are defined by

$$\ell = i\Omega - R_2 + K \tag{23.6}$$

and

$$L = -R_1 + K \tag{23.7}$$

The matrix of the resonance frequencies within the exchange network is

$$\Omega = \begin{pmatrix} \omega_{0,1} & \cdot & \cdot & \cdot & 0 \\ & \cdot & \cdot & & \cdot \\ \cdot & & \cdot & & \cdot \\ \cdot & & & \cdot & \cdot \\ 0 & \cdot & \cdot & \cdot & \omega_{0,N} \end{pmatrix} \tag{23.8}$$

If this were the only finite contribution to the matrix ℓ in the HMM equations, Eqs. 23.3, we would have $dm/dt = i\Omega m$. It is easy to show that this corresponds to a set of the ordinary precession equations for the different m components.

The relaxation-rate matrices,

$$R_1 = \begin{pmatrix} 1/T_{1,1} & \cdot & \cdot & \cdot & 0 \\ & \cdot & \cdot & & \cdot \\ \cdot & & \cdot & & \cdot \\ \cdot & & & \cdot & \cdot \\ 0 & \cdot & \cdot & \cdot & 1/T_{1,N} \end{pmatrix} \;;\quad R_2 = \begin{pmatrix} 1/T_{2,1} & \cdot & \cdot & \cdot & 0 \\ & \cdot & \cdot & & \cdot \\ \cdot & & \cdot & & \cdot \\ \cdot & & & \cdot & \cdot \\ 0 & \cdot & \cdot & \cdot & 1/T_{2,N} \end{pmatrix} \tag{23.9}$$

represent the potential distribution of relaxation rates in the diverse environments. If these matrices were the only contributions to ℓ and L in the HMM equations, Eqs. 23.3, we would have $dm/dt = -R_2 m$ and $dM_z/dt = -R_1[M_z - M_0]$ instead, i.e., component by component, the ordinary rotating-frame Bloch equations.

Finally, the exchange-kinetic matrix, which is of particular interest here, is defined as

$$K = \begin{pmatrix} k_{1,1} & \cdot & \cdot & \cdot & k_{1,N} \\ & \cdot & \cdot & & \cdot \\ \cdot & & \cdot & & \cdot \\ \cdot & & & \cdot & \cdot \\ k_{N,1} & \cdot & \cdot & \cdot & k_{N,N} \end{pmatrix} \tag{23.10}$$

where the elements $k_{f,i}$ are the transfer-rate constants from the environment no. i to environment no. f. Considering the effect of this matrix on the magnetization system vectors separately from the other contributions in the HMM equations, Eqs. 23.3, leads to the kinetic equations $dm/dt = Km$ and $dM_z/dt = K[M_z - M_0]$. The solutions of these equations represent the time dependences of the magnetizations due to exchange between the different environment.

In the equilibrium limit, the populations of the environments and, hence, all magnetizations become stationary. With respect to exchange, this stipulates the condition

$$\frac{d}{dt}\begin{pmatrix} M_{0,1} \\ \cdot \\ \cdot \\ \cdot \\ M_{0,N} \end{pmatrix} = \begin{pmatrix} k_{1,1} & \cdot & \cdot & \cdot & k_{1,N} \\ \cdot & & & & \cdot \\ \cdot & & \cdot & & \cdot \\ \cdot & & & & \cdot \\ k_{N,1} & \cdot & \cdot & \cdot & k_{N,N} \end{pmatrix}\begin{pmatrix} M_{0,1} \\ \cdot \\ \cdot \\ \cdot \\ M_{0,N} \end{pmatrix} = 0 \tag{23.11}$$

or

$$\sum_{i=1}^{N} k_{j,i} M_{0,i} = 0 \tag{23.12}$$

Moreover, the principle of microscopic reversibility requires a detailed balance for each exchange path. The transfer rates back and forth must be the same for each exchange path:

$$k_{i,j} M_{0,j} = k_{j,i} M_{0,i} \tag{23.13}$$

Combining the conditions at Eqs. 23.12 and 23.13 leads to

$$k_{ii} = -\sum_{j \neq i} k_{i,j} \tag{23.14}$$

so that the exchange-kinetic matrix can be rewritten in the form

$$K = \begin{pmatrix} -\sum_{j \neq 1} k_{1,j} & \cdot & \cdot & \cdot & k_{1,N} \\ \cdot & & \cdot & & \cdot \\ \cdot & & & \cdot & \cdot \\ \cdot & & & & \cdot \\ k_{N,1} & \cdot & \cdot & \cdot & -\sum_{j \neq N} k_{N,j} \end{pmatrix} \tag{23.15}$$

In terms of the populations p_i, this result is equivalent to the familiar set of kinetic equations

$$\frac{dp_i}{dt} = \sum_{j=1}^{N} \left(k_{i,j} p_j - k_{j,i} p_i \right) \tag{23.16}$$

The positive sum terms represent the transfer rates increasing the population of environment i, whereas the minus sign indicates depopulation.

23.1.2
HMM Solutions in Terms of Eigenvalues

The general solutions (Eqs. 23.4 and 23.5) of the HMM equations contain exponential factors $\exp\{Pt\}$ where $P = \ell$ or L. The exponential functions are interpreted by their expansions

$$e^{Pt} = \sum_{j} \frac{1}{j!} P^j t^j \tag{23.17}$$

The power of the matrix P is carried out by first diagonalizing the matrix and then applying the power:

$$P^j = U\Lambda^j U^{-1} \tag{23.18}$$

The matrix Λ is the (diagonal) eigenvalue matrix of P, and the transformation matrices are designated by U and its inverse matrix U^{-1}. The eigenvalue matrix is determined by the eigenvalues λ_k and has the form

$$\Lambda = \begin{pmatrix} \lambda_1 & . & . & . & 0 \\ . & . & & & . \\ . & & . & & . \\ . & & & . & . \\ 0 & . & . & . & \lambda_N \end{pmatrix} \quad \text{so that} \quad \Lambda^j = \begin{pmatrix} \lambda_1^j & . & . & . & 0 \\ . & . & & & . \\ . & & . & & . \\ . & & & . & . \\ 0 & . & . & . & \lambda_N^j \end{pmatrix} \tag{23.19}$$

The power series at Eq. 23.17 then reads

$$e^{Pt} = \sum_j \frac{1}{j!} U \begin{pmatrix} \lambda_1^j & . & . & . & 0 \\ . & . & & & . \\ . & & . & & . \\ . & & & . & . \\ 0 & . & . & . & \lambda_N^j \end{pmatrix} U^{-1} t^j = U \begin{pmatrix} e^{\lambda_1 t} & . & . & . & 0 \\ . & . & & & . \\ . & & . & & . \\ . & & & . & . \\ 0 & . & . & . & e^{\lambda_N t} \end{pmatrix} U^{-1} \tag{23.20}$$

Denoting the eigenvalues of ℓ by $\lambda_k^{(\ell)}$, and those of L by $\lambda_k^{(L)}$, the HMM solutions at Eqs. 23.4 and 23.5, can be rewritten as

$$m(t) = U_\ell \begin{pmatrix} e^{\lambda_1^{(\ell)} t} & . & . & . & 0 \\ . & . & & & . \\ . & & . & & . \\ . & & & . & . \\ 0 & . & . & . & e^{\lambda_N^{(\ell)} t} \end{pmatrix} U_\ell^{-1} m(0) \tag{23.21}$$

$$M_z(t) = M_0 - U_L \begin{pmatrix} e^{\lambda_1^{(L)} t} & . & . & . & 0 \\ . & . & & & . \\ . & & . & & . \\ . & & & . & . \\ 0 & . & . & . & e^{\lambda_N^{(L)} t} \end{pmatrix} U_L^{-1} [M_0 - M_z(0)] \tag{23.22}$$

23.1.3
Two-Environment-Exchange Model

Defining two environments "1" and "2" between which exchange occurs with rate constants $k_{1,2}$ and $k_{2,1}$, the exchange model to be considered in the following may be represented by the scheme

$$\boxed{\begin{array}{c} T_{1,1}, \ T_{2,1}, \ \ldots \\ \omega_{0,1}, \ p_1 \end{array}} \quad \begin{array}{c} k_{1,2} \\ \leftarrow \\ \rightarrow \\ k_{2,1} \end{array} \quad \boxed{\begin{array}{c} T_{1,2}, \ T_{2,2}, \ \ldots \\ \omega_{0,2}, \ p_2 \end{array}} \tag{23.23}$$

where the populations obey $p_1 + p_2 = 1$. The dots indicate other relaxation times which, depending on the pulse sequence, potentially play a role.

A typical example of such a situation are hydrogen atoms subject to exchange between OH groups and water molecules in an aqueous alcohol solution. The exchange can also be of a more physical nature such as water molecules exchanging between the hydration shells of macromolecules or other polar surfaces and free water.

The above matrices are now of the format 2×2:

$$\Omega = \begin{pmatrix} \omega_{0,1} & 0 \\ 0 & \omega_{0,2} \end{pmatrix} ; \qquad K = \begin{pmatrix} -k_{2,1} & k_{1,2} \\ k_{2,1} & -k_{1,2} \end{pmatrix} \qquad (23.24)$$

$$R_1 = \begin{pmatrix} 1/T_{1,1} & 0 \\ 0 & 1/T_{1,2} \end{pmatrix} ; \qquad R_2 = \begin{pmatrix} 1/T_{2,1} & 0 \\ 0 & 1/T_{2,2} \end{pmatrix} \qquad (23.25)$$

The general HMM solutions at Eqs. 23.21 and 23.22 thus take the form of linear combinations of two exponential functions:

$$m(t) = \begin{pmatrix} m_1 \\ m_2 \end{pmatrix} = U_\ell \begin{pmatrix} e^{\lambda_1^{(\ell)} t} & 0 \\ 0 & e^{\lambda_2^{(\ell)} t} \end{pmatrix} U_\ell^{-1} \, m(0) \qquad (23.26)$$

$$M_z(t) - M_0 = \begin{pmatrix} M_{z,1} \\ M_{z,2} \end{pmatrix} - \begin{pmatrix} M_{0,1} \\ M_{0,2} \end{pmatrix}$$

$$= -U_L \begin{pmatrix} e^{\lambda_1^{(L)} t} & 0 \\ 0 & e^{\lambda_2^{(L)} t} \end{pmatrix} U_L^{-1} \, [M_0 - M_z(0)] \qquad (23.27)$$

Two limits can be distinguished (compare refs. [8, 535]):

a) **Slow exchange:**

In the limit $|k_{2,1}|, |k_{1,2}| \ll \left| \frac{1}{T_{1,2}} - \frac{1}{T_{1,1}} \right|$, the total longitudinal magnetization, $M_z(t) = M_{z,1}(t) + M_{z,2}(t)$, with the equilibrium value $M_0 = M_{0,1} + M_{0,2} = p_1 M_0 + p_2 M_0$ obeys

$$\frac{M_z(t) - M_0}{M_z(0) - M_0} = p_1 \exp\left\{ -\left(\frac{1}{T_{1,1}} + k_{2,1} \right) t \right\}$$

$$+ (1 - p_1) \exp\left\{ -\left(\frac{1}{T_{1,2}} + k_{1,2} \right) t \right\} \qquad (23.28)$$

Likewise, we find for the limit $|k_{2,1}|, |k_{1,2}| \ll |\omega_{0,1} - \omega_{0,2}|, \left| \frac{1}{T_{2,2}} - \frac{1}{T_{2,1}} \right|$, the total transverse magnetization, $m(t) = m_1(t) + m_2(t)$ the relation

$$\frac{m(t)}{m(0)} = p_1 \exp\left\{ \left(i\omega_{0,1} - \frac{1}{T_{2,1}} - k_{2,1} \right) t \right\}$$

$$+ (1 - p_1) \exp\left\{ \left(i\omega_{0,2} - \frac{1}{T_{2,2}} - k_{1,2} \right) t \right\} \qquad (23.29)$$

b) Fast exchange:

In the limit $|k_{2,1}|, |k_{1,2}| \gg \left| \frac{1}{T_{1,2}} - \frac{1}{T_{1,1}} \right|$, the total longitudinal magnetization turns out to be

$$\frac{M_z(t) - M_0}{M_z(0) - M_0} = \exp \left\{ - \left(\frac{p_1}{T_{1,1}} + \frac{p_2}{T_{1,2}} \right) t \right\} \tag{23.30}$$

The total transverse magnetization in the limit

$$|k_{2,1}|, |k_{1,2}| \gg |\omega_{0,1} - \omega_{0,2}|, \left| \frac{1}{T_{2,2}} - \frac{1}{T_{2,1}} \right|$$

follows

$$\frac{m(t)}{m(0)} = \exp \left\{ \left[i \left(p_1 \omega_{0,1} + p_2 \omega_{0,2} \right) - \frac{p_1}{T_{2,1}} + \frac{p_2}{T_{2,2}} - k_{1,2} \right] t \right\} \tag{23.31}$$

The fast-exchange limit is thus ruled by monoexponential relaxation curves and single-line resonances characterized by the effective parameters[1]

$$\begin{array}{rcl} \dfrac{1}{T_1} & = & \dfrac{p_1}{T_{1,1}} + \dfrac{1 - p_1}{T_{1,2}} \\[2mm] \dfrac{1}{T_2} & = & \dfrac{p_1}{T_{2,1}} + \dfrac{1 - p_1}{T_{2,2}} \\[2mm] \omega_0 & = & p_1 \omega_{0,1} + (1 - p_1) \omega_{0,2} \end{array} \tag{23.32}$$

So far, the above treatment is valid for free evolution of the transverse magnetizations once excited at the beginning of the experiment. This situation is to be discriminated from the behavior arising while CPMG pulse trains are applied. The apparent transverse-relaxation rates which are revealed in the CPMG case have been treated in [211, 306], for instance.

23.2
Two-Dimensional Exchange Spectroscopy

The combined study of the distribution of nuclear environments, exchange rates, and pathways between spin environments which are distinguishable owing to different resonance frequencies, is referred to as two-dimensional (2D) exchange spectroscopy (EXSY).

Two-dimensional exchange spectroscopy was first suggested by Jeener et al. [209]. Subsequently a seemingly endless series of successful applications of this measuring principle followed. The technique has been applied to liquids [139], solids [222], to proton magnetic resonance, to the NMR of other nuclear species such as deuterons [432, 507] and even to electron spin resonance (ESR) [13, 439]. In ref. [480] a sample-turning variant of deuteron 2D exchange spectroscopy was

---

[1]Recall that this behavior stipulates motional averaging as concerns transverse relaxation.

reported. The first nuclear quadrupole resonance (NQR) application was demonstrated in [419].

Dipolar coupling causes an efficient magnetization transfer by cross relaxation. This immaterial spin exchange competes with the real (chemical) exchange, and decoupling pulse sequences or temperature-dependent measurements are necessary for a distinction. On the other hand, there is an interesting technique relying on cross relaxation: **Nuclear Overhauser Effect Spectroscopy (NOESY)** is an extremely powerful and widespread method for the determination of molecular structures [312]. In a sense, this method also belongs to the class of multiple-dimensional exchange experiments.

Further diversification is provided by generalizing the term "exchange" to diffusion and coherent flow (velocity). In [248] two-dimensional experiments are reported probing the position exchange by Brownian motion. The correlation of velocities at time 0 and at a later time may also be interpreted to be equivalent to exchange. A corresponding method termed velocity exchange spectroscopy (VEXSY) was published in Ref. [80].

The idea of multiple-dimensional exchange spectroscopy is to phase encode the coherences of the isochromats in the different environments in evolution intervals that are small compared with the exchange times. The signal is recorded after mixing intervals that are long enough to permit the exchange processes to take place. The lines measured in the detection interval are then correlated in a multiple-dimensional spectrum with those encoded in the evolution intervals. Thus, cross peaks are indicative for nuclei changing the environment in a mixing interval.

The two-dimensional version of this experiments consists of three 90° pulses of equal length separated by intervals t_1 and τ_m as shown in Fig. 23.1. The discussion will be led for uncoupled spins, so that multiple-quantum coherences or spin-order effects need not be considered, although the pulse sequence is identical with that treated in Sects. 4.2 and 7.2 for coupled spins.

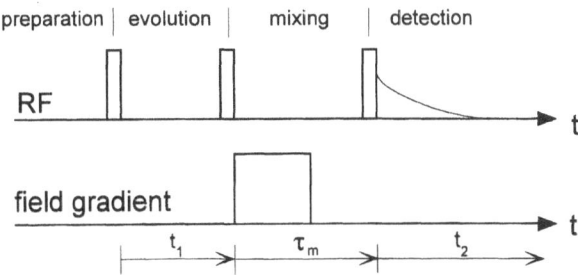

Fig. 23.1. RF pulse sequence of 2D exchange experiments. The sequence consists of three RF pulses. Two time domains, t_1 and t_2, are distinguished. The first evolution interval, t_1, is incremented in subsequent transients, whereas the t_2 domain is probed by acquisition of the FID. The conjugated frequency domains, ω_1 and ω_2, are interrelated to the two time domains by a double Fourier transform. Under normal circumstances, the mixing interval, τ_m, is much longer than t_1 and t_2. Therefore it is this interval which specifies the time scale of the exchange and relaxation processes to be detected via cross peaks.

The exchange is assumed to be slow enough so that the coalescence of the different resonances can be excluded. In other words, exchange during the evolution interval t_1 is assumed to be unlikely. The subsequent mixing interval τ_m, in which part of the coherences are transferred to z magnetization is much longer so that exchange is more likely to become operative. Taking spin-lattice relaxation into account, the τ_m range of particular interest is specified as

$$T_1 \gg \tau_m \gg t_1 \tag{23.33}$$

The coherences remaining after the second RF pulse can be spoiled by a magnetic field gradient pulse. Alternatively one can apply a phase cycle so that no signal contribution arises on this basis.

At the end of the mixing interval, a reading RF pulse is applied generating the free-induction signal to be detected. A data set, $S(t_1, t_2, \tau_m)$, is recorded which depends on the two time domains, t_1, t_2 (during which spin coherences evolve) and the mixing interval τ_m. Note that the signal is recorded, preferably with the aid of the quadrature detection technique, in the t_2 as well as in the t_1 domain. The latter means that two data sets must be recorded with the phase of the second pulse in quadrature [432].

A twofold Fourier transform finally produces the two-dimensional spectrum, $\tilde{S}(\omega_1, \omega_2, \tau_m)$, where the mixing time is an experimental parameter. The diagonal peaks of the spectra represent nuclei having suffered a cyclic or no exchange. Exchange among sites with different resonance frequencies gives rise to cross peaks. Axial peaks eventually appear because of incomplete excitation by the pulses or owing to spin-lattice relaxation in the mixing interval. Typical spectra are shown in Fig. 23.2. This experiment was carried out with 4'-n-pentyl-4-cyanobiphenyl (5CB) filled in a porous glass with a mean pore diameter of 4 nm. At the measuring temperature of 302 K this substance is normally in a nematic phase. However, in the present case, this is prevented by the pore confinement. Rather a pore-specific phase exists with a spectrum of only two resolved resonances. The two-dimensional exchange spectra at Fig. 23.2 illustrate the appearance of cross peaks between the two resonances as a consequence of cross-relaxation. The mean exchange rate was found to be about 100 s^{-1}.

23.2.1
Matrix Formalism

At the beginning of the pulse sequence, the subsystems of the spins are in equilibrium. The transverse and longitudinal magnetization system vectors (see Eqs. 23.2) at time $t = 0-$ are

$$m(0-) = 0 \tag{23.34}$$

$$M_z(0-) = M_0 \tag{23.35}$$

The first 90° pulse is assumed to be applied along the y direction of the frame rotating with the carrier frequency. The magnetization system vectors after the

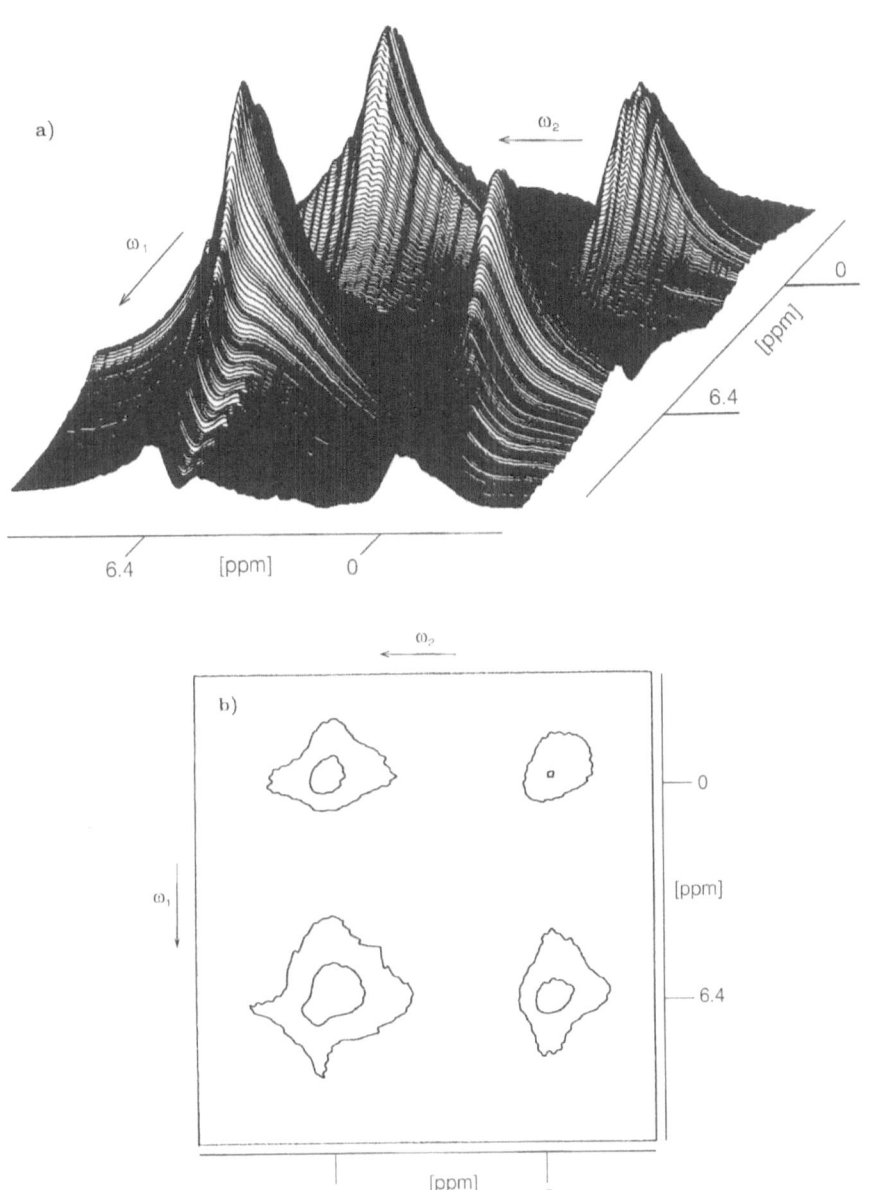

Fig. 23.2. 300 MHz 2D exchange (NOESY) proton spectra of 4'-n-pentyl-4-cyanobiphenyl (5CB) filled in a porous glass with a mean pore diameter of 4 nm. The mixing time was 100 ms: **a)** projection; **b)** contour plot of the magnitude of the two-dimensional Fourier transform. ω_1 and ω_2 are the frequency-domain variables conjugate to the time-domain variables t_1 and t_2 (Fig. 23.1). (Courtesy of F. Grinberg)

pulse are

$$m(0+) = -M_0 \tag{23.36}$$
$$M_z(0+) = 0 \tag{23.37}$$

After the first RF pulse, a free-evolution period follows in which the transverse magnetization components are subject to transverse relaxation, and precession with the resonance frequencies $\omega_{0,k}$ of the respective environments of the exchange network. The magnetization system vectors before the second RF pulse are

$$m(t_1-) = \exp\{\ell t_1\}\, m(0+) \tag{23.38}$$
$$M_z(t_1-) = M_0 - \exp\{L t_1\}\, [M_0 - M_z(0+)] \tag{23.39}$$

The second 90° pulse is again applied along the y direction. This pulse simultaneously transfers the x components of the transverse magnetizations into the z direction whereas the coherence components along the y direction are left untouched. The magnetization system vectors at the beginning of the mixing interval are

$$m(t_1+) = -M_z(t_1-) + i\Im\{m(t_1-)\} \tag{23.40}$$
$$M_z(t_1+) = \Re\{m(t_1-)\} \tag{23.41}$$

The homospoil pulse in the mixing interval is assumed to dephase all coherences completely. The longitudinal magnetization components are subject to exchange and spin-lattice relaxation processes so that we have at the end of the mixing interval

$$m(t_1 + \tau_m-) = 0 \tag{23.42}$$
$$M_z(t_1 + \tau_m-) = M_0 - \exp\{L\tau_m\}\, [M_0 - M_z(t_1+)] \tag{23.43}$$

The third 90° pulse, whose phase is assumed to correspond to the y direction, converts the z magnetizations into transverse magnetizations. That is

$$m(t_1 + \tau_m+) = -M_z(t_1 + \tau_m-) \tag{23.44}$$
$$M_z(t_1 + \tau_m+) = 0 \tag{23.45}$$

The evolution in the detection interval leads to

$$m(t_1 + \tau_m + t_2) = -e^{\ell t_2}\, \underbrace{M_z(t_1 + \tau_m-)} \tag{23.46}$$
$$\overbrace{M_0 - e^{L\tau_m}[M_0 - \underbrace{M_z(t_1+)}]}$$
$$\overbrace{\Re\{\, \underbrace{m(t_1-)}\, \}}$$
$$-e^{\ell t_1} M_0$$

Substituting all underbraced terms by the expressions for the preceding intervals gives

$$m(t_1 + \tau_m + t_2) = -\exp\{\ell t_2\} \left[1 - \exp\{L\tau_m\} \left(1 + \Re\{\exp\{\ell t_1\}\}\right)\right] M_0 \qquad (23.47)$$

All terms not depending on t_1 give rise to axial peaks which are not of interest here, and which can anyway be suppressed by a phase alternating pulse sequence with respect to the first RF pulse and the reference phase for signal acquisition. We therefore omit these terms, i.e., the figures 1 in Eq. 23.47.

The results for the phases of the second pulse in subsequent transients in quadrature, i.e., alternately along the y and x axes, are

$$m(t_1 + \tau_m + t_2)|_y = \exp\{\ell t_2\} \exp\{L\tau_m\} \Re\{\exp\{\ell t_1\} M_0 \qquad (23.48)$$
$$m(t_1 + \tau_m + t_2)|_x = i \exp\{\ell t_2\} \exp\{L\tau_m\} \Im\{\exp\{\ell t_1\} M_0 \qquad (23.49)$$

Ignoring the influence of relaxation and exchange during the evolution periods t_1 and t_2, these equations simplify to

$$m(t_1 + \tau_m + t_2)|_y = \exp\{i\Omega t_2\} \exp\{L\tau_m\} \Re\left\{\exp\{i\Omega t_1\} M_0\right\} \qquad (23.50)$$
$$m(t_1 + \tau_m + t_2)|_x = i \exp\{i\Omega t_2\} \exp\{L\tau_m\} \Im\left\{\exp\{i\Omega t_1\} M_0\right\} \qquad (23.51)$$

The sum of the quadrature signals gives

$$m(t_1 + \tau_m + t_2) = \exp\{i\Omega t_2\} \exp\{L\tau_m\} \exp\{i\Omega t_1\} M_0 \qquad (23.52)$$

23.2.2
Exchange Between Two Environments

The situation to be considered is similar to that represented in the scheme 23.23. For the sake of simplicity, the transfer-rate constants and the spin-lattice relaxation times are assumed to be equal, i.e., $k_{1,2} = k_{2,1} = k = 1/\tau$ and $T_{1,1} = T_{1,2}$. τ is the mean residence time in each of the environments. Exchange and relaxation during the evolution intervals is neglected.

The matrices determining the signal according to Eq. 23.52 are (see Eqs. 23.24 and 23.25)

$$\exp\{i\Omega t_{1,2}\} = \sum_j \frac{1}{j!} \Omega^j t_{1,2}^j = \sum_j \frac{1}{j!} \begin{pmatrix} \omega_{0,1} & 0 \\ 0 & \omega_{0,2} \end{pmatrix}^j t_{1,2}^j$$
$$= \begin{pmatrix} e^{i\omega_{0,1} t_{1,2}} & 0 \\ 0 & e^{i\omega_{0,2} t_{1,2}} \end{pmatrix} \qquad (23.53)$$

and

$$\exp\{L\tau_m\} = \exp\{-R_1 + K\} = \sum_j \frac{1}{j!} \begin{pmatrix} -1/T_1 - k & k \\ k & -1/T_1 - k \end{pmatrix}^j \tau_m^j$$
$$= \begin{pmatrix} \frac{1}{2}\left[1 + e^{-2k\tau_m}\right] e^{-\tau_m/T_1} & \frac{1}{2}\left[1 - e^{-2k\tau_m}\right] e^{-\tau_m/T_1} \\ \frac{1}{2}\left[1 - e^{-2k\tau_m}\right] e^{-\tau_m/T_1} & \frac{1}{2}\left[1 + e^{-2k\tau_m}\right] e^{-\tau_m/T_1} \end{pmatrix} \qquad (23.54)$$

The transverse-magnetization system vector at time $t = t_1 + \tau_m + t_2$ (Eq. 23.52) is thus

$$
\begin{pmatrix} m_1(t = t_1 + \tau_m + t_2) \\ \\ m_2(t = t_1 + \tau_m + t_2) \end{pmatrix} = \begin{pmatrix} e^{i\omega_{0,1}t_1} \frac{1}{2} \left[1 + e^{-2k\tau_m}\right] e^{-\tau_m/T_1} e^{i\omega_{0,1}t_2} \\ \\ e^{i\omega_{0,1}t_1} \tau_m \frac{1}{2} \left[1 - e^{-2k\tau_m}\right] e^{-\tau_m/T_1} e^{i\omega_{0,2}t_2} \end{pmatrix}
$$

$$
\left. \begin{matrix} e^{i\omega_{0,2}t_1} \tau_m \frac{1}{2} \left[1 - e^{-2k\tau_m}\right] e^{-\tau_m/T_1} e^{i\omega_{0,1}t_2} \\ \\ e^{i\omega_{0,2}t_1} \frac{1}{2} \left[1 + e^{-2k\tau_m}\right] e^{-\tau_m/T_1} e^{i\omega_{0,2}t_2} \end{matrix} \right) \begin{pmatrix} M_{0,1} \\ \\ M_{0,2} \end{pmatrix} \tag{23.55}
$$

A twofold Fourier transformation gives **diagonal peaks** at the positions $\omega_{0,1}, \omega_{0,1}$ and $\omega_{0,2}, \omega_{0,2}$ in the two-dimensional ω_1, ω_2 domain. The diagonal-peak intensities are proportional to

$$
\frac{1}{2} \left[1 + \exp\{-2k\tau_m\}\right] \exp\{-\tau_m/T_1\} M_{0,1}
$$

and

$$
\frac{1}{2} \left[1 + \exp\{-2k\tau_m\}\right] \exp\{-\tau_m/T_1\} M_{0,2}
$$

respectively. The intensities of **cross-peaks** arising as a consequence of exchange at the positions $\omega_{0,1}, \omega_{0,2}$ and $\omega_{0,2}, \omega_{0,1}$ are proportional to

$$
\frac{1}{2} \left[1 - \exp\{-2k\tau_m\}\right] \exp\{-\tau_m/T_1\} M_{0,1}
$$

and

$$
\frac{1}{2} \left[1 - \exp\{-2k\tau_m\}\right] \exp\{-\tau_m/T_1\} M_{0,2}
$$

respectively. The maximum cross-peak intensity is achieved for

$$
\hat{t}_m = \frac{1}{2k} \ln\left(2kT_1 + 1\right) \tag{23.56}
$$

23.2.3
Comparison with 2D Exchange NQR Spectroscopy

A complete treatment of the nuclear quadrupole resonance variant of 2D exchange spectroscopy can be found in [365]. In the magnetic-resonance versions of the 2D exchange spectroscopy experiment, the direction of quantization, that is the external magnetic field B_0, remains unaffected by the exchange process. This is in contrast to the NQR case, where the resonance frequency as well as the direction of quantization may change [46]. The latter is given by the local orientation of the electric field gradient (EFG) tensor. That is, it is defined by the local charge distribution determined by the molecular or crystalline structure. In this sense, 2D exchange NQR spectroscopy is more general than the magnetic resonance versions.

Although the 2D exchange NQR experiment is analogous to the NMR case in principle, there are a number of peculiarities to which attention should be paid.

First of all, the efficiency of RF pulses applied in the presence of a powder distribution of the electric field gradient tensors is strongly reduced. The consequence is not only a sensitivity loss which increases multiplicatively with each pulse contributing to the coherence pathway. The excitation is intrinsically imperfect in NQR of powdery materials, so that each of any subsequent pulses gives rise to new coherences originating from the residual equilibrium magnetizations of the subsystems. In the NMR case, by contrast, almost perfect 90° pulses are feasible, so that residual longitudinal magnetizations normally play a minor role.

Moreover, the exchange processes tend to be accompanied not only by a change of the NQR frequency but also by a change of the direction of quantization. Therefore, the cross-peak intensities are reduced by corresponding "projection" losses from the initial to the final direction of quantization of an exchange process. It is obvious that this phenomenon is solely a matter of the molecular structure.

Compounds suitable for NQR studies are prone to resonance line separations larger than the usual bandwidth of the RF pulses. In conventional NQR spectroscopy this is not necessarily a major deficiency. For 2D exchange experiments, however, it is crucial that all resonances of the exchange network are excited at the same time.

A remedy for the problem are phase alternating excitation pulses (PAEP) producing sidebands which can be tuned to several resonances. This was demonstrated in the case of p-chlorobenzotrichloride, for instance, where two resonances separated by 750 kHz had to be excited simultaneously [419]. For exchange networks implying more than two resonances, more complicated pulse modulations may be necessary for the generation of suitable side band patterns.

III

Localization and Imaging

CHAPTER 24

Survey

Most of the fundamental spin-echo phenomena and measuring principles discussed in Parts I and II find their applications in tomography techniques. As a new feature, spatial resolution comes into play. That is, the position dependences of object parameters[1] relevant for NMR are converted into image contrasts.

An **"NMR image"** in a word refers to contrasts determined by several object parameters such as the spin density and the relaxation times. The influence of these quantities need not necessarily be very well specified. As will become clear in subsequent chapters, different experimental parameters, i.e., predominantly the echo time and the repetition time, change the "weight" the individual material parameters possess in the formation of contrasts.

If the experiment is led in such a way that one particular material parameter dominates the contrasts, one speaks of **"filtered"** or **"parameter-weighted imaging."** Examples are relaxation- or diffusion-weighted representations. Nevertheless, the gray shades in all images of this sort are not just reflecting a single parameter. Other material parameters may more or less strongly influence the contrasts as well.

On the other hand, there are measuring protocols providing spatially-resolved data sets of well-defined quantities. Rendering such data of a single well-defined parameter in the form of a picture is termed **"parameter mapping."** Typical examples that will be depicted in subsequent chapters are spin-density maps, transverse-relaxation time maps, spin-lattice relaxation time maps, velocity maps, frequency-offset (or spectral) maps, and diffusivity maps, for instance.

Object-parameter mapping may also imply **multiple-dimensional techniques**. For example, mapping of any spectral parameter such as chemical shift or linewidth requires a further measuring domain in addition to the k space domains from which real-space images are derived. Supplementary measuring domains are also required for velocity (vector) mapping as delineated in subsequent chapters.

"Rendering an image" or a "map" means that the contrast parameter(s) to be displayed are transferred into the shades of a gray (or color) scale. This is what we usually have in mind when we refer to tomography. However, the scope of this technology is much wider. That is, it may be of interest to determine local material parameters quantitatively. The most promising and far-reaching option of tomography is the localized, non-destructive, non-invasive, remotely controlled

---

[1]**"Object parameters"** are determined by the sample whereas **"experimental parameters"** are quantities to be adjusted when implementing a pulse sequence.

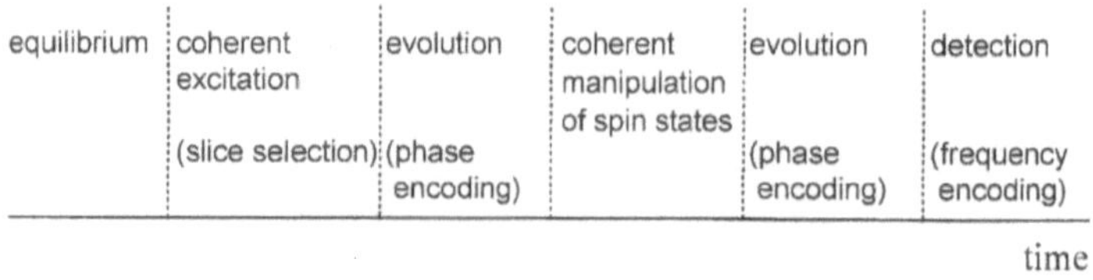

equilibrium	coherent excitation	evolution	coherent manipulation of spin states	evolution	detection
	(slice selection)	(phase encoding)		(phase encoding)	(frequency encoding)

time

Fig. 24.1. Schematic interval sequence of standard NMR imaging experiments.

measurement of an observable via NMR. It is not just the visualization of qualitative properties of an object.

Apart from imaging and parameter mapping, the term tomography therefore addresses all sorts of - potentially image-guided - **localized or volume-selective measurements** of object parameters accessible by NMR techniques. This also includes physical or chemical quantities which can be calibrated to NMR parameters. It may be of interest to measure a local relaxation time, for instance, which is indicative for the local stress influencing the relaxation process in a characteristic manner.

Following the outline of this book, we first consider the elements from which a typical tomography experiment is composed, and turn later to practical schemes suitable for different experimental purposes. Figure 24.1 shows the principal sequel of intervals which may be defined in this context.

An imaging pulse sequence is composed of segments serving the selection of a certain slice or volume region, and the encoding of the spatial information in the signal phase and frequency [288, 289, 295, 323]. The usual selection or encoding principle is based on pulses of a field gradient,

$$G = \nabla B = \begin{pmatrix} \partial B/\partial x \\ \partial B/\partial y \\ \partial B/\partial z \end{pmatrix} \tag{24.1}$$

which modify the external magnetic flux density[2]

$$B(r) \approx \begin{pmatrix} 0 \\ 0 \\ B(r) \end{pmatrix} = \begin{pmatrix} 0 \\ 0 \\ B_0 + G \cdot r \end{pmatrix} \tag{24.2}$$

In the presence of field gradients, the Larmor frequency therefore becomes a function of the position, $\omega_L = \omega_L(r)$. The evolution of coherences, while field gradients are on, consequently leads to position dependent phase shifts, $\varphi = \varphi(r)$. This is the basis of frequency- and phase-encoding of spatial information concerning the signal intensity.

The information is decoded by Fourier transformations of the signals in the respective encoding domains. The principle may be illustrated by considering a

---

[2]In terms of absolute values the field changes caused by practical field gradients are so small that any inclination of the field vectors against the z direction can safely be neglected.

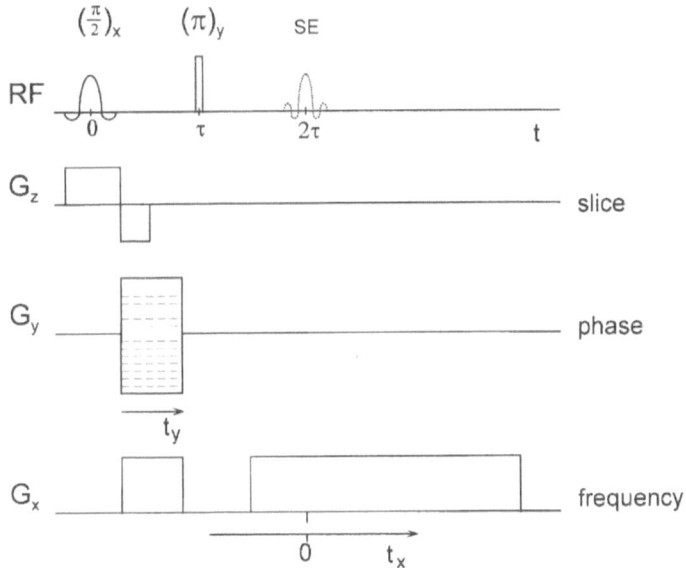

Fig. 24.2. RF and field gradient pulse scheme for two-dimensional Fourier transform (2DFT) NMR imaging also referred to as "spin-warp imaging." A Hahn RF pulse sequence generates the spin echo signal (SE) to be acquired. The first, suitably shaped RF pulse is applied in the presence of a "slice selection gradient." A "phase-encoding gradient" is applied in the first RF pulse interval. This gradient is incremented in a series of subsequent transients. The spin echo is read out in the presence of a "frequency-encoding gradient".

Fourier component of an NMR signal,

$$\xi(r,t) = \xi_0(r) \, \sin[\omega(r)t + \varphi(r)] \tag{24.3}$$

There are three independent parameters which can be used for encoding the information of interest, the amplitude ξ_0, the frequency ω, and the phase φ. It is clear that in a two-dimensional imaging experiment, two of these parameters serve the localization of the third parameter which is supposed to render the image contrast. In a three-dimensional experiment, two-fold phase-encoding must consequently be applied to allow for a third encoding dimension.

The quantities from which the image is reconstructed are normally the frequencies and phases determining the superimposed free-induction signals acquired in the experiment. Thus the amplitude remains as the quantity establishing the contrast. It is determined by object parameters such as the local spin density and the local relaxation times.

Figure 24.3 shows a typical RF and field gradient pulse sequence for NMR imaging of an object slice. The steps of common **(laboratory-frame) imaging** schemes are (a) **slice selection,** (b) (Larmor-precession-) **phase encoding,** (c) (Larmor-) **frequency encoding,** and (d) **image reconstruction**. There is also a **rotating-frame imaging** variant which is based on (e) **(nutation) frequency encoding**, i.e., on

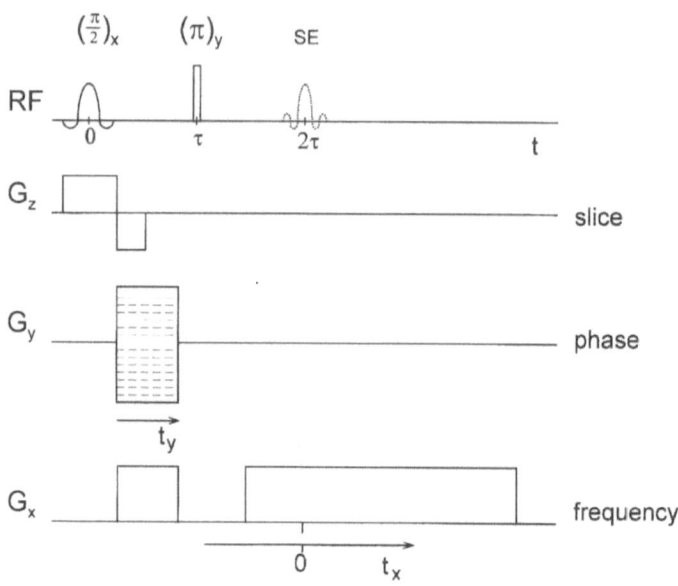

Fig. 24.3. RF and field gradient pulse scheme for two-dimensional Fourier transform (2DFT) NMR imaging also referred to as "spin-warp imaging." A Hahn RF pulse sequence generates the spin echo signal (SE) to be acquired. The first, suitably shaped RF pulse is applied in the presence of a "slice selection gradient." A "phase-encoding gradient" is applied in the first RF pulse interval. This gradient is incremented in a series of subsequent transients. The spin echo is read out in the presence of a "frequency-encoding gradient".

(laboratory-frame) gradients of the RF amplitude

$$G_1 = \nabla B_1 \tag{24.4}$$

Phase- and frequency encoding of spatial information are jointly considered in the "reciprocal space." That is, one introduces an up to three-dimensional wave vector

$$k = \begin{pmatrix} \gamma_n \int_0^{t_x} G_x(t')\,dt' \\ \gamma_n \int_0^{t_y} G_y(t')\,dt' \\ \gamma_n \int_0^{t_z} G_z(t')\,dt' \end{pmatrix} \tag{24.5}$$

as we will see in the subsequent sections. The encoding intervals τ_j ($j = x, y, z$) are defined by the effective period for which the encoding gradient acts on the signal.

It should be noted that the same phase and frequency encoding principles are also used for acquiring additional information on quantities such as frequency offsets (chemical shift, inhomogeneities, magnetic-susceptibility shifts), and velocity components. Pulse sequences serving this purpose will be described in the following after having formulated the principles of spatial encoding. Table 24.1 gives an

Table 24.1. Encoding and localization principles: list of the explicit and implicit dependences on r of the parameters in magnetic-resonance tomography.

	signal: $S \propto \int S_0 \sin(\omega_1 t_w) \exp\{-i(\omega_L t + \varphi)\} \, d^3 r$
$\omega_L = \omega_L(r)$	explicit function of $r \hookrightarrow$ **(Larmor) frequency encoding**
$\omega_L = \omega_L[\sigma(r), \chi(r)]$	implicit function of r
	via chemical shift, magnetic susceptibility
$\varphi = \varphi(r)$	explicit function of $r \hookrightarrow$ **phase encoding**
$\varphi = \varphi[v(r), \sigma(r), \chi(r)]$	implicit function of r
	via flow, chemical shift, magnetic susceptibility
$\omega_1 = \omega_1(r)$	explicit function of $r \hookrightarrow$ **nutation-frequency encoding**
$S_0 = S_0(r)$	explicit function of $r \hookrightarrow$ **slice selection, localization**
$S_0 = S_0[\rho(r), T_1(r), T_{1\rho}(r),$	implicit function of r
$\qquad T_2(r), v(r), D(r), \ldots]$	via spin density, relaxation, inflow/outflow, diffusion, ...

overview of the versatile experimental possibilities NMR imaging and localization offer. The general objective of NMR tomography may be summarized by the formula: *to image the implicit r dependences with the aid of the explicit r dependences.*

CHAPTER 25

Fundamentals of NMR Imaging

The elements of NMR imaging will first be outlined and treated taking the pulse scheme (Fig. 24.3) for two-dimensional Fourier transform imaging as an example. This is basically the **spin-warp technique** described in [134, 212, 354]. Other variants which are related in principle but different in the performance will be depicted thereafter.

25.1
Slice Selection by Soft Pulses

Slice selection by "soft pulses," i.e., narrow-band RF pulses in the presence of a magnetic field gradient [156], is a technique ubiquitous in NMR tomography. The homogeneous flux density of the magnet, $B_0 = B_0 u_z$, is superimposed by a field pulse $\Delta B(r, t) = (G(t) \cdot r) u_z$ which depends on the laboratory-frame position r. Thus the total local magnetic flux density is

$$B(r, t) = B_0 + (G(t) \cdot r) u_z \tag{25.1}$$

The unit vector u_z defines the z direction. The gradient $G = G(t)$ with the components $G_x = \partial B_0/\partial x$, $G_y = \partial B_0/\partial y$, and $G_z = \partial B_0/\partial z$ is chosen to be constant within the sample (i.e., the magnetic flux density varies linearly in the gradient direction). Note that normally only one of the Cartesian components is switched on during selective excitation.

The effective flux density in a frame with the Cartesian coordinates $x', y', z' \equiv z$ rotating with the angular frequency $\omega_0 = -\gamma_n B_0$ about B_0 is

$$B_e(r, t) = (G \cdot r) u'_z + B'_1 \tag{25.2}$$

where $B'_1 = B'_1(t)$ is the RF component relevant in the rotating frame, and where the carrier frequency of the transmitter, ω_c, is assumed to be equal to ω_0 as usual.

Let us first consider a **"hard pulse"** defined by the condition

$$\hbar \gamma_n B_1 \gg \hbar \gamma_n G d_s \gg \langle \mathcal{H}_i \rangle_m \tag{25.3}$$

where d_s is the dimension of the sample and $\langle \mathcal{H}_i \rangle_m$ is the largest expectation value of the spin interaction energy such as those due to chemical shift and spin-spin coupling. In these circumstances the effective field is approximated by

$$B_e(r, t) \approx B'_1(t) \neq f(r) \tag{25.4}$$

During a short intense RF pulse, any influence of spin interactions, relaxation, and diffusion can be neglected so that the expectation value of the local magnetization $M'(r, t)$ obeys the simple equation of motion in the rotating frame

$$\frac{dM'(r, t)}{dt} = \gamma_n M'(r, t) \times B_1'(t) \tag{25.5}$$

The initial condition is $M'(r, 0) = M_0(r)$ where $M_0 = M_0(r) = M_0(r)u_z'$ is the expectation value of the equilibrium magnetization at position r. The RF flux density, $B_1'(t) = B_1 u_x'$, is assumed to be constant during the pulse and directed along the x' axis of the rotating frame. The solution is then a precession about B_1', i.e.,

$$\begin{aligned} M_x'(r, t) &= 0 \\ M_y'(r, t) &= M_0(r) \sin(\gamma_n B_1 t) \\ M_z'(r, t) &= M_0(r) \cos(\gamma_n B_1 t) \end{aligned} \tag{25.6}$$

A hard pulse consequently excites the sample homogeneously with a constant flip angle $\alpha = \gamma_n B_1 t$.

By contrast a **"soft pulse"** is defined by the limit

$$\langle \mathcal{H}_i \rangle_m \ll \hbar \gamma_n B_1 \ll \hbar \gamma_n G d_s \tag{25.7}$$

where the influences of spin interactions, relaxation, and diffusion are again assumed to be negligible during the pulse. The effective field now varies spatially, i.e., $B_e = B_e(r, t)$ (Eq. 25.2). The equation of motion in the rotating frame then reads

$$\frac{dM'(r, t)}{dt} = \gamma_n M'(r, t) \times B_e(r, t) \tag{25.8}$$

The solution for time independent $B_e = B_e(r) \neq f(t)$ is

$$\begin{aligned} M_x'(r, t) &= M_0 \sin\Theta \cos\Theta \, \{1 - \cos(\gamma_n B_e t)\} \\ M_y'(r, t) &= M_0 \sin\Theta \sin(\gamma_n B_e t) \\ M_z'(r, t) &= M_0 \{\cos^2\Theta + \sin^2\Theta \cos(\gamma_n B_e t)\} \end{aligned} \tag{25.9}$$

The angle Θ is spanned by the vectors B_0 and $B_e(r)$ and amounts to

$$\Theta = \Theta(r) = \arctan\left\{\frac{B_1}{G \cdot r}\right\} \tag{25.10}$$

The response of the magnetization to the applied RF field is obviously a function of the position. Only for $\Theta = \pi/2$ is the solution given above for hard RF pulses retained. The situation becomes even more complex if the RF pulses are shaped by an envelope function $B_1 = B_1(t)$. The equation of motion is then no longer linear so that the solution is aggravated.

The purpose of a soft pulse is to produce a spatially dependent magnitude of the transverse magnetization. This is the idea of slice selection. However, as a result of a soft pulse, it is in principle not only the magnitude that becomes a function of the position. The phase of the transverse magnetization vector and the tip angle are inhomogeneously distributed across the sample as well [325].

A rigorous treatment of the response of the magnetization vector to the pulse can be established either by numerically solving the equation of motion (Eq. 25.8), or more advanced analytical techniques for the exact solution of Bloch's equations must be employed [313, 353, 424].

For most applications it suffices to consider approximate solutions. With regard to the instrumental imperfections one is unavoidably facing in tomography experiments, the approach discussed in the following section forms an operational compromise between the rigor of treatment and what can be brought about in practice.

25.1.1
Approximation for Small Tip Angles

For small flip angles an approximate version of the equation of motion can be established, the solution of which suggests Fourier transform responses of the magnetization to soft pulses [19]. This result is readily conceivable because spins can only be excited if the RF pulse contains Fourier components at their resonance frequency. However, the question to be answered is how the local three magnetization components are distributed in space after the soft pulse in the small-tip angle limit.

Let the reference frame rotate with the local resonance frequency $\omega = -\gamma_n (G \cdot r) u_z - \gamma_n B_0$. At a position r, the magnitude of this frequency is $(\omega_0 + \gamma_n G \cdot r)$ rather than the carrier frequency of the RF pulse, $\omega_c = \omega_0$. The consequence is that the effective field, $B_e(r, t)$, is time dependent even if B_1 were constant. It is now directed transversely to the external field, and rotates with an angular frequency $\Omega = -\gamma_n (G \cdot r) u_z'$ about the z direction:

$$B_e(r, t) = B_1(t) \left\{ \cos\left[\gamma_n (G \cdot r)\left(t + \frac{t_w}{2}\right)\right] u_x' + \sin\left[\gamma_n (G \cdot r)\left(t + \frac{t_w}{2}\right)\right] u_y' \right\}$$

$$(25.11)$$

The unit vectors u_x', u_y', and u_z' define the directions of the axes of the local rotating frame. $B_1(t)$ is the envelope function or "pulse shape" assumed to be centered at $t = 0$ and to have a width t_w. That is $B_1(t) = 0$ for $t < -t_w/2$ and $t > t_w/2$ (see Fig. 25.1).

At the beginning of the pulse, i.e., at the time $t = -t_w/2$, the phases of the local rotating frames are the same everywhere. The initial situation corresponds to an alignment of the local effective-field vectors along the coinciding local x' axes.

The equations of motion for the local magnetization components in the their local rotating frames then read

$$\frac{dM_x'(r, t)}{dt} = \gamma_n M_z'(r, t)\, B_1(t) \sin\left[\gamma_n (G \cdot r)\left(t + \frac{t_w}{2}\right)\right]$$

$$\frac{dM_y'(r, t)}{dt} = \gamma_n M_z'(r, t)\, B_1(t) \cos\left[\gamma_n (G \cdot r)\left(t + \frac{t_w}{2}\right)\right]$$

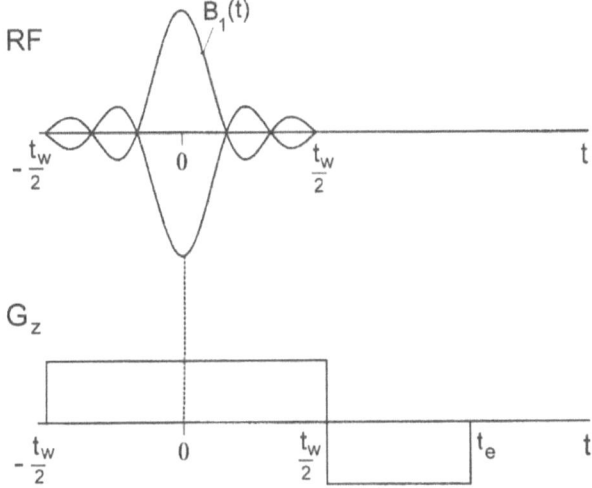

Fig. 25.1. Soft pulse for slice-selective excitation of (gradient refocused) coherences. A narrow-band RF pulse is applied in the presence of a field gradient. For constant magnitude of the gradient the maximum of the gradient echo appears at $t = t_e = t_w$. The envelope of the RF amplitude is proportional to $B_1(t)$, which – in the version shown – is shaped according to a truncated sinc function. The spatially constant magnetic field gradient is arbitrarily assumed in the z direction of the laboratory frame.

$$\frac{dM'_z(r, t)}{ddt} = \gamma_n B_1(t) \left\{ M'_x(r, t) \sin\left[\gamma_n(G \cdot r)\left(t + \frac{t_w}{2}\right)\right] \right.$$
$$\left. - M'_y(r, t) \cos\left[\gamma_n G \cdot r \left(t + \frac{t_w}{2}\right)\right] \right\} \quad (25.12)$$

The transverse components of the magnetization can conveniently be written as a complex quantity with the real and imaginary parts representing the x' and y' components, respectively,

$$m(r, t) = M'_x(r, t) + i M'_y(r, t) \quad (25.13)$$

obeying the combined differential equation

$$\frac{dm(r, t)}{dt} = i\gamma_n M'_z(r, t) \, B_1(t) \, \exp\left\{-i\gamma_n(G \cdot r)(t + t_w/2)\right\} \quad (25.14)$$

For small flip angles we may approximate $M'_z(r, t) \approx M_0(r) = const$. That is, we decouple the evolution of the transverse magnetization from the z component. Equation 25.14 can then be integrated with the initial condition $m(r, t = -t_w/2) = 0$. The transverse magnetization at the end of the pulse is derived on this basis as

$$m(r, t = t_w/2) \approx i\gamma_n M_0(r) \exp\left\{-i\gamma_n(G \cdot r)\frac{t_w}{2}\right\} \int_{-t_w/2}^{t_w/2} B_1(t) \exp\left\{-i\gamma_n(G \cdot r)t\right\} dt$$

$$(25.15)$$

Since $B_1(t) = 0$ for $|t| \geq t_w/2$ the integration limits may be replaced by $\pm\infty$. The complex transverse magnetization $m(r, t_w/2)$ can thus be rewritten as the Fourier-transform expression

$$m(r, t_w/2) \approx \gamma_n M_0(r) e^{-i(\Omega t_w - \pi)/2} \int\limits_{-\infty}^{\infty} B_1(t) e^{-i\Omega t}\, dt \qquad (25.16)$$

where $\Omega = \gamma_n G \cdot r$. In this approximation the distribution of Larmor frequencies of the excited transverse magnetization isochromats is determined by the Fourier transform of the pulse envelope function. The profile of the excited slice thus directly reflects the RF pulse envelope, so that the design of slice-selective excitation pulses is strongly facilitated in this approach.

Assuming a spatially constant gradient arbitrarily chosen along the z axis, the slice profile function, $g(z)$, is implicitly given by

$$m(z, t_w/2) \approx \gamma_n M_0(z) e^{-i(\gamma_n G_z z t_w - \pi)/2} \int\limits_{-\infty}^{\infty} B_1(t) e^{-i\gamma_n G_z z t}\, dt$$

$$= M_0(z)\, g(z)\, e^{-i(\gamma_n G_z z t_w - \pi)/2} \qquad (25.17)$$

That is,

$$g(z) = \gamma_n \int\limits_{-\infty}^{\infty} B_1(t) e^{-i\gamma_n G_z z t}\, dt \qquad (25.18)$$

For example, according to this Fourier-transform relation, a rectangular RF pulse envelope,

$$B_1(t) = \begin{cases} B_1 & \text{for} \quad -\frac{\tau}{2} \leq t \leq \frac{\tau}{2} \\ 0 & \text{otherwise} \end{cases} \qquad (25.19)$$

produces a slice profile described by a sinc function (see Fig. 25.2). Generalized to the case of the carrier frequency resonant at a position $z = z_0$, the normalized

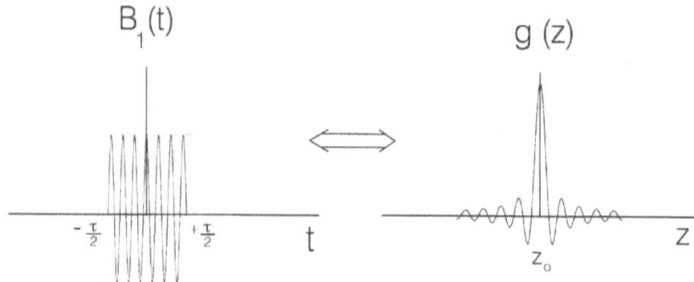

Fig. 25.2. Slice profile excited by a rectangular RF pulse in the presence of a constant field gradient along the z axis. The low-flip angle approximation (Eq. 25.17) is assumed. The zero positions are given by Eq. 25.22.

profile function is

$$\frac{g(z - z_0)}{g(0)} = \frac{\sin\left[\frac{1}{2}(z - z_0)\gamma_n G_z \tau\right]}{\frac{1}{2}(z - z_0)\gamma_n G_z \tau} = \text{sinc}\left[\frac{1}{2\pi}(z - z_0)\gamma_n G_z \tau\right] \qquad (25.20)$$

where

$$\text{sinc}(x) \equiv \frac{\sin(\pi x)}{\pi x} \qquad (25.21)$$

The main maximum at the position z_0 is symmetrically flanked by a sequence of pronounced side-lobes with zero positions at

$$z_n = z_0 + \frac{n}{2\pi\tau\gamma_n G_z} \qquad (25.22)$$

where n is an integer. The shorter the RF pulse, the wider the region excited by the pulse. However this profile does not correspond to a well-defined slice. Other pulse shapes like those described in the following section serve the purpose more appropriately.

25.1.2
Frequently Used Pulseshapes

In the approximation outlined above, a soft pulse with **Gaussian envelope** of the RF,

$$B_1(t) = B_1(0)e^{-at^2} \qquad (25.23)$$

produces a Gaussian slice profile

$$\frac{g(z - z_0)}{g(0)} = e^{-\Omega^2/(4a)} \qquad (25.24)$$

where $\Omega = \gamma_n G_z(z - z_0)$. The parameter a is determined by the full width at half height, $\Delta z_{1/2}$, according to

$$a = \frac{\gamma_n^2 G_z^2}{16 \ln 2}(\Delta z_{1/2})^2 \qquad (25.25)$$

z_0 again indicating the position where the carrier frequency is resonant.

Another pulseshape frequently used has the property of a **sinc function**,

$$B_1(t) = B_1(0)\frac{\sin(bt)}{bt} = B_1(0) \, \text{sinc}\left(\frac{bt}{\pi}\right) \qquad (25.26)$$

In the small-tip-angle approximation, this pulse produces an ideal rectangular profile of the width Δz,

$$\frac{g(z - z_0)}{g(0)} = \begin{cases} 1 & \text{if} \quad |z - z_0| \leq \frac{\Delta z}{2} \\ 0 & \text{otherwise} \end{cases} \qquad (25.27)$$

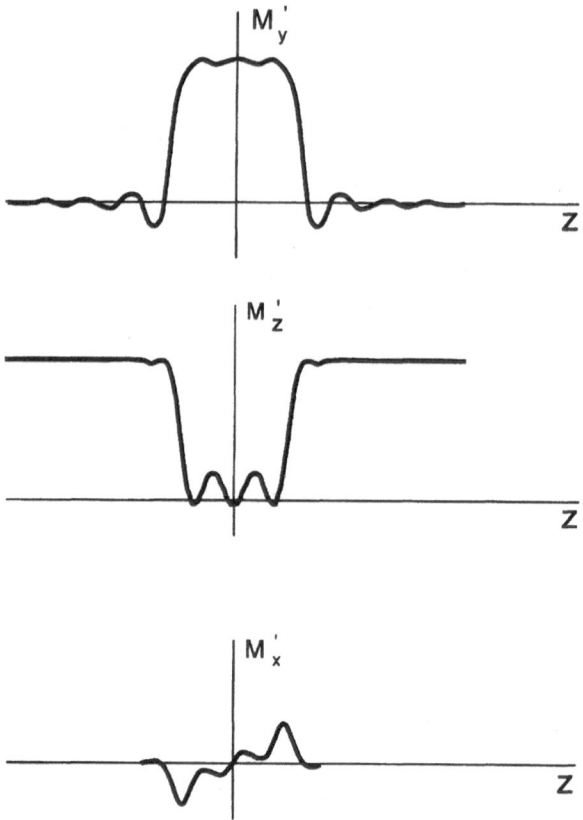

Fig. 25.3. Profiles of the gradient-refocused magnetization components after a double-lobe truncated sinc pulse (data from ref. [76]). The RF phase corresponds to the x' direction of the rotating frame. The nominal flip angle (in the middle of the slice) is 90°. The initial magnetization is assumed to be aligned along the z' direction. The pulse scheme is shown in Fig. 25.1.

The parameter b is related to the slice width by

$$b = \frac{\gamma_n G_z \Delta z}{2} \tag{25.28}$$

The sinc function is characterized by a series of side-lobes on both sides with zero points at $t = \pm n\pi/b$ in between where $n = 1, 2, 3, \dots$. Practically this ideal sinc shape needs be truncated at a finite number of side-lobes. The effect is a slight oscillatory modulation of the rectangular profile and the appearance of small sidebands (Fig. 25.3). These can be reduced substantially by apodizing the **truncated sinc function** with a cosine function [261]. The truncation at $n = 3$, for example, gives the rotating-frame time dependence of the RF amplitude

$$B_1(t) = \begin{cases} \text{sinc}(\frac{bt}{\pi}) \cos(\frac{bt}{6}) & \text{if} \quad \frac{-3\pi}{b} \leq t \leq \frac{3\pi}{b} \\ 0 & \text{otherwise} \end{cases} \tag{25.29}$$

At first sight, the approach so far employed appears to be rather crude but provides amazingly good descriptions of slice profiles for central flip angles up to 90° [76, 451]. For larger flip angles, however, the situation is less favorable. Relative to the transmitter phase, out-of-phase excitations of the transverse magnetization become more and more important. On the other hand, the slice width remains related to the RF bandwidth of the pulse for any nominal flip angle, of course.

A special pulse shape suitable for the slice-selective inversion of z magnetizations was proposed by Silver et al. [451]. This **"hyperbolic secant pulse"** produces almost perfect rectangular profiles.[1] As this pulse requires both amplitude and phase modulation, it is best described using a complex rotating-frame RF field

$$b_1 = B'_{1x} + iB'_{1y} \propto \left(\frac{2}{e^{\beta t} + e^{-\beta t}} \right)^{1+i\mu} = [\mathrm{sech}(\beta t)]^{1+i\mu} \qquad (25.30)$$

where β and μ are real constants. The width of the slice of inverted z magnetization is given by

$$\Delta z = \frac{2\beta\mu}{\gamma_n G_z} \qquad (25.31)$$

25.1.3
Refocusing of the Coherences

Up to now we have not explicitly considered the factor $\exp\{-i(\gamma_n G_z z t_w - \pi)/2\}$ in Eq. 25.17. It produces a distribution of precession phases along the gradient axis. At the end of the pulse, the spin isochromats are therefore dephased entirely, and the total transverse magnetization is canceled. That is, the coherences must be rephased in the form of a spin echo in order to become detectable. A possibility is to apply one or more subsequent RF pulses and a "trimming" gradient pulse in the refocusing interval as suggested in context with coherence transfer pulse sequences [235, 264, 464], for instance.

The simplest compensation method is the generation of a gradient-recalled echo. After the RF pulse the gradient is reversed rather than switched off (see Fig. 25.1) where the magnitude is assumed to be constant. That is, $G(t > t_w/2) = -G(t < t_w/2)$.

The complex transverse magnetization (Eq. 25.17) then evolves as already described in Sect. 2.3. The effective field in the rotating frame is $B_e(z) = -\gamma_n G_z z u_z$ so that the resonant position is in the origin. The transverse magnetization at time t has developed to

$$m(z, t) = m(z, t_w/2)\, e^{i\gamma_n G_z z(t - t_w/2)} = iM_0(z)g(z)e^{i\gamma_n G_z z(t - t_w)} \qquad (25.32)$$

The echo maximum appears at a time $t = t_e$ when the exponent vanishes irrespective of the position z. That is, $t_e = t_w$. The local complex transverse magnetization thus

---

[1]Unfortunately this does not hold good for pulses which are to refocus the transverse magnetization in spin-echo experiments, for instance.

reaches its maximum aligned along the y direction of the rotating frame,

$$m(z, t_e) = i\, M_0(z)\, g(z) \tag{25.33}$$

Note that the slice profile $g(z)$ can be measured directly by leaving the gradient after $t = t_e$ on and Fourier transforming the echo signal of a sample with homogeneous spin density, $M_0(z) = const$,

$$S(t) \propto \int m(z,t)\,\mathrm{d}z = M_0 \int g(z)\, e^{iG_z z(t-t_e)}\,\mathrm{d}z. \tag{25.34}$$

The integration covers the whole extension of the test object.

25.1.4
Variation of the Slice Width and Position

The selected slice is subject to apparative settings and can be defined in different ways. Let us arbitrarily assume a z slice again, i.e., a gradient of the form $G = G_z u_z$. The width of the slice profile can be controlled either by changing the length of the RF pulse (via the parameters a or b in Eqs. 25.23, 25.26, and 25.29 for instance), or by variation of the gradient.

The spatial resolution is obviously a matter of how long the chosen RF pulse is and how strong the field gradient can be produced. The position of the slice can be shifted by readjusting the position z_0 where the carrier frequency is resonant.

25.2
Phase Encoding

The effect of a pulsed, spatially constant phase-encoding gradient assumed in y direction of the laboratory frame,

$$G_p(t) = \begin{cases} G_y & \text{for } 0 \le t \le t_y \\ 0 & \text{otherwise} \end{cases} \tag{25.35}$$

on the evolution of coherences is in principle already implied in the treatment of Hahn echoes outlined in Chap. 2. The local precession-phase shift caused by the field gradient depends on the y coordinate according to

$$\varphi(y, t_y) = \Omega_y t_y = \gamma_n G_y t_y y = k_y y \tag{25.36}$$

where $\Omega_y = \gamma_n G_y y$ is the Larmor frequency offset in the field gradient, and $k_y = \gamma_n G_y t_y$ is the wave vector component generated by the pulse. The local rotating-frame transverse (complex) magnetizations before and after the gradient pulse are then related according to (compare Eq. 2.10)[2]

$$m(r, t_y) = m(r, 0)e^{-i\varphi(y, t_y)} = m(r, 0)e^{-ik_y y} \tag{25.37}$$

---

[2]The minus sign in the exponents originates from the antiparallel alignment of the Larmor frequency and magnetic flux density vectors for positive gyromagnetic ratios.

The local phase in the complex plane of the rotating frame is shifted by the position-dependent angle $\varphi(y, t_y)$, which therefore labels the laboratory-frame y coordinate of the nuclei contributing to the local transverse magnetization. Note that an entirely analogous way of encoding is performed in context with two- or multidimensional spectroscopy as outlined with a series of examples in Chap. 7.

25.3
(Larmor) Frequency Encoding

In imaging experiments, the free-induction signals are usually recorded in the presence of a frequency-encoding (or read) gradient. Assume that a Hahn (Sect. 2.1) or gradient-recalled echo (Sect. 2.3) is generated in the middle of such a gradient pulse, where the gradient is assumed in the x direction of the laboratory frame. The origin of the time scale t_x during the acquisition interval is centered at the maximum of the echo inducing the signal. The local coherence phase shift at time t_x is then

$$\varphi(x, t_x) = \Omega_x t_x = \gamma_n G_x t_x x = k_x x \tag{25.38}$$

where $\Omega_x = \gamma_n G_x x$ is the Larmor frequency offset in the field gradient, and $k_x = \gamma_n G_x t_x$ is the wave vector component generated by the pulse. Note that the phase shift vanishes for $t_x = 0$, i.e., at the maximum of the echo. The local rotating-frame transverse (complex) magnetizations at the echo maximum and before or after it are then related according to (compare Eq. 2.10)

$$m(r, t_x) = m(r, 0)e^{-i\varphi(x, t_x)} = m(r, 0)e^{-i\Omega_x t_x} = m(r, 0)e^{-ik_x x} \tag{25.39}$$

The free-induction signal acquired in this way is composed of contributions evolving according to the local offset frequency Ω_x labeling the nuclei responsible for this local magnetization. Hence the term "frequency encoding." Note that this is the conventional way of encoding in Fourier transform spectroscopy. However, in context with imaging, signals are processed in terms of the wave number k_x rather than of offset frequencies.

25.4
Two- and Three-Dimensional Fourier Imaging

Figures 24.3 and 25.4 show the pulse sequences which, in numerous variants, but based on the same principles, are most frequently applied in tomography. These schemes obviously contain the slice selection and encoding procedures outlined in the previous sections. All resolution and field-of-view aspects of these techniques will be discussed separately in comparison to other digital-acquisition mode techniques (see Chap. 32).

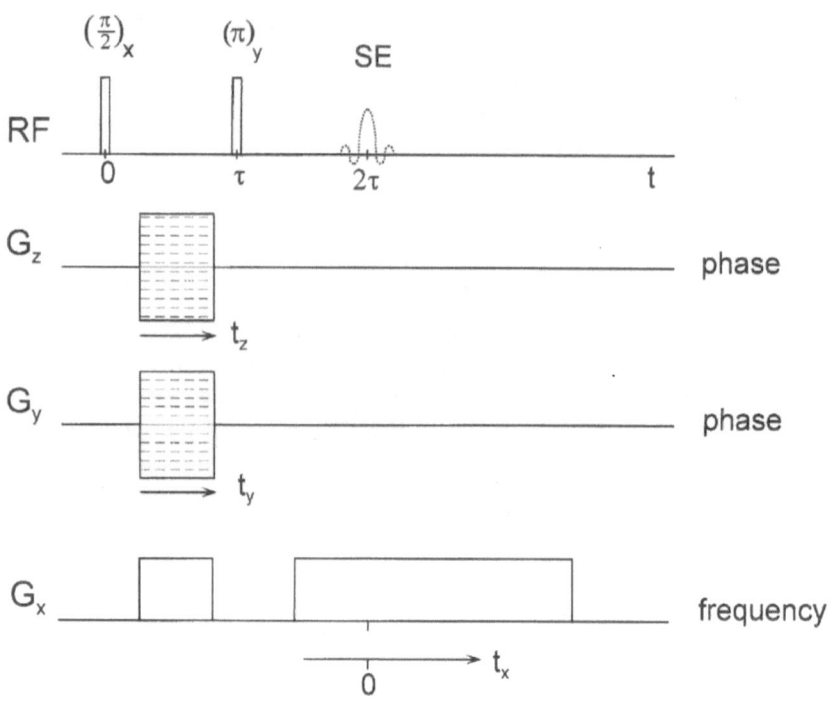

Fig. 25.4. Field gradient and RF pulse sequence for three-dimensional Fourier Transform (3DFT) imaging.

25.4.1
2DFT Imaging

The two-dimensional Fourier transform (2DFT) NMR imaging experiment (Fig. 24.3) begins with the selection of a slice which is then to be imaged. The standard slice selection method is a "soft" RF pulse in the presence of a field gradient as described in Sect. 25.1.

The phase-encoding gradient pulse is applied in the first free-evolution interval. It is incremented in subsequent transients of the pulse sequence in order to vary the wave number k_y. The 180° RF pulse refocuses all frequency offsets which are stationary or symmetrically arranged in both free-evolution intervals before and after this pulse. The phase-encoding gradient pulse does not comply with either of these conditions, so that the phase-encoded spatial information is not spoiled by coherence refocusing. The 180° pulse merely inverts all precession phases relative to the phase of the pulse (see Chap. 2).

The frequency-encoding (or read) gradient pulse is applied in the second free-evolution interval where the coherences are to be refocused as a spin echo. The read gradient tends to dephase the coherences before the spin echo is formed. Therefore, a gradient pulse is applied in the first free-evolution interval in such a way that the first half of the read gradient is compensated.[3] The echo maximum is then reached at $t = 2\tau$ (Fig. 24.3).[4]

The free-induction signal of the echo is composed of the signals of all voxels of the sample. That is, it is represented by an integral over the sample volume,

$$S(k_x, k_y) = \int \rho(r)g(z)e^{-i(k_x x + k_y y)}\, d^3r \approx \Delta z \int \rho(x, y)e^{-i(k_x x + k_y y)}\, dx\, dy \quad (25.40)$$

where $\rho(r)$ is the (parameter-weighted) spin density, $g(z)$ is the (normalized) slice profile, and the wave vector components are $k_x = \gamma_n G_x t_x$ and $k_y = \gamma_n G_y t_y$. The approximation in the above expression means that the slice profile is assumed to be rectangular, and that the spin density is constant across the slice width Δz.

The signal function $S(k_x, k_y)$ represents a "hologram" in complete analogy to the optical case. A graphical representation is shown in Fig. 25.5. The object image becomes accessible to the human eye after two-dimensional Fourier transformation from the reciprocal (k) to the real (r) space. The signal function $S(k_x, k_y)$ is of the form of Eq. 42.14, so that its Fourier transform is given by (see Eq. 42.15)

$$\mathcal{F}_k\{S(k_x, k_y)\} = \rho(x, y)\Delta z \propto \rho(x, y) \quad (25.41)$$

where we have assumed perfectly symmetric echo signals with respect to the origin of the k space and neglected any signal attenuation in the course of the (infinitely long) acquisition interval (compare the Fourier transform pairs listed in Table 42.3 on page 401). In reality a complex function comes out which is usually rendered as an image representing the magnitudes of the complex data.

25.4.2
3DFT Imaging

Three-dimensional image data sets can be recorded with the aid of the pulse sequence shown in Fig. 25.4. Instead of selecting a slice as usual in the two-dimensional case, a further phase-encoding domain probing the third dimension is introduced. The signal encoded in all three dimensions is then represented by the function

$$S(k_x, k_y, k_z) = \int \rho(r)e^{-i(k_x x + k_y y + k_z z)}\, d^3r \quad (25.42)$$

---

[3] Alternatively (but under practical circumstances less favorable) this compensation gradient can be applied in the second free-evolution interval before the read gradient in the form of a gradient pulse in the opposite direction.

[4] If the signal acquisition starts only at the maximum of the spin echo a part of the potential sensitivity is lost because no use is made of the first half of the echo. In other words, only the positive half of the k_x scale is probed. On the other hand, spin echoes tend to have an asymmetric shape owing to transverse relaxation. Therefore the record of both spin-echo halves together may possibly lead to artifacts in the image reconstruction process (compare the Fourier transform pairs for symmetric and asymmetric functions in Table 42.3 on page 401). Also, the longer acquisition times needed for acquiring the full echo stipulates longer pulse intervals, i.e., further T_2 losses. An extensive discussion of the matter can be found in [76].

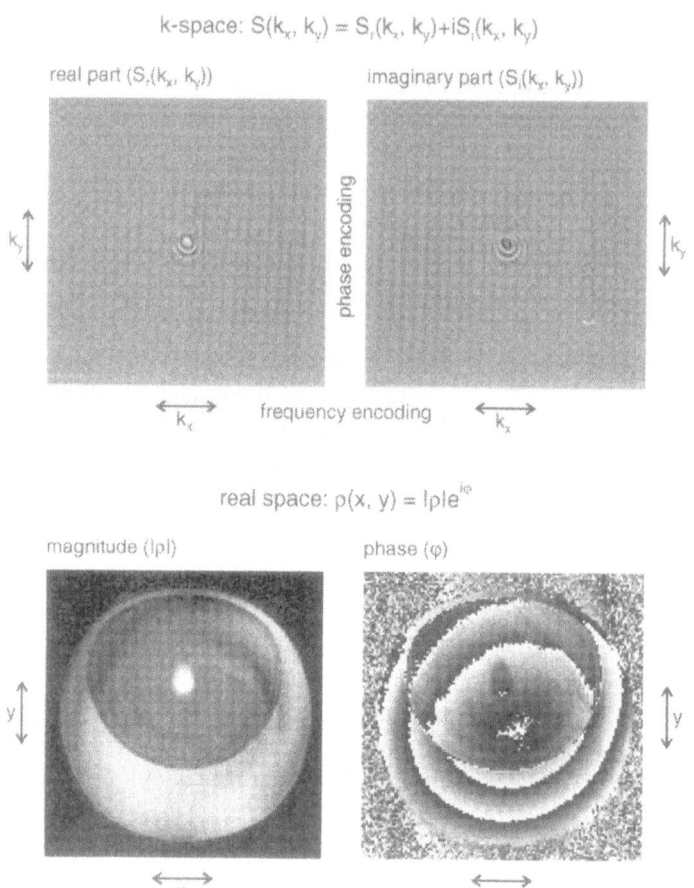

Fig. 25.5. Top: two-dimensional k space plots ("hologram") of the real and imaginary parts of a signal function $S(k_x, k_y)$ recorded in quadrature phase-sensitive detection mode from a quail egg; **bottom:** conjugated magnitude and phase real-space images. (courtesy of U. Görke)

Three-dimensional Fourier transformation leads to the conjugate function (see Eq. 42.15)

$$\mathcal{F}_k\{S(k_x, k_y, kz)\} = \rho(x, y, z) \qquad (25.43)$$

This result is valid for ideal acquisition conditions, i.e., negligible T_2 or T_2^* losses in the (infinitely long) acquisition interval, and perfectly symmetric echo signals with respect to the origin of the k space (see Table 42.3 on page 401). Under real conditions, the result is complex and the data must be rendered in the form of magnitude values.

25.4.3
Gradient-Recalled Spin-Echo Imaging

Spatially encoded signals can also be generated in the form of gradient-recalled echoes instead of the Hahn echoes suggested in Figs. 24.3 and 25.4. The two-dimensional version of corresponding imaging pulse schemes is displayed in Fig. 25.6. The three-dimensional modification can be found as part of the scheme shown in Fig. 33.2. The formalism of k-space data acquisition and image rendering is the same as described above in context with Hahn echo 2DFT and 3DFT pulse sequences.

Such short pulse sequences are of particular interest for **fast imaging**. In this case the flip angle of the excitation pulse is chosen much smaller than 90°. Equation 26.12 shows that much shorter repetition times T_R can be adjusted so that complete sets of image data can be acquired in total times in the order of 100 ms. This version of gradient-recalled spin-echo imaging is referred to as the "fast low-angle shot" (FLASH) protocol [175, 177].

A disadvantage of gradient-recalled spin-echo imaging is, that frequency offsets due to chemical shifts and heterogeneous magnetic susceptibility distributions in the sample are not restricted to the read direction. Rather the phase-encoding domain is affected as well. Remedies are the use of sufficiently strong encoding gradients, or the additional record of the spectroscopic dimension as described in Chap. 26. The latter solution of the problem conflicts with the desire of fast image acquisition, of course.

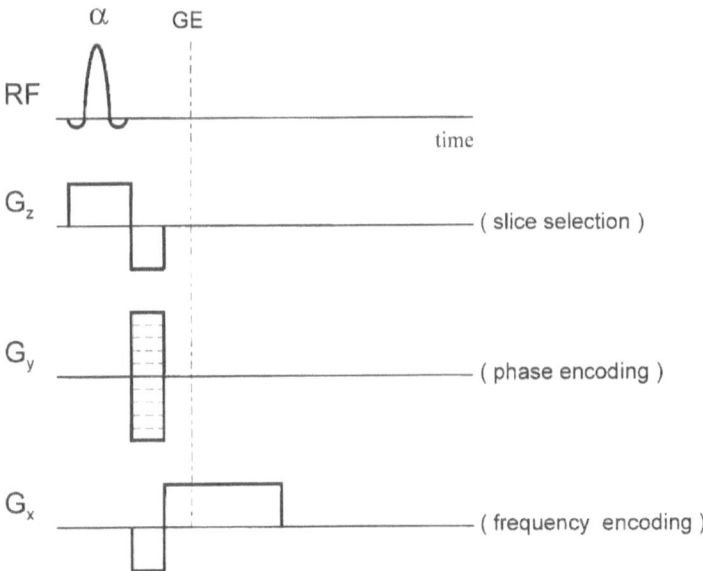

Fig. 25.6. Gradient-recalled echo (GE) pulse sequence for two-dimensional imaging.

25.4.4
Echo-Planar Imaging

The pulse schemes discussed up to now are multi-transient techniques which require at least one transient for each phase-encoding step. Under such circumstances, the time needed for the record of a complete image data set can only be reduced by shortening the repetition time of the transients by exciting only a part of the magnetization in each transient. This is the **low flip angle method** depicted in the previous section. Even faster data acquisition can be achieved by the **multiple signal read-out technique**. That is, the coherences, once excited by a 90° RF pulse, are refocused many times with increasing phase-encoding increments. This is the basic idea of echo-planar imaging [324, 326, 471].

The coherences to be recorded can be generated in a cyclic way with the aid of the CPMG pulse sequence [83, 337], i.e., a train of 180° RF pulses, or using multiply gradient-recalled echoes. The latter version has the advantage that the RF power deposition in the sample is much lower. Figure 25.7 shows the pulse scheme of such experiments. The echo signals are read out in the presence of read gradients of constant magnitude $|G_x|$. The phase-encoding gradient pulses of magnitude $|G_y|$ and of width τ_y lead to phases incremented from echo to echo. The corresponding wave-vector components are

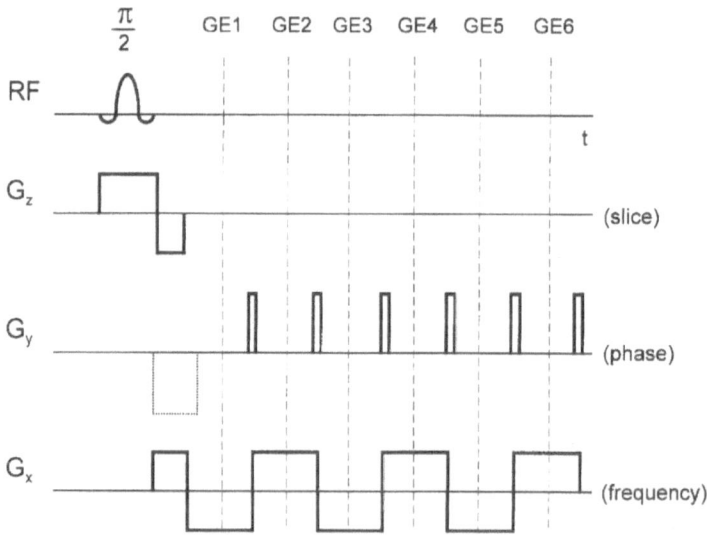

Fig. 25.7. Pulse scheme for echo-planar imaging with the aid of multiply gradient-recalled spin echoes (GE) (in reality more than the six cycles shown are needed, of course). The optional gradient pulse represented by the dotted line serves the precompensation of the phase-encoding gradient pulse train so that the phase increments symmetrically cover negative and positive values.

$$k_x = \gamma_n t_x |G_x|$$
$$k_y = \gamma_n [(n-1)\tau_y - t_c]|G_y|$$

The read-out time t_x is defined in the usual way to be zero at the center of each echo. The total phase-encoding period t_y reached at the n-th echo is composed of $n-1$ additive intervals τ_y owing to the $n-1$ phase-encoding gradient pulses. This pulse train is precompensated at the beginning by a single inverted pulse of width t_c. Its wave-number contribution, $-\gamma_n t_c |G_y|$, is chosen in such a way that the phase increments of the echoes symmetrically cover negative and positive values.

The total image data acquisition time achievable with echo-planar imaging is in the order of 10 ms, so that a good time resolution of the images can be achieved. This is a major advantage if "movies" of temporally varying objects are to be recorded. An example is imaging of flow patterns [281].

CHAPTER 26

Parameter-Weighted Contrasts

26.1
Contrast Parameters of Conventional Images

The gray shades of an ordinary NMR image recorded with the aid of Fourier transform echo pulse sequences such as those shown in Figs. 24.3, 25.4, or 25.6 are the result of several object and experimental parameters. Without paying attention to a special performance of the imaging experiment, one commonly obtains **relaxation-attenuated spin density images** or, in a word, plainly **NMR images**.

This sort of image is to be distinguished from **parameter maps** in which the gray shade is unambiguously related to the quantitative value of a single parameter. Examples are pure spin density or relaxation time (rate) maps which can be evaluated using more elaborate experimental protocols. As further cases, velocity, frequency-offset, $T_{1\rho}$ dispersion, and diffusivity maps are discussed in subsequent chapters.

In an NMR imaging experiment using an ordinary spin echo pulse sequence, the image contrasts are produced by **object parameters** in combination with **experimental parameters**. The latter determine how strong the weight of the former is. The images rendered in this way are object parameter weighted according to the choice of the experimental parameters.

The echo signals are recorded at the **spin-echo time** T_E after the initial excitation RF pulse with the **flip angle** α. The phase-encoding gradient is incremented in subsequent transients with a **repetition time** T_R, that is the interval between two subsequent excitations by the first RF pulse of the pulse sequence. A further experimental parameter of interest may be the **gradient time** t_g which characterizes the extension of dephasing/rephasing gradient pulse pairs (compare Figs. 24.3, 25.4, or 25.6). These are the pulse sequence parameters to be adjusted before the experiment is carried out.

The local signal contributing to the free-induction signal acquired in an imaging experiment is proportional to the number density of nuclei at this position. This material property is usually referred to as the **spin density** $\rho_0 = \rho_0(r)$. It is clear that the spin density may be an effective one. For instance, pulse sequences in standard NMR tomography are usually unsuitable for the record of broadline signals so that solid constituents will not contribute to the signal. In medical diagnostic imaging, for instance, bones appear as black regions although the (proton) spin density is definitely finite, and can in principle easily be imaged by methods appropriate for solids.

The transverse magnetization refocused in the form of a gradient or Hahn echo, measured before in the form of a free-induction signal, decays by transverse relaxation. The corresponding parameter is the local **transverse relaxation time** $T_2 = T_2(r)$ which may vary from voxel to voxel according to the local molecular mobility.

Samples to be imaged by tomography tend to be heterogeneous. Actually, this is the primary motivation for rendering an image, of course. If the sample heterogeneity also manifests itself in a spatial variation of the magnetic susceptibility, internal magnetic field inhomogeneities arise. Coherences dephased by these inhomogeneities are refocused in the form of Hahn echoes but not as gradient echoes as they are intrinsic to many imaging pulse sequences. In this case the local signal is attenuated during the echo time T_E by a local **inhomogeneous dephasing time** $T_2^* = T_2^*(r) \leq T_2(r)$ rather than by the proper transverse relaxation time.

The local signals are further weighted by spin-lattice relaxation if the repetition time T_R is not much longer than the local **spin-lattice relaxation time** $T_1 = T_1(r)$. In this case partial saturation of the local magnetization arises in analogy to the progressive saturation technique mentioned in Table 10.1 on page 94.

As a further parameter specific for the system, the local **diffusivity** $D = D(r)$ may influence image contrasts. Coherence rephasing in echo experiments in which relatively strong gradient pulse pairs are involved tends to be incomplete when self-diffusion (or incoherent flow) becomes perceptible as outlined in Sect. 19.2.

26.2
Contrasts in Gradient-Echo Tomography

Let us consider a gradient-echo pulse sequence (see Fig. 25.6) for simplicity as an example producing typical NMR imaging contrasts. The relaxation curves are assumed to be exponential as predicted by Bloch's equations (see Chap. 10). The potential echo attenuation by translational diffusion (or incoherent flow) is also assumed to be governed by an exponential factor as derived in Sect. 19.2 for ordinary diffusion.

The total gradient echo amplitude is given by the proportionality

$$S(T_E) \propto \int m(r, T_E)\, d^3 r \qquad (26.1)$$

where $m(r, T_E)$ is the local complex transverse magnetization.

The excitation pulse of the scheme at Fig. 25.6 is assumed to have a flip angle α and a phase in x direction of the rotating frame. The magnetization components just before the pulse are then transferred into the magnetizations immediately thereafter,

$$m(r, 0+) = M_z(r, 0-) \sin \alpha \qquad (26.2)$$
$$M_z(r, 0+) = M_z(r, 0+) \cos \alpha \qquad (26.3)$$

where the period from an excitation pulse to the next, T_R, is to obey $T_R \gg T_2^*$. In repetitive applications of the pulse sequence we have

$$M_z(r, 0-) = M_z(r, T_R) \qquad (26.4)$$

The irreversible attenuation of the complex transverse magnetization by transverse relaxation and translational diffusion, or - more likely - incoherent flow, is expressed by

$$m(r, T_E) = m(r, 0+) \exp\left\{-\frac{T_E}{T_2^*(r)}\right\} \exp\left\{-b(t_g)D(r)\right\} \qquad (26.5)$$

The function $b(t_g)$ globally represents the actual shape of the read-gradient in combination with its compensation lobe, and possibly that of the slice-selection gradient. Typical expressions for such functions can be found in Table 19.1 on page 181.

The influence of spin-lattice relaxation is given by the solution of the corresponding Bloch equation in the rotating frame,

$$M_0(r) - M_z(r, T_R) = [M_0(r) - M_z(r, 0+)] \exp\left\{-\frac{T_R}{T_1(r)}\right\} \qquad (26.6)$$

or

$$M_z(r, T_R) = M_z(r, 0+) \exp\left\{-\frac{T_R}{T_1(r)}\right\} + M_0(r)\left[1 - \exp\left\{-\frac{T_R}{T_1(r)}\right\}\right] \qquad (26.7)$$

Combining Eqs. 26.3, 26.4, and 26.7 gives

$$M_z(r, 0-) = M_0(r)\frac{1 - \exp\left\{-T_R/T_1(r)\right\}}{1 - \exp\left\{-T_R/T_1(r)\right\}\cos\alpha} \qquad (26.8)$$

Combining further Eqs. 26.2, 26.5, and 26.8 yields

$$\begin{aligned} m(r, T_E) &= M_0(r)\frac{1 - \exp\left\{-T_R/T_1(r)\right\}}{1 - \exp\left\{-T_R/T_1(r)\right\}\cos\alpha} \\ &\quad \exp\left\{-\frac{T_E}{T_2^*(r)}\right\} \exp\left\{-b(t_g)D(r)\right\}\sin\alpha \end{aligned} \qquad (26.9)$$

Introducing the actual and the effective spin densities,

$$\rho_0(r) \propto M_0(r) \qquad (26.10)$$
$$\rho(r) \propto m(r, T_E) \qquad (26.11)$$

we find the final contrast formula

$$\rho(r) = \rho_0(r)\frac{1 - \exp\left\{-T_R/T_1(r)\right\}}{1 - \exp\left\{-T_R/T_1(r)\right\}\cos\alpha} \exp\left\{-\frac{T_E}{T_2^*(r)}\right\} \exp\left\{-b(t_g)D(r)\right\}\sin\alpha \qquad (26.12)$$

determining the intensity obtained after Fourier transformation of the echo signal data matrix. This function is then rendered in the form of image gray shades.

In this derivation, motional artifacts, e.g., due to inflow or outflow of fluids into or from the sensitive slice, and frequency-offset artifacts were not considered. Contrasts arising on these grounds are discussed in subsequent sections.

26.3
Relaxation-Weighted Contrasts

Equation 26.12 demonstrates that typical imaging pulse sequences such as those displayed in Figs. 24.3 or 25.6 unavoidably produce relaxation-weighted image contrasts. An example is shown in the 2DFT image of a hen egg in Fig. 26.1. The contrasts realistically look like those of a black-and-white photograph: the yolk appears dark whereas the egg-white is rendered white. This amazing similarity with an optical photograph is fortuitous, of course. What we essentially have here is a transverse-relaxation weighted image, and the transverse relaxation times in the yolk are much shorter than those in the egg-white. The in vivo relaxation times at 38 °C and 4.7 T have been reported to be [259]: T_2(yolk, CH_2-signal) = 40 ms; T_2(yolk, water) = 120 ms; T_2(egg-white, water) = 380 ms. For comparison, the spin-lattice relaxation times are T_1(yolk, CH_2-signal) = 14...360 ms; T_1(yolk, water) = 2.4 s; T_1(egg-white, water) = 2.0 s.

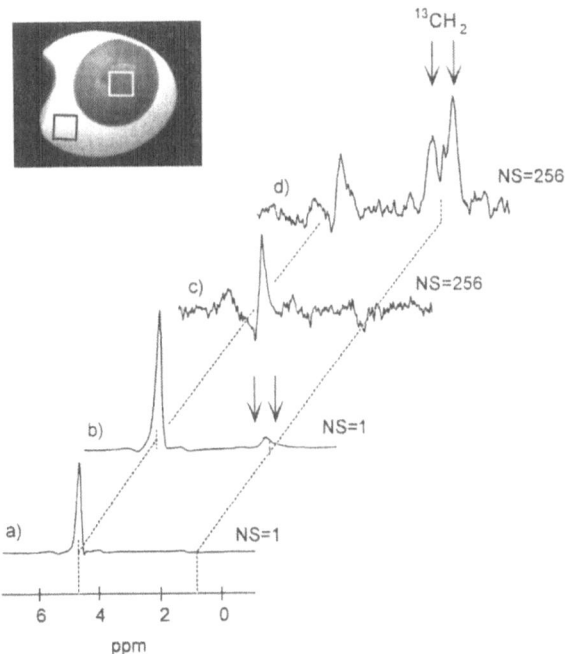

Fig. 26.1. 2DFT image of a hen egg recorded at 4.7 T (top). The pulse sequence used corresponds to Fig. 24.3. The experimental parameters are: slice width 1 cm; field of view 8 cm; echo time 30 ms; repetition time 2 s. The white and black squares indicate the regions to which the localized spectra refer. Spectra (a) and (b) have been recorded with the aid of ^{1}H VOSY (Sect. 37.1) in the egg-white and in the yolk regions, respectively. Spectra (c) and (d) have been recorded in the egg-white and in the yolk, respectively, using the CYCLCROP-LOSY technique for proton detected ^{13}C spectroscopy (Sect. 40.2). (Reproduced by permission from ref. [293])

Equation 26.12 tells us how to choose the experimental parameters if a certain object parameter is to dominate. Of course, imaging pulse schemes such as the one assumed above (Fig. 25.6) can easily be combined with the preparation pulses of conventional relaxation experiments as precursors to the proper imaging pulses in order to enhance the influence of nuclear relaxation.

Non-selective inversion or saturation RF pulses and a certain relaxation interval preceding the proper imaging pulse train may be used as a "filter" for the emphasis of certain relaxation features. In principle most of the experimental schemes discussed in Chap. 10 can be combined with imaging pulse schemes. Saturation/recovery, inversion/recovery, Hahn echo, or CPMG precursors are often employed for preparing relaxation-weighted magnetization distributions in the object prior to the proper imaging experiment. The systematic variation of pulse intervals even permits rendering of pure T_1, T_2, T_2^*, or ρ_0 maps. Typical examples of relaxation time maps recorded from fruits can be found in [79].

Relaxation contrasts can also be produced artificially by treating the object with relaxation contrast agents. These can be electron paramagnetic atoms and molecules, or bio-compatible complexes thereof [277, 358] Applications in medical diagnosis are now routine. Contrasts based on paramagnetic species are also useful in other fields. In [362, 363], for instance, studies of the temporally resolved uptake of heavy metal ions in alginate gels are reported. Figure 26.2 demonstrates that spin-lattice relaxation tends to increase the signal intensity (owing to reduced saturation) whereas transverse relaxation generally causes attenuation.

Modifying the Belousov/Zhabotinsky reaction with Mn^{2+} ions as a catalyst, the spatio/temporal pattern formation ("chemical waves") could impressively be visualized by three-dimensional NMR imaging [486, 487]. In this case the contrasts refer to transverse relaxation because, as a consequence of scalar proton/electron spin coupling, manganese ions reduce T_2 much stronger than T_1 [192].

26.4
Functional Tomography

The coherence attenuation time constant T_2^* is predominantly shortened by inhomogeneities arising from local magnetic susceptibility distributions within the sample. Susceptibility shifts are proportional to the magnetic-flux density and become appreciable in high-field tomography at 2 to 4 T or even more. This is the basis of the so-called "functional imaging" of brain regions active under certain exercises of the test person. The enrichment by (paramagnetic) molecular oxygen changes the local magnetic susceptibility so that high-field T_2^*-weighted images display "blood oxygenation level dependent" (BOLD) contrasts of brain activity [132, 152, 224, 338].

The contrast changes arising in functional imaging are relatively weak and can scarcely be evaluated quantitatively. Based on a multi-parameter weighted image recorded with the aid of a conventional imaging pulse scheme, it may therefore be desirable to obtain quantitative information on a specific object parameter in a certain region of interest. In this case it is more economic to perform an **image-**

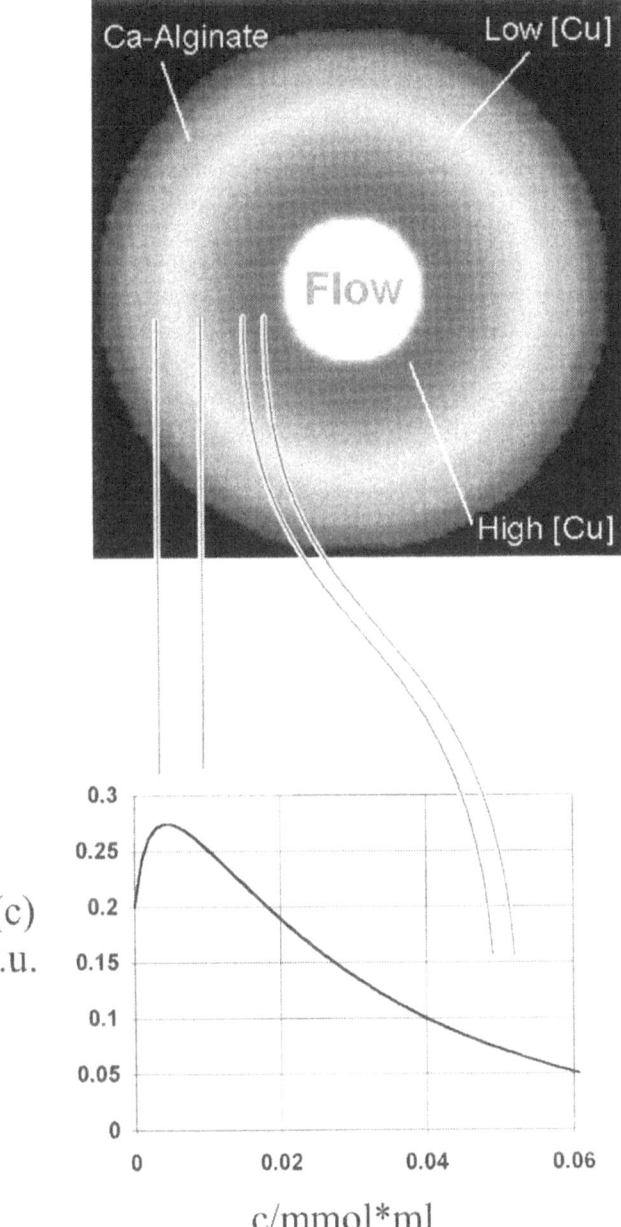

Fig. 26.2. Cross-sectional image of an alginate tube through which a dilute copper ion solution is pumped. A scheme of the experimental set-up is shown in Fig. 28.7. The paramagnetic copper ions are taken up by the alginate via an ion exchange process. The enriched concentration c of paramagnetic species in the alginate leads to transient relaxation-weighted contrasts as long as the equilibrium concentration is not yet reached. Note that the copper ion ingress first leads to an increase (less saturation losses) of the signal intensity $S(c)$ and then to a reduction (more T_2 losses). The plot corresponds to Eq. 26.12. (Reproduced by permission from ref. [363])

guided volume-selective parameter determination experiment rather than modify
the imaging procedure in such a way that the desired information of the region of
interest is provided together with that of regions not of interest. This is just a matter of the measuring time spent to record the desired information. Techniques of
this sort are often combined with volume-selective spectroscopy (VOSY) techniques
[235]. Examples are spectroscopically resolved volume-selective diffusometry [259]
or spectroscopically resolved volume-selective transverse and longitudinal relaxometry [437].

Using the latter technique it was possible to demonstrate a reproducible reduction by about 35 percent of the lipid spin-lattice relaxation time in the human
calf in vivo at 2 T during intense static flexing of the muscles, whereas the transverse relaxation time and both water relaxation times are not affected [438]. This
phenomenon, which occurs independently of the biological variability and which
apparently is not correlated with the muscle temperature may thus be called **"functional volume-selective spectroscopy."**

26.5
Diffusive Attenuation and "Edge Enhancement"

When the root mean squared displacement by self-diffusion or intra-voxel incoherent flow during the application of the field gradients exceeds the pixel resolution, the
signal intensity of the corresponding volume elements will be perceptibly attenuated according to the factor $\exp\left\{-b(t_g)D(r)\right\}$ in Eq. 26.12. This in particular refers
to gradients applied in long periods of the pulse sequence, i.e., to the read gradient
and its compensation counterpart (compare Figs. 24.3 and 25.6). The image contrasts are then weighted by diffusive attenuation according to the local (apparent)
diffusivity.

Under such circumstances bright fringes appear at fluid interfaces to solid structures within the object or to the sample container walls [78, 197, 203, 392]. This
"edge enhancement" effect is attributed to the reduced root mean squared displacements of liquid molecules in the vicinity of regions with strongly reduced diffusivity.
Diffusive attenuation in these edge regions is less effective. That is, boundaries in
the object appear as bright image contrasts.

CHAPTER 27

Relaxation-Dispersion Maps

Apart from the two laboratory-frame relaxation times, T_1 and T_2, the spin-lattice relaxation time effective in the rotating frame, $T_{1\rho}^{(e)}$, is a suitable parameter for modifying image contrasts.[1] Two-dimensional Fourier transform imaging pulse sequences modified for the production of **rotating-frame relaxation-weighted images** have been presented in [411, 442]. Attenuation by rotating-frame spin-lattice relaxation requires a period in the pulse sequence where the magnetization is "spin-locked," i.e., where it is aligned along the effective field in the rotating frame. In [411], this spin-lock pulse is suggested to be applied in the presence of a gradient pulse so that it efficiently serves the production of rotating-frame weighted contrasts and slice selection via the LOSY principle (Chap. 36) at one time. In order to increase the effective frequency and to avoid high RF power deposition while applying the spin-lock pulses, off-resonance experiments have been suggested in [412, 418].

By contrast to the laboratory frame, spin-lattice relaxation in the rotating frame is indicative of slow motions with rates in the frequency regime of $\omega_1 = \gamma_n B_1^{(sl)}$ defined by the spin-lock pulse amplitude $B_1^{(sl)}$. Molecular fluctuations in this range are also relevant for transverse relaxation. Therefore the contrasts of $T_{1\rho}$ and T_2 weighted images tend to be similar.

On the other hand, contrasts of a very peculiar sort can be produced by considering the frequency dependence ("dispersion") of $T_{1\rho}^{(e)}$. In [412, 418] experimental procedures for rendering $T_{1\rho}^{(e)}$ dispersion maps are reported. They were referred to as **"rotating-frame relaxation dispersion imaging"** (RODI). Figure 27.1 shows the pulse schemes employed for this purpose.

RODI techniques refer to the frequency effective in the rotating frame during the spin-lock pulse,

$$\omega_e = \sqrt{\omega_1^2 + \Omega^2} \tag{27.1}$$

where $\omega_1 = 2\pi\nu_1 = \gamma_n B_1^{(sl)}$ and $\Omega = 2\pi(\nu_0 - \nu_c)$. With the aid of the off-resonance technique, the rotating-frame relaxation dispersion is accessible in a relatively wide

---

[1]The superscript "*e*" refers to the magnetization spin-locked by the flux density effective in the rotating frame, $\mathbf{B}_e$. Generally this definition includes off- and on-resonance conditions for the spin-lock pulse. With experiments carried out with resonant spin-lock pulses the superscript is usually omitted (compare Sect. 10.3).

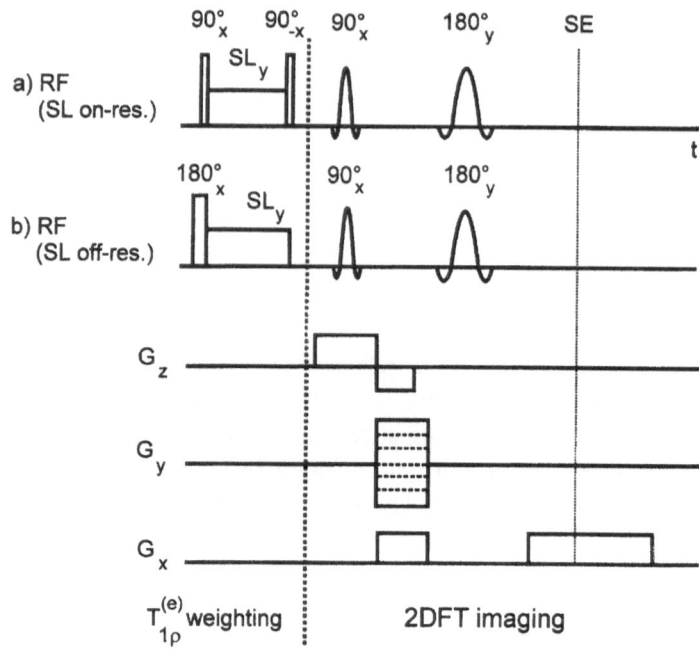

Fig. 27.1. RF and gradient pulse scheme for rotating-frame relaxation dispersion imaging (RODI). The magnetization is first weighted by $T_{1\rho}^{(e)}$ under on-resonance (a) or off-resonance (b) conditions. After these preparation intervals a conventional 2DFT imaging sequence follows. The signals are recorded in the form of a Hahn spin echo (SE). Maps characterizing the dispersion of the rotating-frame spin-lattice relaxation time are rendered by subtracting on-resonance (low effective frequency) (a) from off-resonance (high effective frequency) (b) rotating-frame relaxation weighted signals. The contrast parameter is given in Eq. 27.3.

range (see Sect. 10.3). The problem is now to define a suitable parameter characterizing the slope of the dispersion.

An operational procedure is to represent the dispersion indirectly by the difference of the pixel intensities alternately measured at two different effective frequencies ω_e. The contrast parameter to be mapped is then

$$p_\rho = I(0) \left\{ \exp\left(-\frac{\tau_{sl}}{T_{1\rho}^{(e)}(\omega_e^{(2)})} \right) - \exp\left(-\frac{\tau_{sl}}{T_{1\rho}^{(e)}(\omega_e^{(1)})} \right) \right\} \qquad (27.2)$$

where $I(0)$ is the pixel intensity without spin-lock preparation pulses, and τ_{sl} is the length of the spin-lock pulse.

For short spin-lock intervals, this expression can be approximated linearly, leading to the $T_{1\rho}^{(e)}$ dispersion parameter

$$p_\rho \approx \frac{I(0)\tau_{sl}}{T_{1\rho}^{(e)}(\omega_e^{(1)})T_{1\rho}^{(e)}(\omega_e^{(2)})} \Delta T_{1\rho}^{(e)} \qquad (27.3)$$

where $\Delta T_{1\rho}^{(e)}$ is the difference between the relaxation times at $\omega_e^{(2)}$ and $\omega_e^{(1)}$.

With this parameter representation of the $T_{1\rho}^{(e)}$ dispersion, striking contrast changes have been found with mice tumors [412, 418]. Malignant tissue produced clear bright contrasts, whereas the surrounding tissue (predominantly muscle) was suppressed completely (see Fig. 27.2). This is noteworthy because in normal relaxation-time weighted images one observes changes of contrast but scarcely a complete suppression of certain tissue species.

The contrasts in $T_{1\rho}^{(e)}$ dispersion maps may be modified by contrast agents specifically designed for this purpose. A "contrast agent" of particular interest is ^{17}O water. The natural abundance is about 4×10^{-2}. The modulation of the scalar coupling between protons and ^{17}O by chemical spin exchange leads to a pronounced low-frequency proton spin-lattice relaxation dispersion, especially if the water is isotopically enriched [162]. At pH 7 the mean exchange time is in the order of milliseconds, so that the scalar-coupling relaxation mechanism contributes to rotating-frame relaxation. A strong $T_{1\rho}^{(e)}$ dispersion arises above about 100 Hz. This dispersion can be recorded with the RODI technique, and converted into the gray scale contrasts of a $T_{1\rho}^{(e)}$ dispersion map visualizing the ^{17}O distribution in the sample. Figure 27.3 shows a comparison of a RODI map and a conventional image of water vials for the demonstration of ^{17}O-based contrasts. In the RODI map, ^{17}O enriched water is only visible at neutral pH, i.e., when the $T_{1\rho}^{(e)}$ dispersion occurs in the right frequency range. On the other hand, acidified water has a dispersion shifted outside the frequency window relevant in this experiment. This does not affect conventional imaging, while the corresponding sample is entirely concealed in the RODI map [412].

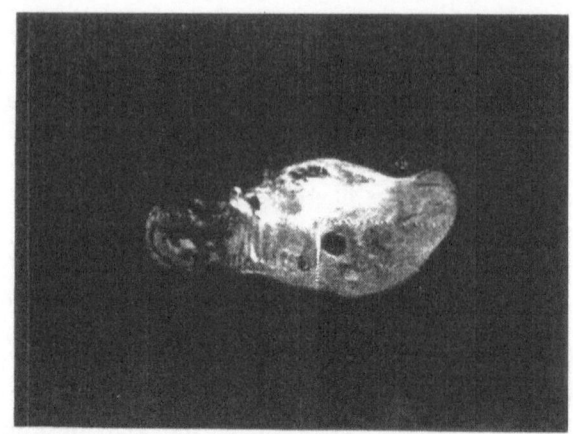

Fig. 27.2. NMR images of a mouse in vivo with an implanted tumor at the right hind leg. The tumor is located at the left side of the images showing cross sections of the abdomen and the upper thighs of the mouse lying on the back. The data were recorded at 4.7 T: **a)** $T_{1\rho}$-weighted two-dimensional spin-echo Fourier transform image (see Fig. 24.3); **b)** rotating-frame dispersion map (RODI) rendering the $T_{1\rho}^{(e)}$ dispersion parameter (Eq. 27.3) as gray scale contrasts. The tumorous tissue on the left-hand side can clearly be distinguished from muscle tissue (right), the signal intensity of which is completely suppressed without application of any gray scale windowing. The dispersion refers to the frequency range 500 - 1670 Hz. The spin-lock time was $\tau_{sl} = 40$ ms. The vertical lines are artifacts due to the superposition of FIDs generated by imperfections of the 180° RF pulse in the 2DFT imaging sequence. These signals are neither phase encoded nor $T_{1\rho}^{(e)}$ weighted. They therefore produce projections of the whole (!) mouse on the read gradient direction even when the signals of the benign tissue are suppressed as is the case in (b). This again demonstrates the efficiency of the $T_{1\rho}^{(e)}$-weighting preparation pulses of the RODI pulse sequence. The structured horizontal line artifact in (a) is a consequence of imperfect baseline corrections of the echo signals. (Reproduced by permission from ref. [412])

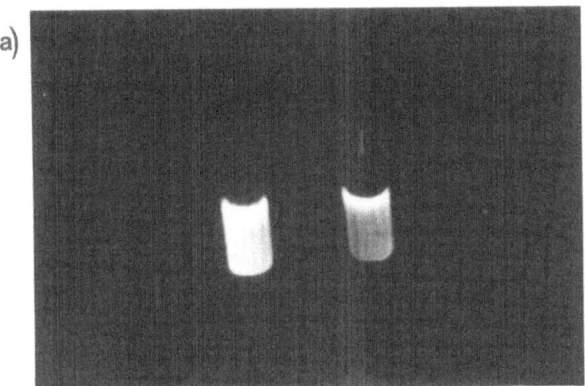

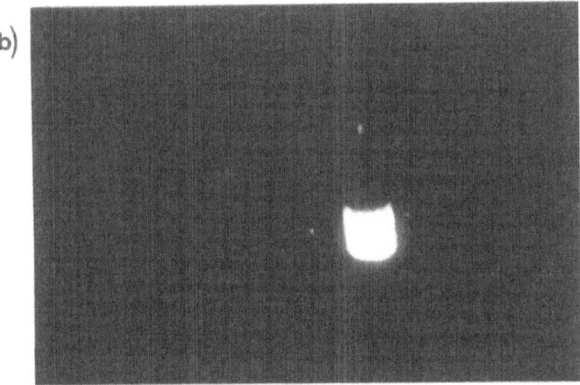

Fig. 27.3. a) $T_{1\rho}$-weighted image of acidified (left) and neutral (right) samples of water enriched with 4 percent ^{17}O. b) $T_{1\rho}^{(e)}$ dispersion (RODI) map of same. The water was doped with $MnCl_2$ in order to reduce the relaxation times (without affecting the low-frequency dispersion). The experimental frequency range was 165 Hz to 620 Hz. The spin-lock time was $\tau_{sl} = 100$ ms. In the RODI map the acidified sample is entirely invisible by contrast to the $T_{1\rho}$-weighted image. No gray-scale windowing was applied. This experiment demonstrates the sensitivity of the RODI technique for the detection of neutral ^{17}O containing water. (Reproduced by permission from ref. [412])

CHAPTER 28

Frequency-Offset Maps

Frequency offsets of any origin can directly and unambiguously be recorded by introducing a further measuring domain, the spectral-encoding domain, in addition to the k space domains. The dimension of the experiment is increased by one. Thus up to four-dimensional pulse schemes are considered. "Mapping" in this context means that the shift of a spectral line is evaluated from the local spectra and rendered in the form of a gray-scale or color map.

One of the most important spectral parameters probed by NMR is the chemical shift of the resonance frequency. This information is closely related to the molecular composition and structure of the compounds bearing the nuclei. Frequency offsets can also arise on a supramolecular length scale by variations of the local magnetic susceptibility or even by inhomogeneities of the magnet.[1] From the experimental point of view, the NMR signals are affected by these frequency offsets in the same way, irrespective of the origin. However, a distinction becomes possible by post-detection processing and considering the different length scales on which these phenomena are relevant. Pulse sequences probing local frequency offsets are referred to as **magnetic resonance spectroscopic imaging** (MRSI) methods.

Local frequency offsets by the inhomogeneous susceptibility and chemical-shift distribution in an object from which an NMR image is to be rendered can cause severe artifacts, especially if the flux density of the main magnetic field is high and the encoding gradients are moderate [73, 95, 389, 407].

However, MRSI provides full information for post-detection processing of the image data, so that artifacts due to local frequency offsets can be corrected whatever the origin. In NMR microscopy, the spatial resolution of such frequency-offset corrected images is merely restricted by the voxel signal-to-noise ratio and the potential influence of diffusion, flow, and motions in the object (see Chap. 32). As a consequence, relatively low read gradients can safely be used even at very high magnetic fields, so that the signal-to-noise ratio benefits from the correspondingly narrow acquisition bandwidth.

The same information background optionally permits mapping of the spatial distribution of frequency offsets in the form of gray shades or colors. This means in particular that the spatial distribution of field offsets due to magnet inhomogeneities or magnetic-susceptibility variations within the sample can be visualized. Moreover,

---

[1] These inhomogeneities must not be confused with the field gradients deliberately applied for spatial encoding purposes.

the spatial distribution of compounds with distinct and resolved chemical shifts can be rendered as chemical-shift selective images.

The first problem to be solved is to have a pulse sequence suitable for encoding the local frequency offsets. That is, a further (spectroscopic) domain is to be added to the two or three spatial dimensions. Another no less important task is to find a suitable post-detection procedure for the correct artifact-free representation of the spectral and spatial information. Experimental data sets achieved in this way permit one to produce maps of certain spectroscopic parameters. The third problem therefore is to define informative quantities which can then be represented in the form of gray-scale or color images.

28.1
MRSI Pulse Sequences

The spatial distribution of frequency offsets can be acquired in imaging pulse sequences either by correspondingly phase-encoding the signals, or by reserving the read-out domain for this purpose. The former, i.e., echo-time encoding of the spectral information, turned out to be particularly favorable [176, 318, 380, 509, 511]. The technique consists of a standard Fourier imaging pulse sequence supplemented by a coherence evolution interval before the echo signals are acquired. The spectral frequency offsets are encoded by incrementing this interval in a series of image records. Figure 28.1 shows a variant based on gradient-recalled echoes.

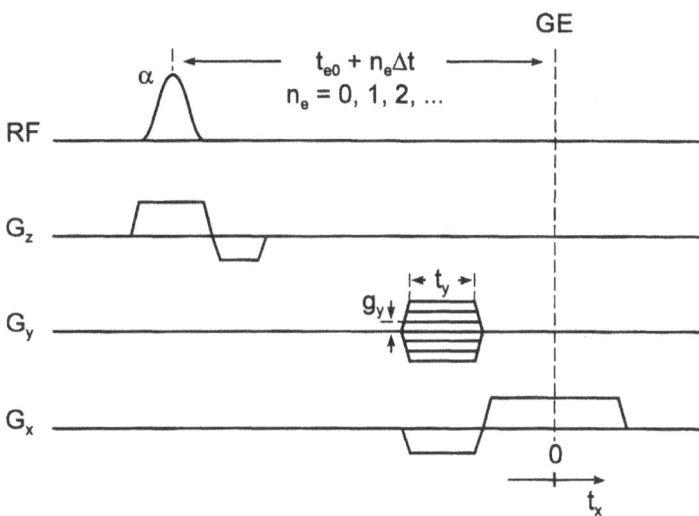

Fig. 28.1. Gradient-recalled echo (GE) pulse sequence (Fig. 25.6) modified for "magnetic resonance spectroscopic imaging" (MRSI). The method is suited for the recording of maps of frequency-offsets due to chemical shifts, magnet inhomogeneities, or magnetic susceptibility. One spectroscopic and two spatial dimensions are probed by phase and frequency encoding. Frequency offsets are phase encoded by incrementing the echo time $T_E = t_{e0} + n_e\Delta t$ in subsequent transients as indicated ("spectroscopic phase encoding").

The complex signal raw data recorded with the aid of this method are processed by three-dimensional Fourier transformation typically after zero-filling an $n \times n \times n$ data matrix for the read, spatial phase-encoding, and spectroscopic phase-encoding dimensions, respectively. The resulting image data matrix provides the full spectral information for each pixel so that the image can be corrected by post-detection processing for frequency-offset artifacts.

28.2
Theory of MRSI

Let us denote the (relaxation and/or diffusion weighted) density of spins with a frequency offset in the range $\Omega \ldots \Omega + d\Omega$ by $\bar{\rho}(x, y, z, \Omega)d\Omega$. The frequency offsets may originate for any of the reasons mentioned before. The signal to be detected is weighted by the slice profile $g(z)$ so that the effective (relaxation/diffusion and slice-profile weighted) offset-specific spin density is

$$\rho(x, y, \Omega) = \int g(z)\bar{\rho}(x, y, z, \Omega)\,dz \tag{28.1}$$

The total signal recorded with the pulse sequences (Fig. 28.1) is then

$$S(k_x, k_y, T_E) = \int \int \int dx\,dy\,d\Omega\; \rho(x, y, \Omega)\; e^{-ik_x x}\; e^{-ik_y y}\; e^{+i\Omega(T_E + t_x)} \tag{28.2}$$

where T_E is the echo time. The wavenumbers, $k_x = \gamma_n G_x t_x$ and $k_y = \gamma_n G_y t_y$ are functions of the encoding gradients and the encoding intervals, G_x, G_y, and t_x, t_y, respectively.

We realize that the Fourier transform $S(k_x, k_y, T_E)$ with respect to k_y yields the right dependence of the spin density to be rendered as an image on the coordinate y, but that the other two dimensions, $k_x \leftrightarrow x$ and $T_E \leftrightarrow \Omega$, are not independent of each other. Rather there is a correlation arising in the course of coherence evolution in the presence of the read gradient. The Fourier transformation without corresponding correction must therefore lead to distorted image data.

The interdependence becomes obvious by the substitution

$$t_x = \frac{k_x}{\gamma_n G_x} \tag{28.3}$$

leading to

$$S(k_x, k_y, T_E) = \int \int \int dx\,dy\,d\Omega\; \rho(x, y, \Omega)\; e^{-ik_x\left(x - \frac{\Omega}{\gamma_n G_x}\right)}\; e^{-ik_y y}\; e^{+i\Omega T_E} \tag{28.4}$$

This relation suggests the substitution

$$x = x' + \frac{\Omega}{\gamma_n G_x} \tag{28.5}$$

$$dx = \left[1 + \frac{1}{\gamma_n G_x}\frac{d\Omega}{dx'}\right]dx' \tag{28.6}$$

The measured signal function can then be represented by

$$S(k_x, k_y, T_E) = \int \int \int dx' \, dy \, d\Omega \, \rho(x' + \frac{\Omega}{\gamma_n G_x}, y, \Omega)$$
$$\left[1 + \frac{1}{\gamma_n G_x} \frac{d\Omega}{dx'}\right] e^{-ik_x x'} \, e^{-ik_y y} \, e^{+i\Omega T_E} \qquad (28.7)$$

The three-dimensional Fourier transformation with respect to the variables T_E, k_x, and k_y yields

$$\rho'(x', y, \Omega) = \rho(x' + \frac{\Omega}{\gamma_n G_x}, y, \Omega) \left[1 + \frac{1}{\gamma_n G_x} \frac{d\Omega}{dx'}\right] \qquad (28.8)$$

The function $\rho(x' + \frac{\Omega}{\gamma_n G_x}, y, \Omega) = \rho(x, y, \Omega)$ is what we want to render as an image whereas $\rho'(x', y, \Omega)$ represents the experimental result after Fourier transform processing. Thus, the function to be mapped in the image is

$$\rho(x, y, \Omega) = \frac{\rho'(x', y, \Omega)}{1 + \frac{1}{\gamma_n G_x} \frac{d\Omega}{dx'}} \qquad (28.9)$$

28.3
Post-Detection MRSI Data Processing

In terms of digital data acquisition and post-detection signal processing, we first acquire a three-dimensional data matrix

$$\left\{S_{l,m,n}\right\} = \left\{S(k_x^{(l)}, k_y^{(m)}, t_e^{(n)})\right\} \qquad (28.10)$$

From this we evaluate the Fourier transformed data set

$$\left\{\rho'_{l,m,n}\right\} = \left\{\rho'(x'_l, y_m, \Omega_n)\right\} \qquad (28.11)$$

The frequency-offset encoding time, i.e., the echo time T_E, can only be varied in the range $0 < t_{e0} \leq T_E \stackrel{<}{\sim} T_2^*$ for experimental reasons. The upper limit, the effective transverse relaxation time T_2^*, restricts the resolution achievable in the spectroscopic dimension (see Chap. 32), whereas the lower limit given by the shortest adjustable echo time t_{e0} leads to frequency-dependent phase shifts. Straightforward Fourier transformation of the measured data representing $S(k_x, k_y, T_E)$ thus provides the function $\rho'(x', y, \Omega) \exp\{i\Omega t_{e0}\}$ in the form of a raw data set. Therefore, the matrix $\{\rho'_{l,m,n}\}$ must be first-order phase corrected prior to further processing.

The resulting three-dimensional data set allocates a local spectrum including any offset relative to the reference frequency to each pixel (x', y). These local spectra permit one to correct the pixel coordinates in the read direction. The principle is to identify a typical resonance line in the spectra of each pixel. In biological objects this may be the water or the lipid ("fat") lines, for instance.

The next step of the procedure is the division of the data by the elements of the scaling matrix

$$\left(1 + \frac{1}{\gamma_n G_x} \frac{d\Omega}{dx'}\right)_{l,m,n} \approx 1 + \frac{1}{\gamma_n G_x} \frac{(\delta\Omega^{(r)})_{l,m}}{\delta x'} \tag{28.12}$$

The quantity $(\delta\Omega^{(r)})_{l,m} = \Omega^{(r)}(x'_{l+1}, y_m) - \Omega^{(r)}(x'_l, y_m)$ is the variation of the offset of the selected reference line at the resonance frequency $\Omega^{(r)}$ between the (apparent) positions x'_{l+1}, y_m and x'_l, y_m. Note that this is a measure of field offsets due to magnet or susceptibility inhomogeneities and therefore a *local* quantity in contrast to the (constant) apparent pixel resolution in x' direction, $\delta x'$.

Finally the resulting data matrix elements for triples l, m, n are assigned to triples $l + \kappa, m, n$ where

$$\kappa = \text{integer nearest to } \frac{\Omega^{(r)}(x'_l, y_m)}{\gamma_n G_x} \tag{28.13}$$

As the quantity $\Omega^{(r)}/(\gamma_n G_x)$ is normally a fractional number, and because the corrected pixel size may deviate from $\delta x'$, the original and new pixel patterns do not necessarily coincide so that the data values of the pixels neighboring the position $l + \kappa$ must be weighted according to the overlap ratio of the original and new pixels [509, 510]. The image data matrix is then given by

$$\{\rho_{l,m,n}\} = \{\rho(x_l, y_m, \Omega_n)\} \tag{28.14}$$

as the corrected effective offset-specific spin density.

This data matrix provides the total spectrum including any inhomogeneity or susceptibility shifts for each pixel (x_l, y_m). A further matrix $\Omega^{(r)}(x_l, y_m)$ representing the local offsets of the reference line from the carrier frequency is readily available. On this basis, different sorts of images can be rendered. For instance, integrated-spectra images are produced by integrating the effective offset-specific spin density over all frequency offsets Ω_n. Resonance-line selective effective spin density images can be generated by integrating over the resonance line of interest. Finally, susceptibility (or more generally, inhomogeneity) images can be rendered by converting the frequency offsets $\Omega^{(r)}(x_l, y_m)$ into gray shades [381, 510].

28.4
Post-Detection Correction of Frequency-Offset Artifacts

Once having an image data set including a frequency-offset dimension, there is a wealth of representation possibilities of different images and maps. In the following we will discuss typical applications to a human finger and an application to time-resolved imaging of heavy ion intrusion in alginate gels. Let us first demonstrate the post-detection correction of frequency-offset artifacts in NMR images. The diverse steps of this procedure are illustrated in Figs. 28.2a-f.

Figure 28.2a,b represent a MRSI data set of a human finger directly after 3D-Fourier transform without any artifact correction. Figure 28.2a shows a cross-sectional image the gray shades of which were calculated as the integral of the

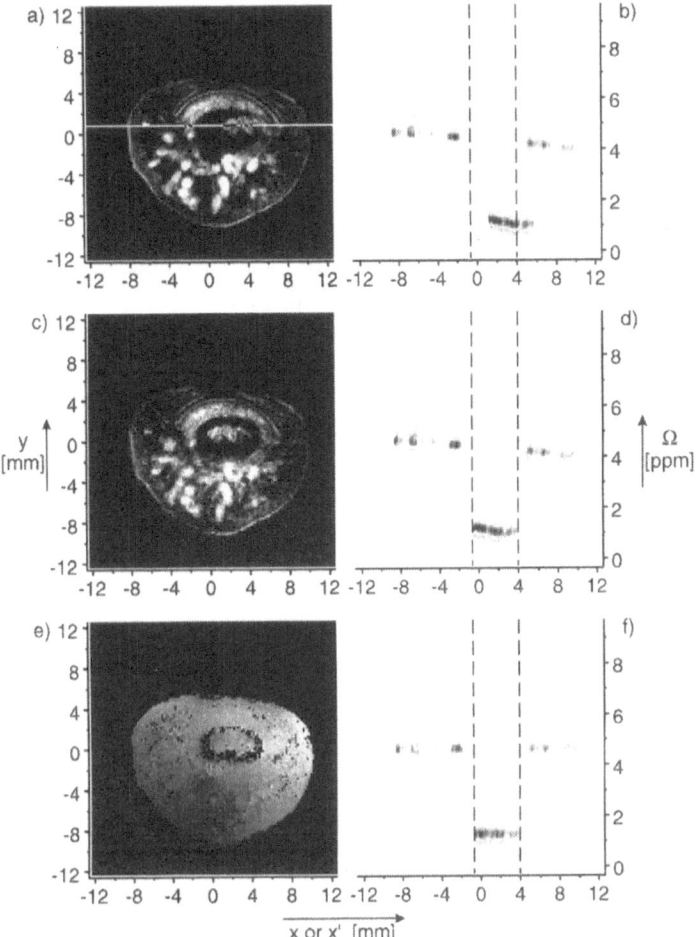

Fig. 28.2. NMR-images and spectroscopic maps of a human finger as recorded and after application of post-detection correction procedures. The data are represented in the magnitude mode. The dashed lines in the spatio/spectral maps indicate the right range of the yellow bone marrow which is dominated by lipid signals. The three-dimensional data matrix recorded in the k_x, k_y and T_E domains of pulse sequence Fig. 28.1 had a format of $128 \times 128 \times 48$. The number of transients was 2. The slice thickness is 1 mm, the in-plane pixel resolution $200\ \mu m \times 200\ \mu m$. The positions of the water or lipid peaks needed for artifact correction procedure were determined from magnitude spectra on a FWHM (full width at half maximum) basis: **a)** image of the integral intensity of the whole spectral axis of a 3D-Fourier transformed MRSI-data set without any post-detection correction. The read gradient was relatively low, so that chemical shift artifacts arose in the form of dislocations up to about 8 to 9 pixels; **b)** uncorrected spatio/spectral intensity map of the spectra of the pixel row marked in (a) by the horizontal stripe in the read-gradient direction. The data are represented in magnitude mode. "Dark" gray shades mean "high intensity", "light" indicates "low intensity". Frequency offset artifacts reveal themselves as dislocations on the x axis, and in the tilted contours of the line ridges; **c)** image (a) corrected for chemical shift but not for susceptibility and inhomogeneity artifacts; **d)** spatio/spectral map (b) after correction for chemical shift artifacts. The influence of susceptibility changes and inhomogeneities are still visible in the form of tilts of the spectroscopic ridge contours; **e)** frequency-offset map of (a) after correction for all shifts Ω. The gray shades represent a range -0.3 ppm $< \Omega/\gamma_n B_0 < 0.6$ ppm where B_0 is the external magnetic flux density; **f)** spatio/spectral map (b) after correction for all frequency-offset artifacts. Resonance ridge contours are now horizontal, and the dislocation of the (lipid) signal of the bone marrow relative to the (water) signal of the muscle tissue has disappeared. (Reproduced by permission from ref. [511])

whole spectrum. The image rendered in this way is equivalent to an ordinary gradient echo 2D image. The white horizontal stripe marks a row of pixels whose spectra are mapped in Fig. 28.2b. The pixel shifts along the read direction due to the chemical-shift and magnetic-susceptibility differences between tissues predominantly consisting of muscles (strong water signal) and bone marrow (strong lipid signal) are obvious (see dashed reference lines). The patterns of the resonance ridges are also tilted in respect to the x' axis as a consequence of inhomogeneous offsets which in the present case predominantly refer to the external magnetic field.

The integrated-spectrum image Fig. 28.2c and the corresponding spatio/spectral intensity map, Fig. 28.2d, represent the data set after correction for chemical shifts. The spatial intensity distribution associated with the lipid line in the bone marrow is now at the correct position relative to the water signal of the tissue outside the bone. However, the susceptibility and inhomogeneity artifacts are still present as indicated by the spectroscopic ridges of the water or lipid lines which are tilted with respect to the baseline.

Figure 28.2e finally shows a map of the frequency offsets of the highest peak of the local spectrum at corrected pixel positions. The offsets are caused by the spatial distribution of the susceptibility of the tissue and the inhomogeneity of the external magnetic field. The spectra obtained after correction for spectral susceptibility or inhomogeneity artifacts are displayed in the spatio/spectral intensity map (Fig. 28.2f) where the x-positions are again corrected as described above.

The distortions by the susceptibility and inhomogeneity artifacts are demonstrated in the finger images Figs. 28.3a,b. The corrected longitudinal cross-section, Fig. 28.3b, is elongated by 6 pixels in the x (read) direction relative to the apparent length in Fig. 28.3a (see dashed reference lines). By contrast, the corresponding shortening in the transverse cross-section (Fig. 28.2c) is only 2 pixels so that it is scarcely perceptible. The larger distortions in the case of the longitudinal cross-section are due to the strong susceptibility step in B_0-direction at the finger tip.

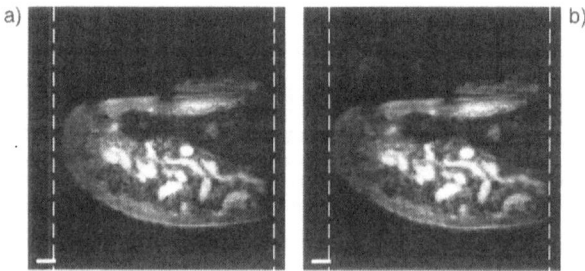

Fig. 28.3. Images of a human finger recorded with the aid of pulse sequence (Fig. 28.1). The experimental parameters are the same as in Fig. 28.2. The white bar represents 2 mm in reality: a) image after correction for chemical shift artifacts; b) image after correction for chemical shift as well as susceptibility and inhomogeneity artifacts. The white lines illustrate the distortion of the read axis length scale by the magnetic susceptibility and inhomogeneity artifacts. Image (a) is apparently shortened in the x direction by 6 pixels relative to the fully corrected image (b). (Reproduced by permission from ref. [511])

28.5
Spectroscopic Maps and Shift-Selective Images

The three-dimensional spatio/spectral data set corrected for artifacts as outlined above can be used for producing maps of different parameters. For instance, Fig. 28.4a-d represents chemical-shift selective images of the integrated intensity of the water and lipid lines in a human finger.

Maps of further spectroscopic parameters are displayed in Fig. 28.5a-d. These are images of different kinds of T_2^* weighted contrasts and direct maps of the percentual contribution of water and lipid signals.

The spectroscopic identification of certain tissue constituents even permits one to render selective images of anatomical details. In Fig. 28.6, different gray shades are attributed to pixels in muscle, fat, blood, and bone marrow.

Analogous to the contrast agents one uses for the artificial enhancement of relaxation contrasts, there is a possibility to enhance chemical shifts with the aid of **lanthanide shift reagents** [334]. In this way, one can separate overlapping resonances of compounds by formation of paramagnetic metal complexes. The chemical shift is then altered either by transfer of electron spin density from the metal ion to the associated nuclei (**"contact shift"**), or by the direct interaction of the unpaired electron with the spins to be observed (**"pseudocontact shift"**).

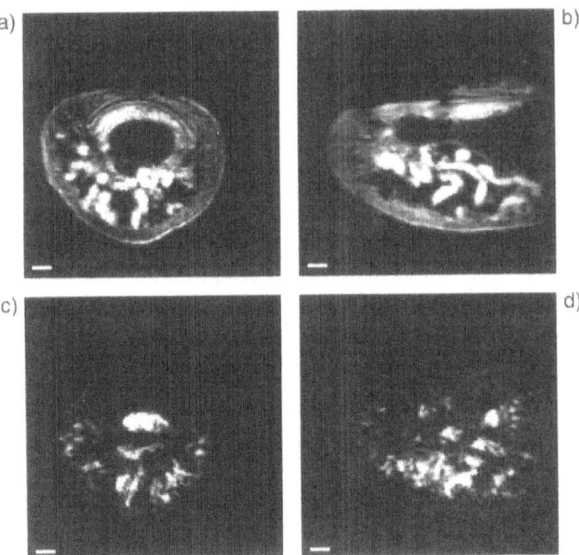

Fig. 28.4. Chemical-shift selective images of: **a)** transverse; **b)** longitudinal; **c)** transverse; **d)** longitudinal cross-sections of a human finger recorded with the aid of pulse sequence (Fig. 28.1). Frequency offset artifacts were corrected as described in Sect. 28.4. The experimental parameters are the same as in Fig. 28.2. The gray shades of (a) and (b) represent the integrated water line, whereas (c) and (d) reflect maps of the integral over the lipid line. The white bar represents 2 mm in reality. (Reproduced by permission from ref. [511])

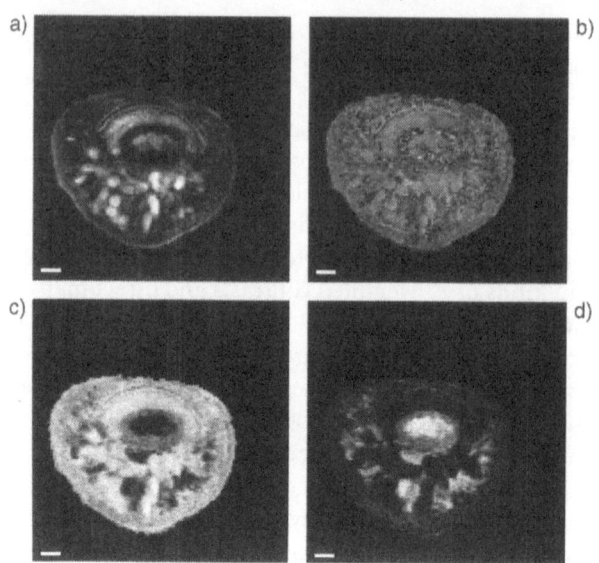

Fig. 28.5. Transverse cross-sections of a human finger recorded with the aid of pulse sequence (Fig. 28.1). The experimental parameters are the same as in Fig. 28.2. The white bar represents 2 mm in reality. The gray shade contrasts represent different spectroscopic parameters evaluated after correction for frequency-offset artifacts (see Sect. 28.4). The image parameters are: **a)** magnitude of the highest peak in each voxel; **b)** full width at half maximum of the highest peak of each voxel (this corresponds to a "T_2^* map"); **c)** quotient of the integrated water peak and the integral of the total spectrum (this essentially corresponds to a "water percentage map" without corrections for the possibly different relaxation losses of the chemical constituents); **d)** quotient of the integrated lipid peak and the integral of the total spectrum (this virtually corresponds to a "lipid or fat percentage map" without corrections for the potentially different relaxation losses of the chemical constituents). (Reproduced by permission from ref. [511])

The frequency offset itself can be a contrast parameter of interest. This was first suggested as a means for imaging field inhomogeneities of the magnet for shimming purposes [332, 333]. In cases where the frequency offset varies spatially due to material or tissue properties, corresponding images may also serve the characterization of the object. A typical application of this sort is "magnetic-susceptibility mapping."

Figure 28.7 represents an experiment where the ion uptake of paramagnetic ions in calcium alginate from a dilute solution is imaged via the local change of the magnetic susceptibility [362]. This ion exchange process leads to a local field shift $\Delta B(r)$ causing a frequency offset $\Omega(r)$. The molar concentration of biosorbed paramagnetic ions is given by

$$c(r) = \frac{3}{\chi_m} \frac{\Delta B(r)}{B_0} = \frac{3}{\chi_m} \frac{\Omega(r)}{\omega_0} \qquad (28.15)$$

where χ_m is the molar magnetic susceptibility of the paramagnetic species. Thus the gray shades in the frequency offset maps directly reflect the local concentration. A series of time-resolved maps of this sort permits one to describe the time evolution

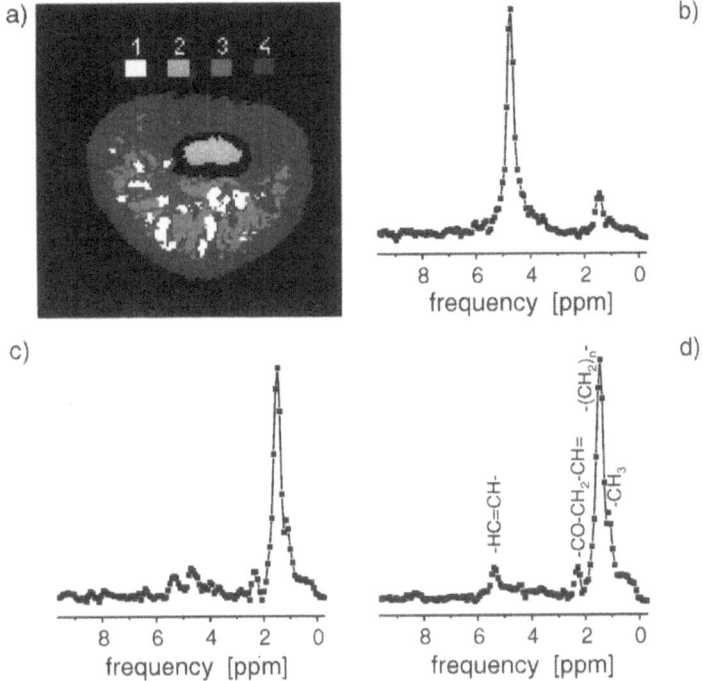

Fig. 28.6. Cross-sectional image and spectra of a human finger recorded using pulse sequence (Fig. 28.1) with the same experimental parameters as in Fig. 28.2. Anatomical details were identified according to the spectroscopic information implied in the three-dimensional spatio/spectral data set after correction for frequency-offset artifacts (see Sect. 28.4): **a)** image composed of gray shades for anatomy-specific spectral properties (key - "1", blood vessels identified as the slowly relaxing component of the water signal (altogether 347 pixels corresponding to a total volume of 13.9 mm^3), "2" yellow bone marrow identified as the lipid signal inside the bone (altogether 211 pixels corresponding to a total volume of 8.4 mm^3), "3" fat identified as the lipid signal outside the bone (altogether 777 pixels corresponding to a total volume of 31.1 mm^3), "4" water outside the blood vessels identified as the fast relaxing component of the water line (altogether 3,967 pixels corresponding to a volume of 158.7 mm^3)); **b)** proton spectrum of all pixels (gray shades "1" to "4" in (a)); **c)** proton spectrum of fat tissue (gray shade "3" in (a)); **d)** proton spectrum of the yellow bone marrow (gray shade "2" in (a)). (Reproduced by permission from ref. [511])

of the diffusion/reaction front which turns out to be remarkably steep. The ingress clearly does not obey Fick's diffusion laws. A better description is possible with a model analogous to Stefan's treatment of a related heat conduction problem [470] This formalism [395] takes into account that the diffusion front proceeds only as fast as the ion exchange reaction permits.

Sometimes the record of the spectroscopic dimension in addition to the two or three spatial domains may turn out to be too time consuming. A favorable strategy in that case is to image the object just in order to be able to define a region where spectral information is of interest. The spectrum can then be recorded in a second experiment using volume-selective spectroscopy. Suitable pulse sequences

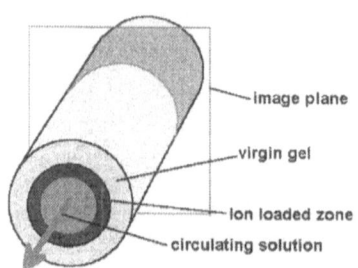
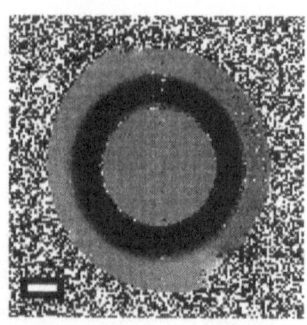

Fig. 28.7. Frequency offset (or magnetic-susceptibility) mapping of the ingress of paramagnetic ions into alginate. A dilute solution (1 mmol/l) of $PrCl_3$ is pumped through a tube prepared from diamagnetic calcium alginate. The cross-sectional image on the right-hand side was recorded after 14 h of exposure to the ion solution. The frequency offsets are mapped in the form of gray shades. In the dilute solution in the middle no perceptible offset is detected. However, the high concentration of the Pr ions reached in the diffusion zone in the alginate leads to a strong offset reflected by the dark gray shade. Note that the diffusion/reaction front is too steep to be represented by Fickian diffusion. The frequency-offset encoding interval (see Fig. 28.1) was incremented in 32 steps. The white bar corresponds to 1 mm in reality. (Reproduced by permission from ref. [362])

for localization are described in Chap. 37. This is particularly rewarding if further experimental parameters such as the local temperature [241] or the local pH value [354] are to be evaluated from single- or double-quantum coherence chemical shifts.

CHAPTER 29

Gradient-Pulse Moments and Motions

Fast coherent displacements of matter by flow or movements of the object to be imaged interfere with spatial phase or frequency encoding and even with slice selection. Artifacts of this origin can partly be avoided by suitable choices of the pulse sequence and the shape of the gradient pulses.

Images of objects with motional components perpendicular to the selected slice are prone to **inflow/outflow artifacts**. Signals of matter displaced out of the initially excited slice can appear with gradient-recalled echoes or Hahn echoes generated with a non-selective refocusing RF pulse. If there are motional components parallel to the slice to be imaged, these signals appear in blurred form at positions shifted with respect to the original location. Hahn echoes refocused with an RF pulse selective to the same slice as the initial excitation pulse avoid such "displaced" signals superimposed on signals from static material. That is, all matter that has moved out or in the selected slice does not contribute to the signal. The corresponding regions appear as black spots in the image. It is clear that echo times as short as possible help to avoid artifacts of this sort.

Spatial phase and frequency encoding of echo signals are also predisposed to motional artifacts. The interval between the phase-encoding and frequency-encoding gradient pulses should be as short as possible so that the k space components addressed by these gradients remain correlated in the presence of motions.

Particular attention must be paid to bipolar gradient pulses used for self-compensation of coherence defocusing by slice-selection or read-out gradient pulses (see Fig. 24.3, for instance). As will be shown in the following, velocity or acceleration dependent phase shifts may arise. A remedy against artifacts based on such phase distortions is to use gradient pulse shapes with vanishing moments in orders relevant for dephasing by position, velocity, and acceleration.

The trajectory of a nucleus, $r = r(t)$, may be expanded according to

$$r(t) = r_0 + v_0 t + \frac{a_0}{2} t^2 + \dots \tag{29.1}$$

where r_0, v_0, a_0 are the initial position, velocity, and acceleration, respectively. Terms of higher order refer to time dependences of the acceleration which will not be examined further here. A field gradient pulse of an arbitrary shape,

$$G = \begin{cases} G(t) & \text{for } 0 \le t \le T \\ 0 & \text{otherwise} \end{cases} \tag{29.2}$$

leads to the accumulative phase shift of a nucleus which is initially located at $r = r_0$:

$$\phi(T) = \int_0^T \Omega[r(t)] \, dt = \gamma_n \int_0^T G(t) \cdot r(t) \, dt \tag{29.3}$$

$\Omega(r) = \gamma_n G \cdot r$ is the angular-frequency offset at the position $r = r(t)$ due to the spatially constant field gradient. Inserting Eq. 29.1 gives the moment series

$$\phi(T) = \gamma_n \left[\underbrace{r_0 \cdot \int_0^T G(t) \, dt}_{m_0 \text{ (0}^{\text{th}} \text{ mom.)}} + \underbrace{v_0 \cdot \int_0^T G(t) \, t \, dt}_{m_1 \text{ (1}^{\text{st}} \text{ mom.)}} + \underbrace{\frac{1}{2} a_0 \cdot \int_0^T G(t) \, t^2 \, dt}_{m_2 \text{ (2}^{\text{nd}} \text{ mom.)}} + \ldots \right]$$

$$= \phi_0(T) + \phi_1(T) + \phi_2(T) + \ldots \tag{29.4}$$

29.1
Bipolar Gradient Pulses

Neglecting any switching ramps, a bipolar gradient pulse may be represented by the equation

$$G_x = \begin{cases} G_0 & \text{for } 0 \le t \le \tau \\ -G_0 & \text{for } \tau + \Delta\tau \le t \le 2\tau + \Delta\tau \\ 0 & \text{otherwise} \end{cases} \tag{29.5}$$

where we have assumed the gradient to be aligned along the x axis of the laboratory frame. A graphical representation is shown in Fig. 29.1 for the case $\Delta\tau = 0$.

The zeroth moment of this pulse obviously vanishes, $m_0 = 0$. That is, all coherence phase shifts depending on the initial position are refocused. This in particular means that static spins provide a completely recovered gradient echo at the end of the pulse (Fig. 29.1). The first moment takes the value $m_1 = -G_0(\tau^2 + \tau \Delta\tau)$. That is, for a stationary velocity component along the gradient axis, a first-moment based phase shift

$$\phi_1(2\tau + \Delta\tau) = -\gamma_n G_0(\tau^2 + \tau \Delta\tau) v_0 \tag{29.6}$$

remains which is proportional to this velocity. Phase shifts of moving nuclei are not refocused at the end of this bipolar pulse (Fig. 29.2). Therefore, bipolar gradient pulses can serve phase encoding of the local velocity as described in Chap. 30.

The second moment is $m_2 = -2G_0(\tau^3 + \tau^2 \Delta\tau + \frac{1}{3}\tau \Delta\tau^2)$. Stationary acceleration components along the gradient axis hence lead to an additional second-moment based phase shift

$$\phi_2(2\tau + \Delta\tau) = -\gamma_n G_0 \left(\tau^3 + \tau^2 \Delta\tau + \frac{1}{3}\tau \Delta\tau^2 \right) a_0 \tag{29.7}$$

Higher-order terms of the moment expansion of the gradient pulse can lead to further additional phase shifts provided that the time-dependent accelerations occur.

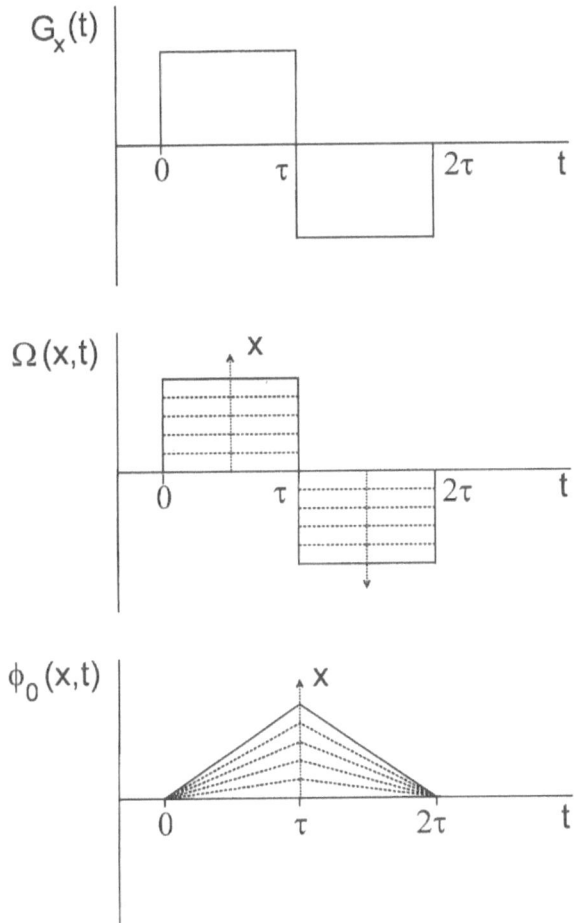

Fig. 29.1. Bipolar gradient pulse, $G_x = G_x(t)$, for $\Delta\tau = 0$. The gradient is assumed along the x axis of the laboratory frame. The angular-frequency offset, $\Omega = \Omega(x, t)$, caused by the gradient is a (preferably linear) function of the position x. The phase shift, $\phi_0 = \phi_0(x, t)$, of static spins is refocused in the form of a gradient echo at the end of the pulse because the zeroth moment vanishes.

It is clear that these motion-induced phase shifts tend to cause severe artifacts in ordinary imaging experiments, and, therefore, are most undesirable. A remedy to a certain degree is provided by so-called "motion-compensated gradient pulses" to be addressed in the following section.

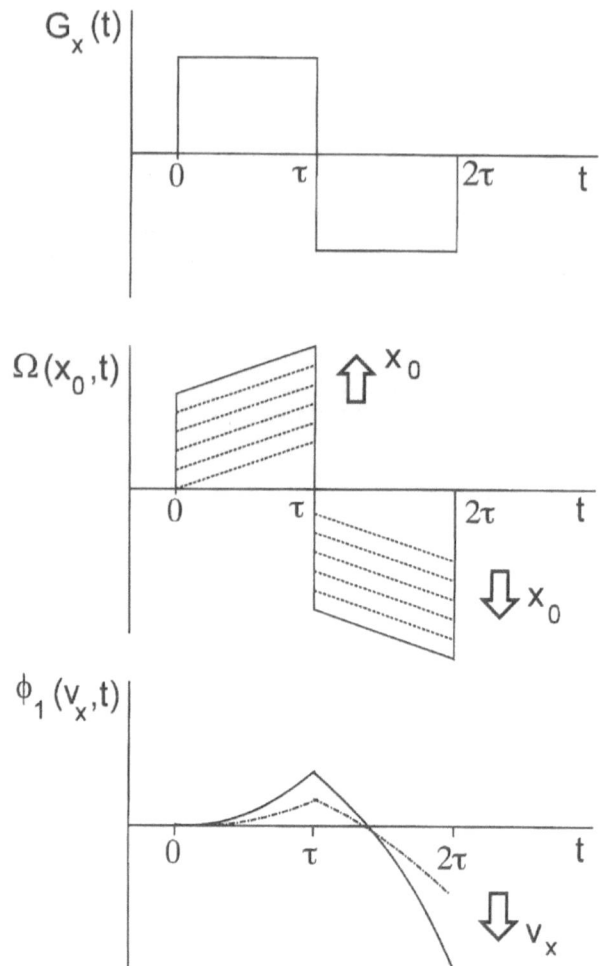

Fig. 29.2. Bipolar x gradient pulse, $G_x = G_x(t)$, with vanishing zeroth moment. The angular-frequency offset, $\Omega = \Omega(x_0, t)$, caused by the gradient is a (preferably linear) function of the position x_0 (linearly) varying with time owing to a (stationary) velocity component along the gradient axis. First-moment based coherence phase shifts, $\phi_1 = \phi_1(v_x, t)$, remain at the end of the pulse, and are proportional to the stationary velocity component v_x along the gradient axis (see Eq. 29.6).

29.2
Velocity-Compensated Gradient Pulses

Motional artifacts[1] arising with gradient pulse pairs can be avoided with the aid of gradient pulse shapes with vanishing first (and potentially also higher) moments.

---

[1] Flow in the presence of bipolar gradient pulse pairs or unipolar gradient pulses separated by a 180° pulse causes phase shifts so that the phase-encoding direction of magnetic-resonance images becomes distorted. Components of **pulsating flow** along gradients therefore result in

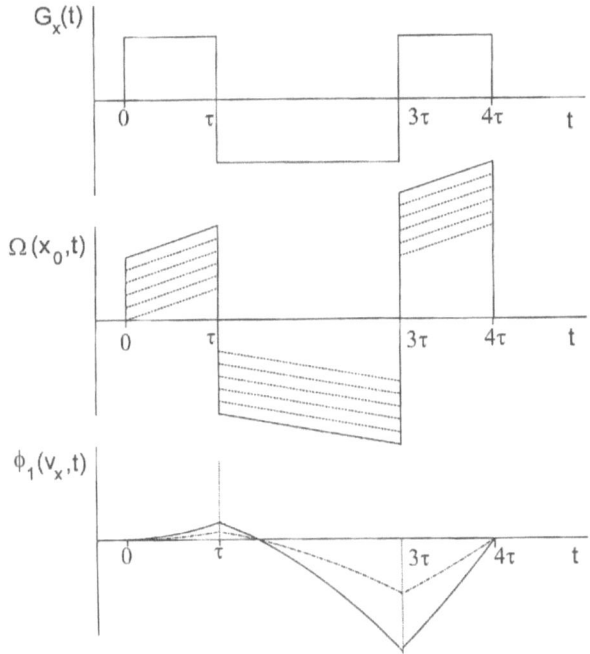

Fig. 29.3. Velocity-compensated x gradient pulse, $G_x = G_x(t)$. The zeroth and first moments are zero. The angular-frequency offset, $\Omega = \Omega(x_0, t)$, caused by the gradient is a (preferably linear) function of the position x (linearly) varying with time owing to a (stationary) velocity component along the gradient axis. At the end of the pulse train, the first-moment based phase shifts, $\phi_1 = \phi_1(v_x, t)$ due to stationary velocity components v_x along the gradient axis are completely refocused. Accelerated motions, on the other hand, lead to second-moment based phase shifts proportional to the acceleration components along the gradient axis.

As an example consider the $(+ - +)$ gradient pulse train (Fig. 29.3):

$$G_x(t) = \begin{cases} G_0 & \text{for} \quad 0 \leq t < \tau \\ -G_0 & \text{for} \quad \tau \leq t < 3\tau \\ G_0 & \text{for} \quad 3\tau \leq t \leq 4\tau \\ 0 & \text{otherwise} \end{cases} \qquad (29.8)$$

The zeroth moment vanishes, i.e., coherences of static spins will be refocused. The first moment vanishes as well, so that coherences of spins moving with a stationary velocity along the gradient axis are also refocused irrespective of the velocity value. However, the second moment takes finite values for accelerated motions:

---

phase-modulated signals. That is, the corresponding sidebands tend to produce "ghost artifacts." Since the series of transients needed for the phase-encoding increments of the gradients and for signal accumulation is normally not coherent with the flow pulsations, an apparent fluctuating signal attenuation occurs. These signal fluctuations may be Fourier analyzed to reveal the frequencies inherent to the pulsations (see ref. [159]).

$m_2 = -4G_0\tau$. The coherence phase shift remaining at the end of the pulse train,

$$\phi_2(4\tau) = -2\gamma_n G_0 \tau^3 a_0 \qquad (29.9)$$

is proportional to the (stationary) acceleration component along the gradient axis. That is, gradient pulses of this shape may in principle be used for phase encoding of the acceleration.

Motion-compensated gradient pulses refocusing phase shifts by the stationary acceleration term or even higher-order terms of the trajectory expansion at Eq. 29.1 are also feasible, of course. Gradient pulse trains with vanishing zeroth, first, and second moment typically consist of alternating gradient lobes, $+G_0, -G_0, +G_0, -G_0$ [366].

For instance, in [259] a pulse sequence for "diffusion and incoherent motion weighted volume-selective NMR spectroscopy" (DICSY) is reported. In order to discriminate molecular self-diffusion from intra-voxel incoherent motions [296], velocity and acceleration compensated pulses were implemented.

CHAPTER 30

Velocimetry and Velocity Maps

Methods for quantitative flow rate measurements (**"velocimetry"**) without spatial resolution have already been suggested before imaging techniques became known [376, 453]. Nowadays phase-encoding techniques [32, 130, 339, 348, 356, 399, 490] are standard for spatially resolved representations. A review of applications can be found in [82].

The effects which - on the one hand - lead to undesired motional artifacts as discussed before, can be used favorably for flow and velocity measurements on the other. One distinguishes techniques merely serving the discrimination of static from flowing material (**"NMR angiography"**) from methods for the quantitative allocation of velocities to the voxels of an image data set. The latter are referred to as **"velocity NMR mapping."**

For example, the former can be based on inflow/outflow effects in the sensitive slice to be imaged [366]. Subtraction of transients recorded with and without gradient pulses for velocity-dependent phase encoding are also in use. In suitably cardiac-cycle triggered versions, i.e., with transients recorded with different instantaneous blood velocity, even the frequency-encoding gradient together with its compensation pulse employed in 2DFT imaging anyway (compare Fig. 24.3) can be utilized for this purpose [491].

Velocity mapping in the proper sense is somewhat more demanding and requires further measuring domains - one per velocity component - in addition to the reciprocal space domains. In each of these additional domains the distribution of the corresponding velocity component is probed. "Mapping" in this context means that, e.g., the average or the most likely velocity of each voxel is evaluated from the local velocity distribution and mapped pixel by pixel. Thus the experiment can in principle become up to six-dimensional if three-dimensional imaging is combined with three-dimensional velocity vector mapping.[1]

There may be situations where the rapid record of a local velocity vector is desirable with a good time resolution. That is, phase-encoding cycles should then

---

[1]From a more operational point of view, three independent, four-dimensional experiments, one for each velocity component, are more favorable than a single six-dimensional conduct of the measurement. A six-dimensional experiment in the proper sense would involve encoding of six quantities in each transient, i.e., five independent phase-encoding increments would be required. Fortunately a series of three four-dimensional experiments is equivalent to such an extremely time-consuming procedure as long as spatial and velocimetrical encoding are independent from each other, as it usually is the case.

be avoided in the measuring process. As a compromise alternative to mapping of the whole velocity vector field in an object, localized velocimetry such as the **"volume-selective measurement of the velocity"** (VOTY) technique has been proposed [258] for this purpose. This method permits the fast, single-transient determination of one or all three velocity components by frequency-encoding in the read-out mode.

30.1
Phase Encoding of the Velocity

The most common technique for velocity mapping is based on phase encoding of the velocity. The phase shift at Eq. 29.6 produced by bipolar gradient pulses such as that displayed in Fig. 29.2 or by unipolar gradient pulse pairs with a 180° RF pulse in between can be favorably employed for this purpose. The local velocity component along the gradient axis generates an unambiguous phase shift of the signal provided that the velocity is stationary.

The combination of such velocity phase-encoding gradient pulses with ordinary imaging pulse schemes permits one to record multidimensional image data sets. Two or three spatial dimensions are supplemented by one, two, or even three velocity dimensions. Thus up to six-dimensional data matrices can be generated [356]. Phase encoding is employed not only for localization but also for spatially resolved velocimetry. All phase-encoding gradients are incremented independently from each other, so that the encoding information can be separately analyzed via Fourier transform processing. That is, the data set does not only refer to the (relaxation or diffusion weighted) spin density in the voxels, but also to the local distributions of the velocity vector components.

30.2
Mapping of Velocity Fields

Figure 30.1 shows a typical pulse scheme for two-dimensional mapping of the velocity vector in a preselected slice of the object. The principle of this technique is known as **"Fourier encoding velocity imaging"** (FEVI) [399]. The velocity components were encoded one by one. In this way three three-dimensional sets of data referring to one velocity component and two spatial dimensions were obtained.

Following Eq. 29.6 the velocity-dependent phase shifts can be expressed in the form

$$\phi_{vj} = k_{vj} v_j \qquad\qquad (j = x, y, z) \qquad\qquad (30.1)$$

where we have introduced a velocity "wave vector"

$$k_v = -\gamma_n G \tau^2 \qquad\qquad (30.2)$$

in analogy to the spatial wave vector k_r (Eq. 24.5). The signal functions acquired in velocity-mapping experiments can then be represented by

$$S(k_x, k_y, k_z, k_{vj}) = \int \int \rho(r) p(r, v_j)\, e^{-i k_r \cdot r}\, e^{-i k_{vj} v_j}\, dv_j\, d^3 r \qquad\qquad (j = x, y, z)$$

$$(30.3)$$

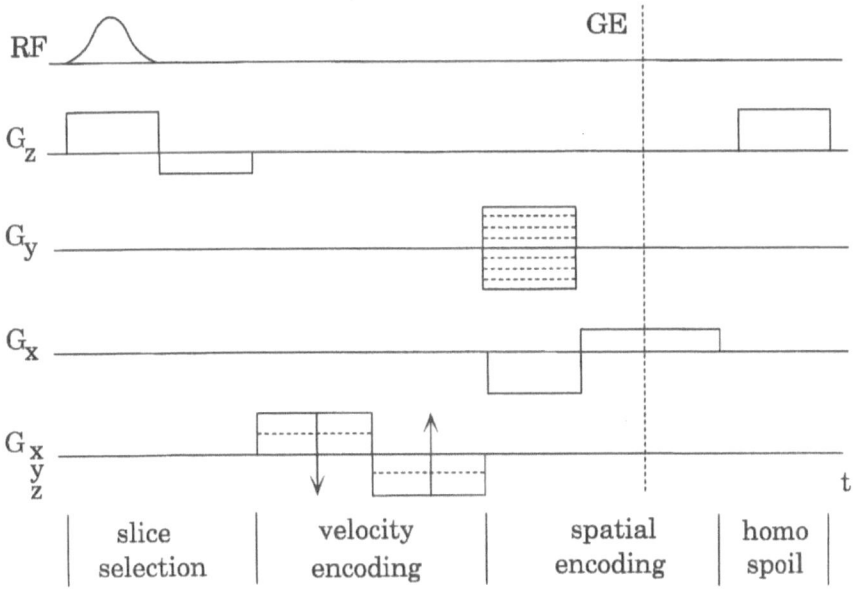

Fig. 30.1. RF and field-gradient pulse scheme for the record of slice-selective velocity maps, the gray scale of which represents a velocity component or the velocity magnitude. The velocity components along the gradient axis are phase-encoded one by one with the aid of bipolar gradient pulses supplemented to a Fourier transform imaging pulse sequence such as that of Fig. 25.6.

In a simplified view one might argue that a certain position can only be connected with a certain velocity at a time. That is, the local velocity distribution would then be given by a δ function,

$$p(r, v_j) = \delta[v_j - v_j(r)] \tag{30.4}$$

so that

$$S(k_x, k_y, k_z, k_{vj}) = \int p(r)\, e^{-ik_r \cdot r}\, e^{-ik_{vj}v_j(r)}\, d^3r \qquad (j = x, y, z) \tag{30.5}$$

The Fourier transform component-by-component with respect to the reciprocal-space variables provides

$$\tilde{S}(x, y, z, k_{vj}) = \mathcal{F}_{k_x}\mathcal{F}_{k_y}\mathcal{F}_{k_z}\left\{S(k_x, k_y, k_z, k_{vj})\right\} = \rho(r)\, e^{-ik_{vj}v_j(r)} \tag{30.6}$$

This expression implies a phase factor merely determined by the local velocity. If the x component of the velocity is encoded in the experiment, for instance, one may evaluate

$$v_x(r) = -\frac{1}{\gamma_n G_x \tau^2} \arctan\left[\frac{\mathcal{I}\{\tilde{S}(r, k_{vx})\}}{\mathcal{R}\{\tilde{S}(r, k_{vx})\}}\right] \tag{30.7}$$

An assessment of a velocity map merely based on the local FT signal phase would be strongly susceptible to artifacts by any inhomogeneous frequency offsets in the object. Therefore it is necessary to probe each velocity component in the form of a whole Fourier dimension by incrementing the velocity phase-encoding gradients in a series of transients.

Furthermore, under experimental conditions, there are two reasons why a δ function is inappropriate for the description of the local velocity distribution. First, one is referring to voxels with a finite volume rather than to point locations in the mathematical sense. That is, a real ("inhomogeneous") velocity distribution may exist within a voxel. Second, measurements are connected with a finite resolution (see Chap. 32), so that a distribution of the local velocity arises anyway for ex-perimental reasons even if the velocity field were strictly homogeneous within the voxel.

Depending on the dimensionality of the wave vectors $\mathbf{k}_r$ and $\mathbf{k}_v$, up to three four-dimensional data matrices for the variable sets (x, y, z, v_x), (x, y, z, v_y), and (x, y, z, v_z) are recorded. After up to four-dimensional Fourier transformation of the signal functions (Eq. 30.3), real-space/real-velocity data sets for

$$\tilde{S}(x, y, z, v_j) = \mathcal{F}_{k_x}\mathcal{F}_{k_y}\mathcal{F}_{k_z}\mathcal{F}_{k_{v,j}}\left\{S(\mathbf{k}_r, k_{vj})\right\} = \rho(x, y, z)\ p(x, y, z, v_j)$$
$$(j = x, y, z) \tag{30.8}$$

are obtained. These data sets imply the full velocity distribution for each voxel, $p(x, y, z, v_j)$. Further evaluation with respect to the average of a velocity component or the most likely velocity component within each voxel gives data sets suitable for rendering as velocity maps. Suitable quantities to be mapped with the aid of gray scales or colors are the velocity components v_x, v_y, v_z or the magnitude of the velocity vector, $v = \sqrt{v_x^2 + v_y^2 + v_z^2}$. Thus, four data sets can be formed, $v_x = v_x(x, y, z)$, $v_y = v_y(x, y, z)$, $v_z = v_z(x, y, z)$, and $v = v(x, y, z)$.

The analogy between the conjugate reciprocal and real spaces concerning the position and the velocity vector also implies that equivalent rules for the "resolution" and the "field of view" apply. This is outlined in more detail in Chap. 32 where these parameters are discussed generally for digital acquisition routines.

The FEVI pulse scheme at Fig. 30.1 permits the record of the velocity vector field in a slice, or, after extension to three spatial dimensions (compare Fig. 25.4), in the whole three-dimensional object to be investigated. The only condition is that the velocity vector field is stationary in the possibly rather lengthy measuring period.

30.3
Typical Applications

As a typical application of the velocity phase-encoding principle, six-dimensional velocity-mapping experiments were employed in experiments serving the deter-mination of the so-called percolation backbone of water flowing through lacunar systems. Examples of such systems are sponges, glass bead agglomerates, or porous rocks [356]. In these experiments, all voxels with velocity magnitudes below the noise level are suppressed, so that only voxels contributing to the transport path-ways are rendered in correspondingly filtered images.

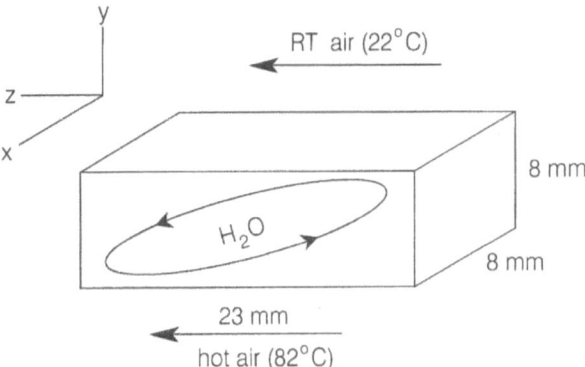

Fig. 30.2. Schematic representation of the convection cell used for the velocity mapping and multiplane/multistripe tagging experiments represented by Figs. 30.3 to 30.5 and Figs. 33.3 to 33.5, respectively. (Reproduced by permission from ref. [512])

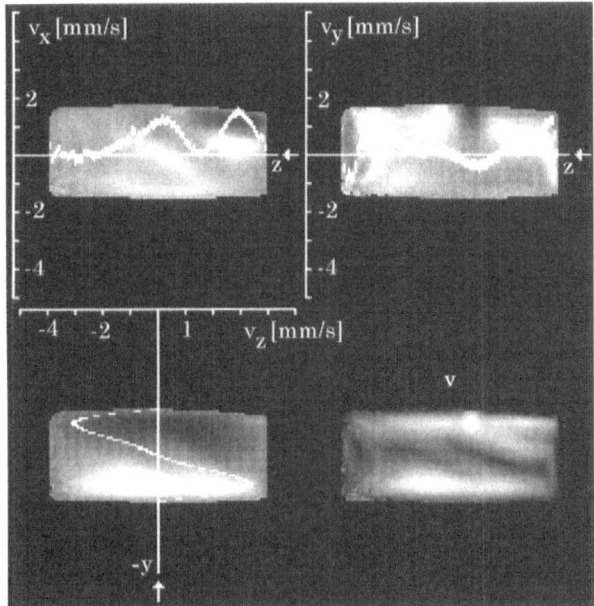

Fig. 30.3. Maps of the maximum local coherent-flow velocity caused by thermal convection in the central y, z plane (temperature gradient along the $-y$ axis) of the cell (Fig. 30.2). The velocity components v_x, v_y, v_z and the magnitude v in the central plane parallel to that spanned by the y and z axes are represented as gray shades (white, high positive velocity; black, high negative velocity). The curves and coordinate axes are superimposed plots of the velocity profiles in the central voxel lines or columns marked by arrows. (Reproduced by permission from ref. [512])

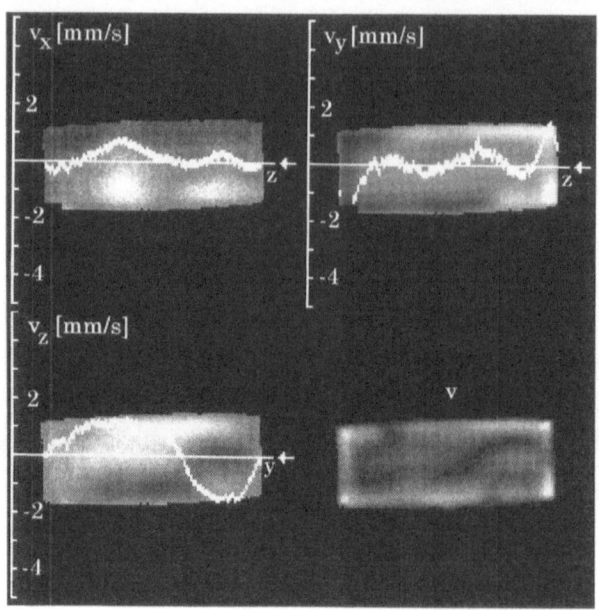

Fig. 30.4. Maps of the maximum local coherent-flow velocity caused by thermal convection in the central x, z plane (temperature gradient along the $-y$ axis) of the cell (Fig. 30.2). The velocity components v_x, v_y, v_z and the magnitude v in the central plane parallel to that spanned by the x and z axes are represented as gray shades (white, high positive velocity; black, high negative velocity). The curves and coordinate axes are superimposed plots of the velocity profiles in the central voxel lines or columns marked by arrows. (Reproduced by permission from ref. [512])

Velocity mapping may also be used for the visualization of thermal convection phenomena including the Rayleigh/Bénard instability of liquids in bulk [512] as well as in porous media [443]. The stationary flow patterns formed by thermal convection in a water-filled cell with a temperature gradient directed from the top to the bottom (see Fig. 30.2) were visualized by the velocity maps shown in Figs. 30.3 to 30.5.

Velocity mapping may also be desirable in situations where the velocity field cannot be kept stationary over prolonged periods. In this case a good time resolution is essential. For this objective, the phase-encoding principle of velocities can be favorably combined with one-transient imaging schemes such as echo-planar imaging (see Fig. 25.7). Using this strategy, velocity fields were mapped in turbulently flowing fluid and in Taylor/Couette flow, e.g., [279, 281].

If the velocity field to be mapped is time dependent and cannot be mapped with sufficiently good time resolution to ensure quasi-stationary conditions, phase encoding of the velocity components as described above fails. In this case signals may even seemingly suggest intra-voxel incoherent motions [296] instead of time dependent coherent motions on a supra-voxel length scale. However, one is dealing in reality with signals with a fluctuating phase due to the instantaneous velocity

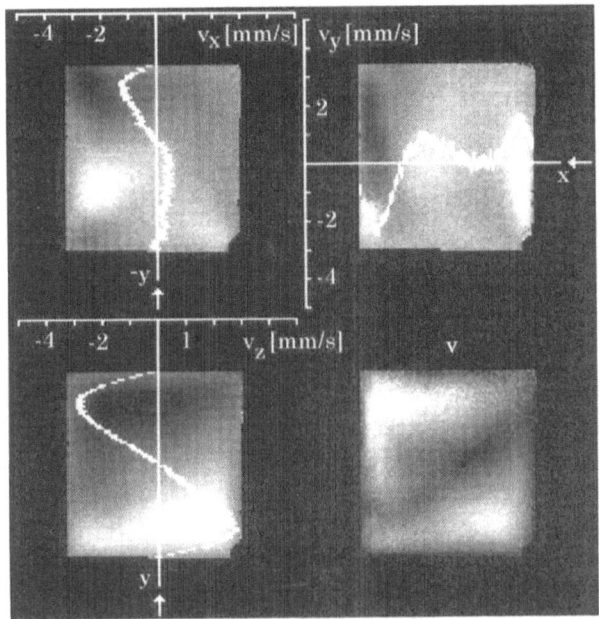

Fig. 30.5. Maps of the maximum local coherent-flow velocity caused by thermal convection in the central x, y plane (temperature gradient along the $-y$ axis) of the cell (Fig. 30.2). The velocity components v_x, v_y, v_z and the magnitude v in the central plane parallel to that spanned by the x and y axes are represented as gray shades (white, high positive velocity; black, high negative velocity). The curves and coordinate axes are superimposed plots of the velocity profiles in the central voxel lines or columns marked by arrows. (Reproduced by permission from ref. [512])

component during the bipolar gradient pulse used for motional phase encoding in the respective transient of the pulse sequence.

The accumulation of signals acquired with fluctuating phases due to periodic motions result in a sort of non-stochastic noise. At first sight this destructive superposition of signals leads to gray shades in the images, giving rise to the impression that incoherent motions or even thermal diffusion have attenuated the signal. However, a Fourier analysis in such cases reveals the frequencies of any coherent components inherent to the motions. This technique was successfully applied for the detection of **pulsatile flow** in fertilized bird eggs at a certain stage of incubation [159]. These motions could not be resolved temporally by ordinary velocity mapping, but in this way were proven to be of a coherent nature.

CHAPTER 31

Diffusivity Maps

Apart from phase encoding of the velocity of coherent motions, bipolar gradient pulses in coherence evolution intervals (or, equivalently, unipolar gradient pulses with a 180° pulse in between) can also serve for probing of incoherent displacements. In particular this refers to self-diffusion. Incrementing the bipolar gradients in subsequent transients, and evaluating the echo attenuation permits one to determine a diffusion coefficient for each voxel (see Sect. 19.2) [24, 31, 79, 482].

As a further step for obtaining as much information as possible, one can combine diffusion imaging with spectroscopic imaging. One then evaluates and maps spatially and spectroscopically resolved diffusion coefficients D as a useful means for tissue characterization. A pulse sequence suitable for this purpose is shown in Fig. 31.1 ("spectroscopic diffusion imaging," SDI).

The relative intensity of the spectral lines is attenuated according to Sect. 19.2.4

$$\frac{A}{A_0} = e^{-bD} \tag{31.1}$$

where

$$b = \gamma_n^2 G_D^2 \delta_D^2 (\Delta_D - \frac{\delta_D}{3}). \tag{31.2}$$

The symbols are explained in Fig. 31.1.

Typical applications of pulse sequence (Fig. 31.1) to the human finger in vivo are shown in Fig. 31.2. Ordinary and diffusion-weighted images referring to the total integrated signal are compared in Fig. 31.2a,b. Several regions with different diffusion coefficients can be distinguished. In Fig. 31.2c-f these are marked by numbers and letters. The numerical values of the local diffusion coefficients are listed in Table 31.1 (compare [107]). Spatially and spectroscopically resolved diffusion coefficients can also be recorded with the aid of (coherent-motion compensated) localization techniques. This was demonstrated in an application to fertilized hen eggs [259], for instance. Localization methods provide accurate information solely from the region of interest rather than from the whole object. Therefore this measuring strategy is particularly economical with respect to measuring time.

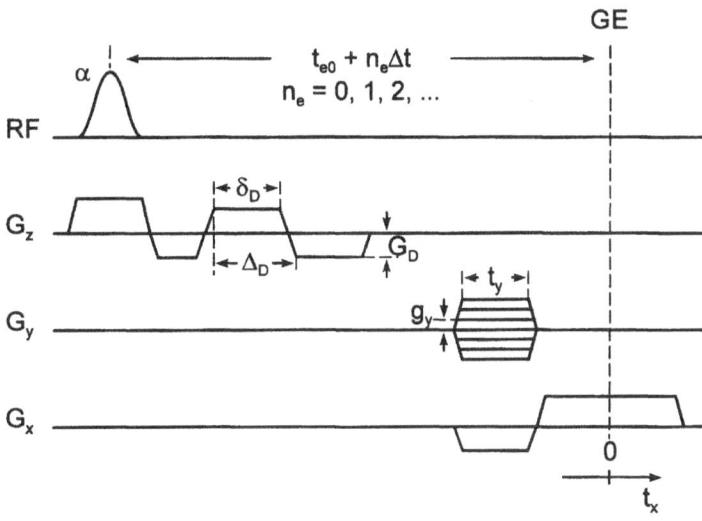

Fig. 31.1. Gradient-recalled echo (GE) pulse sequence for "spectroscopic diffusion imaging" (SDI). The method permits the record of diffusion (or incoherent motion) weighted spectroscopically and spatially resolved data. The echo amplitude is attenuated by incoherent displacements in the presence of a bipolar gradient pulse of magnitude G_D following the slice-selection pulse. The full echo-attenuation curves can be recorded for each pixel by incrementing the bipolar gradient pulse in length or height.

Table 31.1. Diffusion coefficients evaluated from data of the regions illustrated in Figs. 31.2c-f.

region	allocation	spectral line	$D/10^{-9}\,m^2/s$ ($G_D = 449$ mT/m)	$D/10^{-9}\,m^2/s$ ($G_D = 299$ mT/m)
1	below nail		2.0	2.1
2	skin		0.5	0.7
3	skin	water	1.0	1.1
4	skin		0.7	0.6
5	skin		0.5	0.4
6	muscle		0.8	0.9
7	bone marrow		0.7	0.6
8	fatty tissue	fat	0.6	0.7
9	fatty tissue		0.8	0.5

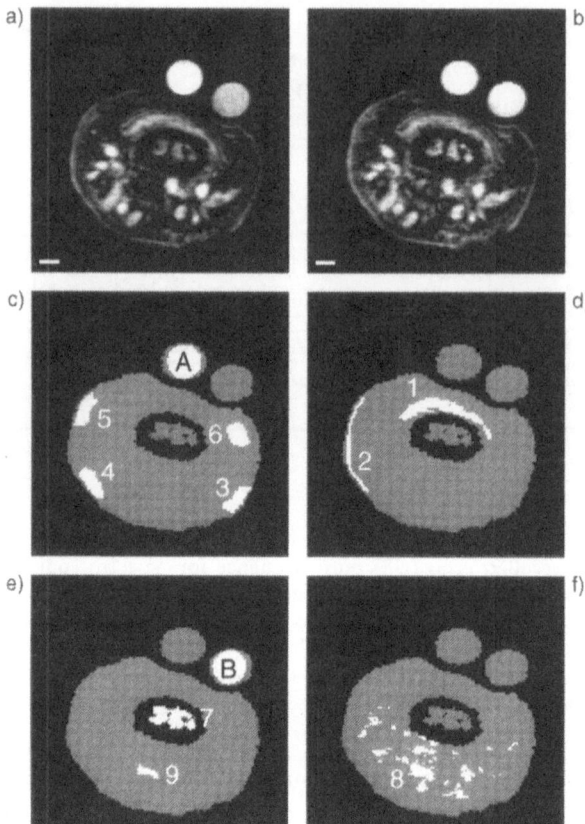

Fig. 31.2. Spectroscopic diffusion images of a human finger recorded with the aid of pulse sequence (Fig. 31.1). The white bar represents 2 mm in reality. The experimental parameters correspond to those specified in the legend of Fig. 28.2: **a)** map of the integrated spectrum after correction for frequency-offset artifacts (see Sect. 28.4) (two reference capillaries filled with water (left) and olive oil (right) are imaged at the same time for comparison); **b)** map as in (a) but with diffusion-weighted intensity (the bipolar gradient pulse used is characterized by $G_D = 449$ mT/m, $\Delta_D = 3.4$ ms, $\delta_D = 3.2$ ms); **c)** **to f)** illustration of the regions "1" to "9" and "A" (water reference sample) and "B" (olive oil reference sample) where distinct diffusion coefficients were measured with the aid of water (c and d) or lipid (e and f) signals (see Table 31.1). (Reproduced by permission from ref. [512])

CHAPTER 32

Resolution

Digital acquisition of signals with the aid of an analog-to-digital converter produces sets consisting of N equidistant data points. Each data point represents an interval on the axis of the independent variable of the signal function. We term this variable the **"measuring variable."** It can be a time, or a component of the reciprocal positional or velocity spaces.

The fast Fourier transform (FFT) procedure [99] common in the numerical analysis of signals requires that N is a power of two. The data may refer directly to the free-induction signal acquired in the experiment, or to the signal phase varied in a series of N transients with incremented evolution intervals or phase-encoding gradients. If echoes are recorded rather than FID-signals after an RF pulse, or if phase-encoded signals of a series of transients are considered, the measuring variable of the signal function from which the digitized data set is obtained takes positive as well as negative values.

Digital acquisition routines are ubiquitous in magnetic resonance. The limits with respect to **resolution** and **range ("field of view")** can generally be formulated for methods in which conjugated domains are linked with primary domains via Fourier transformation. For example, the real-space domains form the counterparts to the k space domains, or the frequency domain is related to the time domain. The primary domains (in which the experiments take place) are generally termed **"measuring domains"** whereas their conjugated counterparts are referred to as **"rendering domains."** The term "rendering" alludes to the data sets to be presented finally after the Fourier transform analysis. With respect to the techniques of interest here, this can be a spectrum, an image, or a velocity (component or vector) distribution.

This chapter depicts the limits of digital acquisition one commonly faces in context with spectroscopy, ordinary imaging, spectroscopic or velocity mapping, and velocimetry. The definitions and limits relevant in the different experiments are compared in Table 32.1.

32.1
Field of View, Spectral Width, Velocity Range

The maximum angular frequency of a **spectrum** that can be detected unambiguously without aliasing effects in the course of digital data processing is given by the Nyquist sampling theorem (see Sect. 42.3):

Table 32.1. Comparison of the limits and parameters of digital acquisition routines in NMR spectroscopy, imaging, and velocimetry. The symbols are the same as in the text, and represent positive quantities throughout (compare Figs. 24.3, 25.4, 25.6, 30.1, 28.1). Abbreviations: DR, digital resolution; SW, spectral width; FOV, field of view; VDR, velocity distribution range.

experiment	conj. var.	information carrier	measuring domain (quad. detection)	rendering domain
spectroscopy and	t_i, ω_i	frequency	$0 \le t_2 \le t_{2,max}$ $t_{2,max} = N\Delta t_2$	DR: $\Delta\omega_2 = \frac{2\omega_{2,max}}{N} = \frac{2\pi}{t_{2,max}}$ SW: $-\omega_{2,max} \le \omega_2 \le \omega_{2,max}$ $\omega_{2,max} = \pi/\Delta t_2$
spectral mapping	$i = 1, 2, ..$	phase	$0 \le t_1 \le t_{1,max}$ $t_{1,max} = N\Delta t_1$	DR: $\Delta\omega_1 = \frac{2\omega_{1,max}}{N} = \frac{2\pi}{t_{1,max}}$ SW: $-\omega_{1,max} \le \omega_1 \le \omega_{1,max}$ $\omega_{1,max} = \pi/\Delta t_1$
2DFT	k_i, r_i	frequency	$-k_{x,max} \le k_x \le k_{x,max}$ $k_{x,max} = \frac{1}{2}N\Delta k_x$ $\Delta k_x = \gamma_n G_x \Delta t_x$	DR: $\Delta x = \frac{2x_{max}}{N} = \frac{\pi}{k_{x,max}}$ FOV: $-x_{max} \le x \le x_{max}$ $x_{max} = \pi/\Delta k_x$
imaging	$i = x, y, ..$	phase	$-k_{y,max} \le k_y \le k_{y,max}$ $k_{y,max} = \frac{1}{2}N\Delta k_y$ $\Delta k_y = \gamma_n t_y \Delta G_y$	DR: $\Delta y = \frac{2y_{max}}{N} = \frac{\pi}{k_{y,max}}$ FOV: $-y_{max} \le y \le y_{max}$ $y_{max} = \pi/\Delta k_y$
velocimetry and velocity mapping	k_v, v	phase	$-k_{v,max} \le k_v \le k_{v,max}$ $k_{v,max} = \frac{1}{2}N\Delta k_v$ $\Delta k_v = \gamma_n \tau^2 \Delta G$	DR: $\Delta v = \frac{2v_{max}}{N} = \frac{\pi}{k_{v,max}}$ VDR: $-v_{max} \le v \le v_{max}$ $v_{max} = \pi/\Delta k_v$

$$\omega_{max} = \frac{\pi}{\Delta t} \tag{32.1}$$

where Δt is the "dwell time" of the analog-to-digital converter with which the time-domain signal is acquired in the form of N data points. This condition means that at least two data points (two dwell-time intervals) per period are needed for the identification of a Fourier component. For quadrature detection[1] the total **spectral width** is then

$$- \omega_{max} \leq \omega \leq \omega_{max} \tag{32.2}$$

In complete analogy, this statement can be translated to the terminology of other pairs of conjugated variables. In the case of **k** space vs real space digital transform pairs, the Nyquist theorem refers to "waves" which are to be unambiguously identified with respect to wave numbers. With ordinary **imaging**, one obtains the "**field of view**"

$$- r_{i,max} \leq r_i \leq r_{i,max} \qquad (i = x, y, z) \tag{32.3}$$

where

$$r_{i,max} = \frac{2\pi}{2\Delta k_i} = \frac{\pi}{\Delta k_i} \qquad (i = x, y, z) \tag{32.4}$$

The frequency-encoding domain is probed in N steps

$$\Delta k_x = \gamma_n G_x \Delta t \tag{32.5}$$

where Δt is again the dwell time of the analog-to-digital converter. Instead of a dwell time, the phase-encoding domains are sampled with the aid of gradient-pulse increments

$$\Delta k_i = \gamma_n G_i t_i \qquad (i = y, z) \tag{32.6}$$

Finally, the Nyquist sampling theorem also defines "field-of-view" of **velocimetry**, i.e., the range in which the velocity distribution is examined. As with all other digital Fourier transformation methods, the maximum range of the variables conjugated to those of the measuring domain is given by the increment of the latter. In the FEVI case (see Fig. 30.1) this refers to the bipolar gradient pulses which define a "reciprocal velocity space." The maximum magnitude of the velocity component along the gradient axis is

$$v_{max} = \frac{\pi}{\Delta k_v} \tag{32.7}$$

where the increment is given by (compare Eq. 30.2)

$$\Delta k_v = \gamma_n \tau^2 \Delta G \tag{32.8}$$

---

[1]Quadrature detection of the signal and its 90° phase-shifted counterpart permits the distinction of positive and negative frequencies relative to the carrier (or reference) frequency of the system. The complex transverse magnetization is then unambiguously defined in the complex plane. Quadrature detection with respect to phase-encoding domains is possible by recording signals from two series of transients acquired with correspondingly phase shifted RF pulses.

The larger the increment the more narrow is the **velocity field-of-view**,

$$- v_{max} \leq v \leq v_{max} \qquad (32.9)$$

If velocimetry is combined with imaging for velocity mapping purposes, a further limit of a more practical nature must be taken into account. Analogous to the inflow/outflow artifacts discussed above, the voxels may be "washed out" if the displacements by coherent motions in the echo time, T_E, exceed the length scale of the spatial resolution, Δx. That is, the maximum velocity must also comply to the condition

$$v_{max} \ll \Delta x / T_E \qquad (32.10)$$

The number of increments in multi-dimensional experiments is kept as small as possible, of course, in order to save measuring time. This is of particular importance when all three velocity dimensions are to be examined. For instance, in the three-dimensional velocity vector mapping scheme of Fig. 30.1, only two phase-encoding increments are suggested for probing the velocity distribution. One of the three encoding levels corresponds to no bipolar gradient at all.[2] That is, the signals acquired in this case can be used for the evaluation of all three components of the velocity vector.

32.2
Digital Resolution

The digital resolution is generally defined as the separation of two neighboring data points in the rendering domain. This is to be distinguished from the physical resolution limits to be discussed afterwards. Let us first consider the resolution that can be achieved at best in terms of the digital signal acquisition parameters. The resolution will generally be better the more strongly the quantities to be resolved affect the signal, i.e., the larger the range of the measuring variable chosen.

With the time domains considered in ordinary **spectroscopy** or in spectroscopic imaging, the limiting measuring variables are the longest acquisition time $t_{2,max}$ or the maximum evolution period $t_{1,max}$. The former is a "frequency-encoding" time according to the very definition, the latter is the interval in which the spectral information is "phase encoded."

In terms of angular frequencies the **spectral digital resolution** is

$$\Delta \omega_i = \frac{2\omega_{i,max}}{N} = \frac{2\pi}{N\Delta t_i} = \frac{2\pi}{t_{i,max}} \qquad (i = 1, 2) \qquad (32.11)$$

where $t_{i,max} = N\Delta t_i$ is the maximum encoding time, and N is the respective number of dwell-time intervals, Δt_i, in the ith measuring domain. Recall that the data acquired with quadrature detection are complex and each consists of a real and an imaginary part.

---

[2]The FFT procedure [99] common in the numerical analysis of signals requires that the number N of data points in the measuring domain is a power of two. Recall that this can readily be fulfilled by zero-filling the data matrix as usual (see below).

Imaging experiments are related to k space domains. Hence, the maximum measuring variables determining the digital resolution in this case is the maximum component along the x axis, $\pm k_{x,max}$, in the frequency-encoding domain, and the maximum component along the y (or z) axis, $\pm k_{y,max}$ (or $\pm k_{z,max}$) in the phase-encoding domain(s). Provided that the measuring domains are probed on the positive as well as on the negative half axes, the **spatial digital ("pixel") resolution** is

$$\Delta r_i = \frac{2r_{i,max}}{N} = \frac{2\pi}{N\Delta k_i} = \frac{\pi}{k_{i,max}} \qquad (i = x, y, z) \qquad (32.12)$$

where $k_{x,max} = N\Delta k_x = \gamma_n G_x N\Delta t_x$, $k_{y,max} = N\Delta k_y = \gamma_n t_y N\Delta G_y$, and $k_{z,max} = N\Delta k_z = \gamma_n t_z N\Delta G_z$. The number N counts the respective (complex) data corresponding to the dwell-time intervals or increments of the phase-encoding gradients by which the measuring domains are probed.

Finally, the record of the **velocity distribution** is again based on phase encoding as outlined above. The resolution of the velocity component along the gradient axis is limited by the maximum wave number in the reciprocal velocity space, $\pm k_{v,max}$. The **velocity digital resolution** concerning the component along the gradient axis is

$$\Delta v = \frac{2v_{max}}{N} = \frac{2\pi}{N\Delta k_v} = \frac{\pi}{k_{v,max}} \qquad (32.13)$$

where $k_{v,max} = \frac{N}{2}\Delta k_v = \frac{1}{2}\gamma_n \tau^2 N\Delta G$. The number N again refers to the (complex) data set acquired in the measuring domain with the aid of N increments of the phase-encoding gradient.

For a given dwell time Δt_i or phase-encoding increment Δk_i, fixed as needed for the spectral width or the field of view, the digital resolution is improved with the number N of collected data points. However, high numbers N stipulate the record of signals up to regimes of the measuring variable which may be burdened with strong noise compared with the decaying signal amplitude in this variable range. In this case it is more favorable to skip the acquisition of all further data points recorded under too "noisy" conditions. Instead the acquisition data matrix can be **"zero-filled."** Thus, the rendering-domain signal becomes less susceptible to noise.

This of course is a compromise to be formed between signal-to-noise ratio on the one hand and spectral or pixel information content on the other. The expansion of the acquisition data matrix by zero-filling improves the digital resolution but not the sample-related (physical) resolution to be discussed in the subsequent section.

Zero-filling is particularly useful in phase-encoding domains. In order to save measuring time, it may be desirable to combine a certain field-of-view with a certain digital resolution while the total of increments of the phase-encoding gradients is lower than the required number N (see Table 32.1). The problem can then easily be solved by post-detection zero-filling the data matrix.

32.3
Physical Resolution Limits

The digital resolution is as good as its adjustment. Without considering the physical limits of the encoding or acquisition intervals, the digital resolution can be made arbitrarily good. However, there are several experimental, signal-processing, and sample-related restrictions which must be matched [33, 95, 285].

In imaging, for instance, improving the pixel resolution only makes sense as far as the gray shades of two neighboring pixels can be distinguished if principally different. That is, the physical resolution of superimposed signals such as the intensities of neighboring voxels means that the information inherent to these signals can be evaluated separately. The limit is given by the maximum measuring variable which is permitted under acceptable signal-to-noise conditions. This commonly holds true in all dimensions of spectroscopy, tomography, and velocimetry.

For discussion we restrict ourselves to a one-dimensional experiment, e.g., frequency-encoding imaging. The principles, thereby well illustrated, can easily be applied to other measuring schemes.

32.3.1
Spatial In-Plane Resolution

Phase- and frequency-encoding imaging are formally equivalent, so that it suffices to consider signals recorded as a function of any k space component[3]. Artifacts of the imaging procedure possibly spoiling the resolution, as already discussed in context with image contrasts in the previous chapters, are only marginally regarded here. We rather examine the restrictions unavoidably connected with the signal generation and with data processing. Relevant mechanims are

a) signal attenuation by transverse relaxation
b) signal attenuation by inhomogeneous intra-voxel spin interactions
c) signal attenuation by self-diffusional or incoherent flux intra-voxel displacements

Displacements by diffusion or incoherent flux in the measuring time may also lead to displacements beyond the voxel length scale to be resolved, so that

d) fading of the voxel contrast by incoherent displacements

is of potential importance. In general, resolution is intimately related to the signal-to-noise ratio achievable in the experiment. This ratio may be improved by digital filtering via exponential multiplication of the acquired data set. However, this in turn leads to a further restriction of the maximum measuring variable owing to

e) digital filtering by exponential multiplication

---

[3]With 2DFT imaging, apart from the in-plane resolution, the resolution in the third spatial dimension is also of interest as concerns the selection of the slice to be imaged. The factors influencing the slice thickness are discussed in Sects. 25.1 and 36.3 with respect to soft-pulse and spin-lock pulse slice selection, respectively.

Let us first discuss the limitation of the maximum measuring variable by exponential attenuation factors of the above list.

32.3.1.1
Limitation by Exponential Attenuation Factors

The influence of items (a), (b), and (e) may be assumed to be governed by mono-exponential decays, for simplicity. Consider two neighboring voxels of equal spin density at positions x_a and x_b. The signals originating from these voxels can be expressed by

$$\tilde{S}_a(k_x) = S_0\, e^{-\bar{c}t_x}\, e^{-ik_x x_a} \tag{32.14}$$
$$\tilde{S}_b(k_x) = S_0\, e^{-\bar{c}t_x}\, e^{-ik_x x_b} \tag{32.15}$$

The first exponential functions on the right-hand sides represent the echo signal attenuation by transverse relaxation, by intra-voxel frequency distributions due to inhomogeneous spin interactions, and by exponential multiplication for digital-filtering purposes. Thus

$$\bar{c} = \frac{1}{T_2^*} + \frac{1}{\tau_f} \tag{32.16}$$

where T_2^* is the combined time constant of transverse relaxation and inhomogeneous intra-voxel defocusing of the coherences, and τ_f is the time constant of digital filtering. The second exponential functions are the usual frequency-encoding factors depending on the wave number $k_x = \gamma_n G_x t_x$. The substitution $t_x = k_x/(\gamma_n G_x)$ leads to

$$\tilde{S}_a(k_x) = S_0\, e^{-ck_x}\, e^{-ik_x x_a} \tag{32.17}$$
$$\tilde{S}_b(k_x) = S_0\, e^{-ck_x}\, e^{-ik_x x_b} \tag{32.18}$$

where $c = \bar{c}/(\gamma_n G_x)$. The total signal of the two voxels is

$$\tilde{S}(k_x) = \tilde{S}_a(k_x) + \tilde{S}_b(k_x) \tag{32.19}$$

The Fourier transform (see Eq. 42.15)

$$S(x) = \frac{1}{2\pi} \int\limits_{-\infty}^{\infty} \tilde{S}(k_x) e^{ik_x x}\, dx \tag{32.20}$$

leads to a superposition of two Lorentzians,

$$S(x) = S_0 \left(\frac{c}{c^2 + (x - x_a)^2} + \frac{c}{c^2 + (x - x_b)^2} \right) \tag{32.21}$$

where we have taken advantage of the shifting property of Fourier transforms. The full width at half maximum of each of these Lorentzians is

$$(\Delta x)_{FWHM} = 2c \tag{32.22}$$

The signals of the two voxels can be evaluated separately if

$$\Delta x = |x_b - x_a| \geq (\Delta x)_{FWHM} \tag{32.23}$$

The physical resolution limit on grounds of mechanisms (a), (b), and (e) hence is

$$(\Delta x)_{phys} = (\Delta x)_{FWHM} = \frac{2}{\gamma_n G_x} \left(\frac{1}{T_2^*} + \frac{1}{\tau_f} \right) \tag{32.24}$$

It is therefore reasonable to adjust the digital resolution, Δx_d, in such a way that

$$(\Delta x)_d \approx (\Delta x)_{phys} = \frac{2}{\gamma_n G_x} \left(\frac{1}{T_2^*} + \frac{1}{\tau_f} \right) \tag{32.25}$$

That is, the acquisition time $T_a = N\Delta t_x$ should obey the condition

$$\frac{1}{T_a} \approx \left(\frac{1}{T_2^*} + \frac{1}{\tau_f} \right) \tag{32.26}$$

From the point of view of the sample properties and the data processing, good resolutions require long T_2^* filtering and little digital filtering. The latter is a matter of the acceptable noise level, of course. On the other hand, the best experimental means for resolution enhancement is the use of strong field gradients. For instance, a field gradient of 70 mT/m and an acquisition time of 1 s lead to a proton image resolution of 2 μm.

32.3.1.2
Influence of Diffusion and Incoherent Motions

The influence of diffusion [72] and incoherent intra-voxel flow, i.e., mechanism (c) in the list given in Sect. 32.3.1, has been skipped in the discussion up to now, because these phenomena cannot be described by monoexponential attenuation functions. Anyway, there is a potential signal attenuation on these grounds. For example, self-diffusion with the self-diffusion coefficient D in the readout time, t_x, causes an attenuation factor $\exp\{-\gamma_n^2 G_x^2 t_x^3 D/12\}$ (see Sect. 19.2.4).

Note that the readout time t_x is relevant for mechanism (c) rather than the interval between the read gradient and its compensation pulse (compare Figs. 24.3 or 25.6). Diffusional displacements in that interval and those of any other field-gradient pulse pairs also attenuate the echo signal amplitude. This "precursor" attenuation leads to sensitivity losses but, in the first instance, not to a deterioration of the resolution provided that the root mean squared displacements are less than the length scale of the voxels. Otherwise a second sort of diffusive limitation of the resolution arises, namely mechanism (d) (Sect. 32.3.1).

The relevant time scale of mechanism (d), fading of the voxel contrasts by inter-voxel diffusion, may crudely be identified with the echo time T_E. The root mean squared displacement due to self-diffusion in this period is

$$\sqrt{\langle r^2 \rangle} = \sqrt{6DT_E} \tag{32.27}$$

For water at room temperature, i.e., $D = 2 \times 10^{-9}$ m^2/s, and an echo time of typically $T_E = 10$ ms, one estimates the maximum length scale of this effect to be $\sqrt{\langle r^2 \rangle} \approx 10$ μm.

32.3.1.3
Frequency-Offset Artifact Compensation

Frequency offsets due to field inhomogeneities, $\Omega_i(r) = \gamma_n \Delta B_0(r)$, magnetic-susceptibility variations in the object, $\Omega_\chi(r) = \chi_m(r)B_0/3$, or chemical shifts, $\Omega_{cs}(r) = \gamma_n \sigma(r)B_0$ make voxels appear in the image at shifted positions, and may result in a certain limitation of the spatial resolution [73]. However, as outlined in Sect. 28.4, such artifacts can be cured by post-detection correction provided that the spectral dimension has been sampled in the experiment. A drawback is the reduction of the sensitivity owing to the additional measuring time needed for probing the spectral dimension.

32.3.2
The Sensitivity Limit

We have already referred to the importance of the signal-to-noise ratio, S/N, for the resolution one can achieve. Lack of sufficient sensitivity forms the ultimate limitation of resolution from which, at the end of the day, all restrictions mentioned above ensue [71]. The general means for the improvement of the signal-to-noise ratio are accumulation of N_t transients of the pulse sequence for each phase-encoding step, and the use of a high magnetic-flux density. As the accumulated signal increases linearly with the number of transients, whereas the sum of random noise voltages is proportional only to the square root of that number, the signal-to-noise ratio in total varies as $S/N \propto \sqrt{N_t}$. It depends on the magnetic flux density according to [198] $S/N \propto B_0^{7/4}$.

The most important factor in this context is the filling factor of the RF probe. With imaging, signals from each voxel of the object must be identified independently. That is, the filling factor with respect to a single voxel is relevant, $S/N \propto (\Delta x)^3/V_{probe}$. That is, in order to gain extreme spatial resolutions, the RF probe volume and, hence, that of the object to be imaged, must be reduced as far as possible. The limiting proton sensitivity in this respect amounts at present to about 10^{13} protons[4] per voxel permitting an in-plane resolution in the order of micrometers (e.g., [157]) and approximately matches the limit given by diffusion in liquids discussed before. With solid-like samples new perspectives arise on the basis of **magnetic-resonance force microscopy** (MRFM), which will be discussed in Sect. 35.3.

Another non-inductive detection scheme of extreme sensitivity is β-**detected NMR** [133, 206]. The principle is to spin-polarize β-active nuclei and to probe the longitudinal magnetization via the parity violating forward/backward asymmetry of β emission.

---

[4]According to Curie's law this corresponds to about 10^7 uncompensated proton dipoles aligned along the magnetic field at room temperature.

CHAPTER 33

Multi-Stripe/Plane Tagging

The purpose of tagging experiments is to visualize the displacements by flow or movements during a preselected "time-of-flight." The principle is to "tag" certain space regions by saturating the local spin populations at least partially, and to image the magnetization distribution after the time-of-flight with any fast NMR imaging method [17, 530]. The marker lines then appear in black superimposed on the image. Their contours are deformed according to coherent displacements of the nuclei reached in the time-of-flight. If the motions are of a more incoherent nature, the black contrast of the marker lines will fade relative to the contrast of the adjacent material. With two-dimensional imaging, the technique is referred to as "multi-stripe tagging" or "marker line" imaging. In context with three-dimensional image data one speaks of "multi-plane tagging."

33.1
DANTE Pulse Combs

The marker grid is prepared with the aid of DANTE pulse combs [350] in the presence of magnetic-field gradients. Figure 33.1 shows a suitable version of such a pulse sequence.

Ideally a DANTE RF pulse comb with pulses separated by delays τ may be represented by the factorized function

$$h(t) = f(t)g(t) \tag{33.1}$$

It consists of a δ pulse comb,

$$f(t) = \sum_{n=-\infty}^{\infty} \delta(t - n\tau) \tag{33.2}$$

modulated by the sinc function,

$$g(t) = \frac{A}{\pi} \frac{\sin(\omega_m t)}{t} \equiv 2v_m A \mathrm{sinc}(2v_m t) \tag{33.3}$$

In linear (low flip-angle) approximation, the excitation profile generated in the presence of a constant field gradient G_x in the sample is given by the Fourier

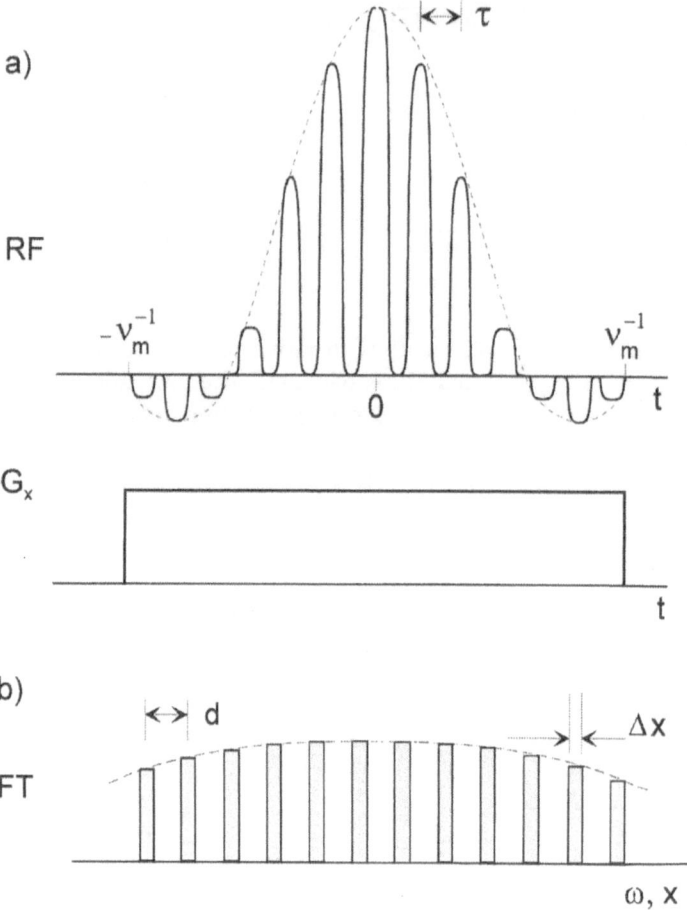

Fig. 33.1. a) Schematic DANTE RF pulse comb in the presence of a constant magnetic-field gradient for the preparation of a magnetization grid. **b)** Schematic Fourier transform of the RF pulse comb. The Fourier transform refers to the conjugated-variable pair t, ω. In the presence of a constant field gradient the spectral coordinate of the Fourier transform corresponds to the spatial coordinate $x = \omega/(\gamma_n G_x)$. The RF pulse amplitudes are modulated by a (truncated) sinc function (schematically represented by the dotted envelope in (a)) in order to approach rectangular profiles of the slices forming the grid. The homogeneity of the excitation in these slices depends on how short the RF pulses are (see the schematic dashed envelope line in (b). The spacing of the grid slices is given by $d = 2\pi/(\gamma_n G_x \tau)$ where τ is the RF pulse delay.

transform (see Sect. 25.1)

$$h(t) \rightarrow H(\omega) = \frac{1}{2\pi} F(\omega) \otimes G(\omega) \equiv \int\limits_{-\infty}^{\infty} F(\omega')G(\omega - \omega')\,d\omega' \qquad (33.4)$$

where $\omega = \gamma_n G_x x$. That is, in a sense, the angular frequency ω is synonymous with the spatial coordinate x. This convolution integral is evaluated with the aid of the transform pairs

$$f(t) \rightarrow F(\omega) = \frac{2\pi}{\tau} \sum_{n=-\infty}^{\infty} \delta\left(\omega - n\frac{2\pi}{\tau}\right)$$

$$g(t) \rightarrow G(\omega) = \begin{cases} A & \text{for} \quad |\omega| \leq \omega_m \\ 0 & \text{otherwise} \end{cases}$$

where $G(\omega)$ represents the rectangular profile of a grid slice.
The result is

$$H(\omega) = \frac{1}{\tau} \sum_{n_{min}}^{n_{max}} G\left(\omega - n\frac{2\pi}{\tau}\right) \qquad \left(\frac{\tau}{2\pi}[\omega - \omega_m] \leq n \leq \frac{\tau}{2\pi}[\omega + \omega_m]\right) \quad (33.5)$$

This is a sum of rectangular profile functions of constant height A/τ in the range $|\omega - n2\pi/\tau| \leq \omega_m$.

In the presence of a constant gradient G_x, only spins located in equidistant slices are excited. The spacing of the grid slices is given by

$$d = \frac{2\pi}{\gamma_n \tau G_x} \tag{33.6}$$

where one chooses $\tau < \pi/\omega_m$ in order to avoid overlap. Gradients of 15 mT/m and pulse delays of 1 ms lead for protons to grid spacings of 1.6 mm, for instance. The thickness of the slices is determined by the sinc modulation frequency,

$$\Delta\omega = \gamma_n G_x \Delta x = 2\omega_m \tag{33.7}$$

In reality, DANTE combs neither consist of δ pulses nor are they infinite. Finite RF pulse widths lead to tagging-slice profile heights decaying towards the fringes of the field of view, while the truncation of the sinc modulation function causes deviations from rectangular tagging-slice profiles (compare Fig. 33.1). However, these limitations are not critical. One even obtains good results with unmodulated DANTE pulse trains. Test experiments with and without sinc modulation show that the stripe edges effectively appear in the image even sharper with constant amplitudes of the DANTE pulses [512].

33.2
Imaging Pulse Scheme and Applications

Figure 33.2 shows a gradient and radio frequency pulse scheme of the three-dimensional multi-plane tagging method for the three-dimensional visualization of motions. The pulse sequence consists of a preparation and an imaging section. The initial magnetization saturation grid is prepared by three DANTE combs in the presence of gradients applied in the three space directions. Each DANTE train consists of several hard RF pulses. The number depends on how many marker slices are to be prepared in the field of view. The width and the spacing of the tagging

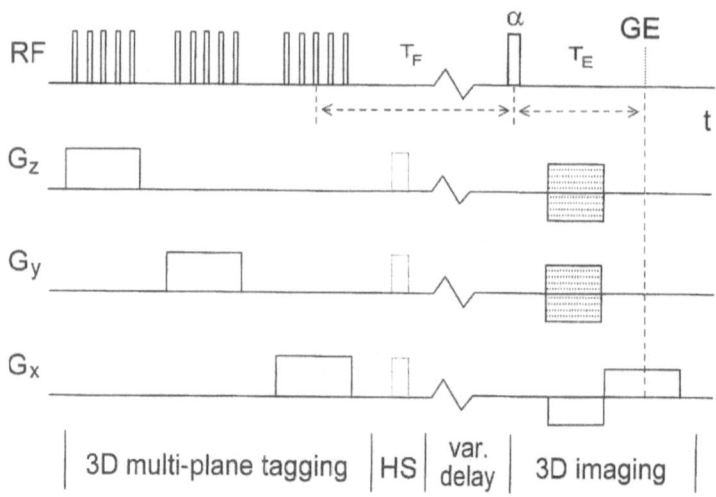

Fig. 33.2. Pulse sequence for three-dimensional multi-plane tagging three-dimensional NMR imaging. The imaging part is the three-dimensional version of the gradient-recalled echo sequence (Fig. 25.6). It is separated from the tagging-plane pulses by the time-of-flight T_F. In this period, optional homospoil (HS) gradient pulses may be applied in order to prevent refocusing of coherences by the subsequent RF pulses. The amplitudes of the DANTE pulses may optionally be modulated according to a sinc function.

stripes depend on the parameters of the DANTE comb as described in the previous section.

The imaging part of the pulse sequence should be as short as possible. Unlike the gradient-recalled echo sequence shown in Fig. 33.2, echo-planar imaging is also in use [280]. It is clear that the echo time should be much shorter than the time-of-flight in order to measure correspondingly fast motions under well-defined conditions. Extremely fast motions may require velocity-compensated gradient pulses in order to avoid artifacts on these grounds. The upper limit of the time-of-flight is determined by the spin-lattice relaxation time of the flowing material, so that extremely slow displacements may also escape detection by the tagging method. Typical orders of magnitude of the velocities suitable for this technique are mm/s to m/s.

Applications of two-dimensional multi-stripe tagging methods were reported for medical purposes such as the examination of heart wall motion [17, 355, 530], the visualization of streamlines in laminar [204] or turbulent [280] flow of liquids in technical systems, and in context with the study of transport properties in bird eggs [159]. A three-dimensional variant of this measuring principle was employed for the visualization of thermal convection patterns [512].

Figure 30.2 shows the thermal convection box used in the latter experiment. The thermal convection rolls appearing under the competitive action of gravity and buoyancy resemble the situation encountered with well known Rayleigh/Bénard instability problem [129, 255]. By contrast to the situation originally described by

Bénard [27], we are dealing here with a three- rather than two-dimensional problem. That is, the boundary conditions at the side walls are important as well. The three-dimensional multi-plane tagging three-dimensional imaging method (Fig. 33.2) is therefore well suited for the visualization of the complicated convectional flow patterns arising under such circumstances.

Figures 33.3 to 33.5 represent images recorded after different times-of-flight in the central voxel planes along the x, y, and z direction of the water-filled convection cell (Fig. 30.2). The time-of-flight was varied from 6 to more than 800 ms. The deformations of the tagging stripes visible in the images reflect the displacements attained during the time-of-flight. Figure 33.6 finally shows perspective/cross-sectional views of the time evolution of the three-dimensional convection patterns. These representations are the illustrative counterpart to the more quantitative velocity maps of the same convection cell discussed in Chap. 30.

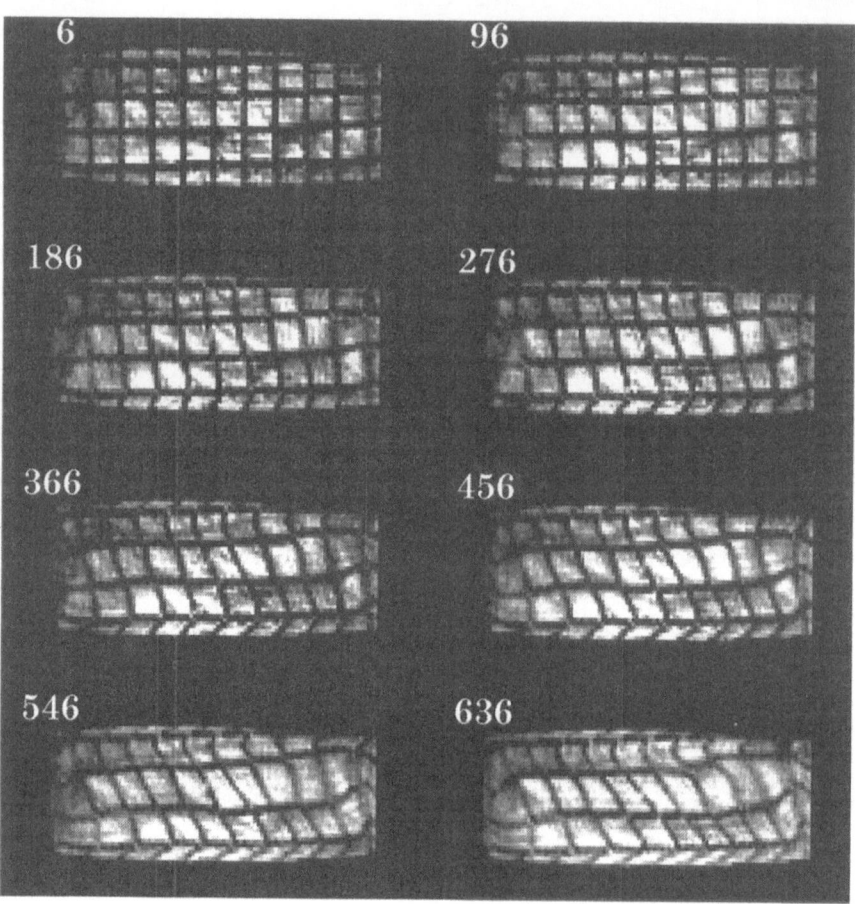

Fig. 33.3. Multi-plane tagging NMR images of thermal convection in the central y, z plane (temperature gradient along the $-y$ axis) of the cell (Fig. 30.2). The parameters of the pulse sequence (Fig. 33.2) were $T_E = 11$ ms and $T_R = 800$ ms. The measured k space data matrix was 128×128, the slice thickness 0.5 mm. The pixel resolution is 0.2×0.2 mm^2. The background gray shades correspond to an ordinary gradient echo image. The initial stripe spacing is 2 mm. The numbers indicate the time-of-flight, T_F, in ms. (Reproduced by permission from ref. [512])

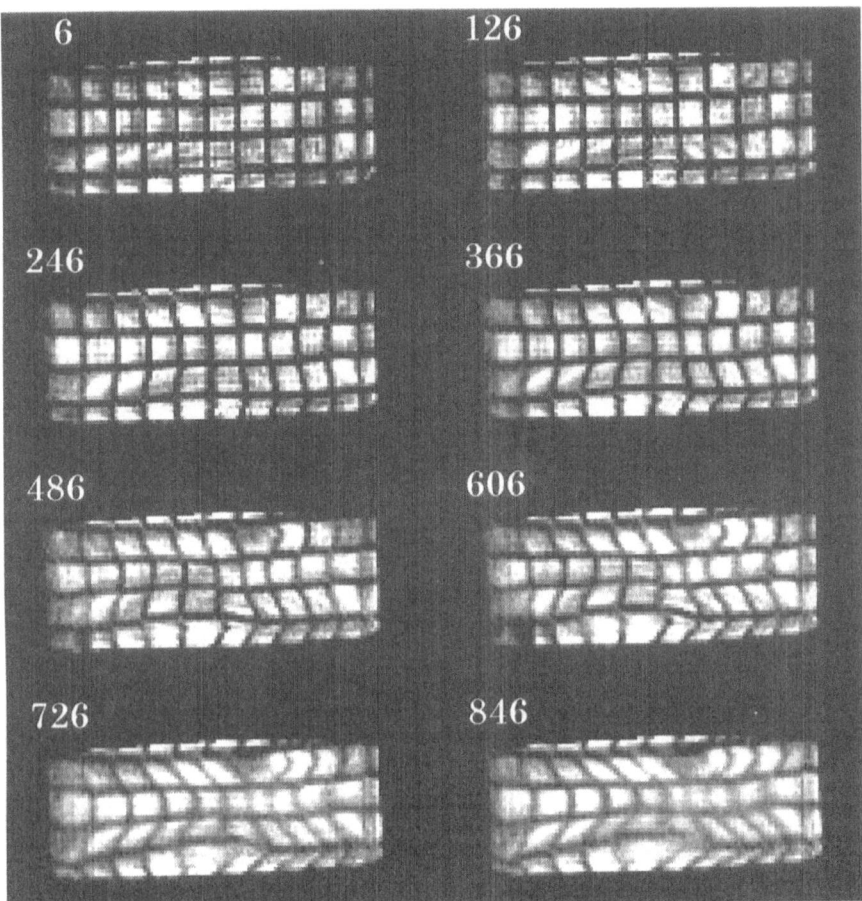

Fig. 33.4. Multi-plane tagging NMR images of thermal convection in the central x, z plane (temperature gradient along the $-y$ axis) of the cell (Fig. 30.2). The experimental parameters are the same as given in the legend of Fig. 33.3. The numbers indicate the time-of-flight, T_F, in ms. (Reproduced by permission from ref. [512])

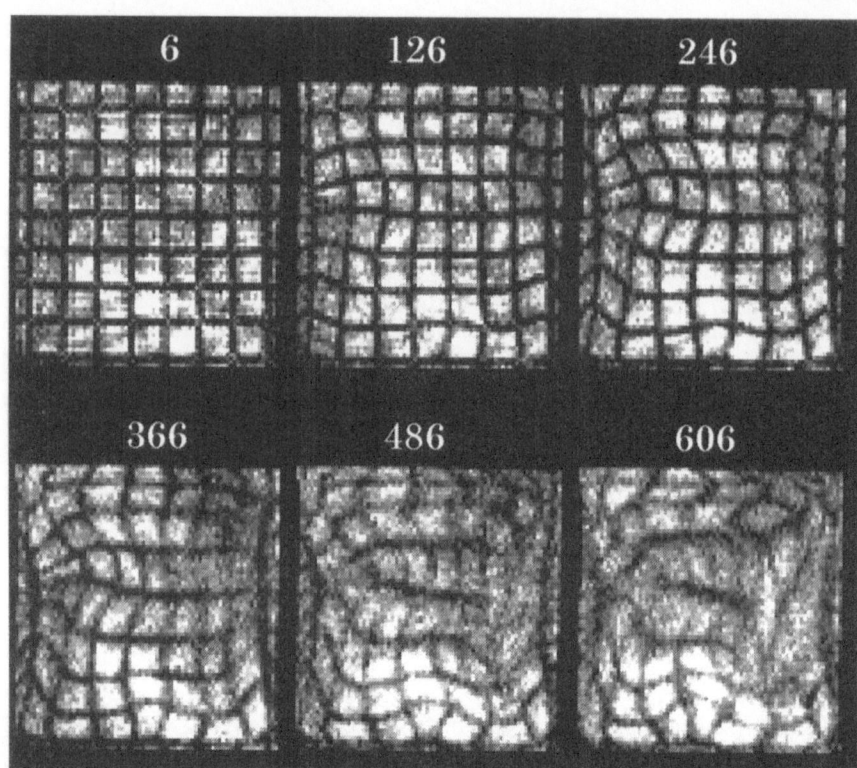

Fig. 33.5. Multi-plane tagging NMR images of thermal convection in the central x, y plane (temperature gradient along the $-y$ axis) of the cell (Fig. 30.2). The experimental parameters are the same as given in the legend of Fig. 33.3. The numbers indicate the time-of-flight, T_F, in ms. (Reproduced by permission from ref. [512])

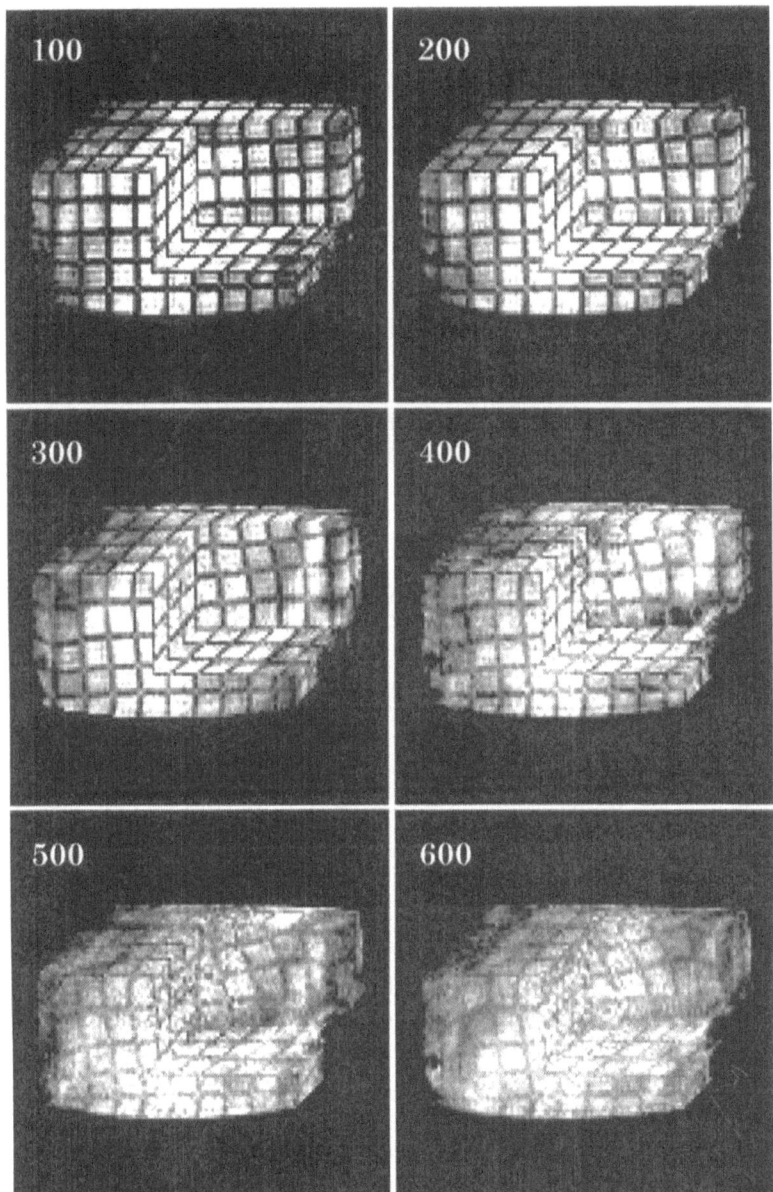

Fig. 33.6. Perspective (cross sectional) representations of the three-dimensional multi-plane tagging three-dimensional image data of thermal convection in the cell (Fig. 30.2). The experimental parameters are the same as given in the legend of Fig. 33.3. The measured k space data matrix was $128 \times 96 \times 96$, the zero-filled data matrix was $128 \times 128 \times 128$. A voxel resolution of $0.2 \times 0.2 \times 0.2$ mm^3 was achieved in this way. The background gray shades correspond to an ordinary gradient echo image of the surface. Partially black shades are visible which must be attributed to tagging planes intersecting the image plane. The numbers indicate the time-of-flight, T_F, in ms. (Reproduced by permission from ref. [512])

CHAPTER 34

Rotating-Frame Imaging

34.1
Nutation Frequency Encoding

The rotating-frame counterpart to (Larmor) frequency encoding in laboratory-frame imaging experiments is **"nutation frequency encoding"** [199]. Instead of gradients of the external magnetic field, rotating-frame imaging is based on gradients of the RF amplitude B_1. That is, the effective field in the resonantly rotating frame has the form

$$B_1 = B_1(r) = B_1^{(0)} + G_1 \cdot r \tag{34.1}$$

where we have neglected the influence of local fields by spin interactions. The B_1 gradient is defined by[1]

$$G_1 = \nabla B_1 \tag{34.2}$$

Below we will assume that the B_1 gradient is aligned along the x axis of the laboratory frame. Furthermore it is to be constant in the range of interest for simplicity. The effective field can then be written in the form

$$B_1 = B_1(x) = B_1^{(0)} + G_1 x \tag{34.3}$$

The nutation frequency thus becomes a function of the position:

$$\omega_1 = \omega_1(x) = \gamma_n B_1^{(0)} + \gamma_n G_1 x \tag{34.4}$$

The signal, i.e., the amplitude of an FID following the B_1 gradient pulse (see Fig. 34.1), is a superposition of signals of all volume elements along the B_1 gradient:

$$S \propto \int m(t_1, x) \, dx \tag{34.5}$$

where $m(t_1, x)$ is the local transverse magnetization at the position x after the RF pulse of length t_1. The local flip angles generated by that pulse are a function of the position as well:

$$\alpha = \alpha(x) = \alpha^{(0)} + \Delta\alpha(x) = \gamma_n B_1^{(0)} t_1 + k_x x \tag{34.6}$$

where we have introduced the wavenumber,

$$k_x = \gamma_n G_1 t_1 \tag{34.7}$$

---

[1]Note that this gradient and the nabla differential operator refer to the laboratory frame, of course.

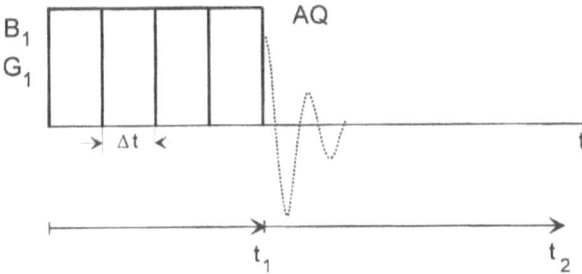

Fig. 34.1. RF pulse sequence for rotating-frame imaging. The RF-field amplitude B_1 is subject to a gradient G_1. The RF pulse width is incremented in a series of transients with recycle delays in between permitting equilibration of the magnetization. The initial amplitude of the FID just after the pulse is recorded after each step as a function of t_1. These data form the pseudo-FID which is evaluated with respect to the spatial information. Optionally the proper FID in the time scale t_2 can be acquired (AQ) as a second (spectral) dimension.

in analogy to the B_0 gradient techniques. The "k-space signal" recorded as the FID amplitudes immediately after the $B_1(x)$ pulse thus is

$$S(k_x) \propto \int \int \int \rho(x, y, z) \, \sin(\alpha^{(0)} + k_x x) \, dx \, dy \, dz \qquad (34.8)$$

The wavenumber k_x is varied by incrementing the RF-pulse length or, equivalently, the RF-pulse amplitude. The Fourier transform of the "pseudo-FID" recorded in this way eventually yields the projection of the spin-density of the sample on the gradient direction,

$$\tilde{S}(x) = \mathcal{F}_{k_x}\{S(k_x)\} \propto \int \int \rho(x, y, z) \, dy \, dz \qquad (34.9)$$

This represents a one-dimensional image of the object. In principle the same rules for the digital resolution and the field of view apply as with the laboratory-frame imaging methods described above.

Gradients of the RF amplitude naturally arise in the fringe field of RF probes such as simple surface coils [3, 174, 364]. The gradients can be considerable. The fringe field of an ordinary 300 MHz solenoid for 5-mm sample tubes, for instance, reaches gradients more than 3 T/m [452]. Of course, gradients produced in this way can be taken as constant only in a rather limited volume range. In this respect anti-Helmholtz coil geometries are more favorable [416]. Extremely strong RF gradients follow from the attenuation of electromagnetic waves at conducting surfaces. This phenomenon was employed by Skibbe and Neue in a depth-resolved study of molecular hydrogen in palladium foils [454]. Using this "skin-effect-enhanced imaging" (SEEING) method a resolution in the order of μm in the direction perpendicular to the metal surface was achieved.

A prominent merit of nutation frequency encoding is the insensitivity to chemical shift or magnetic-susceptibility artifacts due to the low B_1 magnitudes compared with B_0 [396]. On the other hand, the weak influence of B_0 inhomogeneity distortions is opposed by problems arising from inhomogeneously distributed B_1 gradient magnitudes or even gradient vector directions. However, to a certain degree, artifacts of this origin can be corrected by post-detection processing.

34.2
Multi-Dimensional Representations

It may be difficult to produce two or even three independent B_1 gradients for probing two or three space directions in a rotating-frame imaging experiment. A more operational procedure is to reconstruct two- or three-dimensional data sets with the aid of backprojection methods [295]. One records a series of projections of the object on different gradient directions relative to the object. This can be performed by either turning the sample step by step in the presence of one fixed B_1 gradient produced by a surface coil, for instance, or, vice versa, by gradually moving the surface coil around the object.

The former was applied in the NQR version of the technique, rotating-frame NQR imaging (ρNQRI) [242], for instance. Note, however, that a Fourier transform analysis is inadequate for the analysis of a NQR pseudo FID. Rather special deconvolution, Hankel transform, or maximum-entropy data manipulation procedures have been suggested for this purpose [404, 420].

One-dimensional rotating-frame imaging can be favorably extended by considering the spectroscopic dimension (i.e., the time scale t_2 in Fig. 34.1). Such a frequency domain is readily provided by the FID itself. Fourier transforming the pseudo-FID as well as the proper FIDs recorded in full, while incrementing the pulse length, provides a two-dimensional data set from which a spatio/spectral map can be evaluated. This method was suggested for spatially resolved ^{31}P spectroscopy at tissue surfaces [173] or for depth-resolved measurements of temperature or pressure via NQR [253], for instance.

34.3
Rapid Rotating-Frame Imaging

Instead of incrementing the pulse length in subsequent transients a single-shot multiple-pulse sequence[2] can be used [314, 340]. A series of short B_1-gradient pulses is applied with narrow intervals in between, permitting the "stroboscopic" acquisition of data points of the pseudo-FID (Fig. 34.2). Each RF pulse is understood as an increment of the previous RF pulses taken together. Thus the whole k-space information needed for the FT evaluation of the projection of the object on the gradient direction can be recorded in one transient of the pulse train.

The advantage of extremely fast access to image data of an object is opposed by the drawback that the option to record spectral information is sacrificed in this way. In the short pulse intervals no full FID can evolve. That is, there is no longer

---

[2]An application of this principle has already been outlined in Sect. 21.1.1 in context with the MAGROFI diffusometry method.

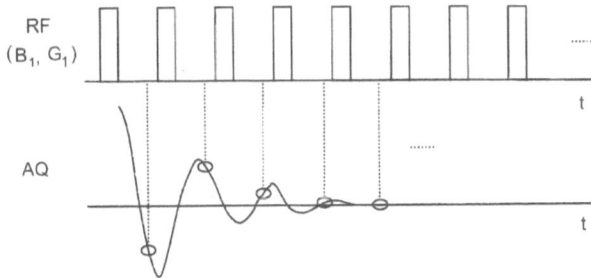

Fig. 34.2. B_1-gradient multiple-pulse sequence for the "stroboscopic" acquisition (AQ) of the pseudo-FID in a single transient permitting fast rotating-frame imaging.

the possibility to probe the spectral dimension. The same applies to the fast ρNQRI variant for rapid rotating-frame NQR imaging [405].

CHAPTER 35

Imaging of Solid Samples

A "solid" in this context means that the sample is far away from complete motional averaging of secular spin interactions. This situation certainly arises in rigid materials like crystalline or glassy substances. However, it also occurs in non-glassy macromolecular systems, elastomers or liquid crystals, for instance.

The origin of spectral line broadening in solids is the anisotropy of spin couplings, i.e., in particular, the dipolar, quadrupolar, and chemical shift interactions. What may be disagreeable for the high-resolution spectroscopist is most desirable for solid-state imaging. This is a major source of information on local material properties one would like rendered in the form of images or parameter maps.[1] On the other hand, the full exploitation of this potential requires special experimental procedures which can be more demanding than the imaging schemes delineated above.[2] Experiments for imaging solid-like materials must be designed according to which of the following three features is to be accentuated at the expense of the others:

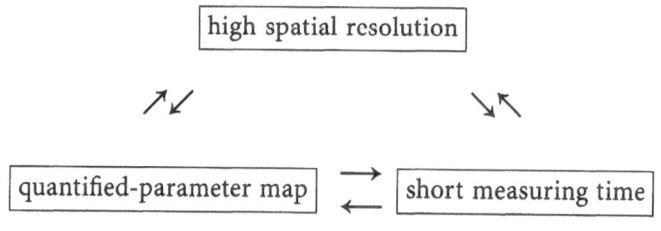

---

[1]U Haeberlen once proclaimed in his book [179] "narrow is beautiful!" In this context we may counter: sure, however, broad is meaningful!

[2]To a certain degree the conventional imaging pulse sequences can be adapted to the requirements of very viscous systems. Typical modifications of this sort are "oscillating gradient" methods [146, 150, 317], the "back-projection low-angle shot" (BLAST) technique [184], and constant-time imaging procedures of the simplest form [164]. In the following, techniques are envisaged which, at least in principle, are suitable for imaging of materials of any degree of rigidity or softness.

35.1
Experimental Strategies of Materials Imaging

35.1.1
The Reservoir of Contrast Parameters

Magnetic resonance imaging of materials offers interesting perspectives because of the wealth of different NMR parameters that can be exploited for the generation of image contrasts visualizing material properties in a non-destructive way with reasonable spatial, temporal, and - by circumstance - spectral resolutions [49]. The general objective is to transfer a parameter field of the object into a pixel/voxel data matrix, from which a gray-shade image can be produced. The parameter-fields of interest here do not only imply those already mentioned in context with conventional magnetic resonance imaging. The variety is much wider. Typical examples are:

a) spin density related quantities such as the concentration and the porosity, $\rho = \rho(r)$,
b) spin-lattice relaxation times in the laboratory frame, $T_1 = T_1(r)$, in the rotating frame, $T_{1\rho} = T_{1\rho}(r)$, or in local fields produced by dipole-dipole interaction, $T_{1d} = T_{1d}(r)$,
c) lineshape measures such as the linewidth, $\Delta\Omega_{1/2} = \Delta\Omega_{1/2}(r)$, the second (or higher) moments, $M_2 = M_2(r)$, the effective transverse relaxation time under multiple-pulse irradiation, $T_{2e} = T_{2e}(r)$, or resonance offsets due to chemical shifts, $\sigma = \sigma(r)$, or the magnetic susceptibility, $\chi = \chi(r)$,
d) displacements by flow, $v = v(r)$, or due to the local diffusion coefficient, $D = D(r)$, in soft or fluid areas of the object.

35.1.1.1
Parameter Mapping

Based on the systematic variation of the experimental parameters of the pulse sequence, or with the aid of the frequency-encoding dimension kept free from gradients, a whole set of local parameter values can in principle be evaluated and attributed to each pixel or voxel of an image data matrix. Based on such parameter maps, one can conveniently evaluate histograms of the second line-moment M_2 or of the transverse relaxation time T_2, for instance, for the global characterization of an object such as a cross-linked polymer sample [286]. Furthermore, the directly accessible NMR parameter fields may reflect fields of other parameters of interest to which they are related in such a way as to permit calibration. Examples are stress, polymer chain orientation, crystallinity, polymer cross-link density, temperature, and so on.

35.1.1.2
Parameter-Weighted Spin Density Images

In a more qualitative but less time-consuming way, the spatial distribution of material properties can also be rendered as parameter-weighted spin-density images instead of mapping the NMR parameters themselves quantitatively. The pulse sequences commonly used for imaging of solids tend to provide some intrinsic parameter-weighting of the signal intensity. The situation is very similar to that depicted in Chap. 26 in context with spin-echo FT imaging of liquid-like samples.

A Hahn-type spin echo, for instance, comprises only signals from coherences the dephasing process of which is not dominated by secular spin interactions. Any imaging sequence based on Hahn echo formation per se is a "filter" suppressing signals from solid-like constituents. Vice versa, magic echoes generated by pulse sequences like those shown in Fig. 6.1 selectively represent the rigid parts of a sample underlying secular spin interactions. That is, magic-echo pulse sequences act as filters in the complementary sense. An example was demonstrated with rubbers in [23]. In a rubber in the glassy state the magic-echo signals start to disappear when the temperature approaches the glass transition where the material gets soft.

The use of **"filters"** can be cultivated with the aid of precursor pulse sequences modifying the spatial distribution of the magnetization before the proper imaging pulse sequence begins. This turned out to be a very successful strategy for the fast visualization of qualitative material properties. For instance, on this basis it was possible to picture the spatial extension of shearbands in a solid polymer sample [508].

35.1.2
The Spatial-Resolution Problem

The k space signal (or "hologram") generally needed in any NMR FT imaging experiment is given by

$$S = S(k) = \int \tilde{\rho}(r)e^{-ik\cdot r}\, d^3r \tag{35.1}$$

where $\tilde{\rho}(r)$ is the effective, i.e., possibly "filtered" spin density, and k is the wave vector assumed to be three-dimensional and defined as usual by

$$k = \gamma_n \begin{pmatrix} G_x t_x \\ G_y t_y \\ G_z t_z \end{pmatrix} = \gamma_n \begin{pmatrix} t_x \partial B_0/\partial x \\ t_y \partial B_0/\partial y \\ t_z \partial B_0/\partial z \end{pmatrix} \tag{35.2}$$

The wave vector is determined by the gradients of the external magnetic field, B_0, and the encoding times t_x, t_y, t_z, during which the respective field gradients are effective. The threefold Fourier transform of Eq. 35.1 leads to the image data $\tilde{\rho} = \tilde{\rho}(r)$ (usually in the form of a magnitude representation).

The three-dimensional digital voxel resolution (see Chap. 32) is

$$\Delta x \, \Delta y \, \Delta z = \frac{\pi}{k_x^{max}} \frac{\pi}{k_y^{max}} \frac{\pi}{k_z^{max}} = \frac{\pi}{\gamma_n G_x^{max} t_x^{max}} \frac{\pi}{\gamma_n G_y^{max} t_y^{max}} \frac{\pi}{\gamma_n G_z^{max} t_z^{max}} \quad (35.3)$$

It is determined by the maximum encoding gradients and the maximum encoding times available for probing the spatial information in the three space directions.

The maximum gradients that can be applied depend on technical limits given by the performance of the gradient system, i.e., the power and the switching rates achievable.[3] They are also limited by the bandwidth of the receiver system, and by signal-to-noise requirements. Strong field gradients accelerate dephasing of the coherences, so that signal acquisition requires a higher bandwidth. This unavoidably degrades the signal-to-noise ratio [198].

A crucial condition for high-resolution FT imaging is therefore to ensure encoding times as long as possible. The physical limit of the encoding intervals that can be adjusted is given by the decay time of coherences, i.e., in particular by transverse relaxation. Naturally this tends to be rather unfavorable with solid materials because of the dominance of secular spin interactions. Every effort must therefore be made to increase the effective lifetime of coherences with the aid of spin manipulation techniques.

35.1.2.1
Coherence Refocusing

A possibility for reducing the effect of secular spin interactions in spatial encoding intervals while maintaining the full solid-state spectrum information content in the acquired signals is to refocus spin coherences as far as possible by spin-echo techniques. In this sense, solid echo [116, 311], Jeener/Broekaert echo [116, 415], and magic-echo [117, 181, 244] pulse sequences have been suggested. In this way, an undistorted spectral dimension can be probed in addition to the **k** space domains.

The spectral information is then frequency-encoded[4]. in the echo signals which are read out in the absence of field gradients. That is, the spatial allocation is performed solely via phase encoding and, potentially, slice selection. The most efficient

---

[3] The highest - although stationary - field gradients in use are those readily available in the stray field of superconducting magnets. On this basis a special scanning technique, the stray-field imaging (STRAFI) method, has been suggested [429, 533]. With gradients in the order of 50 T/m, even extremely short RF pulses take the character of "soft" pulses. Therefore they act slice-selectively. Instead of evaluating frequency-encoded *k*-space signals, the object is scanned slice by slice. The signals are solid echoes or accumulated echoes of a multiple-solid-echo train (see Sect. 4.1). The echo amplitudes are acquired as a function of the position on the gradient axis by shifting the object step by step along this axis. In this way a data set representing the projection of the object on the gradient direction is obtained. Gradually turning the sample and repeating the scans along the gradient direction provides a complete set of projections from which 2D or 3D image data can be reconstructed using the backprojection procedure [295]. Note that STRAFI is a scanning rather than an FT imaging method.

[4] Note that spectral parameters such as line moments can also be evaluated directly from the curvature of solid echoes (compare Eq. 4.24).

refocusing procedures are based on magic echoes or, more generally, mixed echoes. The outstanding feature of magic echoes is that coherences can be refocussed even after delays much longer than the transverse relaxation time. This is what significantly improves the spatial encoding potential with rigid solids.

35.1.2.2
Averaging of Secular Bilinear Spin Interactions

The strategy to refocus coherences in the form of echoes can be extended to a situation where the secular bilinear spin interactions are largely averaged. Based on any of the solid-echo pulse sequences mentioned before, suitable **multi-pulse trains** can be composed for this objective. In the pulse intervals, the signals are strobo-scopically acquired. They then evolve under the action of the average Hamiltonian. That is, in principle there is no coherence-dephasing effect by secular bilinear spin interactions such as quadrupolar or homonuclear dipolar couplings, whereas linear spin interactions due to chemical shift or any field-gradient dependent offset are scaled down by a certain constant factor. Famous pulse cycles serving this purpose are the Waugh/Huber/Haeberlen (WAHUHA or WHH-4) sequence [506],

$$
\left[\underbrace{(\pi/2)_{-x} - \tau - (\pi/2)_y - \tau - (\bullet) - \tau -} \right.
$$

$$
\left. \underbrace{(\pi/2)_{-y} - \tau - (\pi/2)_x - \tau - (\bullet) - \tau -} \right]_n \qquad (35.4)
$$

the Mansfield/Rhim/Elleman/Vaughan (MREV-8) sequence [321, 402],

$$
\left[\underbrace{(\pi/2)_x - \tau - (\pi/2)_{-y} - \tau - \tau -} \quad \underbrace{(\pi/2)_y - \tau - (\pi/2)_{-x} - \tau - (\bullet) - \tau -} \right.
$$

$$
\left. \underbrace{(\pi/2)_{-x} - \tau - (\pi/2)_{-y} - \tau - \tau -} \quad \underbrace{(\pi/2)_y - \tau - (\pi/2)_x - \tau \, \tau -} \right]_n \qquad (35.5)
$$

and the time reversal spin echo sequence (TREV-8) [331, 479] which consists of a series of magic-sandwich pulses for the production of a train of magic or mixed echoes in the intervals (see Figs. 6.1 and 6.2)[5] The repetition number of the pulse cycles is n. Other means for averaging secular spin interactions are **magic-angle spinning** (MAS) [11] or CRAMPS, a combination of MAS with multi-pulse line-narrowing techniques [427, 473].

Imaging experiments were suggested in combination with multi-pulse sequences [94, 102, 331, 508], with magic-angle spinning [493] or with CRAMPS [476]. On the same basis, spatio/spectral maps [196] or T_{2e}-filtered images [508] were recorded.

However, line-narrowing strategies change the nature of the signals which are acquired for rendering the images. Coherences in solids which are normally governed by anisotropic secular spin interactions are converted into liquid-like signals

---

[5]The dots indicate the intervals where the signal is sampled.

with all anisotropies more or less averaged out. Therefore, spectral information based on anisotropic secular spin interactions is sacrificed unless filter techniques are employed as mentioned above. That is, frequency encoding is preferably employed as a k-space dimension rather than for spectroscopic purposes. Note that the field gradient effective under multi-pulse irradiation is scaled down by the same factor as the chemical shift or any other frequency offset by linear spin interactions. On the other hand, the intervals available for encoding are increased.

35.2
Magic- and Mixed-Echo Phase-Encoding Imaging

Pulse sequences used for imaging of solid objects [185] are favorably based on magic echoes or on mixed echoes (see Chap. 6). Magic echoes which are formed under secular dipolar or quadrupolar interactions selectively render signals of rigid constituents of the object, whereas mixed echoes are sensitive to rigid as well as soft materials.

If the full solid-state spectroscopic information is to be maintained, one phase- or frequency-encoding domain must be reserved for spectral purposes. This strategy led to the magic- or mixed-echo phase-encoding solid imaging (MEPSI) technique [181] which embraces two or three phase-encoding k-space dimensions and one spectral (frequency-encoding) domain. The corresponding pulse sequences are shown in Figs. 35.1 and 35.2.[6]

The total length of the phase-encoding intervals is 2τ. That is, the wave-vector is given by

$$k = 2\gamma_n G\tau \tag{35.6}$$

The time-reversed evolution of coherences during the magic-sandwich pulses is not affected by phase encoding (compare the treatment in Chap. 6 or [117]). The signals recorded in quadrature as magic- or mixed echoes are thus given by

$$S = S(k, t_1) = e^{i\varphi} \int \int \bar{\rho}(r, \Omega) e^{-ik\cdot r} \, e^{-i\Omega t_1} \, \mathrm{d}^3 r \, \mathrm{d}\Omega \tag{35.7}$$

where φ is the (arbitrary) phase difference between the preparation and the sandwich pulses. Frequency offsets Ω originating from secular spin interactions are probed in the spectral time-domain, t_1.

Note that frequency offsets due to magnetic-susceptibility heterogeneities or magnet inhomogeneities are refocused in the mixed-echo case. However, they may contribute in the spectral time-domain, of course. A post-detection processing scheme like that outlined in Sect. 28.4 may be feasible for correction of corresponding effects. On the other hand, echo signals of solids may be evaluated directly

---

[6]The switching times of the phase-encoding gradient pulses should be much shorter than the encoding intervals in order to maintain the full encoding efficiency in the free-evolution periods. If this turns out to be a technical problem the gradients may also be left on during the whole pulse sequence including the magic sandwich. The artifacts arising on these grounds are minor [117]. Merely a sort of blurred inherent LOSY slice selection mechanism is produced by spin-locking in the presence of gradients.

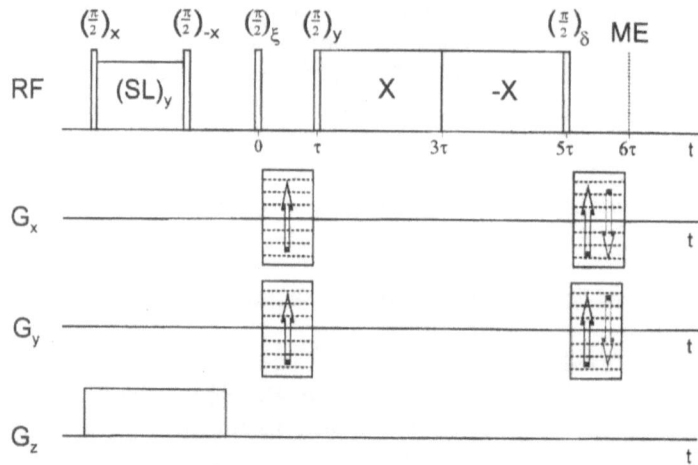

Fig. 35.1. RF and field-gradient pulse sequence for two-dimensional magic- or mixed-echo (ME) phase encoding solid imaging (2D-MEPSI). The frequency-encoding domain t_1 (or $t > 6\tau$) is suggested for spectroscopic purposes. A LOSY pulse is suggested for slice selection prior to the proper imaging sequence. **Magic-echo version:** $\delta \equiv -y$; left and right gradient pulses are incremented in the same sense (black arrows). **Mixed-echo version:** $\delta \equiv y$; left and right gradient pulses are incremented in opposite directions (left, black arrows; right, gray arrows). The phase difference between the LOSY pulse and the imaging pulses is arbitrary. The phase direction ξ is also uncritical.

in the spectral time-domain t_1 with respect to the linewidth or the line moments irrespective of any line shifts by inhomogeneities.

As mentioned before, the spectral frequency-encoding in the time domain t_1 may be sacrificed for the sake of spatial frequency-encoding, preferably in the form of multiple magic/mixed echo pulse trains [185]. This replaces one of the spatial phase-encoding dimensions. Spectral information may nevertheless be implied in a more qualitative manner by filtering precursor pulse sequences.

Figure 35.3 shows an application of a 3D-MEPSI pulse sequence to a hexamethylbenzene object. From such data sets arbitrary cross-sectional views may be rendered.

35.3
Magnetic Resonance Force Microscopy

The technology developed in context with atomic force microscopy (AFM) can be exploited for noninductively detecting magnetic resonance by coupling the nuclear magnetization to a mechanical oscillator [447]. To the very extreme this technology would refer to relatively few spins so that it would be possible to scan a sample with spatial resolution. This "magnetic resonance force microscopy" (MRFM) [425, 426] is a method for probing magnetic resonance of unpaired electrons or nuclei in a way

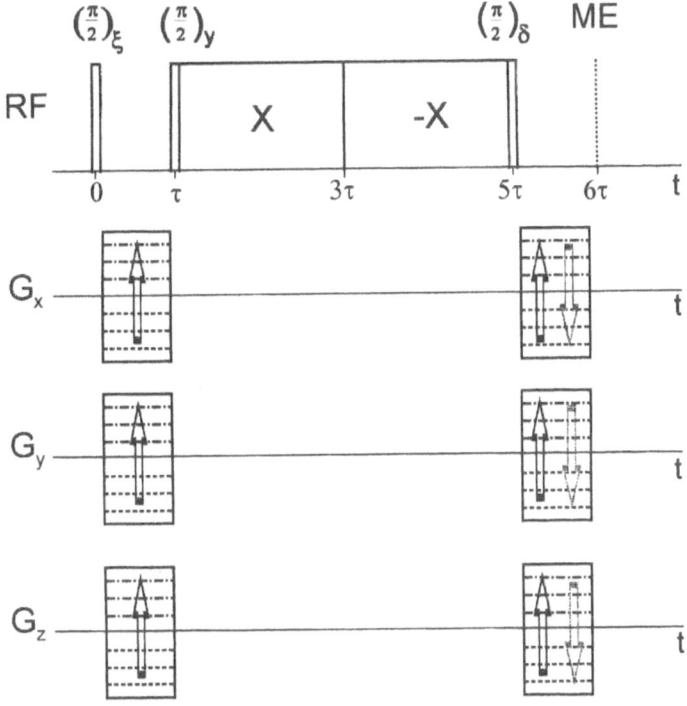

Fig. 35.2. RF and field-gradient pulse sequence for three-dimensional magic- or mixed-echo phase encoding solid imaging (3D-MEPSI). The frequency-encoding domain t_1 (or $t > 6\tau$) is suggested for spectroscopic purposes. **Magic-echo version:** $\delta \equiv -y$; left and right gradient pulses are incremented in the same sense (black arrows). **Mixed-echo version:** $\delta \equiv y$; left and right gradient pulses are incremented in opposite directions (left, black arrows; right, gray arrows). The phase difference between the LOSY pulse and the imaging pulses is arbitrary. The phase direction ξ is also uncritical.

alternative to the conventional FID acquisition or continuous wave (CW) absorption experiments.[7]

The sample, a tiny quantity of typically 10 ng, is mounted on a microscopic cantilever that is part of a mechanical oscillator with an eigenfrequency in the order of 10^3 Hz. This device is arranged close to an RF transmitter coil in a strong magnetic field with a strong gradient of the order of almost 100 T/m. The magnetization of those particles which are at resonance in the magnetic field gradient can be modulated by the RF irradiation. The principle can be either adiabatic inversion or nutation. This in turn modulates the force the sample experiences in the magnetic field gradient. If the modulation frequency is tuned to the mechanical eigen-frequency, the cantilever starts to oscillate as detected by a sensitive optical

---

[7]MRFM is to be distinguished from experiments where isolated electron spins are detected by scanning tunnelling microscopy (STM) [30, 160] via the modulated tunnelling current [319].

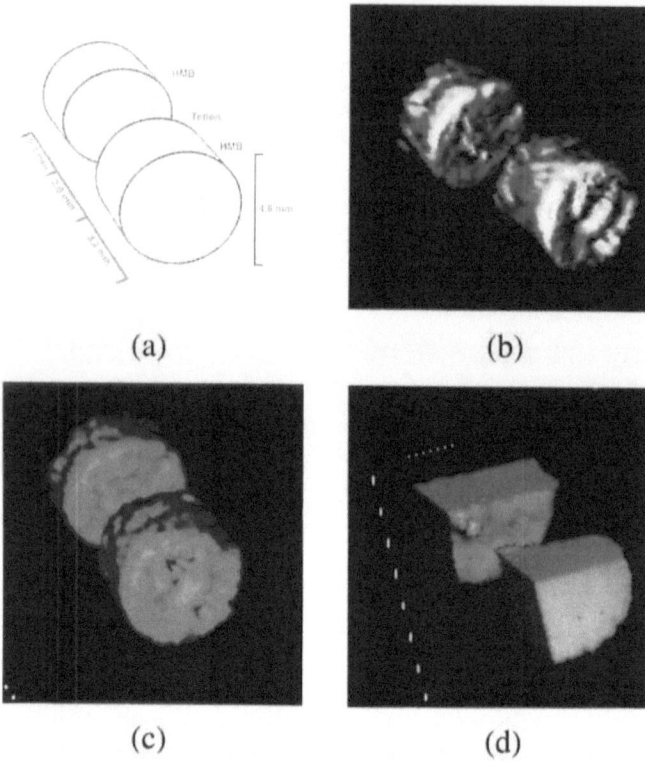

(a)   (b)

(c)   (d)

Fig. 35.3. 3D-MEPSI surface and cross-sectional views of a powdery hexamethylbenzene object. The spectral dimension was not exploited: a) schematic representation of the object and its dimensions; b), c) perspective MEPSI surface pictures; d) cross-sectional view calculated from the same data set. The phase-encoding time was $2\tau = 200$ μs, the repetition time 0.8 s. 64 × 64 × 64 voxels with an edge length of 300 μm were recorded. (Reproduced by permission from ref. [183])

interferometry device. Variation of the radio frequency or shifting the sample in the field gradient permits one to probe different parts of the sample. Spatial resolution is obtained by gradually scanning the resonant volume element across the sample. The mechanical oscillator serves as the proper detector monitoring the local magnetic resonance. Its high sensitivity makes it possible to detect an order of magnitude of 10^{13} protons per resolved voxel. That is, a micrometer or even submicrometer resolution becomes feasible with this point-by-point spin-density imaging technique.

CHAPTER 36

Slice-Selective Homonuclear Spin-Locking

36.1
Review of Slice-Selection Principles

Slice-selective excitation of an object is of interest in context with two-dimensional imaging as already delineated in Sect. 25.1. It also forms the crucial element of volume-selection procedures providing localized free-induction signals (Chap. 37). Basically one can distinguish three principles.

a) The technique most common in magnetic resonance imaging is the direct excitation of the region of interest by **"soft" pulses**, i.e., the application of narrow-band RF pulses in the presence magnetic field gradients (see Sect. 25.1).

b) An indirect means complementary to a) is **to spoil or presaturate the magnetization outside the region of interest** so that the subsequent pulse sequence can only generate free-induction signals originating from the selected slice. This can be achieved with the aid of correspondingly "tailored" shapes of the RF pulses again applied in the presence of magnetic field gradients. The whole volume is then excited but the desired slice is left (or recovered) in the initial state [15, 125, 305], A second method of this category, which is of particular interest here, is the application of spin-lock pulses in the presence of field gradients [115, 180, 244, 413, 515].

c) With heteronuclear spin experiments a slice can also be directly addressed by coherent or incoherent **slice-selective cross-polarization** from one spin species to the other [119, 120, 121, 182, 283, 292, 293].

The spatial resolution of slice-selective excitation according to principle **a)** becomes more restrictive with the shorter lifetimes of the spin coherences one normally has to expect with viscous liquids or solids. The efficiency may then be drastically reduced compared with liquid-like samples. Fast transverse relaxation decays of the spin coherences restrict the RF pulse length and, hence, permit only bandwidths which are by circumstance not narrow enough for the selection of the desired slice in a given field gradient. The problem can possibly be circumvented with the aid of DANTE combs by placing one narrow peak of the Fourier spectrum at the desired position of the slice to be selected [101]. With solid-like materials, "soft" single-pulse excitation is only feasible with extremely strong gradients in the order of 10 T/m such as those employed with the "stray-field imaging" (STRAFI) technique [429, 533].

With handier gradients the problem of slice-selection in solid-like materials can be solved by indirectly addressing the desired slice so that the direct selective excitation is not necessary any longer. This is the idea of principle b). Spin-locking in the presence of a field gradient turned out to be most useful in this respect. Slice selection by a spin-locking pulse sequence was actually the first method used for "localized spectroscopy" (LOSY) in solids [180]. Therefore the term "LOSY pulse" is used for brevity for this type of slice selection. The slice profiles obtained on this basis will be treated in the subsequent sections of this chapter.

Finally, localization and imaging experiments with heteronuclear spin systems suggest profit from cross-polarization methods. The corresponding formalism will be presented in Chap. 38. Imaging and localization pulse schemes can be streamlined by combining cross-polarization and slice selection to a single element of the sequence (category c)). Applications will be discussed in Chaps. 40 and 41.

36.2
Theory of Slice Selection by Spin-Locking

The magnetization component along the magnetic field that is effective in the rotating frame tends to be "locked", i.e., is prevented from getting spoiled by minor inhomogeneities or local fields. If spin coherences are locked in the presence of a strong enough field gradient, this phenomenon leads to a slice-selection effect.

The locked spins relax according to their longitudinal relaxation time in the rotating frame, $T_{1\rho}$. Non-resonant spins tend to remain unlocked and, hence, are dephased with a (much shorter) time constant T_2^*. With solids, T_2^* is practically determined by transverse relaxation caused by the secular spin-interaction terms. This process is intrinsically fast and dominates the selection process. Under motional narrowing conditions in liquids T_2^* also comprises the spoiling effect of the field gradient.

Spin-locking in the presence of field-gradients thus selectively conserves the magnetization of a slice about the mid-resonance plane whereas that of the residual volume is spoiled. Figure 36.1 shows the RF and field-gradient pulse sequence for production of slice-selective magnetization on this basis.

This composite pulse consists of a preparation 90° pulse, a 90° phase-shifted spin-lock pulse, and an optional 90° storage pulse. The lengths of the preparation and spin-lock pulses are denoted by t_p and t_{sl}, and the rotating-frame RF amplitudes by $B_1^{(p)}$ and $B_1^{(sl)}$, respectively. The RF amplitudes are constant during the pulses. The spin-locking condition is

$$|\omega - \omega_c| \overset{<}{\approx} |\gamma_n B_1^{(sl)}| \tag{36.1}$$

where $\omega = \omega(z)$ is the local Larmor frequency and ω_c is the angular carrier frequency of the pulse.

The RF pulses are applied in the presence of a constant field gradient arbitrarily assumed to be aligned along the z direction. That is, $G = G_z u_z = u_z \partial B_0 / \partial z = \text{const}$. Provided that the spin-locking interval t_{sl} obeys

$$T_2^* \ll t_{sl} \ll T_{1\rho} \tag{36.2}$$

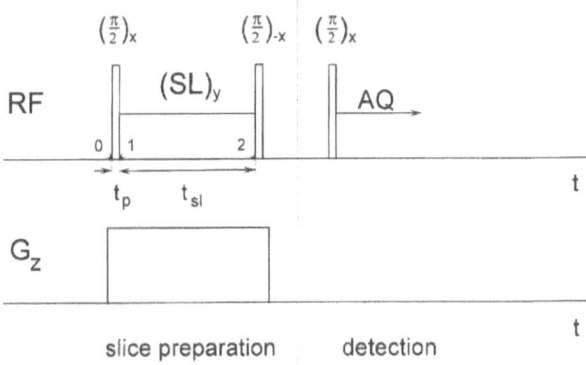

Fig. 36.1. LOSY pulse scheme for the production of slice-selective z magnetization. The sequence consists of a (preferably "hard") preparation RF pulse, $(90°)_x$, a ("soft") spin-lock RF pulse, $(SL)_y$, and a storage RF pulse, $(90°)_{-x}$. The slice-selective signal can be acquired (AQ) after switching off the field gradient by a reading 90° pulse. The numbers indicate times relevant for the treatment presented in the text.

the magnetization surviving the spin-lock pulse is that of the spins in the selected slice.

Under normal conditions, the preparation pulse can be assumed to be "hard," and its flip angle is 90°. That is, it obeys

$$\omega_1^{(p)} t_p = \pi/2 \tag{36.3}$$

and

$$\hbar \gamma B_1^{(p)} \gg \hbar \gamma G_z d_s \gg \langle \mathcal{H}_i' \rangle_m \tag{36.4}$$

where $\omega_1^{(p)} = \gamma_n B_1^{(p)}$, d_s is the dimension of the sample, and $\langle \mathcal{H}_i' \rangle_m$ is the largest expectation value of the spin interaction energy (in the rotating frame) such as those owing to chemical shift and spin-spin coupling.

By contrast, the spin-lock pulse must be "soft." The corresponding condition is

$$B_1^{(sl)} \ll G_z d_s \tag{36.5}$$

where the magnitude of the local fields caused by spin interactions is not yet specified. In the latter respect, the "liquid-" and "solid-state limits" will be distinguished for the formal treatment below [115, 244].

36.2.1
Liquid-State Limit

Let us first consider the case that $B_1^{(sl)}$ is much greater than the local fields arising from the spin interactions, $\langle \mathcal{H}_i' \rangle_m / (\hbar \gamma_n)$, but less than the total field variation across the sample:

$$\langle \mathcal{H}_i' \rangle_m \ll \hbar \gamma_n B_1^{(sl)} \ll \hbar \gamma_n G_z d_s \tag{36.6}$$

Any spin couplings can then be neglected. This case is denoted the "liquid-state" limit. The equation of motion of the local magnetization is given by Eq. 25.8 applied to the effective field of the spin-lock pulse,

$$B_e = G_z z\, u_z + B_1^{(sl)} \tag{36.7}$$

with the local tilt angle

$$\Theta_{sl} = \Theta_{sl}(z) = \arctan \frac{B_1^{(sl)}}{G_z z} \tag{36.8}$$

The position $z = 0$ at which the selected slice is centered is defined as the position where the carrier frequency is resonant.

The solution of the equation of motion represents a precession of the magnetization $M'(z,t)$ about $B_e = B_e(z)$ in the frame rotating with the carrier frequency. The component of $M' = M'(z,t)$ along B_e,

$$M'_{sl}(z,t) = M_0 \sin \Theta_{sl}\, e^{-t/T_{1\rho}} \tag{36.9}$$

is spin-locked and therefore conserved (apart from the attenuation by spin-lattice relaxation with the possibly spatially varying time constant $T_{1\rho}$). The perpendicular components, on the other hand, are spoiled by the field gradient and transverse relaxation. M_0 is the magnitude of the magnetization at the beginning of the spin-lock pulse.

The transverse magnetization just after the spin-lock pulse is given as the projection on the transverse plane [115, 180, 414],

$$M'_y(z,t_{sl}) = M_0 \sin^2 \Theta_{sl}\, e^{-t_{sl}/T_{1\rho}} \tag{36.10}$$

This is the component of the spin-locked magnetization which is eventually stored in the z direction by the 90° pulse terminating the LOSY sequence. In a homogeneous medium the slice profile thus has a Lorentzian shape,

$$g(z) = \frac{M'_y(z,t_{sl})}{M_0 \exp\{-t_{sl}/T_{1\rho}\}} = \sin^2 \Theta_{sl} = \frac{1}{1 + \left(G_z z / B_1^{(sl)}\right)^2} \tag{36.11}$$

The dependence of the profile function on the gradient strength is plotted in Fig. 36.2.

36.2.2
Solid-State Limit

The opposite limit, the "solid-state" limit, is defined by

$$\langle \mathcal{H}'_i \rangle \approx \hbar \gamma_n B_1^{(sl)} \ll \hbar \gamma_n G_z d_s \tag{36.12}$$

In this case, spin couplings are relevant at least during spin locking, and a more extended treatment must be carried out.

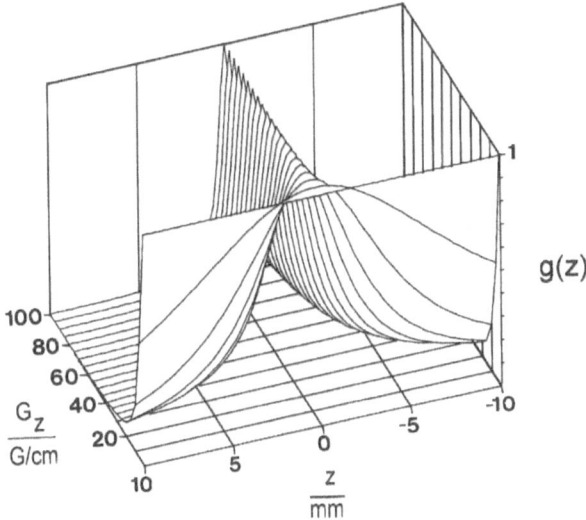

g(z)

Fig. 36.2. Slice profiles generated by a LOSY pulse in the liquid-state limit (Eq. 36.11). The amplitude of the spin-lock pulse was set at $B_1^{(sl)} = 5$ G. (Reproduced by permission from ref. [115])

36.2.2.1
General Procedure

Consider a spin ensemble at the position z in a constant magnetic field gradient $G = G_z u_z$. The local density operator, $\rho(z,t)$, obeys the Liouville/von Neumann equation

$$\frac{\partial \rho(z,t)}{\partial t} = -\frac{i}{\hbar}[\mathcal{H}(z,t), \rho(z,t)]. \tag{36.13}$$

The (laboratory frame) Hamiltonian is

$$\mathcal{H}(z,t) = -\hbar\gamma_n B_0 I_z - \hbar\gamma_n G_z z I_z - 2\hbar\gamma_n B_1 I_x \cos(\omega_c t) + \mathcal{H}_i^{(0)} \tag{36.14}$$

where we have assumed that the RF field oscillates along the x direction. $\mathcal{H}_i^{(0)}$ represents the possibly spatially varying laboratory-frame secular part of the local dipolar or quadrupolar spin interactions. The laboratory-frame nonsecular contributions cause spin-lattice relaxation but do not affect the evolution of coherences. They are therefore irrelevant in this context and can be ignored. The z dependence of the spin interaction term takes into account that the atomistic structure of the sample possibly varies with the position z referred to. The slice to be excited is again centered at $z = 0$ where the carrier frequency is resonant.

The total-spin vector operator, I, of a local many-spin system at position z has the laboratory-frame components I_x, I_y, I_z, where $I_k = \sum_j I_k^{(j)}$ ($k = x, y, z$). The sum refers to all interacting spins within the local spin ensemble.

The standard solution of the Liouville/von Neumann equation,

$$\rho(z,t) = e^{-(i/\hbar)\mathcal{H}(z)t}\rho(z,0)e^{(i/\hbar)\mathcal{H}(z)t} \tag{36.15}$$

stipulates that $\mathcal{H}(z)$ does not explicitly depend on time. The laboratory frame for which Eq. 36.14 is valid is therefore not suitable for such evaluations.

Rather a unitary transformation to a frame of reference is required so that the explicit time dependence vanishes. In particular, this applies to the RF field term which is made stationary by a transformation to a frame rotating with the carrier frequency ω_c about the z axis (see Sect. 48.8).

The solution (Eq. 36.15) is further simplified by another transformation leading to the "tilted rotating frame" in which the axis of quantization coincides with the z axis, i.e., these directions should be aligned along the effective field. In this particular frame, no spin transitions are induced by the RF field because no transverse field component is left in the effective Hamiltonian.

As the RF amplitude may be different with the preparation and spin-lock pulses, and because of the RF phase shift between the pulses, the transformations must be carried out interval by interval with constant RF amplitudes B_1 and phases (see Fig. 36.1). The spin system is assumed to be in thermodynamic equilibrium at the beginning of the LOSY pulse. The initial density operator in the laboratory frame is then in linear high-temperature approximation

$$\rho(z,0) \approx \left(\mathcal{E} + \beta\hbar\omega_0 I_z\right)/\text{Tr}\{\mathcal{E}\} \tag{36.16}$$

where $\omega_0 = \omega(z=0) = \gamma_n B_0$, $\beta = 1/(k_B T_l)$, with k_B Boltzmann's constant, T_l the absolute temperature of the lattice, and $\mathcal{E}$ the unity operator. This initial density operator is uniform across the sample for $\omega_0 \gg \gamma_n G_z d_s$.

The most general procedure for the treatment of the time evolution of the density operator in the course of a LOSY pulse sequence is

a) transformation from the laboratory frame (coordinates x, y, z) to the frame rotating with the carrier frequency of the preparation pulse (coordinates x', y', z' and index R)

b) transformation to the tilted rotating frame (coordinates x'', y'', z'' and index TR) so that the z'' axis is aligned along the effective field of the preparation pulse

c) derivation of the evolution of coherences in the tilted rotating frame during the preparation pulse (interval t_p)

d) back transformation to the rotating frame for the preparation pulse

e) transformation to the rotating frame of the spin-lock pulse (coordinates $\tilde{x}', \tilde{y}', \tilde{z}'$ and index $\tilde{R}$) which is phase shifted by 90°

f) transformation to the tilted rotating frame of the spin-lock pulse (coordinates $\tilde{x}'', \tilde{y}'', \tilde{z}''$ and index $\tilde{T}\tilde{R}$)

g) derivation of the evolution during the spin-lock pulse (interval t_{sl}) using the spin-temperature concept

h) back transformation to the rotating frame of the spin-lock pulse

Schematically this procedure may be written as

$$\rho(z,0) \xrightarrow{R} \rho_R(z,0) \xrightarrow{T} \rho_{TR}(z,0) \xrightarrow{P_p} \rho_{TR}(z,1)$$

$$\xrightarrow{T^{-1}} \rho_R(z,1) \xrightarrow{\tilde{R}} \rho_{\tilde{R}}(z,1) \xrightarrow{\tilde{T}} \rho_{\tilde{T}\tilde{R}}(z,1) \xrightarrow{P_{sl}} \rho_{\tilde{T}\tilde{R}}(z,2) \xrightarrow{\tilde{T}^{-1}} \rho_{\tilde{R}}(z,2) \quad (36.17)$$

The arrows and their superscripts represent the transformations to different frames of reference and the propagators acting during the RF pulses. The numbers in the arguments refer to the times indicated in Fig. 36.1.

36.2.2.2
Transformation to the Rotating Frame of the Preparation Pulse

The first step is the transformation to the frame rotating with the carrier frequency ω_c of the preparation pulse about the z' axis. It is carried out with the aid of the rotation operator (see Sect. 48.5)

$$R = e^{-i\omega_c t I_z}, \tag{36.18}$$

The density operator in the laboratory frame (Eq. 36.16) is not affected by this transformation,

$$\rho_R(z,0) = R \, \rho(z,0) \, R^{-1} = \rho(z,0) \tag{36.19}$$

The laboratory frame Hamiltonian at Eq. 36.14, on the other hand, is transformed to[1]

$$\mathcal{H}_R(z) = R \, \mathcal{H}(z,t) \, R^{-1} + i\hbar \dot{R} R^{-1}$$
$$= -\hbar \Delta\Omega(z) I_z - \hbar \omega_1^{(p)} I_x + R \, \mathcal{H}_i^{(0)}(z) \, R^{-1} \tag{36.20}$$

The local offset of the rotation frequency from resonance is

$$\Delta\Omega = \Delta\Omega(z) = (\omega_0 + \gamma_n G_z z) - \omega_c \tag{36.21}$$

36.2.2.3
Transformation to the Tilted Rotating Frame of the Preparation Pulse

The RF part of the Hamiltonian still depends on transverse spin components, i.e., nonsecular terms. These are removed by the transformation to the tilted rotating frame which is reached by turning the rotating frame about its y'' axis by the tilt angle

$$\Theta_p = \Theta_p(z) = \arctan \frac{\omega_1^{(p)}}{\Delta\Omega(z)} \tag{36.22}$$

The corresponding transformation operator is

$$T = T(z) = e^{i\Theta_p(z) I_y} \tag{36.23}$$

---

[1]Remember that we always use spin operators without labels specifying the reference frame they are supposed to refer to.

With this operator, Eqs. 36.16 and 36.19 are transformed to the new initial density operator

$$
\begin{aligned}
\rho_{TR}(z,0) &= T(z)\, \rho_R(z,0)\, T^{-1}(z) \\
&= \left[\mathcal{E} + \beta\hbar\omega_0 \left(I_z \cos\Theta_p + I_x \sin\Theta_p \right) \right] / \mathrm{Tr}\{\mathcal{E}\}
\end{aligned}
\tag{36.24}
$$

The Hamiltonian at Eq. 36.20 is transformed to

$$
\begin{aligned}
\mathcal{H}_{TR}(z) &= T\, \mathcal{H}_R\, T^{-1} \\
&= -\hbar\omega_e^{(p)} I_z + T\, R\, \mathcal{H}_i^{(0)}\, R^{-1}\, T^{-1}
\end{aligned}
\tag{36.25}
$$

with the effective frequency

$$
\omega_e^{(p)} = \omega_e^{(p)}(z) = \sqrt{\omega_1^{(p)2} + \Delta\Omega^2}
\tag{36.26}
$$

The situation is strongly simplified in the hard-pulse limit which reads in the present nomenclature

$$
\hbar\omega_1^{(p)} \gg \hbar\gamma_n G_z d_s \gg \langle T\, R\, \mathcal{H}_i^{(0)}\, R^{-1}\, T^{-1} \rangle
\tag{36.27}
$$

In this case the whole spectrum and the whole sample are excited homogeneously. Irrespective of the position, the tilt angle and the effective angular frequency are

$$
\Theta_p \approx \pi/2 \quad \text{and} \quad \omega_e^{(p)} \approx \omega_1^{(p)} = \text{const}
\tag{36.28}
$$

respectively. Dipolar, quadrupolar, or chemical-shift interactions may be neglected, and the Hamiltonian in the tilted rotating frame becomes

$$
\mathcal{H}_{TR}^{(p)} \approx -\hbar\omega_e^{(p)} I_z \approx -\hbar\omega_1^{(p)} I_z
\tag{36.29}
$$

The initial density operator of the preparation pulse interval, Eq. 36.24, is approached in this case by

$$
\rho_{TR}(z,0) \approx \left[\mathcal{E} + \beta\hbar\omega_0 I_x \right] / \mathrm{Tr}\{\mathcal{E}\}
\tag{36.30}
$$

36.2.2.4
Evolution During the Preparation Pulse

The initial density operator evolves in the tilted rotating frame of the preparation pulse according to the transformed Liouville/von Neumann equation,

$$
i\hbar\frac{\partial\rho_{TR}(z,t)}{\partial t} = [\mathcal{H}_{TR}^{(p)}(z), \rho_{TR}(z,t)]
\tag{36.31}
$$

The solution is

$$\rho_{TR}(z,t) = P_p(z,t)\,\rho_{TR}(z,0)\,P_p^{-1}(z,t)$$
$$\equiv e^{-(i/\hbar)\mathcal{H}_{TR}^{(p)}(z)t}\,\rho_{TR}(z,0)\,e^{(i/\hbar)\mathcal{H}_{TR}^{(p)}(z)t} \tag{36.32}$$

where

$$P_p(z,t) = e^{-(i/\hbar)\mathcal{H}_{TR}^{(p)}(z)t} \tag{36.33}$$

is the propagator during the RF irradiation. Dipolar, quadrupolar, or chemical shift interactions also may be neglected with respect to the evolution during the preparation pulse, so that Eq. 36.32 yields for the time "1" at the end of the pulse (see Fig. 36.1):

$$\rho_{TR}(z,1) = \left\{\mathcal{E} + \beta\hbar\omega_0 \left(I_z\cos\Theta_p - \left[I_x\cos(\omega_e^{(p)}t_p)\right.\right.\right.$$
$$\left.\left.\left. - I_y\sin(\omega_e^{(p)})\right]\sin\Theta_p\right)\right\}/\mathrm{Tr}\{\mathcal{E}\}$$
$$\approx \left\{\mathcal{E} + \beta\hbar\omega_0\,I_y\right\}/\mathrm{Tr}\{\mathcal{E}\}$$
$$\text{for}\quad \Theta_p \approx \pi/2; \quad \omega_e^{(p)}t_p \approx \omega_1^{(p)}t_p = \pi/2 \tag{36.34}$$

36.2.2.5
Back Transformation to the Rotating Frame of the Preparation Pulse

The back transformation to the rotating frame gives

$$\rho_R(z,1) = e^{-i\Theta_p I_y}\,\rho_{TR}(z,1)\,e^{i\Theta_p I_y}$$
$$= \left(\mathcal{E} + \beta\hbar\omega_0\left\{\left(I_z\cos\Theta_p + I_x\sin\Theta_p\right)\cos\Theta_p\right.\right.$$
$$- \left[\left(I_x\cos\Theta_p - I_z\sin\Theta_p\right)\right.$$
$$\left.\left.\cos(\omega_e^{(p)}t_p) - I_y\sin(\omega_e^{(p)}t_p)\right]\sin\Theta_p\right\}\right)/\mathrm{Tr}\{\mathcal{E}\}$$
$$\approx \left(\mathcal{E} + \beta\hbar\omega_0\,I_y\right)/\mathrm{Tr}\{\mathcal{E}\}$$
$$\text{for}\quad \Theta_p \approx \pi/2; \quad \omega_e^{(p)}t_p \approx \omega_1^{(p)}t_p = \pi/2 \tag{36.35}$$

36.2.2.6
Transformation to the Rotating Frame of the Spin-Lock Pulse

The subsequent spin-lock pulse is phase shifted by 90°. Therefore the rotating frame must be transformed to a correspondingly phase shifted frame:

$$\rho_{\bar{R}}(z,1) = e^{iI_z\pi/2}\,\rho_R(z,1)\,e^{-iI_z\pi/2}$$
$$= \left(\mathcal{E} + \beta\hbar\omega_0\left\{\left(I_z\cos\Theta_p - I_y\sin\Theta_p\right)\cos\Theta_p\right.\right.$$
$$- \left[\left(-I_y\cos\Theta_p - I_z\sin\Theta_p\right)\right.$$
$$\left.\left.\cos(\omega_e^{(p)}t_p) - I_x\sin(\omega_e^{(p)}t_p)\right]\sin\Theta_p\right\}\right)/\mathrm{Tr}\mathcal{E}$$
$$\approx \left(\mathcal{E} + \beta\hbar\omega_0\,I_x\right)/\mathrm{Tr}\{\mathcal{E}\}$$
$$\text{for}\quad \Theta_p \approx \pi/2; \quad \omega_e^{(p)}t_p \approx \omega_1^{(p)}t_p = \pi/2 \tag{36.36}$$

36.2.2.7
Transformation to the Tilted Rotating Frame of the Spin-Lock Pulse

The density operator in the tilted rotating frame of the spin-lock pulse reads

$$
\begin{aligned}
\rho_{\tilde{T}\tilde{R}}(z,1) &= e^{i\Theta_{sl}I_y}\,\rho_{\tilde{R}}(z,1)\,e^{-i\Theta_{sl}I_y} \\
&= \big(\mathcal{E} + \beta\hbar\omega_0\,\{\,[\,(I_z\cos\Theta_{sl} - I_x\sin\Theta_{sl})\cos\Theta_p \\
&\quad -I_y\sin\Theta_p]\,\cos\Theta_p - \{\,[-I_y\cos\Theta_p - (I_z\cos\Theta_{sl} \\
&\quad -I_x\sin\Theta_{sl})\sin\Theta_p]\,\cos(\omega_e^{(p)}t_p) - (I_x\cos\Theta_{sl} \\
&\quad +I_z\sin\Theta_{sl})\,\sin(\omega_e^{(p)}t_p)\,\}\,\sin\Theta_p\,\}\big)\,/\mathrm{Tr}\{\mathcal{E}\} \\
&\approx \big(\mathcal{E} + \beta\hbar\omega_0\,\{\,I_x\cos\Theta_{sl} + I_z\sin\Theta_{sl}\,\}\big)\,/\mathrm{Tr}\{\mathcal{E}\} \\
&\quad\text{for}\quad \Theta_p \approx \pi/2;\quad \omega_e^{(p)}t_p \approx \omega_1^{(p)}t_p = \pi/2
\end{aligned}
\tag{36.37}
$$

The polar tilt angle of the spin-lock pulse is

$$
\Theta_{sl} = \arctan\left(\omega_1^{(sl)}/\Delta\Omega\right)
\tag{36.38}
$$

where $\omega_1^{(sl)} = \gamma_n B_1^{(sl)}$. Equation 36.37 may be written as

$$
\rho_{\tilde{T}\tilde{R}}(z,1) = \{\,\mathcal{E} + \beta\hbar\omega_0\,U\cdot I\,\}\,/\mathrm{Tr}\{\mathcal{E}\}
\tag{36.39}
$$

The vector U has the form

$$
U = \begin{pmatrix} U_1 \\ U_2 \\ U_3 \end{pmatrix}
\tag{36.40}
$$

where

$$
\begin{aligned}
U_1 &= s_p\Sigma_p c_{sl} - c_p^2 s_{sl} - s_p^2\gamma_p s_{sl} \\
U_2 &= s_p c_p\gamma_p - c_p s_p \\
U_3 &= c_p^2 c_{sl} + s_p^2\gamma_p c_{sl} + s_p\Sigma_p s_{sl}
\end{aligned}
\tag{36.41}\tag{36.42}\tag{36.43}
$$

and

$$
\begin{aligned}
\Sigma_p &= \sin(\omega_e^{(p)}t_p) & (36.44) \\
\gamma_p &= \cos(\omega_e^{(p)}t_p) & (36.45) \\
s_p &= \sin\Theta_p & (36.46) \\
c_p &= \cos\Theta_p & (36.47) \\
s_{sl} &= \sin\Theta_{sl} & (36.48) \\
c_{sl} &= \cos\Theta_{sl} & (36.49)
\end{aligned}
$$

In the hard 90° preparation pulse limit, $\Theta_p \approx \pi/2;\ \omega_e^{(p)}t_p \approx \omega_1^{(p)}t_p = \pi/2$, we have

$$
U \approx \begin{pmatrix} \cos\Theta_{sl} \\ 0 \\ \sin\Theta_{sl} \end{pmatrix}
\tag{36.50}
$$

This is the situation reached at the beginning of the spin-lock pulse. The next step is to consider the evolution under spin-locking conditions.

36.2.2.8
Evolution During the Spin-Lock Pulse

The spin-lock pulse is "soft," and the influence of spin interactions can no longer be neglected. That is

$$\left\langle \tilde{T}(z)\, \tilde{R}\, R\, \mathcal{H}_i^{(0)}(z)\, R^{-1}\, \tilde{R}^{-1}\, \tilde{T}^{-1}(z) \right\rangle \lessgtr \hbar\omega_1^{(sl)} \gtrless \hbar\gamma_n G_z d_s \tag{36.51}$$

The laboratory-frame Hamiltonian, Eq. 36.14, modified according to the phase shift of the spin-lock pulse is

$$\mathcal{H}(z,t) = -\hbar\gamma_n B_0 I_z - \gamma_n \hbar G_z z I_z - 2\hbar\gamma_n B_1 I_x \sin(\omega_c t) + \mathcal{H}_i^{(0)}(z) \tag{36.52}$$

The transformation to the tilted rotating frame of the spin-lock pulse leads to

$$\begin{aligned}
\mathcal{H}_{\tilde{T}\tilde{R}}(z) &= \tilde{T}(z)\, \tilde{R}\, \mathcal{H}(z,t)\, \tilde{R}\, \tilde{T}^{-1}(z) \\
&= -\omega_e^{(sl)} I_z + \tilde{T}\, \tilde{R}\, \mathcal{H}_i^{(0)}(z)\, \tilde{R}^{-1}\tilde{T}^{-1}
\end{aligned} \tag{36.53}$$

where

$$\omega_e^{(sl)} = \omega_e^{(sl)}(z) = \sqrt{\omega_1^{(sl)2} + \Delta\Omega^2} \tag{36.54}$$

The transformed Hamiltonians of dipolar or quadrupolar spin interactions are derived in Sects. 48.9 and 48.10. As an example, the dipolar-coupling case will be treated in the following. The local secular dipolar Hamiltonian in the tilted rotating frame is

$$\begin{aligned}
\mathcal{H}_d^{\tilde{T}\tilde{R}}(z) &= P_2(\cos\Theta_{sl})\, \mathcal{H}_d^{(0)} + \frac{3}{4} P \sin^2\Theta_{sl} \\
&\quad + \frac{3}{2} Q \sin\Theta_{sl} \cos\Theta_{sl}
\end{aligned} \tag{36.55}$$

where the second-order Legendre polynomial is defined by

$$P_2(\cos\Theta_{sl}) = \frac{1}{2}(3\cos^2\Theta_{sl} - 1) \tag{36.56}$$

The nonsecular contributions $P = P(z)$ and $Q = Q(z)$ arise from the unitary transformations and are given in Eqs. 48.113 and 48.114.

Immaterial spin diffusion in solids is fast enough to guarantee the establishment of a quasiequilibrium with the local spin temperature $T_s(z) < T$ on a time scale much faster than the spin-lattice relaxation time $T_{1\rho}$ in the tilted rotating frame. As soon as the quasi-equilibrium is established all transverse magnetization components have disappeared in the tilted rotating frame. That is, the "locked" magnetization is aligned along the $\tilde{z}''$ axis. The time scale on which this state is reached is given by the transverse relaxation time in the tilted rotating frame, $T_{2\rho}$.

Thus the final local density operator for $T_{2\rho} \ll t_{sl} \ll T_{1\rho}$ is

$$\rho_{\tilde{T}\tilde{R}}(z,2) = \left(\mathcal{E} + \beta_s \hbar \omega_e^{(sl)} I_z - \beta_s \hbar \mathcal{H}_d^{\tilde{T}\tilde{R}} \right) / \text{Tr}\{\mathcal{E}\} \tag{36.57}$$

where

$$\beta_s = \beta_s(z) = \frac{1}{k_B T_s} \tag{36.58}$$

is proportional to the reciprocal spin temperature. As spin-lattice relaxation is negligible in the time scale considered, we may assume that the spin energy is conserved during the spin-lock pulse. That is

$$\langle \mathcal{H}_{\tilde{T}\tilde{R}}(z) \rangle = \text{Tr}\{ \mathcal{H}_{\tilde{T}\tilde{R}}(z) \rho_{\tilde{T}\tilde{R}}(z,1) \} = \text{Tr}\{ \mathcal{H}_{\tilde{T}\tilde{R}}(z) \rho_{\tilde{T}\tilde{R}}(z,2) \} \tag{36.59}$$

The right-hand equation,

$$\text{Tr}\{ \mathcal{H}_{\tilde{T}\tilde{R}}(z) \rho_{\tilde{T}\tilde{R}}(z,1) \} = \text{Tr}\{ \mathcal{H}_{\tilde{T}\tilde{R}}(z) \rho_{\tilde{T}\tilde{R}}(z,2) \} \tag{36.60}$$

links the beginning and the end of the spin-lock pulse. Inserting the initial and final density operators, Eqs. 36.39 and 36.57, respectively, and the Hamiltonians, Eqs. 36.53 and 36.55 leads to traces of products of the Cartesian components of the spin operators. These can be evaluated using the rules and formulae given in Chap. 43.

The only non-vanishing terms are obviously $\text{Tr}\{I_z^2\}$, $\text{Tr}\{\mathcal{H}_d^{(0)2}\}$, $\text{Tr}\{P^2\}$, and $\text{Tr}\{Q^2\}$. Evaluating these leads to

$$\frac{\text{Tr}\{ \mathcal{H}_d^{(0)2} \}}{\hbar^2 \text{Tr}\{ I_z^2 \}} = \frac{3}{4} \frac{\text{Tr}\{ P^2 \}}{\hbar^2 \text{Tr}\{ I_z^2 \}} = \frac{3}{4} \frac{\text{Tr}\{ Q^2 \}}{\hbar^2 \text{Tr}\{ I_z^2 \}} \equiv \omega_{loc}^2 = (\gamma_n B_{loc})^2 \tag{36.61}$$

where we have introduced the laboratory frame local field B_{loc} (compare refs [158, 523]).

Solving Eq. 36.60 with respect to the reciprocal spin temperature leads to the slice profile to be expected in a homogeneous sample

$$g(z) = \frac{\beta_s(z)}{\beta} = \frac{\omega_0 \omega_e^{(sl)} U_3}{\omega_e^{(sl)2} + L \omega_{loc}^2} \tag{36.62}$$

where

$$L = L(z) = P_2^2(\cos \Theta_{sl}) + \frac{3}{4} \sin^4 \Theta_{sl} + 3 \sin^2 \Theta_{sl} \cos^2 \Theta_{sl} \tag{36.63}$$

and (for resonant irradiation)

$$\sin \Theta_{sl}(z) = \frac{\omega_1^{(sl)}}{[\omega_1^{(sl)2} + (\gamma_n G_z z)^2]^{1/2}} \tag{36.64}$$

$$\cos \Theta_{sl}(z) = \frac{\gamma_n G_z z}{[\omega_1^{(sl)2} + (\gamma_n G_z z)^2]^{1/2}} \tag{36.65}$$

For a hard preparation pulse, i.e., $\Theta_p \approx \pi/2;\; \omega_e^{(p)} t_p \approx \omega_1^{(p)} t_p = \pi/2$, the profile function reduces to

$$g(z) \approx \frac{\omega_1^{(sl)2} + \omega_{loc}^2}{\omega_e^{(sl)2} + L(z)\omega_{loc}^2} \tag{36.66}$$

or

$$g(z) \approx \frac{1 + \left(\frac{B_{loc}}{B_1^{(sl)}}\right)^2}{1 + \left(\frac{G_z z}{B_1^{(sl)}}\right)^2 + L(z)\left(\frac{B_{loc}}{B_1^{(sl)}}\right)^2} \tag{36.67}$$

Finally, in the limit $B_1^{(sl)} \gg B_{loc}$, Eq. 36.67 becomes identical to Eq. 36.11, of course. Representative graphical representations of slice profiles in the solid-state limit are shown in Fig. 36.3.

Note that there is a pair of "magic" z slices symmetric to the central plane at $z = 0$, in which the secular contribution vanishes. This magic z value is given by

$$z_m = \frac{B_1^{(sl)}}{G_z \tan \Theta_m} \tag{36.68}$$

where the magic angle is $\Theta_m = \arccos(1/\sqrt{3})$. At the corresponding positions, the secular dipolar interactions do not contribute any more to the local field, whereas the nonsecular terms are still effective. Hence, this situation is intermediate between the solid and liquid limits and, strictly speaking, a spin-temperature concept is not applicable to these particular positions.

The treatment of the solid-state limit also shows that the spin temperature can be established *locally*. The spin temperature is therefore a function of the position. In the time scale of slice selection, spin diffusion fortunately is not able to equilibrate the temperature field.

Finally it should be mentioned that the field gradient must not be switched during the spin-lock pulse. Otherwise the magnetization is possibly transferred adiabatically in the direction of the effective field (compare Sect. 15.2.1). The slice selection effect then disappears or remains incomplete.

36.3
Variation of the Slice Width and Position

The profile of the slice is symmetrical to the position $z = 0$ where the carrier frequency is resonant. In the liquid-state limit and also in the solid-state case for spin-lock amplitudes exceeding the local field, the full width at half height is given by

$$\Delta z_{1/2} = \frac{2\, B_1^{(sl)}}{G_z} \tag{36.69}$$

It is determined by the ratio $G_z/B_1^{(sl)}$ and varies linearly with the spin-lock RF amplitude.

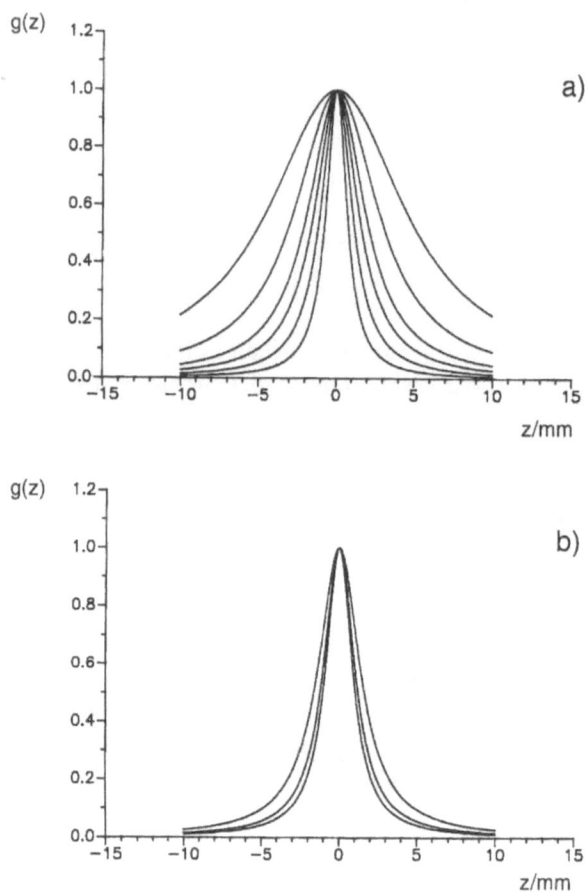

Fig. 36.3. Slice profiles generated by a LOSY pulse in the solid-state limit at Eq. 36.67. The 90° preparation pulse was assumed hard. The field gradient was set $G_z = 9.6$ G/cm. **a)** Profiles for different spin-lock amplitudes (beginning with the inner curve the parameters are $B_1^{(sl)} = 0.5$ G; 1.0 G; 1.5 G; 2.0 G; 3.0 G; 5.0 G, respectively; the local field was assumed to be $B_{loc} = 0.6$ G). **b)** Profiles for different local fields (beginning with the inner curve the parameters are $B_{loc} = 0$ G (liquid-state limit); 0.66 G ($< B_1^{(sl)}$); 1.26 G ($> B_1^{(sl)}$), respectively; in these cases, the spin-lock amplitude was set at $B_1^{(sl)} = 1$ G). (Reprinted by permission from ref. [115])

Thus the thickness of the slice can be adjusted by changing $B_1^{(sl)}$ while the gradient can be kept constant. The RF pulse shape need not be varied, and the excitation is independent of the pulse length in contrast to soft-pulse techniques. The slice width is determined by the spin-locking conditions rather than by the RF bandwidth of the pulse.

The only criterion for the length of the spin-lock pulse is the period needed for spoiling the magnetization outside the locked slice. With solid-like materials, this is an extremely fast process. Under practical circumstances, the condition for the spin-lock period,

$$T_2^* \ll t_{sl} \ll T_{1\rho} \tag{36.70}$$

allows much shorter pulse durations than would be needed with the soft-pulse technique. This is the reason that makes LOSY slice-selection feasible for applications to solids.

The slice can conveniently be positioned while keeping the adjusted thickness constant. The variation of the carrier frequency shifts the resonance position $z = 0$ to the desired region of interest.

CHAPTER 37

Homonuclear Localized NMR

The full spatially resolved spectral information of an object can be recorded with the aid of magnetic resonance spectroscopic imaging (MRSI) as described in Chap. 28. This of course may be a time-consuming procedure. Therefore one often prefers first to record an ordinary NMR image using a fast-imaging pulse sequence, for instance, and then to define a region of interest from which a localized spectrum can selectively be acquired with the aid of a volume-selective spectroscopy technique. One thus gains fast and accurate access to local spectral parameters at the expense of information from other object regions.

With the transients of an MRSI experiment the spectral information is recorded from the whole object within the field-of-view. However, in order to be able to discriminate the spectral information of the region of interest from other regions by post-detection Fourier data processing, the complete k space domains must be probed in addition to the spectral dimension (see Eq. 28.2). By contrast, localized spectroscopy is not based on a k space scanning rule. In principle, a single transient, i.e., one FID, suffices for addressing the spectral properties of the region of interest. Accumulation of multiple signals for the improvement of the signal-to-noise ratio may nevertheless be necessary, of course.[1] On the other hand, if spectral information from the whole object is of interest, it is much less efficient than MRSI to record the volume elements one after the other, of course.

Since the main motivation for localized-NMR experiments is to get rapid access to material properties of a well-defined region determined beforehand, the technique is not only of interest for spectroscopic purposes. Volume-selection can also be combined with the determination of parameters like relaxation times, diffusivities, or velocities. Moreover it is feasible to intertwine spectroscopy with relaxometry or diffusometry, so that such parameters can be determined, localized, and spectroscopically resolved.

When we speak here of localization we mean that the region of interest is determined beforehand under software control. This is to be distinguished from hardware-based methods such as those using surface coils [3] or "topical NMR" magnets [161]. The elements of volume selection procedures are rather set up on

---

[1] Accumulation of coherent-phase signals must not be confused with the acquisition of multiple signals each phase-encoded with different increments of the encoding gradient. Transients of the latter sort do not strengthen the signal-to-noise ratio because the phases of signals originating from different voxels are incoherent (compare Eq. 28.2).

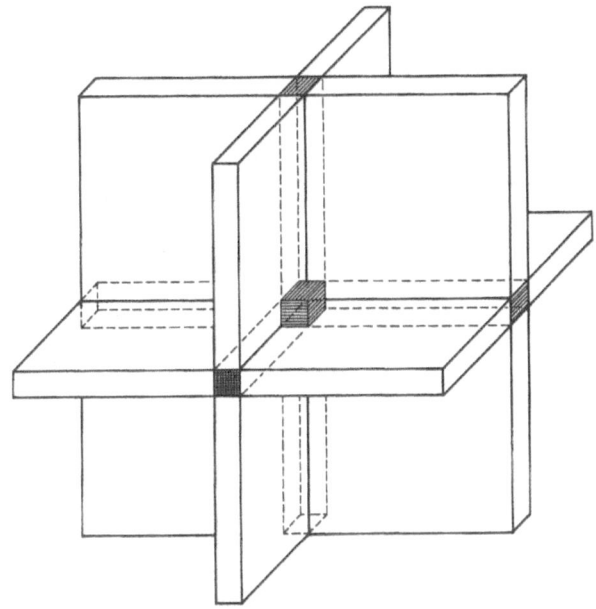

Fig. 37.1. Volume-selective NMR: schematic representation of the volume regions to be excited or saturated in a sequence of three slice-selective pulses. The central cube is common to all three orthogonal slices, and is placed in the region of interest.

the slice selection methods described above (Sect. 25.1 and Chap. 36). It is obvious that the simultaneous selection of three orthogonal slices in all three space directions defines a volume element that can then be positioned in the region of interest. Figure 37.1 illustrates the regions excited one after another by three slice-selection steps. Two principles serving this purpose may be distinguished.

a) The three orthogonal slices are subsequently excited with the aid of soft pulses applied in the presence of gradients along the three space directions. Three-pulse echoes are recorded so that the volume element common to all three slices is the only one which contributes to the acquired signal. This class of methods will be referred to as **"VOSY" (volume-selective spectroscopy)** or **"VOSING" (volume-selective spectral editing)**. The latter implies that the certain resonances are rendered in their full intensity or even enhanced, whereas others are suppressed. This is particularly of interest when a line of interest is superimposed by a (stronger) signal from another compound. As will be shown below, volume selection and spectral editing can be performed in an integrated way by the appropriate choice of the coherence pathways of the pulse sequence.

b) The regions outside the three orthogonal slices are saturated one after another with the aid of spin-lock pulses in the presence of field gradients in the three space directions.[2] As the only region with virginal magnetization the volume

---

[2]The same situation can also be produced with the aid of combined hard and soft-pulse excitation sequels [15, 125, 374].

element common to all three slices remains. It can be examined by an RF pulse or a pulse sequence applied after the localization pulses. This procedure is denoted "LOSY" (localized or locking spectroscopy) techniques.

37.1
The Homonuclear VOSY/VOSING Family

Principle a) for volume selection comprises a large variety of echo signals that can be exploited for localization and spectral editing. If the volume selection is to refer to all three space directions, the minimum number of pulses needed for the generation of VOSY signals consists of three slice-selective RF pulses as shown in Fig. 37.2. Dephasing of the coherences by the slice gradients is balanced by compensation gradient pulses in evolution intervals before or after the slice-gradient interlude.

The signals of J coupled spin systems acquired in this way may generally be classified as **coherence and spin state transfer echoes** [235, 236]. In Sects. 2.2 and 7.2 we have already treated the respective response of uncoupled spins and of a simple AX spin system to three RF pulses. Restricting ourselves to three-pulse echoes having their center at time $t = 2\tau_1 + \tau_2$, the following phenomena must be distinguished:

(i)   the stimulated Hahn echo of uncoupled spins
(ii) the longitudinal-magnetization transfer echo of coupled spins
(iii) the scalar-order transfer echo of coupled spins

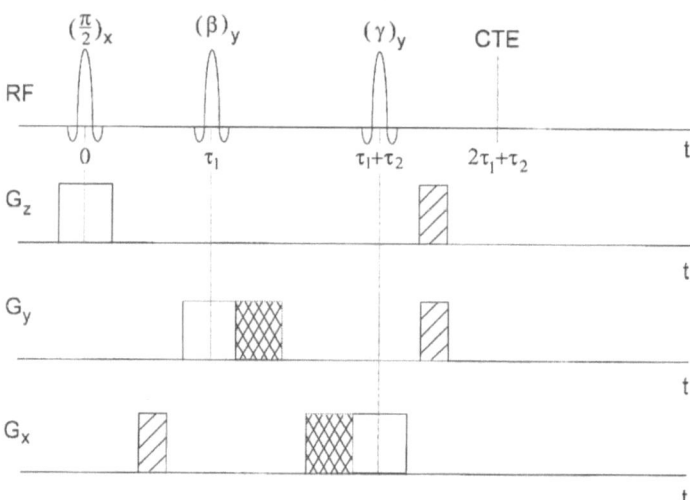

Fig. 37.2. Pulse sequence for volume-selective spectroscopy (VOSY) using three slice-selection pulses for the generation of coherence and spin-state transfer echoes (CTE). The single-hatched gradient pulses serve for the compensation of dephasings by the slice-selection gradients. The optional extensions of the gradient pulses (doubly hatched) in the second pulse interval are supposed to spoil unwanted coherences. The RF pulse phases are indicated according to the treatment in Sect. 7.2. However, owing to coherence dephasing by the gradients, they are arbitrary in principle.

(iv) the zero-quantum coherence transfer echo of coupled spins

(v) the double-quantum coherence transfer echo of coupled spins

Apart from the secondary Hahn echo (see Sect. 2.2), it is clear that uncoupled spins are restricted to the three-pulse echo of the first sort. By contrast, coupled spins are subject to the full scale of coherence and spin-state transfer echoes.[3] The signals then tend to be J modulated owing to coherence evolution unless the "low-coupling limit," $\tau_1 \ll J^{-1}$, is fulfilled.

The different echoes for coupled and uncoupled spins permits the recording of "edited spectra." The special choice of the flip angles β, γ and of the pulse delay τ_1 (see Fig. 37.2) optimizes or suppresses certain coherence pathways [236]. This can be shown with the aid of the product operator treatment described in Sect. 7.2.

Consider a weakly-coupled two-spin 1/2 system, where the spins are subject to offsets Ω_1 and Ω_2 due to the local gradient field and the chemical shift. As the field gradients are pulsed, the offsets Ω_1 and Ω_2 are time dependent quantities. On the other hand, the field gradients are trimmed for coherence rephasing in a subsequent interval. Therefore the offsets Ω_1 and Ω_2 can effectively be set constant during the whole pulse sequence without severe loss of generality.

Soft RF pulses in the presence of field gradients as they are used for the excitation of slices must generally be classified as "semiselective pulses" obeying the condition

$$2\pi|J| < |\gamma_n B_1| < |\Omega_1|, |\Omega_2| \tag{37.1}$$

Consequently a treatment using tilted RF-pulses [457] would be appropriate. In order to avoid expressions too complicated, we restrict ourselves to non-selective RF pulses, i.e.,

$$|\gamma_n B_1| \gg |\Omega_1|, |\Omega_2| \gg 2\pi|J| \tag{37.2}$$

In other words, all spins outside of the center plane of the selected slice are neglected, so that the resonance offsets of the considered spins fulfill the above condition.

The first RF pulse excites single-quantum coherences, while the second pulse transfers them into different spin states and coherences. The third RF pulse produces the observable single-quantum coherences eventually to be acquired. The density operator of these coherences at time $t = 2\tau_1 + \tau_2$ (see Fig. 37.2) consists of three terms,

$$\sigma_{1Q}(2\tau_1 + \tau_2) = \sigma_{(0Q)} + \sigma_{(so)} + \sigma_{(lm)} \tag{37.3}$$

where the subscripts in parentheses indicate the states in the τ_2 interval from which the coherences have been converted. The states in that interval comprise single-quantum coherences (1Q), zero-quantum coherences (0Q), longitudinal scalar order (so), and longitudinal magnetization (lm). In principle further contributions originating from single or double-quantum coherences during τ_2 might be of inter-

---

[3]Volume-selective spectroscopy is normally performed in order to examine coupled spin systems. Therefore the term "stimulated-echo acquisition mode" (STEAM) originally suggested for imaging schemes [151] and occasionally encountered in the literature is not appropriate for experiments with coupled spins.

est. However, in the pulse sequence shown in Fig. 37.2, these would immediately be spoiled by the field gradients in this interval.[4]

The signal contributions considered above are three-pulse effects as required for true volume selection with respect to all three space directions. These must be discriminated from coherences generated by less pulses. That is, the field-gradient pulses in Fig. 37.2 must be adjusted in such a way that undesired signals are spoiled. The acquired signal is proportional to the transverse magnetization which can be derived directly from σ_{1Q}. The result provides the crucial dependences on the flip angles and on the pulse delays.

A first step towards spectral editing is to suppress the three contributions partially and simultaneously to optimize the parameters of the pulse sequence for the record of a signal with the desired coherence pathway. In this way a kind of spectral editing is integrated in the volume-selection pulse sequence. A number of parameter sets of interest for this purpose are listed in the following.

Case (a) $\beta = \gamma = \pi/2$; $\tau_1 = J^{-1}$:
The coherences described by $\sigma_{(lm)}$ and originating from longitudinal magnetization in the τ_2 interval show their maximum values. In a sense the situation is analogous to the "stimulated echo" appearing with uncoupled spins. If these are present, as for instance the water protons in an aqueous solution, they contribute in full to the acquired signal. The other contributions vanish in this case, i.e., $\sigma_{(0Q)} = \sigma_{(so)} = 0$.

Case (b) $\beta = \gamma = \pi/4$; $\tau_1 = (2J)^{-1}$:
Here we have a situation opposite to case (a): $\sigma_{(so)}$ adopts its maximum value while $\sigma_{(lm)} = 0$ for the coupled spins (uncoupled spins still contribute to the signal!). The zero-quantum transfer signal has a finite amplitude. The order transfer echo can be used to edit simultaneously lines with coupling constants defined by $J = (2\tau_1)^{-1}$.

Case (c) $\beta = \gamma = \frac{\pi}{4}$; $\tau_1 = \frac{(2j+1)\pi}{\Omega_1+\Omega_2}$ (with $j = 0, 1, 2, 3, ...$) :
The special choice of τ_1 leads to $\sigma_{(0Q)} = 0$ provided that we may assume that $\Omega_1+\Omega_2$ has a value homogeneous in the whole volume element of interest.

Case (d) $\beta = \gamma = \frac{\pi}{4}$; $\tau_1 = \frac{(2j+1)\pi}{|\Omega_1-\Omega_2|}$ (with $j = 0, 1, 2, 3, ...$):
The offset difference comes in here so that any influence of field gradients cancel. The pulse spacing τ_1 is now chosen to fulfill $\sigma_{(so)} = 0$ while the contribution from $\sigma_{(0Q)}$ is particularly large.

In its basic form (Fig. 37.2) the VOSY three-pulse sequence already provides manifold localized-spectroscopy experiments. After slight modifications it is well suited for spectral editing of overlapping resonance lines on the basis of J coupling. For instance, case (b) can be employed for **scalar-order filtered volume-selective spectral editing (SOF-VOSING)** [262].

The principle of this method is that the scalar order is alternately reversed in subsequent transients of the experiment with the aid of a frequency-selective 180° pulse in the middle of the τ_2 interval. The carrier frequency and the bandwidth

---

[4]A modified pulse sequence for "double-quantum filtered volume-selective editing" (DQF-VOSING) will be discussed below.

of this pulse is adjusted to act only on one of the coupling partners of the compound to be edited. As a consequence, the sign of $\sigma_{(so)}$ and, hence, the signal of interest takes turns as well. Alternately adding and subtracting the echo signals recorded with and without 180° pulse accumulates the signal of interest in the form of scalar-order transfer echoes whereas signals from other sources and in particular from uncoupled spins cancel. The coherence pathway of the pay-signal is

single-quantum coherences → scalar order → single-quantum coherences.

A formal treatment of the scalar-order transfer echo for an AX spin system can be found in Sect. 7.2.2.

The SOF-VOSING procedure is of the "signal-add/subtract type." This is to be distinguished from single-transient techniques. That is, the complete editing effect is implied in each transient of any accumulation or phase-cycling series. In the following, two methods of this sort will be discussed in more detail.

37.1.1
Double-Quantum Volume-Selective Spectral Editing

In Sect. 7.2.5 the principle of double-quantum coherence transfer echoes is described, and a treatment is given for an AX spin system. Also, double-quantum filtered COSY is known as a standard technique in high-resolution NMR (see Fig. 7.4). This sort of phenomenon anticipates the basis of **double-quantum filtered volume-selective editing (DQF-VOSING)** spectroscopy put forward in [264]. The same principle was also suggested for spectroscopic imaging purposes [202] and for the unlocalized determination of lactate [464].

The essence of the DQF-VOSING measuring protocol is to suppress all coherence pathways other than

single-quantum coherences → double-quantum coherences → single-quantum
coherences.

This goal can already be achieved with the three-pulse VOSY pulse sequence by appropriate timing of the RF pulse intervals and of the gradient pulse lengths (compare Fig. 7.6). However, the pay-signal intensity achieved with the simple VOSY sequence is reduced because the second and third 90° pulses are applied while all coherences are completely dephased so that the coherence transfer by each of the pulses is less efficient.

Therefore additional 180° pulses are employed in the single-quantum coherence evolution intervals. Figure 37.3 shows a corresponding pulse scheme. The purpose of the 180° pulses is to refocus the single-quantum coherences in the relevant pulse intervals in full with respect to offsets by field inhomogeneities and chemical shifts. The transfer efficiency thus becomes maximal for double-quantum coherences.[5]

---

[5]Note that the phases of the two 90° pulses of the preparation pulse sequence should be in phase in this case when all offsets by gradients, inhomogeneities, and chemical shifts are refocused. The transfer from single-quantum to double-quantum coherences will only then be complete.

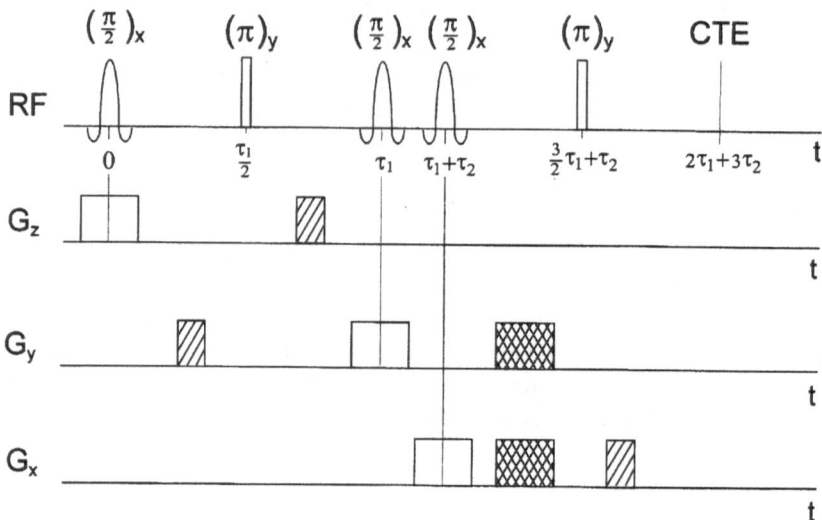

Fig. 37.3. Pulse scheme for double-quantum filtered volume-selective spectral editing (DQF-VOSING) spectroscopy. The coherence-transfer echo (CTE) results from a pathway consisting of the sequel single-quantum coherences ($0 < t < \tau_1$), double-quantum coherences ($\tau_1 < t < \tau_1 + \tau_2$), single-quantum coherences ($\tau_1 + \tau_2 < t < 2\tau_1 + 3\tau_2$). In the middle of each single-quantum interval a refocusing π RF pulse is applied. The three $\pi/2$ RF pulses are irradiated in the presence of slice gradients. The hatched gradient pulses serve for the refocusing of the slice gradients as concerns the first single-quantum coherence interval. The doubly hatched gradient pulses selectively compensate the phase shifts of the double-quantum coherence by single-quantum coherence phase shifts in the subsequent interval. That is, their "areas" are doubled relative to what would be needed for the restitution of phase shifts of single-quantum coherences.

On the other hand, the double-quantum coherences in the interval $\tau_1 \ldots (\tau_1 + \tau_2)$ are deliberately refocused only after converting them to single-quantum coherences at the end of this interval. The editing effect of this method is based on a double-quantum coherence dephasing and single-quantum rephasing cycle which distinguishes this coherence pathway from any other potential echo formation sequel.

The interval $\tau_1 \ldots (\tau_1 + \tau_2)$ is kept as short as needed for the slice-selection gradients (which simultaneously serve for the defocusing of the double-quantum coherences). The coherence-transfer echo for the selected coherence pathway is centered finally at time $t = 2\tau_1 + 3\tau_2$. The 180° pulse in this latter echo-formation interval is placed in such a manner that only dephasings of the single-quantum coherences accumulated in that interval are compensated, whereas restitution of the phase shifts originating from offsets in the double-quantum coherence evolution interval is made by "double-area" gradient pulses and by extending the echo-formation interval by a further delay of $2\tau_2$ (see Fig. 37.3).

37.1.1.1
Formal Treatment for A_3X Spin Systems

We consider a (weakly coupled) A_3X spin system as it is formed by the non-exchanging hydrogen nuclei of lactate $CH_3CH...$, for instance. The RF pulse sequence to be treated is

$$\left(\frac{\pi}{2}\right)_x - \frac{\tau_1}{2} - (\pi)_y - \frac{\tau_1}{2} - \left(\frac{\pi}{2}\right)_x - \tau_2 - \left(\frac{\pi}{2}\right)_x - \frac{\tau_1}{2} - (\pi)_y - \frac{\tau_1}{2} - (\text{echo})$$

where $\tau_2 \ll \tau_1$ is assumed so that dephasings in that interval by field inhomogeneities can be neglected. The formal treatment is carried out using Cartesian product operators the rules of which are summarized in Table 51.3 on page 487.

The reduced equilibrium density operator of the A_3X system in the high-temperature approximation is

$$\sigma(0-) = \sum_{k=1}^{4} I_{kz}. \tag{37.4}$$

The subscripts $k = 1, 2, 3$ refer to the three A spins, and $k = 4$ to the X spin. The first RF pulse, $(\pi/2)_x$, produces

$$\sigma(0+) = \sum_{k=1}^{4} I_{ky} \tag{37.5}$$

which corresponds to in-phase single-quantum coherences.

In the interval $0 < t < \tau_1$ the coherences evolve according to spin-spin coupling with the coupling constant J. On the other hand, frequency offsets by the field gradients, chemical shifts and field inhomogeneities need not be considered explicitly because the phase shifts arising on this basis are refocused by the 180° pulse in the middle of the interval (note that the field gradient pulses are balanced symmetrically to the 180° pulse).

A general treatment shows that the single-quantum coherences of the first free-evolution interval are most efficiently converted to double-quantum coherences by the second 90° pulse if τ_1 is adjusted relative to the spin-spin coupling constant according to

$$\tau_1 = \frac{1}{2J} \tag{37.6}$$

For the sake of simplicity we will therefore immediately replace τ_1 in the calculation by $(2J)^{-1}$.

The net evolution of the reduced density operator in the interval $0 < t < \tau_1 = (2J)^{-1}$ follows from the respective operator relations for the A spins ($k = 1, 2, 3$),

$$I_{kx} \xrightarrow{\pi J \tau_1 2 I_{kz} I_{4z}} I_{kx} \cos(\pi J \tau_1) + 2 I_{ky} I_{4z} \sin(\pi J \tau_1) = 2 I_{ky} I_{4z} \tag{37.7}$$

$$I_{ky} \xrightarrow{\pi J \tau_1 2 I_{kz} I_{4z}} I_{ky} \cos(\pi J \tau_1) - 2 I_{kx} I_{4z} \sin(\pi J \tau_1) = -2 I_{kx} I_{4z} \tag{37.8}$$

and for the X spin $(k = 4)$,

$$I_{4x} \xrightarrow{\pi I_{4z}I_{1z}} 2I_{4y}I_{1z} \xrightarrow{\pi 2I_{4z}I_{2z}} -4I_{4x}I_{1z}I_{2z} \xrightarrow{\pi 2I_{4z}I_{3z}} -8I_{4y}I_{1z}I_{2z}I_{3z} \tag{37.9}$$

$$I_{4y} \xrightarrow{\pi I_{4z}I_{1z}} -2I_{4x}I_{1z} \xrightarrow{\pi I_{4z}I_{2z}} -4I_{4y}I_{1z}I_{2z} \xrightarrow{\pi I_{4z}I_{3z}} 8I_{4x}I_{1z}I_{2z}I_{3z} \tag{37.10}$$

The reduced density operator thus becomes

$$\sigma(\tau_1-) = -\sum_{k=1}^{3} 2I_{kx}I_{4z} + 8I_{4x}I_{1z}I_{2z}I_{3z} \tag{37.11}$$

The second 90° pulse, $(\pi/2)_x$, results in

$$\sigma(\tau_1+) = -\sum_{k=1}^{3} 2I_{kx}I_{4y} + 8I_{4x}I_{1y}I_{2y}I_{3y} \tag{37.12}$$

The last term cannot lead to single-quantum coherences in the detection interval, and is therefore discarded in the further treatment. The sum terms can be analyzed according to

$$\begin{aligned} -2I_{kx}I_{4y} &= (I_{ky}I_{4x} - I_{kx}I_{4y}) - (I_{kx}I_{4y} + I_{ky}I_{4x}) \\ &= \frac{1}{2i}(I_k^+ I_4^- - I_k^- I_4^+) - \frac{1}{2i}(I_k^+ I_4^+ - I_k^- I_4^-) \\ &= \sigma_{0qc,y}^{k,4} - \sigma_{2qc,y}^{k,4} \end{aligned} \tag{37.13}$$

The density operators $\sigma_{0qc,y}^{k,4}$ and $\sigma_{2qc,y}^{k,4}$ stand for zero- and double-quantum coherences between one of the A spins $(k = 1, 2, 3)$ on the one hand, and the X spin $(k = 4)$ on the other.

The gradient pulses of the DQF-VOSING scheme are balanced in such a way that no coherence order other than double-quantum coherences can lead to refocused single-quantum coherences in the detection interval. Therefore the zero-quantum coherence terms can be dropped. During the τ_2 interval the double-quantum coherences experience the J coupling to "passive" spins, and frequency offsets by field gradients, inhomogeneities and chemical shifts.

The angular-frequency offset of the A spins is denoted by Ω_A, that of the X spin by Ω_X. These offsets are functions of time owing to the field-gradient pulses in the free-evolution intervals (see Fig. 37.3). The total phase angles accumulated by the double-quantum coherences due to the field gradients or inhomogeneities, and - to a minor degree - chemical shifts are defined by

$$\varphi_{A2} \equiv \int_{\tau_1}^{\tau_1+\tau_2} \Omega_A(t')\,dt' \tag{37.14}$$

$$\varphi_{X2} \equiv \int_{\tau_1}^{\tau_1+\tau_2} \Omega_X(t')\,dt' \tag{37.15}$$

In terms of these phase shifts the double-quantum coherences evolve according to

$$\sigma_{2qc,y}^{k,4} \xrightarrow{(\varphi_{A2}I_{kz}+\varphi_{X2}I_{4z})} \sigma_{2qc,y}^{k,4} \cos(\varphi_{A2} + \varphi_{X2}) + \sigma_{2qc,x}^{k,4} \sin(\varphi_{A2} + \varphi_{X2}) \tag{37.16}$$

where $k = 1, 2, 3$.

The passive A spins do not take part in double-quantum coherences between one A and the X spin, but influence their evolution. The evolution during the τ_2 interval under the J coupling of the active spins, i.e., one A spin (say $k = 1$) and the X spin ($k = 4$), to the two passive A spins ($k = 2, 3$), for instance, gives rise to the transformations

$$\sigma_{2qc,x}^{1,4} \xrightarrow{\pi J\tau_2 2I_{2z}I_{4z}} \sigma_{2qc,x}^{1,4} \cos(\pi J\tau_2) + 2I_{2z}\sigma_{2qc,y}^{1,4} \sin(\pi J\tau_2)$$

$$\xrightarrow{\pi J\tau_2 2I_{3z}I_{4z}} \sigma_{2qc,x}^{1,4} \cos^2(\pi J\tau_2) + (I_{2z} + I_{3z})\sigma_{2qc,y}^{1,4} \sin(2\pi J\tau_2)$$

$$- 4I_{2z}I_{3z}\sigma_{2qc,x}^{1,4} \sin^2(2\pi J\tau_2) \tag{37.17}$$

Only the first term of the last expression can lead to observable single-quantum coherences in the detection period. The other terms are therefore dropped in the following.

At the end of the τ_2 interval just before the third 90° pulse the relevant density operator terms read

$$\sigma(\tau_1 + \tau_2-) = -\sum_{k=1}^{3} \left[\sigma_{2qc,y}^{k,4} \cos(\varphi_{A2} + \varphi_{X2}) \right.$$

$$\left. + \sigma_{2qc,x}^{k,4} \sin(\varphi_{A2} + \varphi_{X2}) \right] \cos^2(\pi J\tau_2) + \ldots$$

$$= -\sum_{k=1}^{3} \left[(I_{kx}I_{4y} + I_{ky}I_{4x}) \cos(\varphi_{A2} + \varphi_{X2}) \right.$$

$$\left. + (I_{kx}I_{4x} - I_{ky}I_{4y}) \sin(\varphi_{A2} + \varphi_{X2}) \right] \cos^2(\pi J\tau_2) + \ldots \tag{37.18}$$

The third 90° pulse produces

$$\sigma(\tau_1 + \tau_2+) = \frac{1}{2}\sum_{k=1}^{3} (2I_{kx}I_{4z} + 2I_{kz}I_{4x}) \cos(\varphi_{A2} + \varphi_{X2}) \cos^2(\pi J\tau_2) + \ldots \tag{37.19}$$

where we have dropped multiple-quantum coherence and longitudinal scalar-order terms. The relevant terms correspond to antiphase single-quantum coherences which evolve in the subsequent interval $\tau_1 = (2J)^{-1}$ according to the transformations

$$2I_{kx}I_{4z} \xrightarrow{\pi I_{kz}I_{4z}} I_{ky} \xrightarrow{\varphi_{A3}I_{kz}} I_{ky} \cos\varphi_{A3} + I_{kx} \sin\varphi_{A3} \tag{37.20}$$

$$2I_{4x}I_{1z} \xrightarrow{\pi I_{1z}I_{4z}} I_{4y} \xrightarrow{\pi J I_{2z}I_{4z}} -2I_{4x}I_{2z} \xrightarrow{\pi I_{3z}I_{4z}} -4I_{4y}I_{2z}I_{3z}$$

$$\xrightarrow{\varphi_{X3}I_{4z}} -4I_{4y}I_{2z}I_{3z} \cos\varphi_{X3} - 4I_{4x}I_{2z}I_{3z} \sin\varphi_{X3} \tag{37.21}$$

where we have already taken into account the 180° pulse in the middle of this interval which refocuses all dephasings by chemical shifts and field inhomogeneities, and the gradient pulses. The doubly hatched gradient pulses in Fig. 37.3 are matched to the dephasings by the double-quantum coherence evolution in the τ_2 interval. That is, the phase shift each nuclear species experiences via single-quantum coherences must be equal to the total phase shift accumulated by the double-quantum coherence in the τ_2 interval. That is,

$$\varphi_{A3} \approx \varphi_{X3} = \varphi_{A2} + \varphi_{X2} \equiv \varphi \tag{37.22}$$

Thus the reduced density operator effectively evolves according to

$$\sigma(2\tau_1 + \tau_2) = \frac{1}{2} \sum_{k=1}^{3} \big[I_{ky} \cos\varphi + I_{kx} \sin\varphi - 2I_{4y}I_{kz}$$
$$\cos\varphi - 2I_{4x}I_{kz} \sin\varphi \big] \cos\varphi \cos^2(\pi J \tau_2) + \dots \tag{37.23}$$

The antiphase single-quantum coherences of the X nucleus are of minor interest and tend to cancel anyway under low-resolution conditions. They are therefore ignored. Assuming an equipartition of the phase shifts φ due to the strong gradients employed in this method, the average density operator is formed as

$$\langle \sigma(2\tau_1 + \tau_2) \rangle = \frac{1}{2\pi} \int_{0}^{2\pi} \sigma(2\tau_1 + \tau_2) \, d\varphi$$
$$= \frac{1}{4} \sum_{k=1}^{3} I_{ky} \cos^2(\pi J \tau_2) + \dots \tag{37.24}$$

The cosine term may be approximated by 1 in the limit $\tau_2 \ll J^{-1}$ so that the amplitude of the coherence-transfer echo signal of the A spins finally is proportional to the complex transverse rotating-frame magnetization

$$m(2\tau_1 + \tau_2) = \frac{n}{4} \gamma_n \hbar b \text{Tr} \left\{ \sum_{k=1}^{3} (I_{kx} + iI_{ky})\langle \sigma(2\tau_1 + \tau_2) \rangle \right\} = \frac{1}{4} m_A(0+) \tag{37.25}$$

where $n/4$ is the number density of A_3X spin systems and $m_A(0+)$ the complex transverse magnetization of the A_3 spins immediately after the first RF pulse. The other symbols are defined in context with Eq. 47.7.

The sensitivity is 1/4 compared with the unedited (but potentially concealed) signal of the same number of nuclei. A factor of 1/2 is sacrificed with the conversion from single-quantum to double-quantum coherences by the second 90° pulse. Another factor of 1/2 is due to the transfer from completely dephased double-quantum coherences to single-quantum coherences by the third 90° pulse.

An appealing feature of the DQF-VOSING technique is its simplicity in terms of the RF pulse number needed. At the expense of a further factor of 1/2 in the sensitivity relative to the version treated above, volume-selective editing can be managed already with three pulses [264], i.e., the minimum number needed for volume selection with respect to all three space directions.[6]

37.1.2
Cyclic Polarization Transfer Volume-Selective Spectral Editing

A further method of interest in this context is based on laboratory frame homonuclear polarization transfer of the INEPT type [351] which is performed in a cyclic way. The pulse sequence is shown in Fig. 37.4. The editing principle of this

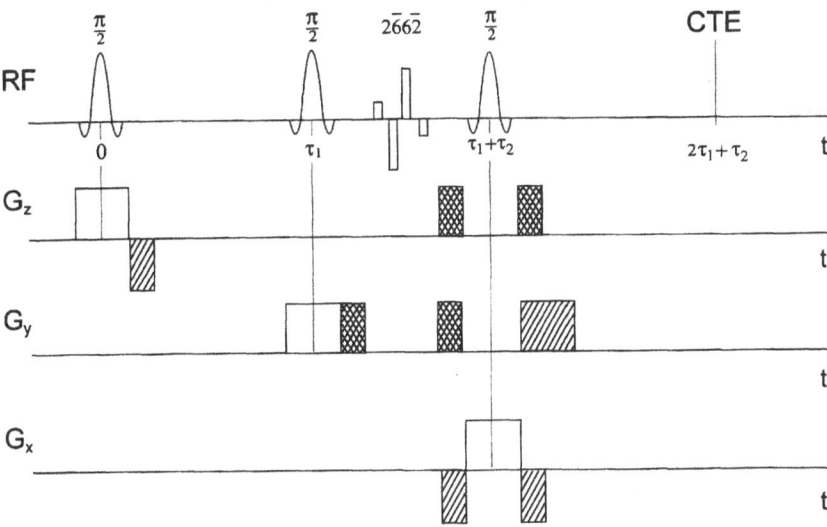

Fig. 37.4. RF and gradient pulse sequence for the homonuclear CYCLPOT-VOSING technique. A coherence transfer echo (CTE) of a preselected group of a coupled spin system is produced with the aid of three slice-selective 90° pulses the phases of which are arbitrary. The frequency window of the binomial pulse ($2\bar{6}6\bar{2}$) is adjusted selectively to the resonance of the spin group to be edited. The hatched gradient pulses serve for the compensation of dephasings by the slice gradients. Homospoil gradient pulses are doubly hatched.

---

[6]Employing considerably more RF pulses another method based on multiple-quantum coherences was suggested in [265] for the same purpose. However, this "split-pathway volume-selective editing" (SP-VOSING) spectroscopy technique requires many more pulses. The principle is to split intermittently the coherence pathways of coupled as well as uncoupled spins. Each pathway leads eventually to single-quantum coherences so that all echo signals appear in superimposed form. Coupled spins follow pathways implying single-quantum and multiple-quantum coherences in parallel. The coherence-transfer echoes into which these pathway terminate are constructively superimposed. This is in contrast to uncoupled spins, which produce stimulated and secondary Hahn echoes cancelling each other.

"cyclic polarization-transfer filtered volume-selective spectral editing (CYCLPOT-VOSING)" technique [268] is to transfer the spin polarization between two spin groups of the molecule in a cyclic way. The frequency region of the group of interest is selectively spoiled after having transferred the polarization to another group. This removes all signals from other compounds potentially concealing the line of interest. After that the spins to be detected are repolarized. This pathway is specifically adjusted for the J coupling and the chemical-shift of the envisaged spin system. Spin systems with other spectral parameters, in particular uncoupled spins, cannot follow the cyclic transfer and, hence, are suppressed.

The selection of the volume element to be probed is again performed with three slice-selective 90° pulses. The pulse interval τ_1 is adjusted according to the J coupling constant of the spin system of interest, i.e., ideally but not stringently $\tau_1 = (2J)^{-1}$. In the middle of the τ_2 interval a "binomial pulse," $(2\bar{6}6\bar{2})$, is applied [398]. It is composed of four broadband pulses with alternating phases and an intensity ratio corresponding to the binomial coefficients. As a whole it acts like a 180° pulse on the entire frequency range with the exception of a narrow frequency window in which the spins are left untouched. In the present application this pulse is adjusted in such a way that the resonance of one group of the coupled spin system is centered in the middle of the frequency window whereas that of the other is subject to the nominal 180° flip angle in full. The consequence is that the coherence evolution experienced by J coupling is refocused.

The phases of the RF pulses of the pulse scheme at Fig. 37.4 are arbitrary because they are applied deliberately to dephased coherences throughout. This is a particularly strong point of the technique although twice a factor of 1/2 in the polarization transfer efficiency is sacrificed relative to the CYCLPOT-VOSING variant using refocused coherences [263].

37.1.2.1
Formal Treatment for A₃X Spin Systems

For the formal treatment using Cartesian product operators (see Table 51.3 on page 487) a sequence of RF pulses with well defined phases is needed, e.g.,

$$\left(\frac{\pi}{2}\right)_x - \tau_1 - \left(\frac{\pi}{2}\right)_x - \frac{\tau_2}{2} - (2\bar{6}6\bar{2})_y - \frac{\tau_2}{2} - \left(\frac{\pi}{2}\right)_x - \tau_1 - (\text{echo})$$

Let us first treat the coherence evolution during this sequence for the set of pulse phases specified in this way, and for well-defined phase angles accumulated in the pulse intervals. The result will then be averaged over all coherence-phase angles assuming equipartition as a consequence of the field gradient pulses illustrated in Fig. 37.4. It is clear that under such circumstances the pulse phases become irrelevant for the average signal.

As in the previous section, a (weakly coupled) A₃X spin system is considered such as formed by lactate $CH_3CH....$ The initial reduced equilibrium density operator is

$$\sigma(0-) = \sum_{k=1}^{4} I_{kz} \tag{37.26}$$

where the subscripts $k = 1, 2, 3$ label the three A spins having angular-frequency offsets $\Omega_A(t)$, and $k = 4$ the X spin which is subject to the offset $\Omega_X(t)$.

The length of the first pulse interval optimal for our purpose can be shown to be again

$$\tau_1 = \frac{1}{2J} \tag{37.27}$$

In order to simplify the formalism we restrict ourselves to this value. That is, τ_1 is immediately replaced by $(2J)^{-1}$ where relevant.

The phase angles accumulated in the free-evolution intervals by the field gradients or inhomogeneities, and - to a minor degree - by chemical shifts are defined by

$$\varphi_{A1} \equiv \int_0^{\tau_1} \Omega_A(t') \, dt' \tag{37.28}$$

$$\varphi_{X1} \equiv \int_0^{\tau_1} \Omega_X(t') \, dt' \tag{37.29}$$

$$\varphi_{A2} \equiv \int_{\tau_1}^{\tau_1+\tau_2} \Omega_A(t') \, dt' \tag{37.30}$$

$$\varphi_{A3} \equiv \int_{\tau_1+\tau_2}^{2\tau_1+\tau_2} \Omega_A(t') \, dt' \tag{37.31}$$

The first RF pulse, $(\pi/2)_x$, produces

$$\sigma(0+) = \sum_{k=1}^{4} I_{ky} \tag{37.32}$$

During the first pulse delay the operator terms of the A_3 spins evolve with respect to the frequency offset Ω_A according to

$$\sum_{k=1}^{3} I_{ky} \xrightarrow{\varphi_{A1} I_{kz}} \sum_{k=1}^{3} \left[I_{ky} \cos \varphi_{A1} + I_{kx} \sin \varphi_{A1} \right] \tag{37.33}$$

and with respect to spin-spin coupling according to

$$I_{kx} \xrightarrow{\pi J \tau_1 2 I_{kz} I_{4z}} I_{kx} \cos(\pi J \tau_1) + 2 I_{ky} I_{4z} \sin(\pi J \tau_1) = 2 I_{ky} I_{4z} \tag{37.34}$$

$$I_{ky} \xrightarrow{\pi J \tau_1 2 I_{kz} I_{4z}} I_{ky} \cos(\pi J \tau_1) - 2 I_{kx} I_{4z} \sin(\pi J \tau_1) = -2 I_{kx} I_{4z} \tag{37.35}$$

The corresponding relations for the X spin are

$$I_{4y} \xrightarrow{\varphi_{X1}I_{4z}} I_{4y}\cos\varphi_{X1} + I_{4x}\sin\varphi_{X1} \tag{37.36}$$

and for $\tau_1 = (2J)^{-1}$

$$I_{4x} \xrightarrow{\pi 2I_{4z}I_{1z}} 2I_{4y}I_{1z} \xrightarrow{\pi 2I_{4z}I_{2z}} -4I_{4x}I_{1z}I_{2z} \xrightarrow{\pi I_{4z}I_{3z}} -8I_{4y}I_{1z}I_{2z}I_{3z} \tag{37.37}$$

$$I_{4y} \xrightarrow{\pi I_{4z}I_{1z}} -2I_{4x}I_{1z} \xrightarrow{\pi I_{4z}I_{2z}} -4I_{4y}I_{1z}I_{2z} \xrightarrow{\pi I_{4z}I_{3z}} 8I_{4x}I_{1z}I_{2z}I_{3z} \tag{37.38}$$

At the end of the first free-evolution interval the density operator thus becomes

$$\sigma(\tau_1-) = \sum_{k=1}^{3}\left[-2I_{kx}I_{4z}\cos\varphi_{A1} + 2I_{ky}I_{4z}\sin\varphi_{A1}\right]$$
$$+ 8I_{4x}I_{1z}I_{2z}I_{3z}\cos\varphi_{X1} - 8I_{4y}I_{1z}I_{2z}I_{3z}\sin\varphi_{X1} \tag{37.39}$$

The second RF pulse, $(\pi/2)_x$, results in

$$\sigma(\tau_1+) = \sum_{k=1}^{3}\left[-2I_{kx}I_{4y}\cos\varphi_{A1} - 2I_{kz}I_{4y}\sin\varphi_{A1}\right]$$
$$+ 8I_{4x}I_{1y}I_{2y}I_{3y}\cos\varphi_{X1} + 8I_{4z}I_{1y}I_{2y}I_{3y}\sin\varphi_{X1} \tag{37.40}$$

The operator terms in this expression represent antiphase single-quantum co-herences ($2I_{kz}I_{4y}$), zero- and double-quantum coherences ($2I_{kx}I_{4y}$), and multiple-quantum coherences of higher order ($8I_{4x}I_{1y}I_{2y}I_{3y}$ and $8I_{4z}I_{1y}I_{2y}I_{3y}$). Merely the antiphase single-quantum coherence term can lead to observable echo signals. The multiple-quantum coherences either are not converted into single-quantum coher-ences by the third 90° pulse, or the single-quantum coherences descending from them are not refocused in the presence of the special gradient pulse sequel of the pulse scheme (Fig. 37.4). The corresponding terms can therefore be omitted from the further treatment.

The binomial pulse selectively imposes a 180° flip angle on the X spins in the middle of the τ_2 interval while the A_3 spins remain untouched. For that reason the coherence evolution with respect to spin-spin coupling will be refocused at the end of the interval, and need not be considered explicitly. The antiphase single-quantum coherence term thus evolves effectively according to

$$\sigma(\tau_1 + \tau_2-) = -\sum_{k=1}^{3}\left[2I_{4y}I_{kz}\sin\varphi_{A1}\cos\varphi_{A2} + 2I_{4x}I_{kz}\sin\varphi_{A1}\sin\varphi_{A2}\right] + \ldots \tag{37.41}$$

The last RF pulse, $(\pi/2)_x$, generates

$$\sigma(\tau_1 + \tau_2+) = \sum_{k=1}^{3}\left[2I_{4z}I_{ky}\sin\varphi_{A1}\cos\varphi_{A2} + 2I_{4x}I_{ky}\sin\varphi_{A1}\sin\varphi_{A2}\right] + \ldots \tag{37.42}$$

The second operator term in the sum representing zero- and double-quantum coherences is unable to produce any detectable signals. It can therefore be left out in the further consideration. The first term stands for antiphase single-quantum coherences of the A_3 spins which evolve in the interval $(\tau_1 + \tau_2) \ldots (2\tau_1 + \tau_2)$ according to

$$2I_{ky}I_{4z} \xrightarrow{\varphi_{A3}I_{kz}} 2I_{ky}I_{4z} \cos\varphi_{A3} + 2I_{kx}I_{4z} \sin\varphi_{A3} \tag{37.43}$$

$$2I_{kx}I_{4z} \xrightarrow{\pi I_{kz}I_{4z}} I_{ky} \tag{37.44}$$

$$2I_{ky}I_{4z} \xrightarrow{\pi I_{kz}I_{4z}} -I_{kx} \tag{37.45}$$

The resulting reduced density operator is

$$\sigma(2\tau_1 + \tau_2) = \sum_{k=1}^{3} \left[-I_{kx} \sin\varphi_{A1} \cos\varphi_{A2} \cos\varphi_{A3} + I_{ky} \sin\varphi_{A1} \cos\varphi_{A2} \sin\varphi_{A3} \right] + \ldots \tag{37.46}$$

The gradient pulses in the pulse scheme (Fig. 37.4) are placed in such a way that

$$\varphi_{A3} = \varphi_{A1} + \varphi_{A2} \tag{37.47}$$

This is the condition for refocusing of the coherences in the form of an echo. Thus

$$\sigma(2\tau_1 + \tau_2) = \sum_{k=1}^{3} \left[I_{kx} \left(\sin^2\varphi_{A1} \sin\varphi_{A2} \cos\varphi_{A2} - \sin\varphi_{A1} \cos\varphi_{A1} \cos^2\varphi_{A2} \right) \right.$$

$$\left. + I_{ky} \left(\sin^2\varphi_{A1} \cos^2\varphi_{A2} + \sin\varphi_{A1} \cos\varphi_{A1} \sin\varphi_{A2} \cos\varphi_{A2} \right) \right] + \ldots \tag{37.48}$$

Owing to the large gradients applied in the pulse intervals an equipartition of the phase angles φ_{A1} and φ_{A2} can be assumed. That is why the ensemble average over all phase angles can be performed as

$$\langle \sigma(2\tau_1 + \tau_2) \rangle = \frac{1}{4\pi^2} \int_0^{2\pi} \int_0^{2\pi} \sigma(2\tau_1 + \tau_2) \, d\varphi_{A1} \, d\varphi_{A2} = \frac{1}{4} \sum_{k=1}^{3} I_{ky} + \ldots \tag{37.49}$$

The amplitude of the coherence-transfer echo signal finally is proportional to the complex transverse rotating-frame magnetization

$$m(2\tau_1 + \tau_2) = \frac{n}{4} \gamma_n \hbar b \, \mathrm{Tr} \left\{ \sum_{k=1}^{3} (I_{kx} + iI_{ky}) \langle \sigma(2\tau_1 + \tau_2) \rangle \right\} = \frac{1}{4} m_A(0+) \tag{37.50}$$

where the symbols have the same meaning as in Eq. 37.25.

As a result, there is no refocused-coherence contribution other than from the spin-system to which the pulse sequence is adjusted, neither from uncoupled spins

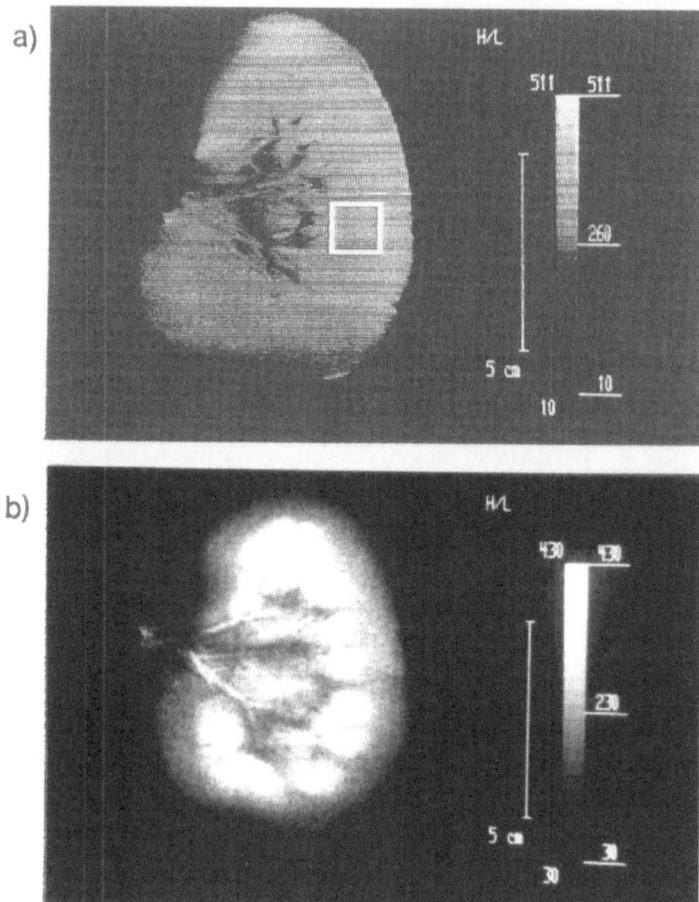

Fig. 37.5. a) Proton images of a porcine kidney perfused with a perfluorocarbon emulsion as a blood substitute. **b)** Fluorine-19 images of the same. The images were recorded using the hydrogen/fluorine retuning tomography (HYFY) technique (see Chap. 41. The white square indicates the volume in which the ^{19}F spectrum and the ^{19}F relaxation curves shown in Fig. 37.6 were measured. The fluorine image parameters are: repetition time 1.2 s; echo time 34 ms; number of transients 8; slice width 1 cm; pixel matrix 256 × 256. The length scale is given by the bars. (Reproduced by permission from ref. [437])

nor from spin systems not taking part in the cyclic polarization transfer and semiselective refocusing scheme outlined above. The amplitude of the edited signal retains 1/4 of the maximal value. In this a factor of 1/2 is due to each of the two 90° pulses applied to completely dephased coherences.

In the above treatment, relaxation has not been taken into account. Provided that $2\tau_1 \gg \tau_2$, losses by transverse relaxation are mainly due to the two τ_1 intervals. It may therefore be favorable to reduce these intervals to minimize the T_2 losses at the expense of the polarization transfer efficiency.

The sensitivity of the phase-insensitive CYCLPOT-VOSING variant is half as high as that of the phase-sensitive method which completely recovers the coherences after the polarization transfer cycle as suggested in [263]. Compared with the DQF-VOSING pulse scheme (Fig. 37.3) the same signal strength can be expected. However, the CYCLPOT variant (Fig. 37.4) is entirely insensitive to the pulse phases, and therefore tends to be more robust under operational measuring conditions. This was demonstrated in [268] where the in vivo detection of lactate in the human brain was reported. On the other hand, owing to its simplicity the DQF-VOSING pulse sequence has the clear advantage of being particularly easy to implement on a tomograph.

37.1.3
Volume-Selective Relaxometry, Diffusometry, and Velocimetry

Volume-selective measuring techniques are of interest when precise and quantitative information is needed from a certain region of interest. It is then a matter of economy first to record a fast image serving the definition of a volume element from which precise information is to be recorded in a second, more demanding step. The VOSY and VOSING experiments described above permit the precise measurement of spectral parameters such as chemical shifts or line intensities even if the compounds to be detected are concealed under otherwise overlapping spectra of other substances. Furthermore, chemical shifts may be indicators of the local pH value [155], the local temperature [207, 241], and the local concentration of shift reagents [171]. The above volume-selection pulse sequences in principle imply intervals which are subject to coherence evolution as well as relaxation at the same time. This suggests the volume and spectral-line selective determination of relaxation times [235, 263, 265].

Because of the superposition of different coherence pathways in the course of the VOSY and VOSING pulse sequences, the use of special precursor pulse trains for the unambiguous determination of relaxation times may be advisable. Corresponding VOSY modifications were reported in [437] in context with hydrogen/fluorine retuning tomography (see Chap. 41). Figure 37.5 shows proton (top) and fluorine (bottom) images of a porcine kidney perfused with a perfluorocarbon emulsion as a blood substitute. The white square in the proton image indicates the volume element to which a VOSY pulse sequence was adjusted. Part of the ^{19}F spectrum recorded from this region is inserted in Fig. 37.6. It shows the CF_3 and CF_2 lines of the perfluorocarbon emulsion. Furthermore the transverse and longitudinal relaxation curves of these lines have been recorded as plotted in the same figure. This demonstrates that volume-selective and spectroscopically resolved relaxometry is well feasible.

Apart from volume-selective laboratory-frame relaxometry a number of methods for the localized determination of the rotating-frame spin-lattice relaxation time $T_{1\rho}$ have been suggested in the literature [411, 412, 418]. As already discussed in Chap. 27, the volume-selective measurement of the frequency dispersion of $T_{1\rho}$ thus becomes possible.

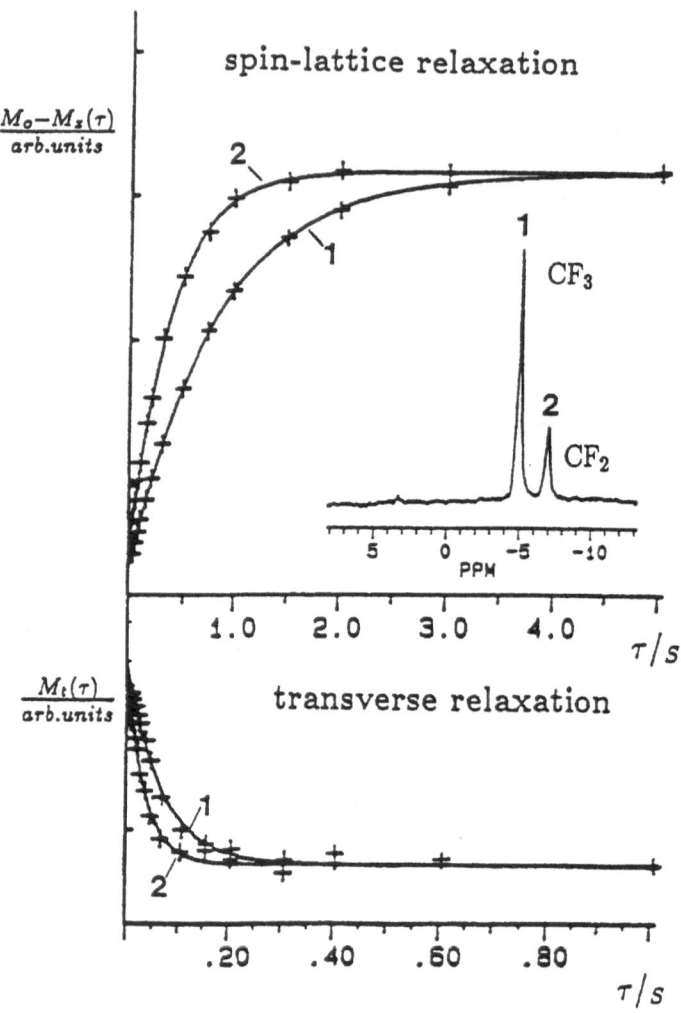

Fig. 37.6. Volume-selective and spectroscopically resolved ^{19}F relaxometry in a porcine kidney perfused with a perfluorocarbon emulsion. The transverse and longitudinal relaxation curves were measured with the indicated resonances in the volume element defined in Fig. 37.5. The experimental parameters are: flux density 4.7 T; room temperature; volume element $(1.5 \text{ cm})^3$. (Reproduced by permission from ref. [437])

Finally, volume-selective spectroscopy can be combined with velocimetry or diffusometry. In [259] "diffusion and incoherent motion weighted volume-selective NMR spectroscopy (DICSY)" was suggested and applied for the localized and spectrocopically resolved determination of diffusivities in fertilized hen eggs. The coherent motion analogue, the "volume-selective measurement of velocity (VOTY)," permits the single-transient determination of the mean velocity vector from a preselected volume element [258].

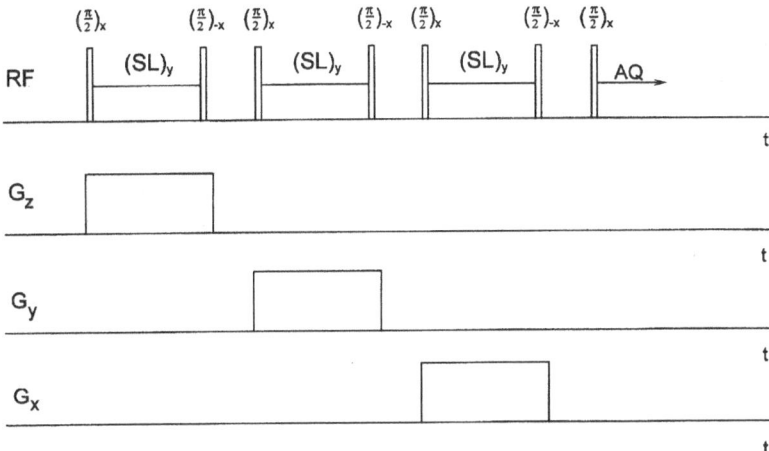

Fig. 37.7. RF and field-gradient pulse scheme for localized spectroscopy (LOSY) based on spin-locking (SL) in the presence of field gradients. The signal detected in the acquisition (AQ) interval originates from the volume element common to the three orthogonal slices chosen by the three spin-lock/restore pulses (compare Fig. 36.1).

37.2
The Homonuclear LOSY Pulse Sequence

The general localization method to be discussed here is based on principle b) mentioned above in the introduction of this chapter. The theory of slice selection by spin-locking pulses in the presence of field gradients is outlined in Chap. 36. Applying one after another three spin-locking/storing pulses of the type shown in Fig. 36.1 in combination with three orthogonal field gradient pulses forms a complete scheme for localized spectroscopy (LOSY) [413]. The pulse sequence is schematically shown in Fig. 37.7.

This method is particularly efficient if the unlocked coherences, i.e., coherences in volume regions outside of the chosen slice, are rapidly attenuated by relaxation and field inhomogeneity effects. Such a situation is given in solid-like materials. An application to polymers was demonstrated in [180].

CHAPTER 38

Cross-Polarization Principles

38.1
Categorization of Cross-Polarization Techniques

We distinguish **"laboratory-frame polarization transfer"** such as the INEPT technique [352] from **"rotating-frame polarization transfer"** or more commonly **"cross-polarization"** [189].[1] Both classes of double-resonance experiments refer to coupled non-equivalent spins. The spin-bearing particles can be homo- or hetero-nuclei or, in principle, even unpaired electrons. To the INEPT procedure we have already referred in the treatment outlined in Sect. 37.1.2. Cross-polarization was first demonstrated by Hartmann and Hahn in 1962. Therefore it may synonymously be called the **"Hartmann/Hahn (HH) experiment."** This chapter is mainly devoted to the theoretical principles on which HH procedures are based.

A typical example are coupled ^{1}H and ^{13}C nuclei. In context with solids, one often speaks of **"abundant"** spins (e.g., ^{1}H with spins I) and **"rare"** spins (e.g., ^{13}C with spins S). Referring to liquids, where the systems of coupled spins tend to be restricted to certain chemical groups, one should better classify the cross-polarization partners according to **"sensitive"** and **"less sensitive"** nuclei, i.e., "large" and "small" magnitudes of the gyromagnetic ratios.

The essence of cross-polarization is to bring two sorts of coupled spins into "contact." This is normally performed by RF irradiating simultaneously to both resonance frequencies. The effective splittings of the spin energies of the two interacting spin species I and S are "matched." That is, they are adjusted to be equal, so that mutual spin-energy exchange is facilitated. The **Hartmann/Hahn matching condition** is

$$\omega_{e,I} = |\gamma_I| B_{e,I} = |\gamma_S| B_{e,S} = \omega_{e,S} \tag{38.1}$$

where $\omega_{e,I}$ and $\omega_{e,S}$ are the effective angular frequencies of spins I and S in their respective rotating frames.[2]

---

[1]The main objective of cross-polarization experiments is to increase the detection sensitivity. In this context, **optical pumping methods** [219] are also of interest. Nuclear spins in noble gases such as ^{129}Xe or ^{3}He can be polarized by irradiating circularly polarized light on hyperfine split atomic transitions. In this way, NMR sensitivity enhancements by five orders of magnitude may be achievable. Furthermore, the polarization of the noble gas nuclei may be transferred to other nuclear species by bubbling the hyperpolarized gas through liquids [361].

[2]The energy splittings referred to need not necessarily be connected by single-quantum transitions, and the quantizing field is not necessarily based on the application of RF. **Double-quantum cross-polarization** and cross-polarization involving one spin species subject to local

Cross-polarization is to be contrasted on the one hand from the **(nuclear) Over-hauser effect** [368] which arises irrespective of any RF matching if one spin species is subject to permanent or pulsed RF irradiation so that non-equilibrium populations are produced. On the other hand, cross-polarization must not be confused with laboratory-frame polarization transfer (e.g., INEPT) as mentioned before. Let us differentiate the three double-resonance/coupled-spin phenomena in the following way.

- **Cross-polarization** (see below):
  a) HH matching permits spin-energy conserving zero-quantum transitions of I and S spins in the doubly-rotating frame;
  b) the spin-state populations are predominantly changed by **coherence evolution in the doubly-rotating frame** if the secular interactions are motionally averaged, and the spectral density at frequency zero is small (liquids);
  c) the spin-state populations are predominantly changed by **cross relaxation in the doubly-rotating frame** if the secular interactions contribute in full, i.e., the spectral density at frequency zero is large (solids).
- **INEPT** (see Sect. 37.1.2):
  a) a sequence of unmatched RF pulses serves the manipulation of spin states;
  b) the spin-state populations are changed as the combined result of **free coherence evolution in the laboratory frame, and spin manipulation by RF pulses.**
- **Overhauser effect:**
  a) unmatched RF pulses or permanent RF irradiation to the coupling partners bring these into a non-equilibrium state; in NOESY experiments [312, 526], the mutual effect is considered at one time;
  b) the populations of the observed spins are changed by **laboratory-frame cross relaxation** with the coupling partners; the full laboratory-frame Zeeman splitting pattern is relevant.

A list of typical cross-polarization/polarization-transfer methods is given in Table 38.1.

The original Hartmann/Hahn experiment follows the pulse scheme displayed in Fig. 38.1a. The I spins are spin locked by an RF pulse phase shifted by $90°$ against the initial excitation pulse. The S spins are then brought into contact with the I spins with the aid of a contact RF pulse, the RF amplitude of which is HH matched to the spin-lock pulse of the I spins. The HH matching condition is given in Eq. 38.1. The RF amplitudes produced by the preferably resonant RF channels are adjusted in such a way that the rotating-frame Zeeman splittings "match" each other and spin-energy conserving flip-flop transitions can take place.

Such transitions arise in solids predominantly by dipolar interaction whereas in liquids, in the motional averaging limit, indirect spin-spin coupling with the

---

fields relevant in **dipolar** and J **ordered states** have been reported in [377, 400, 436], for instance, where the order state is produced either by adiabatic demagnetization in the rotating frame (ADRF) or with the aid of the Jeener/Broekaert pulse sequence.

Table 38.1. Typical cross-polarization / polarization-transfer methods.

frame	spatial selectivity	solids (large spin systems, incoherent transfer)	liquids (small spin systems, coherent transfer)
lab.	none		INEPT [351] DEPT [124]
lab.	localized		spat. res. INEPT, DEPT [16, 477] CYCLPOT-VOSING [266, 267, 268, 270]
rot.	none	CP [189, 385, 386] SLOPT, DOPT [67, 436]	JCP [92, 189, 357], AJCP [91, 93] MOIST [298], RJCP [90]
rot.	localized	SLOPT-LOSY [120, 121, 182, 244] DOPT-LOSY [120, 121, 182]	JCP-LOSY, CYCLCROP-LOSY [292, 293] AJCP-LOSY [283], VJCP [282] adiabatic CYCLCROP-LOSY [283]

coupling constant J prevails. In the latter case one speaks of J cross polarization or, in short, JCP.

38.2
Spatially Selective HH-Matching

Cross-polarization methods were originally designed as tools for the **enhancement of the detection sensitivity**. However, as will be outlined in the following, they can also serve for **localization** and **spectral editing**. In the presence of gradients of the main magnetic field or of the RF field amplitudes, cross-polarization becomes localized, so that a third class of localization methods becomes feasible with heteronuclear spin systems.

For instance, if a spatially constant field gradient G_x is applied along the x axis of the laboratory frame, the resonance frequencies become functions of the coordinate x. Moreover, in AJCP variants, the RF amplitudes are time-dependent so that the effective fields are functions of time t. The effective spin-energy splittings in the rotating frames are then matched in a double-resonance experiment according to the Hartmann/Hahn condition

$$\gamma_I B_{e,I}(x, t) = \gamma_S B_{e,S}(x, t) \tag{38.2}$$

or

$$\omega_{e,I}(x, t) = \omega_{e,S}(x, t) \tag{38.3}$$

where

$$B_{e,I}(x, t) = \sqrt{B_{1,I}^2(t) + (B_I(x) - \omega_I/\gamma_I)^2} \tag{38.4}$$

$$B_{e,S}(x, t) = \sqrt{B_{1,S}^2(t) + (B_S(x) - \omega_S/\gamma_S)^2} \tag{38.5}$$

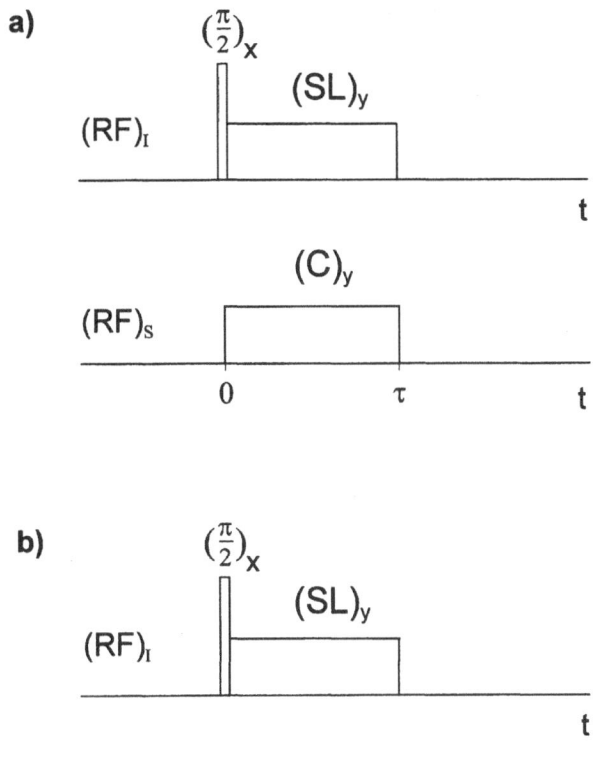

Fig. 38.1. Double-resonance RF pulses for cross polarization (or rotating-frame polarization transfer). The spin-locking pulse (SL), which is 90° phase shifted to the 90° preparation pulse of the I spins, is matched to the contact pulse (C) according to the Hartmann/Hahn condition: **a)** stationary matching - Hartmann/Hahn cross polarization; **b)** adiabatic crossing of the matching point – adiabatic cross-polarization.

38.3
Adiabatic J Cross-Polarization

Cross-polarization can also be conducted by adiabatically crossing the HH matching point (see Fig. 38.1b). In an ordinary JCP experiment the Hartmann/Hahn condition is kept for a certain contact time. With the adiabatic variant, at least one of the RF amplitudes is varied according to a ramp function so that the spin energy levels in the tilted doubly-rotating frame cross one another sufficiently slowly, i.e., "adiabatically." In the pulse sequence shown in Fig. 38.1b, $B_{1,I}$ is kept constant while

$B_{1,S}(t)$ is a ramp function. Equally well a pair of rising and decaying ramp functions can be used for $B_{1,S}(t)$ and $B_{1,I}(t)$, respectively.

The ramp shape of the contact pulse leads to a level crossing of the rotating-frame energies depending on the local magnetic field. This corresponds to transient Hartmann/Hahn matching according to Eq. 38.1. The variation of the Hamiltonian by the ramp is assumed to be "adiabatic." This is a term which is well established in quantum mechanics, and which has already been discussed in Sect. 15.2.1 in context with relaxometry. Let us briefly review it for the present purpose.

38.3.1
Adiabatic Level-Crossing Condition

The criterion for adiabatic level crossing is that no transitions in the spin system take place. The system remains all the time in eigenstates of the *instantaneous* Hamiltonian, the time dependence of which must consequently be sufficiently weak so that no transitions are induced. The condition for adiabatic conduct of a process was derived in Sect. 15.2.1 (see Eq. 15.16):

$$\left| \frac{dE}{d\tau} \right| \ll \frac{\hbar}{\tau^2} \qquad (38.6)$$

where E is the energy splitting varied in the process, and τ is the minimum time needed for a transition of the system according to Heisenberg's energy/time uncertainty relation, $E\tau \geq \hbar$.

In the present case, the variation of the spin Hamiltonian is due to the ramp of $B_{1,S}(t)$ (see Fig. 38.1b), i.e., in the doubly resonantly rotating frame, $E = \hbar\omega_{1,S}$. On the other hand, the lowest energy splitting in this frame is due to J coupling causing an energy splitting $E_J = 2\pi\hbar J$. The minimum transition time follows from the uncertainty relation as $\tau = \hbar/E_J = (2\pi J)^{-1}$. The condition for adiabatic level crossing consequently is

$$\left| \frac{d(\gamma_S B_{1,S})}{dt} \right| \ll (2\pi J)^2 \qquad (38.7)$$

This is the most stringent criterion for adiabatic level crossing in the presence of a field gradient. At locations other than the double-resonance position, we are dealing with off-resonance conditions. The energies must then be considered in the tilted doubly resonant reference frame (in which the z axis is aligned along the effective field). The minimum energies occurring in this frame imply energy contributions from the off-resonance fields and, hence, are larger than that due to J coupling alone. The variation rate of the spin Hamiltonian, i.e., of $B_{1,S}(t)$ in our case, may therefore be greater.

38.3.2
The Principle of Adiabatic J Cross-Polarization

According to Eq. 48.134, the Hamiltonian effective in the rotating doubly-rotating frame is

$$\mathcal{H}'_e = -\hbar\frac{\Delta\omega_1}{2}(I_y - S_y) + \frac{hJ}{2}[I_z S_z + I_x S_x] \tag{38.8}$$

where $\Delta\omega_1 = \omega_{1,I} - \omega_{1,S}$. This expression can be rewritten in the form

$$\mathcal{H}'_e = -\hbar\frac{\Delta\omega_1}{2}(I_y - S_y) + \frac{hJ}{2}\left[I_z S_z + \frac{1}{4}(I^+S^+ + I^-S^- + I^+S^- + I^-S^+)\right] \tag{38.9}$$

Consider now an AX spin system with spins $I = S = 1/2$. We denote the spin up and down states with respect to the y'' axis of the rotating doubly-rotating frame by α and β, respectively. Cross polarization basically implies an interchange of spin up and down states. Therefore, the $m = 0$ states of the spin system, i.e., the product wave functions $|\alpha\beta\rangle$ and $|\beta\alpha\rangle$ are of particular interest. In these kets the first position refers to the I spin, the second to the S spin. The Hamiltonian, Eq. 38.9, superimposes the product states, and the corresponding eigenfunctions are of the type

$$|\psi\rangle = c_1|\alpha\beta\rangle + c_2|\beta\alpha\rangle \tag{38.10}$$

Under adiabatic control of the level-crossing process, the spin system always stays in an eigenstate of the instantaneous Hamiltonian, so that the time-independent Schrödinger equation

$$\mathcal{H}'_e|\psi\rangle = E|\psi\rangle \tag{38.11}$$

is valid. Multiplying Eq. 38.11 on the left with $\langle\alpha\beta|$ and $\langle\beta\alpha|$ produces a set of two linear equations for the coefficients c_1 and c_2 of the wavefunction at Eq. 38.10. This can be written in matrix form:

$$\begin{pmatrix} \langle\alpha\beta|\mathcal{H}'_e|\alpha\beta\rangle - E & \langle\alpha\beta|\mathcal{H}'_e|\beta\alpha\rangle \\ \langle\beta\alpha|\mathcal{H}'_e|\alpha\beta\rangle & \langle\beta\alpha|\mathcal{H}'_e|\beta\alpha\rangle - E \end{pmatrix}\begin{pmatrix} c_1 \\ c_2 \end{pmatrix} = \begin{pmatrix} 0 \\ 0 \end{pmatrix} \tag{38.12}$$

This equation is soluble if the secular determinant vanishes, i.e.,

$$\begin{vmatrix} -\frac{\hbar}{2}\Delta\omega_1 - \frac{hJ}{8} - E & \frac{hJ}{8} \\ \frac{hJ}{8} & \frac{\hbar}{2}\Delta\omega_1 - \frac{hJ}{8} - E \end{vmatrix} = 0 \tag{38.13}$$

The solutions are the eigenvalues

$$E_{2,3} = -\frac{hJ}{8} \mp \frac{1}{2}\left((\hbar\Delta\omega_1)^2 + \frac{(hJ)^2}{16}\right)^{1/2} \tag{38.14}$$

From Eq. 38.12 we obtain the coefficients for the corresponding eigenfunctions. The result is

$$|\psi_2\rangle = c_1|\alpha\beta\rangle - c_2|\beta\alpha\rangle \tag{38.15}$$
$$|\psi_3\rangle = c_1|\beta\alpha\rangle + c_2|\alpha\beta\rangle \tag{38.16}$$

where

$$c_1 = \frac{1}{\sqrt{2}} \left(1 + \frac{\Delta\omega_1}{\sqrt{(\pi J)^2 + (\Delta\omega_1)^2}} \right)^{1/2} \tag{38.17}$$

$$c_2 = \frac{1}{\sqrt{2}} \left(1 - \frac{\Delta\omega_1}{\sqrt{(\pi J)^2 + (\Delta\omega_1)^2}} \right)^{1/2} \tag{38.18}$$

In the pulse scheme (Fig. 38.1b), $\Delta\omega_1$ is varied by the ramp of the $(RF)_S$ amplitudes from a large positive value $\Delta\omega_1 \gg +|\pi J|$ at the beginning to a small negative value $\Delta\omega_1 \ll -|\pi J|$ at the end. This means that the eigenfunctions experience a crossover between an initial and a final limit (see Fig. 38.2)

$$|\psi_2\rangle_{initial} \approx |\alpha\beta\rangle \quad\longrightarrow\quad |\psi_2\rangle_{final} \approx |\beta\alpha\rangle \tag{38.19}$$

$$|\psi_3\rangle_{initial} \approx |\beta\alpha\rangle \quad\longrightarrow\quad |\psi_3\rangle_{final} \approx |\alpha\beta\rangle \tag{38.20}$$

At the Hartmann/Hahn matching instant, i.e., for $\Delta\omega_1 = 0$, we have

$$|\psi_{2,3}\rangle = \frac{1}{\sqrt{2}}(|\alpha\beta\rangle + |\beta\alpha\rangle) \tag{38.21}$$

In the course of adiabatic level crossing, the spin up and down states of the $m = 0$ levels are interchanged, while the populations remain unaffected.

The population differences for the single-quantum I and S spin transitions are consequently also interchanged. The signal achievable with the S spins is thus enhanced by a factor of γ_I/γ_S.

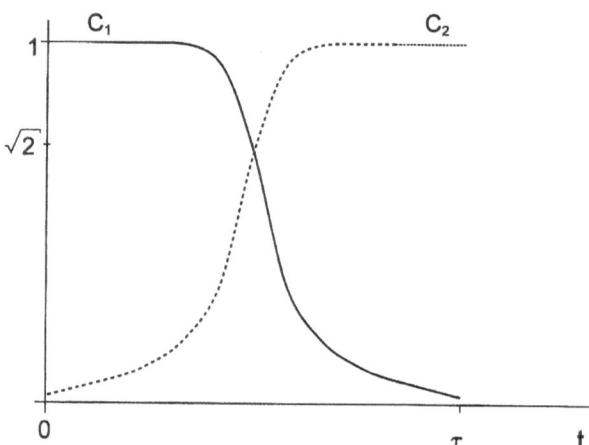

Fig. 38.2. Coefficients of the eigenfunctions for an AX spin-1/2 system (Eqs. 38.15 and 38.16) as functions of the contact time during adiabatic crossing of the Hartmann/Hahn condition in an AJCP experiment with linear time dependence of at least one of the RF amplitudes. The spin-spin coupling constant was assumed to be 150 Hz. $\Delta\omega_1$ was increased from $2\pi \cdot 200$ Hz to $2\pi \cdot 400$ Hz with linear time dependence.

CHAPTER 39

Single-Transition Operator Theory of Cross-Polarization

We consider typical cross-polarization pulse sequences such as those shown in Fig. 38.1. The first objective of this chapter is to treat the evolution of the density operator in the course of such pulse schemes in far-reaching generality. In a second step, we will give the solutions for special cases such as ordinary Hartmann/Hahn cross-polarization, the adiabatic variant, and slice-selective cross-polarization.

39.1
Weakly Coupled AX Spin Systems

The theory to be outlined in the following is a generalized **single-transition operator theory** [119] valid for weakly-coupled two-spin 1/2 systems in the cases

- J as well as dipolar coupling
- liquid as well as solid samples
- stationary HH matching as well as adiabatic crossing of the matching point
- matched as well as mismatched spin-lock and contact pulses
- on- as well as off-resonance RF irradiation
- with as well as without magnetic field gradients

This catalogue is quite general so that all phenomena occurring in context with cross-polarization experiments can be discussed on this basis. The consideration of multiple-spin systems and eventually solids will then be the subject of a later section.

The general strategy is to solve the Liouville/von Neumann equation for the density operator,

$$\frac{\partial \rho}{\partial t} = -\frac{i}{\hbar} \left[\mathcal{H}, \rho \right] \tag{39.1}$$

where relaxation is neglected. The resulting density operator permits the derivation of the expectation values of all observables of interest. The solution is established by transforming to reference frames where the time dependence of the Hamiltonian either disappears or assumes a form easy to deal with. Under suitable circumstances, a "precession-type solution" can be achieved in this way (compare Sect. 48.5.3). As we will see, the first step toward this goal is a transformation to the tilted doubly-rotating frame. The second step will be a transformation referring to the so-called zero-quantum Liouville space.

39.1.1
Laboratory-Frame Hamiltonian

The two-spin system is supposed to consist of a nucleus A (typically ^{1}H) with spin I and a nucleus X (typically ^{13}C) with spin S. The Zeeman Hamiltonian in the laboratory frame is

$$\mathcal{H}_0 = -\hbar\,\underbrace{\gamma_I B_0}_{\omega_{0,I}}\,I_z - \hbar\,\underbrace{\gamma_S B_0}_{\omega_{0,S}}\,S_z \qquad (39.2)$$

The double-irradiation RF fields are assumed to oscillate along the laboratory-frame y axis, so that the term contributing to the total Hamiltonian is

$$\mathcal{H}_{rf} = -2\hbar\,\underbrace{\gamma_I B_{1,I}}_{\omega_{1,I}}\,I_y\,\cos(\omega_I t) - 2\hbar\,\underbrace{\gamma_S B_{1,S}}_{\omega_{1,S}}\,S_y\,\cos(\omega_S t) \qquad (39.3)$$

The carrier frequencies are denoted by ω_I for the I-spin channel, and by ω_S for the S-spin channel. The interaction Hamiltonian effective for the coherence evolution of two weakly J or dipolar coupled spins is (Eq. 51.24)[1]

$$\mathcal{H}_{J,d}^{(ev)} = h c_{IS} I_z S_z \qquad (39.4)$$

where

$$c_{IS} \equiv \begin{cases} J & \text{for } J \text{ coupling} \\ \frac{\mu_0}{4\pi}\frac{\gamma_I\gamma_S\hbar}{2\pi r^3}(1 - 3\cos^2\vartheta) & \text{for dipolar coupling} \end{cases}$$

The total Hamiltonian,

$$\mathcal{H} = \mathcal{H}(r,t) = \mathcal{H}_0(r) + \mathcal{H}_{rf}(r,t) + \mathcal{H}_{J,d}^{(ev)} \qquad (39.5)$$

will be considered as a function of the position in the presence of main- or RF-field gradients, and as a function of time in the case of time varying RF field amplitudes (compare Fig. 38.1).

39.1.2
Transformation to the Tilted Doubly-Rotating Frame

A solution of the Liouville/von Neumann equation can be obtained by transforming the density operator as well as the Hamiltonian to the tilted doubly-rotating frame. That is, we subsequently perform unitary transformations

---

[1]Note that in the high-field limit the coupling between heteronuclei is always "weak." In the homonuclear case, the situation is less clear. In context with J cross-polarization one rather uses the complete coupling Hamiltonian

$$\mathcal{H}_J = hJ(I_x S_x + I_y S_y + I_z S_z)$$

The coherence evolution subject to this "isotropic" Hamiltonian during the cross-polarization RF pulses is referred to as "isotropic mixing" [60, 86, 278].

- with respect to the I spins to the frame rotating with the carrier frequency ω_I
- with respect to the S spins to the frame rotating with the carrier frequency ω_S
- with respect to the I spins to the frame tilted by the angle Θ_I against B_0
- with respect to the S spins to the frame tilted by the angle Θ_S against B_0

The transformation principles are delineated in Sect. 48.7.2. The first two steps leading to the doubly rotating frame results in the secular effective Hamiltonian given in Eq. 48.126. The total transformation equations are

$$\rho_{TR} = T R \rho R^{-1} T^{-1} \tag{39.6}$$

$$\mathcal{H}_{TR} = T R \mathcal{H} R^{-1} T^{-1} + i [T \dot{R}] [T R]^{-1} + i \dot{T} T^{-1} \tag{39.7}$$

where the transformation operators are

$$R = e^{-i\omega_I t I_z} e^{-i\omega_S t S_z} \tag{39.8}$$

$$T = e^{i\Theta_I I_y} e^{i\Theta_S S_y} \tag{39.9}$$

The instantaneous local rotating-frame tilt angles are

$$\Theta_I = \arctan\left\{ \frac{\omega_{1,I}}{\Delta\omega_I} \right\} \tag{39.10}$$

$$\Theta_S = \arctan\left\{ \frac{\omega_{1,S}}{\Delta\omega_S} \right\} \tag{39.11}$$

where $\Delta\omega_I = \omega_{0,I} - \omega_I$ and $\Delta\omega_S = \omega_{0,S} - \omega_S$. Discarding all rapidly oscillating terms (compare Sect. 48.11) we obtain

$$\begin{aligned} \mathcal{H}_{TR} = {} & -\hbar\omega_{e,I} I_z - \hbar\omega_{e,S} S_z + hc_{IS} \cos\Theta_I \cos\Theta_S I_z S_z \\ & + \frac{hc_{IS}}{4} \sin\Theta_I \sin\Theta_S (I^+ S^- + I^- S^+) \end{aligned} \tag{39.12}$$

where the local instantaneous effective frequencies are

$$\omega_{e,I} = \{\omega_{1,I}^2 + \Delta\omega_I^2\}^{1/2} \tag{39.13}$$

$$\omega_{e,S} = \{\omega_{1,S}^2 + \Delta\omega_S^2\}^{1/2} \tag{39.14}$$

39.1.3
Single-Transition Operator Representation

The Hamiltonian at Eq. 39.12 may be rewritten using zero- and double-quantum transition operators defined in Sect. 42.6.7,

$$\begin{aligned} \mathcal{H}_{TR} = {} & -\hbar\left(\omega_{e,I} + \omega_{e,S}\right) I_z^{(2qt)} - \hbar\left(\omega_{e,I} - \omega_{e,S}\right) I_z^{(0qt)} \\ & + hc_{IS} \cos\Theta_I \cos\Theta_S \left[\left(I_z^{(2qt)}\right)^2 - \left(I_z^{(0qt)}\right)^2 \right] \\ & + \frac{hc_{IS}}{2} \sin\Theta_I \sin\Theta_S I_x^{(0qc)} \end{aligned} \tag{39.15}$$

where

$$
\begin{aligned}
I_x^{(0qt)} &= \tfrac{1}{2}(I^+S^- + I^-S^+)\\
I_z^{(0qt)} &= \tfrac{1}{2}(I_z - S_z)\\
I_z^{(2qt)} &= \tfrac{1}{2}(I_z + S_z)
\end{aligned}
\tag{39.16}
$$

The zero-quantum transition operators $I_x^{(0qt)}, I_y^{(0qt)}, I_z^{(0qt)}$ may formally be interpreted as the unit "vectors" of the coordinate axes of the so-called **"zero-quantum Liouville space."** This frame can be transformed in such a way that the Hamiltonian expressed in the new zero-quantum Liouville space coordinates merely depends on z components. The solution of the Liouville/von Neumann equation for such a Hamiltonian can then readily be established in the form of a precession-type expression analogous to that derived in Sect. 48.5.3.

In this sense, the zero-quantum Liouville space vector

$$
-\hbar\left(\omega_{e,I} - \omega_{e,S}\right) I_z^{(0qt)} + \frac{hc_{IS}}{2}\sin\Theta_I\sin\Theta_S\, I_x^{(0qc)}
$$

as a part of the Hamiltonian at Eq. 39.15 is tilted by the angle

$$
\varphi = \arctan\left\{\frac{\pi c_{IS}\sin\Theta_I\sin\Theta_S}{\omega_{e,I} - \omega_{e,S}}\right\}
\tag{39.17}
$$

against the $-I_z^{(0qt)}$ direction. Under Hartmann/Hahn matching conditions,

$$
\omega_{e,I} = \omega_{e,S}
\tag{39.18}
$$

the angle φ takes the value $\pi/2$. Deviations are indicative for mismatch. One therefore speaks of the "mismatch parameter."

This tilt angle suggests a transformation rotating the $I_z^{(0qt)}$ axis about the $I_y^{(0qt)}$ axis into the direction of the above zero-quantum Liouville space vector. The Hamiltonian is then subject to the unitary transformation

$$
\tilde{\mathcal{H}}_{TR} = e^{-i\varphi I_y^{(0qt)}}\,\mathcal{H}_{TR}\,e^{i\varphi I_y^{(0qt)}}
\tag{39.19}
$$

The individual single-transition operators are transformed just as ordinary spin operators (see Table 48.1 on page 442). Moreover employing the first of the equations 42.72, the commutator relations 42.78, and the anti-commutator relation 44.17, we find

$$
\begin{aligned}
e^{-i\varphi I_y^{(0qt)}}\, I_x^{(0qt)}\, e^{i\varphi I_y^{(0qt)}} &= I_x^{(0qt)}\cos\varphi - I_z^{(0qt)}\sin\varphi\\
e^{-i\varphi I_y^{(0qt)}}\, I_z^{(0qt)}\, e^{i\varphi I_y^{(0qt)}} &= I_z^{(0qt)}\cos\varphi + I_x^{(0qt)}\sin\varphi\\
e^{-i\varphi I_y^{(0qt)}}\, I_z^{(2qt)}\, e^{i\varphi I_y^{(0qt)}} &= I_z^{(2qt)}\\
e^{-i\varphi I_y^{(0qt)}}\left(I_z^{(2qt)}\right)^2 e^{i\varphi I_y^{(0qt)}} &= \left(I_z^{(2qt)}\right)^2 = \tfrac{1}{4}\mathcal{E}^{(2qt)}\\
e^{-i\varphi I_y^{(0qt)}}\left(I_z^{(0qt)}\right)^2 e^{i\varphi I_y^{(0qt)}} &= \tfrac{1}{4}\mathcal{E}^{(0qt)}
\end{aligned}
\tag{39.20}
$$

where $\mathcal{E}^{(0qt)}$ and $\mathcal{E}^{(2qt)}$ are the unit operators in the zero-quantum and double-quantum transition Liouville spaces, respectively. The resulting Hamiltonian is

$$\tilde{\mathcal{H}}_{TR} = -\hbar\left(\omega_{e,I} + \omega_{e,S}\right) I_z^{(2qt)} + \frac{hc}{4} \cos\Theta_I \cos\Theta_S \, \mathcal{E}^{(2qt)} - \frac{hc}{4} \cos\Theta_I \cos\Theta_S \, \mathcal{E}^{(0qt)}$$

$$- \underbrace{\hbar\left(\omega_{e,I} - \omega_{e,S}\right)}_{A} \left[I_z^{(0qt)} \cos\varphi + I_x^{(0qt)} \sin\varphi \right]$$

$$- \underbrace{\frac{hc_{IS}}{2} \sin\Theta_I \sin\Theta_S}_{B} \left[I_z^{(0qt)} \sin\varphi - I_x^{(0qt)} \cos\varphi \right] \qquad (39.21)$$

The last two terms of the right-hand side may be expressed by the identity

$$C \equiv -A \left(I_z^{(0qt)} \cos\varphi + I_x^{(0qt)} \sin\varphi \right) + B \left(I_z^{(0qt)} \sin\varphi - I_x^{(0qt)} \cos\varphi \right)$$

$$= -\left\{ \left[A \left(I_z^{(0qt)} \cos\varphi + I_x^{(0qt)} \sin\varphi \right) \right. \right.$$

$$\left. \left. + B \left(I_z^{(0qt)} \sin\varphi - I_x^{(0qt)} \cos\varphi \right) \right]^2 \right\}^{1/2} \qquad (39.22)$$

Carrying out the square and again using the spin-operator relations mentioned above leads to

$$C = -\left\{ A^2 + B^2 \right\}^{1/2} I_z^{(0qt)}$$

$$= -\left\{ \hbar^2 \left(\omega_{e,I} - \omega_{e,S}\right)^2 + \frac{h^2 c_{IS}^2}{4} \sin^2\Theta_I \sin^2\Theta_S \right\}^{1/2} I_z^{(0qt)} \qquad (39.23)$$

The terms with the Liouville-subspace unit operators $\mathcal{E}^{(2qt)}$ and $\mathcal{E}^{(0qt)}$ commute with all other operators and do not contribute to the evolution of the spin coherences. These terms therefore can be discarded. The correspondingly truncated Hamiltonian thus reads

$$\tilde{\mathcal{H}}_{TR} = -\hbar p \, I_z^{(2qt)} - \hbar q I_z^{(0qt)} \qquad (39.24)$$

where the abbreviations

$$p \equiv \omega_{e,I} + \omega_{e,S} \qquad (39.25)$$

$$q \equiv \left\{ \left(\omega_{e,I} - \omega_{e,S}\right)^2 + \pi^2 c_{IS}^2 \sin^2\Theta_I \sin^2\Theta_S \right\}^{1/2} \qquad (39.26)$$

are of the dimension of angular frequencies.

39.1.4
Solution of the Liouville/von Neumann Equation

The Hamiltonian at Eq. 39.24 is of the desired form. It merely depends on longitudinal terms. The transformed Liouville/von Neumann equation reads

$$\frac{\partial \tilde{\rho}_{TR}}{\partial t} = \frac{i}{\hbar} \left[\tilde{\rho}_{TR}, \tilde{\mathcal{H}}_{TR} \right] \qquad (39.27)$$

The formal solution of this equation is

$$\tilde{\rho}_{TR}(t) = \mathcal{T} \exp\left\{ -\frac{i}{\hbar} \int_0^t \tilde{\mathcal{H}}_{TR} \, dt' \right\} \tilde{\rho}_{TR}(0) \, \mathcal{T} \exp\left\{ \frac{i}{\hbar} \int_0^t \tilde{\mathcal{H}}_{TR} \, dt' \right\} \tag{39.28}$$

where $\mathcal{T}$ is the Dyson time-ordering operator.

In the adiabatic variant of cross-polarization, the prefactors of the single-transition operators in Eq. 39.24 are (adiabatically varying) functions of time whereas all operators remain the same during the spin-lock/contact RF pulses. Therefore the integral in Eq. 39.28 can be evaluated explicitly, and the solution of the Liouville/von Neumann equation takes the precession-type form

$$\tilde{\rho}_{TR}(t) = \exp\left\{ i \int_0^t p(t') \, dt' \, I_z^{(2qt)} \right\} \exp\left\{ i \int_0^t q(t') \, dt' \, I_z^{(0qt)} \right\}$$

$$\tilde{\rho}_{TR}(0) \exp\left\{ -i \int_0^t q(t') \, dt' \, I_z^{(0qt)} \right\} \exp\left\{ -i \int_0^t p(t') \, dt' \, I_z^{(2qt)} \right\} \tag{39.29}$$

39.1.4.1
The Preparation RF Pulse

The preparation pulse shown in the schemes at Fig. 38.1 is nominally a 90° pulse applied in the I-spin channel with the B_1 field along the x direction of the I-spin rotating frame. Since the theory is supposed to be valid for all situations including the presence of gradients of the main magnetic field or of the RF field amplitude, the flip angle specified in the figure should not be taken too literally. Rather this is a nominal value thought to be valid at a certain (resonant) position within the sample.

Depending on the local flip angle and on the local resonance offset, the preparation pulse produces I magnetization components transverse to the local field effective in the rotating frame in the presence of the spin-lock pulse. For the sake of simplicity, we assume that these rotating-frame coherences are so short-lived owing to RF field inhomogeneities that they can be neglected on the cross-polarization time scale. Moreover, according to the different gyromagnetic ratios, the initial magnetization of the S spins is assumed to be negligible compared to that of the I spins.[2]

By stripping off all constant terms and factors, the reduced density operator in the tilted doubly-rotating frame just after the preparation pulse at the beginning of the spin-lock/contact-pulse pair is then

$$\sigma_{TR}(0) = I_z \sin \Theta_I(0) = \left(I_z^{(0qt)} + I_z^{(2qt)} \right) \sin \Theta_I(0) \tag{39.30}$$

---

[2]Actually, this situation can be set up in the rigorous sense by selectively saturating the S spins with the aid of an RF pulse comb prior to the cross-polarization pulses.

where I_z represents the longitudinal component along the field effective in the rotating frame. Applying the transformation represented by Eqs. 39.20 gives

$$\tilde{\sigma}_{TR}(0) = \left(I_z^{(0qt)} \cos\varphi(0) + I_x^{(0qt)} \sin\varphi(0) + I_z^{(2qt)} \right) \sin\Theta_I(0) \tag{39.31}$$

39.1.4.2
Evolution During the Spin-Lock/Contact Pulse Pair

Inserting the initial reduced density operator at Eq. 39.31, into Eq. 39.29 in its reduced-density-operator form gives

$$\begin{aligned}
\tilde{\sigma}_{TR}(t) &= \exp\left\{ iP(t)\, I_z^{(2qt)} \right\} \exp\left\{ iQ(t)\, I_z^{(0qt)} \right\} \\
&\quad \tilde{\sigma}_{TR}(0) \exp\left\{ -iQ(t)\, I_z^{(0qt)} \right\} \exp\left\{ -iP(t)\, I_z^{(2qt)} \right\} \\
&= \left\{ I_z^{(2qt)} + I_x^{(0qt)} \sin\varphi(0) \cos Q(t) - I_y^{(0qt)} \sin\varphi(0) \sin Q(t) \right. \\
&\quad \left. + I_z^{(0qt)} \cos\varphi(0) \right\} \sin\Theta_I(0)
\end{aligned} \tag{39.32}$$

where

$$P(t) \equiv \int_0^t p(t')\, dt' \tag{39.33}$$

$$Q(t) \equiv \int_0^t q(t')\, dt' \tag{39.34}$$

The quantities p and q are defined in Eqs. 39.25 and 39.26. Furthermore the commutator relations 42.78 and the zero-quantum and double-quantum unitary-transformation rules (see Table 48.1 on page 442),

$$\begin{aligned}
e^{i\alpha I_z^{(2qt)}} I_z^{(nqt)} e^{-i\alpha I_z^{(2qt)}} &= I_z^{(nqt)} & (n = 0, 2) \\
e^{i\alpha I_z^{(0qt)}} I_x^{(0qt)} e^{-i\alpha I_z^{(0qt)}} &= I_x^{(0qt)} \cos\alpha - I_y^{(0qt)} \sin\alpha & (39.35) \\
e^{i\alpha I_z^{(2qt)}} I_\xi^{(0qt)} e^{-i\alpha I_z^{(2qt)}} &= I_\xi^{(0qt)} & (\xi = x, y, z)
\end{aligned}$$

have been taken into account in the derivation of this expression.

To evaluate the transferred polarization, the density operator in the tilted doubly-rotating frame will be needed. The transformation reverse to that applied in Eq. 39.19, i.e.,

$$\tilde{\sigma}_{TR}(t) = e^{i\varphi(t)\, I_y^{(0qt)}} \tilde{\sigma}_{TR}(t)\, e^{-i\varphi(t)\, I_y^{(0qt)}} \tag{39.36}$$

gives

$$\sigma_{TR}(t) = \sin \Theta_I(0) \left\{ I_z^{(2qt)} \right.$$
$$+ \left[-\cos \varphi(0) \sin \varphi(t) + \sin \varphi(0) \cos \varphi(t) \cos Q(t) \right] I_x^{(0qt)}$$
$$- \sin \varphi(0) \sin Q(t) I_y^{(0qt)}$$
$$\left. + \left[\cos \varphi(0) \cos \varphi(t) + \sin \varphi(0) \sin \varphi(t) \cos Q(t) \right] I_z^{(0qt)} \right\} \quad (39.37)$$

The density operators obviously evolve only in the zero-quantum Liouville subspace. The double-quantum Liouville space component $\sin \Theta_I(0)\, I_z^{(2qt)}$ is stationary and commutes with all other operators. In other words, the sum of the spin-locked magnetizations of the I and S spins represented by the operator sum $I_z + S_z$ in the tilted doubly-rotating frame is a constant of time. However, the distribution of the polarization among the two spin species is determined by the zero-quantum Liouville space component

$$\sin \Theta_I(0) \sin \varphi(0) \sin \varphi(t) \cos Q(t) I_z^{(0qt)}$$

which is a time-dependent function.

If the excitation of the I spins is performed by a "hard" 90° RF pulse and an initially "hard" spin-lock pulse, so that the flip angle is everywhere the same within the sample, the initial tilt angle $\Theta_I(0)$ in Eq. 39.37 is also uniformly equal to 90°.

39.1.5
The Cross-Polarized Magnetizations

The transverse-magnetization amplitudes at time $t = \tau$ just after the cross-polarization pulses (see Fig. 38.1) are proportional to the ensemble expectation values of the transverse spin components in the doubly-rotating frame,

$$\langle I_\alpha \rangle (\tau) = \mathrm{Tr}\{I_\alpha \sigma_R(\tau)\} \quad (39.38)$$
$$\langle S_\alpha \rangle (\tau) = \mathrm{Tr}\{S_\alpha \sigma_R(\tau)\} \quad (39.39)$$

where $\alpha = x, y$, and $\sigma_R(\tau)$ is the reduced doubly-rotating frame density operator at time τ. The traces are invariant against transformations so that the arguments may be transformed to the tilted doubly-rotating frame,

$$\langle I_\alpha \rangle (\tau) = \mathrm{Tr}\left\{ T(\tau)\, I_\alpha\, T^{-1}(\tau)\, \sigma_{TR}(\tau) \right\} \quad (39.40)$$
$$\langle S_\alpha \rangle (\tau) = \mathrm{Tr}\left\{ T(\tau)\, S_\alpha\, T^{-1}(\tau)\, \sigma_{TR}(\tau) \right\} \quad (39.41)$$

Using Eq. 39.9 and Table 48.1 on page 442, we obtain

$$\langle I_x \rangle (\tau) = \mathrm{Tr}\{ [I_x \cos \Theta_I(\tau) + I_z \sin \Theta_I(\tau)]\, \sigma_{TR}(\tau) \} \quad (39.42)$$
$$\langle S_x \rangle (\tau) = \mathrm{Tr}\{ [S_x \cos \Theta_S(\tau) + S_z \sin \Theta_S(\tau)]\, \sigma_{TR}(\tau) \} \quad (39.43)$$
$$\langle I_y \rangle (\tau) = \mathrm{Tr}\left\{ I_y\, \sigma_{TR}(\tau) \right\} \quad (39.44)$$
$$\langle S_y \rangle (\tau) = \mathrm{Tr}\left\{ S_y\, \sigma_{TR}(\tau) \right\} \quad (39.45)$$

Note that the spin operator components in these equations refer to the tilted doubly-rotating frame. The complex transverse rotating-frame magnetizations relative to the equilibrium magnetizations are (compare Eqs. 47.7 and 47.24)

$$
\begin{aligned}
a_I(\tau) &\equiv \frac{m_I(\tau)}{M_{I,0}} \\
&= \frac{n_I \gamma_I \hbar b_I \left(\langle I_x \rangle (\tau) + i \langle I_y \rangle (\tau) \right)}{n_I \gamma_I \hbar b_I \langle I_z \rangle} \\
&= \frac{1}{2} \sin \Theta_I (\tau) \sin \Theta_I (0) \left[1 + \cos \varphi (0) \cos \varphi (\tau) \right. \\
&\quad \left. + \sin \varphi (0) \sin \varphi (\tau) \cos Q (\tau) \right]
\end{aligned}
\tag{39.46}
$$

$$
\begin{aligned}
a_S(\tau) &\equiv \frac{m_S(\tau)}{M_{S,0}} \\
&= \frac{n_S \gamma_S \hbar b_I \left(\langle S_x \rangle (\tau) + i \langle S_y \rangle (\tau) \right)}{n_S \gamma_S \hbar b_S \langle S_z \rangle} \\
&= \frac{1}{2} \frac{\gamma_I}{\gamma_S} \sin \Theta_S (\tau) \sin \Theta_I (0) \left[1 - \cos \varphi (0) \cos \varphi (\tau) \right. \\
&\quad \left. - \sin \varphi (0) \sin \varphi (\tau) \cos Q (\tau) \right]
\end{aligned}
\tag{39.47}
$$

The quantities n_I and n_S are the spin number densities of the I and S spins, respectively. The respective constants b_I and b_S are defined in Eq. 47.27. Note that a_S depends on the ratio $b_I : b_S$, that is, on the quotient $\gamma_I : \gamma_S$. This factor limits the maximum polarization enhancement of the S spins. For $^1H/^{13}C$ pairs, this amounts to a factor of about four.

As long as the experiment is carried out adiabatically and relaxation is negligible, Eqs. 39.46 and 39.47 are generally valid for J or dipolar-coupled two-spin-1/2 systems in liquids or solids for any temporal or spatial dependence of the experimental parameters. In the following we will discuss a number of special cases often encountered in double-resonance NMR.

39.1.6
HH-Matched Resonant Cross-Polarization

Cross-polarization of two-spin systems is normally considered for liquids under motional averaging conditions, so that it refers to J coupling. In the following we therefore equate $c_{IS} = J$. Resonant J cross-polarization (JCP) can be performed with the pulse sequence shown in Fig. 38.1a. The spin-lock and contact pulses are both resonant and have constant amplitudes. The B_0 and B_1 fields are assumed to be homogeneous in the sample. Since the RF carrier frequencies are resonant to the two spin species, all offsets are zero,

$$
\Delta \omega_I = 0 = \Delta \omega_S
\tag{39.48}
$$

In this case, the HH-matching condition is

$$\omega_{e,I} = \omega_{1,I} = \omega_{1,S} = \omega_{e,S} \tag{39.49}$$

The rotating-frame tilt angles Θ_I and Θ_S (Eqs. 39.10 and 39.11) hence take the values 90° independently of time. The same applies to the zero-quantum Liouville frame tilt angle φ (Eq. 39.17), so that

$$\Theta_I = \Theta_S = \varphi = \frac{\pi}{2} \tag{39.50}$$

The quantity q (Eq. 39.26) thus becomes $q = \pi J$. For $t = \tau$, the integral at Eq. 39.34 results in

$$Q(\tau) = \pi J \tau \tag{39.51}$$

Equations 39.46 and 39.47 then read

$$
\begin{array}{rcl}
a_I(\tau) & = & \cos^2(\pi J \tau / 2) \\[2ex]
a_S(\tau) & = & (\gamma_I / \gamma_S)\, \sin^2(\pi J \tau / 2)
\end{array}
\tag{39.52}
$$

The polarization oscillates between the I and the S spins. The first maximum of the polarization transfer from the I to the S spins is reached after the period

$$\tau_{op} = \frac{1}{J} \tag{39.53}$$

The relative magnetizations are then

$$a_I(\tau_{op}) = 0 \tag{39.54}$$

$$a_S(\tau) = \frac{\gamma_I}{\gamma_S} \tag{39.55}$$

That is, the polarization is completely transferred to the S spins. It is then a factor of γ_I / γ_S higher than in equilibrium. After a period twice as long, the total polarization is allocated to the I spins, and so on. In reality, the oscillatory polarization transfer between the two coupled spins is attenuated by relaxation processes, field inhomogeneities, and interactions to more distant spins, of course. It is also clear that any mismatch causing $\omega_{e,I} \neq \omega_{e,S}$ leads to angles $\varphi \neq \pi/2$ and, hence, to incomplete, but still oscillatory polarization transfer (see Fig. 39.1). The maximum cross-polarization efficiency is achieved for perfect HH matching (see Fig. 39.2).

39.1.7
HH-Mismatch Losses of Resonant Cross-Polarization (VJCP)

The parameters ω_{1I}, ω_{1S}, Θ_I, Θ_S are again assumed to be independent of time during the contact, and the external magnetic field is considered to be homogeneous. However, the RF amplitudes are now assumed to be a function of the position. This

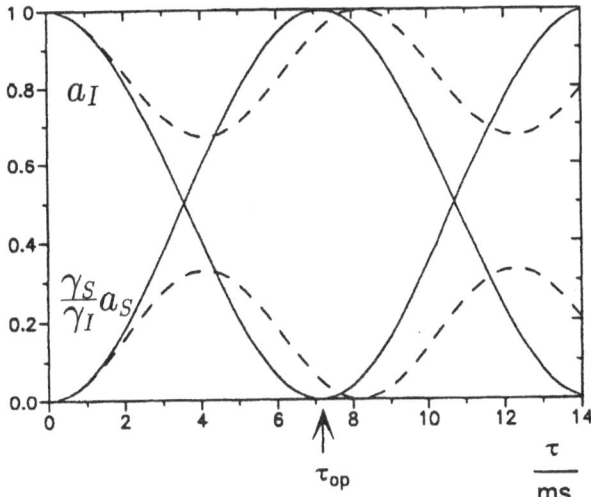

Fig. 39.1. Normalized relative magnetization amplitudes a_I and $\gamma_S a_S/\gamma_I$ of an AX-spin 1/2 system as function of the contact time τ assuming resonant irradiation ($\Theta_I = \Theta_S = \pi/2$; see Eqs. 39.10 and 39.11) under stationary conditions. The coupling constant was assumed to be $J = 140$ Hz. The contact time for optimal cross-polarization of the S spins is $\tau_{op} = 1/J$. The solid lines refer to matched spin lock/contact pulses ($\varphi = \pi/2$), the dashed curves were calculated for $\varphi = 0.6$ on the basis of Eqs. 39.46 and 39.47. This corresponds to a HH mismatch of $|\nu_{1,I} - \nu_{1,S}| = 100$ Hz. (Courtesy of C. Kunze)

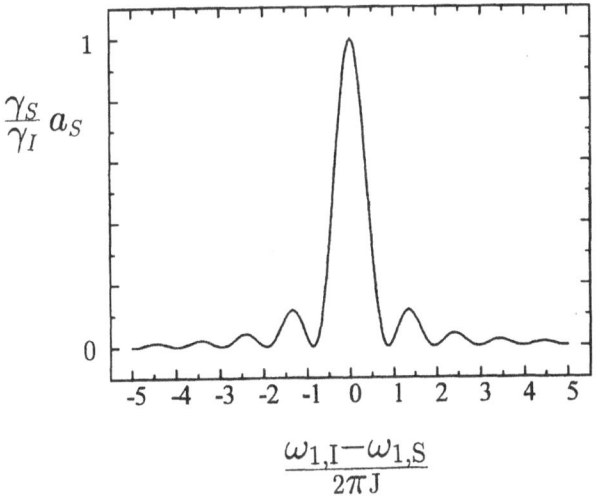

Fig. 39.2. Normalized relative magnetization amplitude $\gamma_S a_S/\gamma_I$ of an AX spin-1/2 system as a function of the HH mismatch $\omega_{1,I} - \omega_{1,S}$ (see Eq. 39.18) for the optimal contact time τ_{op} and resonant irradiation under stationary conditions. The curve was calculated with the aid of Eq. 39.47. (Courtesy of H. Köstler)

situation arises with the combination of a surface RF coil with a bird-cage resonator, for instance. In [282] it was suggested to employ HH-mismatch losses outside the region of interest for volume-selective J cross-polarization (VJCP).

The RF amplitudes are then functions of the position, i.e., $B_{1I} = B_{1I}(r)$, $B_{1S} = B_{1S}(r)$, $\omega_{1I} = \omega_{1I}(r)$, $\omega_{1S} = \omega_{1S}(r)$. The Hartmann/Hahn condition, Eq. 39.18, is fulfilled only locally, and a certain region is selectively addressed by cross polarization. The consequences of HH mismatch are illustrated in Figs. 39.1 and 39.2. Losses on this basis can be reduced with the aid of the MOIST variant of JCP [298] or with the RJCP modification [90], for instance.

39.1.8
Off-Resonance Losses of HH Cross-Polarization (JCP-LOSY)

The external magnetic field is assumed to be superimposed by a gradient whereas B_{1I}, B_{1S}, and, hence, ω_{1I} and ω_{1S} are homogeneously distributed. During the spin contact, all parameters are independent of time. The cross-polarization process can then take place only in a certain slice around the resonance position. Since the tilt angles Θ_I (Eq. 39.10), Θ_S (Eq. 39.11), and φ (Eq. 39.17) are now functions of the position, all parameters in Eq. 39.47 are also spatial functions. Figure 39.3 shows the profile of the relative normalized magnetization amplitude of the S spins as a function of the magnetic flux density as calculated on the basis of Eq. 39.47 for a constant field gradient.

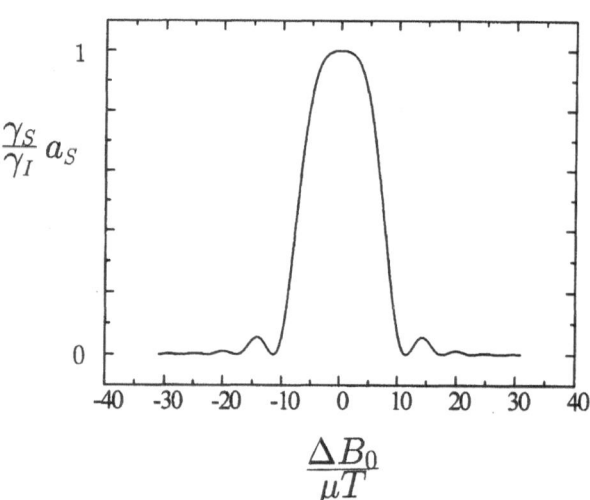

Fig. 39.3. Normalized relative ^{13}C magnetization amplitude $\gamma_S a_S / \gamma_I$ of a ^{1}H^{13}C system as a function of the magnetic-field offset ΔB_0 for JCP. Such a system is formed by formic acid ($J=220$ Hz), for instance. At the positions where $\Delta B_0 = 0$, both channels are resonant. The contact time was assumed to be equal τ_{op}. The curve was calculated with the aid of Eq. 39.47. (Courtesy of H. Köstler)

Off-resonance losses can be used for suppressing signals from outside a region of interest. A corresponding method for J cross-polarization localized spectroscopy (JCP-LOSY) was suggested in [292].

39.1.9
Resonant Adiabatic J Cross-Polarization (AJCP)

In the pulse scheme at Fig. 38.1b, the RF amplitude of the S spin channel is assumed to vary linearly with time so that the HH matching point is crossed adiabatically (see Eq. 38.7). The two RF channels are assumed to be resonant, i.e.,

$$\Delta\omega_I = 0 = \Delta\omega_S \tag{39.56}$$

The effective frequencies are given by the RF amplitudes,

$$\omega_{e,I} = \omega_{1,I} \tag{39.57}$$

$$\omega_{e,S}(t) = \omega_{1,S}(t)$$
$$= \omega_{1,S}^{(0)}(1 + \kappa t) \tag{39.58}$$

where $\omega_{1,S}^{(0)} < \omega_{1,I}$ and κ are constant parameters defining the $\omega_{1,S}$ ramp (Fig. 38.1b). The rotating-frame tilt angles Θ_I (Eq. 39.10) and Θ_S (Eq. 39.11) take the values 90° independently of time. The zero-quantum Liouville frame tilt angle φ (Eq. 39.17) adopts the time-dependent value

$$\varphi(t) = \arctan\left\{\frac{\pi J \sin\Theta_I \sin\Theta_S}{\omega_{1,I} - \omega_{1,S}(t)}\right\} \tag{39.59}$$

The quantity q (Eq. 39.26) is also a function of time:

$$q(t) = \left\{\left[\omega_{1,I} - \omega_{1,S}(t)\right]^2 + \pi^2 J^2\right\}^{1/2} \tag{39.60}$$

The integral at Eq. 39.34 is of the type

$$Q(\tau) = \int_0^\tau q(t')\,dt' = \int_0^\tau \left(\kappa^2 t'^2 + c_1 t' + c_2\right)^{1/2} dt' \tag{39.61}$$

where

$$c_1 \equiv 2\kappa(\omega_{1,S}^{(0)2} - \omega_{1,I}\,\omega_{1,S}^{(0)})$$
$$c_2 \equiv (\omega_{1,I} - \omega_{1,S}^{(0)})^2 + \pi^2 J^2$$

There are different analytical solutions depending on the values of the coefficients relative to each other. At the end of the contact pulse the complex transverse rotating-frame S magnetization relative to the equilibrium magnetizations (Eq. 39.46) is

$$a_S(\tau) = \frac{1}{2}\frac{\gamma_I}{\gamma_S}\left[1 - \cos\varphi(0)\cos\varphi(\tau) - \sin\varphi(0)\sin\varphi(\tau)\cos Q(\tau)\right] \tag{39.62}$$

39.1.10
Adiabatic *J* Cross Polarization Localized Spectroscopy (AJCP-LOSY)

If the external magnetic field is superimposed by a constant gradient G, the RF channels are resonant only at positions were $r \cdot G = 0$. Far away from this region the HH matching point is not reached so that the cross-polarized signals are restricted to a localized volume. A corresponding technique, adiabatic J cross-polarization localized spectroscopy (AJCP-LOSY) was suggested in [283].

The RF amplitudes B_{1I} and B_{1S}, i.e., ω_{1I} and ω_{1S}, may be assumed to be homogeneously distributed in space at any time. During the spin contact, the RF parameters depend, however, in an adiabatic way on time. At off-resonance positions the cross-polarization parameters in Eq. 39.47 become functions of the position and the contact time. This in particular applies to the tilt angles $\Theta_I, \Theta_S, \varphi$ (Eqs. 39.10, 39.11, 39.17) and the integral Q (Eq. 39.34). Figure 39.4 shows the numerically evaluated profile resulting in the presence of a constant gradient.

39.2
From Two-Spin to Multi-Spin Systems and Solids

The above single-transition operator treatment refers to AX two-spin 1/2 systems. A remarkable outcome is the fact that cross-polarization occurs in a completely coherent way. The polarization oscillates between the two spin species without any losses apart from relaxation (see Fig. 39.1). This applies to liquids where J coupling dominates owing to motional averaging of the dipolar-coupling constant. It

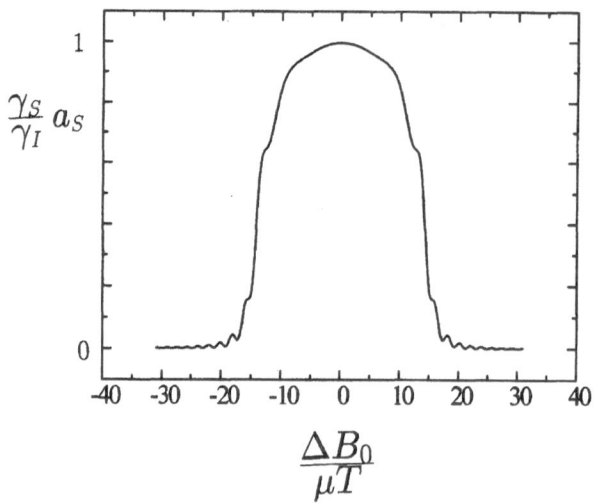

Fig. 39.4. Normalized relative ^{13}C magnetization amplitude $\gamma_S a_S/\gamma_I$ of a $^1H^{13}C$ system as a function of the magnetic-field offset ΔB_0 for AJCP. An example of such a system is formic acid ($J=220$ Hz), for instance. At the positions where $\Delta B_0 = 0$, both channels are resonant. The curve was calculated with the aid of Eq. 39.47. (Courtesy of H. Köstler)

would also hold true in the somewhat artificial situation of isolated dipolar-coupled two-spin systems in a solid single-crystal. With the exception of the magic-angle orientation, cross-polarization is then governed by dipolar coupling leading in this particular case to exactly the same formalism.

The situation changes with larger spin systems with more than two interacting spins, though J coupled AX_2 spin systems still show the completely periodical transfer of polarization among the A and X spins [85, 92, 357]. With increasing number of spins participating in the cross-polarization process the transfer adopts a more and more incoherent character. That is, the polarization is no longer oscillating periodically between the two spin species. A chaotic element enters into the phenomenon so that the process is directed toward the more likely distribution of polarizations. The evolution of many-spin coherences is subject to quasi-random statistics even without participation of the lattice degrees of freedom.

Under motional-averaging conditions, the number of interacting spins increases with the size of the molecule. In the opposite case, i.e., in solid-like materials, dipolar coupling links practically all spins present in the sample. There are two major consequences as concerns cross-polarization. First, the coherences in the doubly-rotating frame decay as a consequence of incoherent influences by the many coupling partners. Any oscillatory transfer patterns back and forth vanish. Second, cross-relaxation comes into play because the spin-interaction spectral density at frequency zero is particularly large in solids. Relaxation means spin transitions and coherence evolutions by randomly fluctuating perturbations, i.e., an incoherent process by nature.

Randomized coherences permit the establishment of a spin temperature in the thermodynamic sense. This means that the spin system adopts an equilibrium state much faster than spin-lattice relaxation. The spin coherence phases are random, corresponding to a maximum of entropy. A spin temperature deviating from the lattice temperature may be defined. Cross-relaxation in solids may then be considered as an equilibration of the spin temperatures of the two spin species after bringing them into contact [114, 386].

CHAPTER 40

Proton-Detected Localized ^{13}C NMR

The homonuclear volume-selective spectral-editing principles outlined in Chap. 37 are also suited very well for heteronuclear applications in modified form. Of particular interest are techniques exploiting the slice-selectivity of cross-polarization pulses as described in Chap. 38. A list of cross-polarization and laboratory-frame polarization transfer methods is summarized in Table 38.1 on page 362.

The prominent nuclear species to which such techniques can be successfully applied are ^{1}H and ^{13}C which are ubiquitous in the organic world. Other nuclides of interest such as ^{31}P or ^{15}N either are rarely associated to other nuclei with high gyromagnetic ratio, or occur in rather low natural abundance. Quadrupole nuclei such as ^{2}H or ^{14}N are suitable for **slice-selective double-quantum coherence transfer** [121] but may be subject to unfavorably short relaxation times (see Sect. 12.2.2).

As already outlined in Sect. 38.2, cross-polarization becomes volume-selective in the presence of gradients of the main field or if RF fields with different gradients are superimposed. In principle, this volume-selective cross-polarization is feasible with coupled spins in liquids as well as in solids, i.e., for J and dipolar coupling. The implications of the size of the spin system are discussed in Sect. 39.2. Treatments for dipolar-coupled spins in solids can be found in [120, 182, 244].

The most promising realm of such techniques is that of small ^{13}C$_n$^{1}H$_m$ spin systems in liquids or liquid-like materials such as tissue or elastomers. The methodological objectives are

- to **edit** the spectral lines of the chemical groups of interest,
- to **localize** the signals relative to an ordinary NMR image, and
- to **enhance** the detection sensitivity

As will be depicted below, there is a wealth of means to achieve all three goals at one time.

Sensitivity enhancement refers to either laboratory-frame polarization transfer (compare the treatment of INEPT based pulse sequences in Sect. 37.1.2) or rotating-frame cross-polarization (compare the JCP and AJCP theory delineated in Chap. 38) in one-way or cyclic versions. In the case of cyclic polarization pathways ^{1}H$\rightarrow$^{13}C$\rightarrow$^{1}H, the maximum gyromagnetic-ratio based enhancement relative to direct ^{13}C detection[1] theoretically is

---

[1]The nuclear Overhauser effect associated with proton decoupling is not considered here.

$$\left(\tfrac{\gamma_H}{\gamma_C}\right)^3 \approx 64 \qquad (40.1)$$

The exponent 3 indicates that the quotient of the gyromagnetic ratios relates the dipole moments, the polarizations, and the detection frequencies in the same way. Furthermore, the signal-accumulation rate limiting spin-lattice relaxation time tends to be much shorter on the proton side. Polarization transfer techniques stipulate the existence of hydrogens directly bound to a carbon atom, of course. If there are more than one bound hydrogen atom, a further enhancement is expected. In principle, the ^{13}C containing chemical groups can be detected with the same sensitivity as would be possible by conventional proton NMR in the absence of all other proton signals.

Before this background, a number of heteronuclear volume selective spectral editing techniques have been suggested and applied. These imply heteronuclear cyclic polarization transfer volume-selective editing spectroscopy (heteronuclear CYCLPOT-VOSING spectroscopy) [266, 270], i.e., INEPT-based pulse sequences, and heteronuclear multiple-quantum filtered cyclic polarization transfer spectroscopy (heteronuclear MQF-VOSING spectroscopy) [266] in analogy to homonuclear double-quantum filter spectroscopy discussed in Sect. 7.3.1.

Cross-polarization, i.e., rotating-frame polarization transfer, also offers numerous ways for heteronuclear spectral editing, localization, and signal enhancement. The Hartmann/Hahn matching is restricted to a certain volume if different spatial distributions of the RF fields are superimposed (Sects. 38.2 and 39.1.7). This permits volume-selective J cross-polarization (VJCP) without any gradient of the main field. For example, it is possible to combine a ^1H bird-cage resonator with a ^{13}C surface coil. This volume-selective cross-polarization can also be applied in a cyclic way, and can be combined with spectral editing [282].

Another means for rendering cross-polarization volume selective is to apply a gradient of the main field during the contact pulses (compare Sect. 39.1.8). A method of this sort is cyclic cross-polarization localized spectroscopy (CYCLCROP-LOSY) [292, 293] which is based on the MOIST variant of J cross-polarization. The adiabatic cross-polarization version of it is called adiabatic CYCLCROP-LOSY [283]. In the following, we will restrict ourselves to the discussion of the MQF-VOSING and the CYCLCROP-LOSY techniques.

40.1
Heteronuclear MQF-VOSING Spectroscopy

The pulse sequence in Fig. 40.1 consists of only four RF pulses. Any three of these are slice-selective so that the final signal originates from a well-defined volume element (see Fig. 37.1). In contrast to the homonuclear VOSY sequence (Sect. 37.1), the slice selection pulses are applied in different RF channels. The editing principle is to spoil all ^1H coherences, then to produce the heteronuclear multiple-quantum coherences, which eventually are converted into new ^1H single-quantum coherences. "New" means that there is no pure ^1H coherence through-pathway to the signal to be detected.

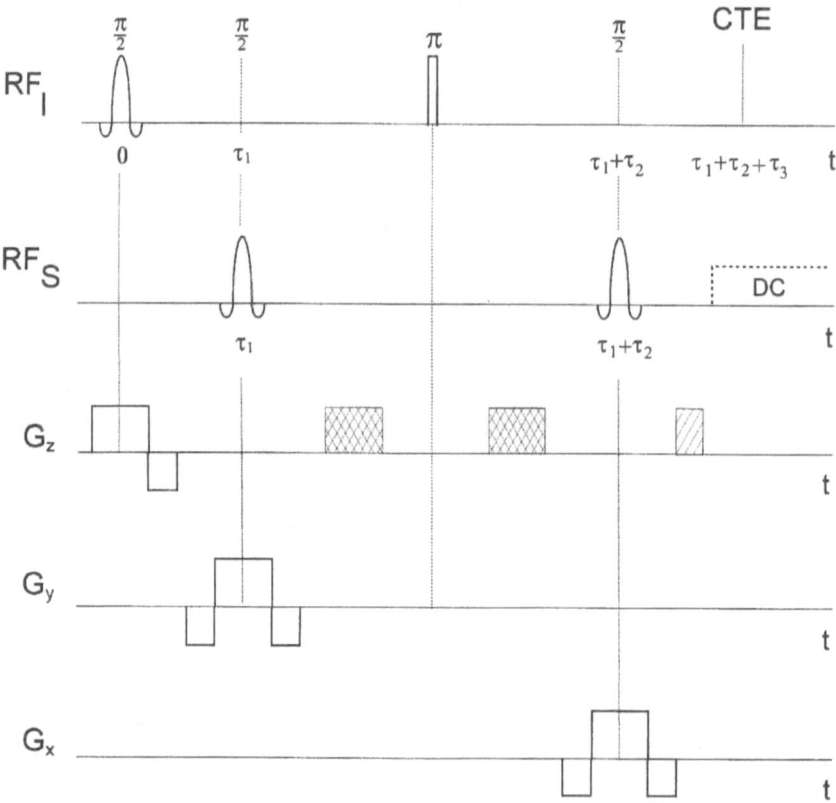

Fig. 40.1. Radio frequency and field gradient pulse sequence for heteronuclear multiple-quantum filtering volume-selective editing (MQF-VOSING) spectroscopy [266]. This method produces volume-selective I spin coherence-transfer echoes (CTE) from coupled spins I (e.g., ^{1}H) and S (e.g., ^{13}C). The I spin signal is preferably recorded under decoupling (DC) of the S spins. The sequence consists of four RF pulses in the two channels. Any three of these pulses are supposed to be slice-selective. The hatched gradient pulse serves for compensation of the phase shifts originating from multiple-quantum coherences in the τ_2 interval in the presence of the doubly hatched gradient pulses.

The first pulse interval τ_1 shown in the scheme (Fig. 40.1) is optimized for the conversion from single-quantum coherences in the first to multiple-quantum coherences in the second interval. With an AX spin system, for instance, the condition to be met is (see Eq. 7.48)

$$\tau_1 = \frac{2k+1}{2J} \qquad (k = 0, 1, 2, \ldots) \qquad (40.2)$$

where J is the heteronuclear coupling constant. From the heteronuclear coherences produced by the second pulse, merely those referring to zero- and double-quantum transitions are of interest. Single-quantum coherences, in particular those of uncoupled spins, cannot lead to echo signals at the time when the final coherence-transfer echo is recorded.

The 180° pulse in the middle of the τ_2 interval is selectively applied to the proton spins. Therefore, the zero- and double-quantum coherences are interchanged at this instant. A heteronuclear zero-quantum coherence continues to evolve as double-quantum coherence after the 180° pulse, and vice versa. Another effect of the 180° pulse is that the proton part of the multiple-quantum coherences is phase-inverted so that the doubly hatched gradient pulse pair (Fig. 40.1) has no impact on it. This is in contrast to the ^{13}C part which is dephased in full.

The fourth pulse which is applied in the ^{13}C channel converts the multiple-quantum coherences again. From the resulting coherences merely the single-quantum proton coherences are able to contribute to the echo signal after the refocusing time τ_3. This period and the gradient pulse therein (drawn hatched in Fig. 40.1) must be adjusted in such a way that the phase shifts of the ^{13}C coherence part caused by the doubly hatched gradient pulses is compensated. This spectral-editing condition means that the "area" of the hatched compensation pulse is related to that of the two doubly hatched gradient pulses according to the quotient of the gyromagnetic ratios γ_C/γ_H.

Secondly, any coherence phase shifts caused by inhomogeneities and chemical shifts in the τ_1 and τ_2 intervals must be accounted for. The phase shifts arising in the τ_1 interval are due to proton coherences, whereas those related to the τ_2 interval effectively[2] refer to ^{13}C. That is, the τ_3 interval needed for compensation of the phase shifts accumulated on the relevant coherence pathway in the total period $\tau_1 + \tau_2$ is

$$\tau_3 = \tau_1 + \frac{\gamma_C}{\gamma_H}\frac{(1-\sigma_C)}{(1-\sigma_H)}\tau_2 \tag{40.3}$$

where σ_C and σ_H are the shielding constants of the ^{13}C and ^1H nuclei, respectively. In other words, proton single-quantum coherence evolution under the influence of field offsets is $\gamma_H/\gamma_C \approx 4$ as efficient as ^{13}C single-quantum coherence evolution under the same field offsets.

40.2
Cyclic Cross-Polarization Localized Spectroscopy (CYCLCROP-LOSY)

The second example for heteronuclear localized spectroscopy techniques to be depicted is cyclic cross-polarization localized spectroscopy (CYCLCROP-LOSY). Figure 40.2 shows the pulse scheme. Apart from the detection interval, this method completely avoids free-evolution intervals. For biological applications this is a rather important feature because any free-evolution period that is essential for the spectral-editing process may cause severe artifacts if the object under examination moves [293].

The volume selectivity of cross-polarization in the presence of field gradients has already been discussed in Chaps. 38 and 39. Compensation for Hartmann-Hahn mismatch can be achieved to a certain degree with the aid of the MOIST variant

---

[2]Remember that protons are subject to a 180° pulse in this interval.

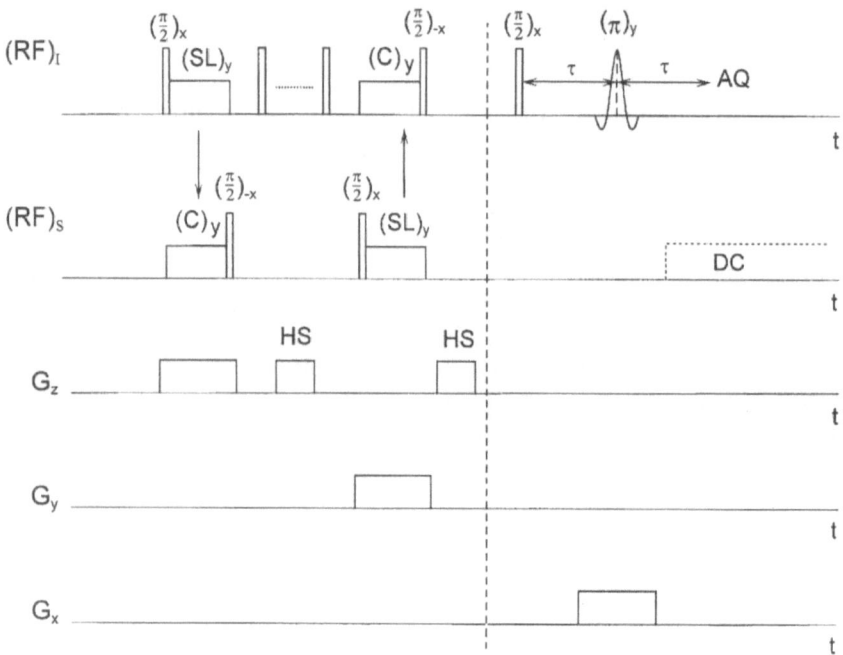

Fig. 40.2. Pulse sequence for localized proton-detected ^{13}C spectroscopy (cyclic J cross polarization localized spectroscopy **CYCLCROP-LOSY**) on the basis of cyclic and slice-selective J cross-polarization (left-hand side). The shown Hahn echo sequence (right-hand side) or alternatively a slice-selective 90° ^{1}H reading pulse serves for the selection of a slice in the third space direction. The spin-locking/contact-pulse pairs are matched according to the Hartmann/Hahn condition. Mismatch can favorably be compensated to a certain degree with the MOIST variant. Abbreviations are: SL, spin-locking pulse; C, contact pulse; HS, homospoil gradient pulse; AQ, acquisition; DC, decoupling.

by reversing the phases of the spin-locking and contact pulses in intervals short compared with the reciprocal coupling constant J^{-1}.

In the simplest case, i.e., a two-spin-$\frac{1}{2}$ system, AX, and under perfect Hartmann/Hahn matching conditions, the reduced density operator after the contact time τ and for forward polarization transfer is given by (see Chap. 39)

$$\sigma = S_y \sin^2\left(\frac{\pi J\tau}{2}\right) - I_z S_x \sin(\pi J\tau), \tag{40.4}$$

The first term represents single-quantum in-phase coherences of the S spins and the second antiphase coherences. With the optimal contact time for complete polarization transfer, it follows $\sigma = S_y$.

After having transferred the polarization from the I to the S spins, the in-phase coherences of the S spins are stored as z magnetization with the aid of a $90°_{-x}$ pulse, $\sigma = S_z$. All I spin populations are subsequently saturated by a comb of pulses followed by homospoil (HS) gradients (Fig. 41.3).

The S spin magnetization is then spin-locked again and the polarization is transferred back to the I spins, selectively producing I spin coherences of the coupled spins, whereas the abundant uncoupled spins still remain saturated. At the end of the backward polarization-transfer pulse pair, the reduced density operator is

$$\sigma = I_y \sin^2\left(\frac{\pi J \tau}{2}\right) - S_z I_x \sin(\pi J \tau). \tag{40.5}$$

The optimum contact time is the same as in the case of the forward transfer. As before, the coherences are stored as z magnetization by the aid of a $90°_{-x}$ pulse, i.e., the density operator $\sigma = I_y$ is transferred to $\sigma = I_z$.

The application of this heteronuclear editing procedure to $^{13}CH_n$ systems thus selectively produces z magnetization of the protons coupled to the ^{13}C nuclei of interest. All other and, in particular, uncoupled protons are saturated at this instant.

A typical application of the CYCLCROP-LOSY method (Fig. 40.2) in this modification is shown in Fig. 26.1. Indirect ^{13}C spectra were recorded in preselected volumes in the yolk and egg-white of a hen's egg with ^{13}C in natural abundance. Conventional volume-selective ^1H spectra recorded with a VOSY sequence (Fig. 37.2) (voxel size: 1 x 1 x 1 cm^3) are plotted in the same figure for comparison. An estimation of the RF power deposition in tissue can be found in [293].

CHAPTER 41

Heteronuclear Imaging

The pulse sequences for magnetic resonance imaging outlined in Chap. 25 are in principle applicable to any nuclide providing detectable NMR signals. The number of potential applications is therefore innumerable. However, lack of sensitivity limits the versatility in this respect. This may be a consequence of low natural abundance of the isotope, low abundance of the element, or low gyromagnetic ratios. Other factors such as extremely short relaxation times may also play a restrictive role.

The nuclide ^{19}F is a relatively favorable species for NMR experiments. It is of special interest for biomedical applications (e.g., [214, 329, 330, 437]). As the natural abundance in tissue is negligible, it suggests itself as a label of substances, the distribution and the local properties of which are to be examined. The gyromagnetic ratio is only slightly less than that of protons, so that an attractive sensitivity can be achieved in principle.

Hydrogen/fluorine retuning tomography (HYFY) is a probehead setup permitting the change from ^{1}H to ^{19}F resonance and vice versa without removing the object from the magnet [437]. A bird-cage resonator is retuned to ^{19}F resonance by changing the shortening capacitors of the antennas with the aid of copper foils placed adjacent to them. This method can be used for the record of correlated ^{1}H and ^{19}F images from the same object. A typical example is shown in Fig. 37.5 on page 356. The images represent an animal kidney perfused with a perfluorocarbon emulsion. In the ^{19}F image, the network of blood capillaries is clearly visible.

Another nuclide suitable for labeling is ^{2}H. As concerns hydrogen-bond forming species, this may be critical for in vivo applications. However, material properties can be examined in this way quite favorably. This includes double-quantum filtering variants for the discrimination of oriented from isotropic regions [170, 260] (compare Sect. 5.2). As a promising option double-quantum filtering can be carried out slice-selectively by double-quantum coherence transfer between ^{1}H and ^{2}H [121].

The most versatile potential of heteronuclear imaging applications is, certainly offered by ^{13}C NMR. The situation is quite analogous to that of conventional spectroscopy. The reason is that ^{13}C in natural abundance as well as in enriched substances provides access to so many organic compounds in general, and to hydrocarbons in particular. In the following we therefore consider special schemes for indirect ^{13}C imaging[1] suitable for overcoming the sensitivity problem.

---

[1]This is to be distinguished from direct ^{13}C imaging methods [26, 359, 477, 488, 527] in combination with laboratory-frame polarization transfer for signal enhancement.

41.1
Proton-Detected ^{13}C Imaging

Indirect, i.e., proton-detected imaging of ^{13}C requires spectral editing suppressing all coherences of protons other than those coupled to ^{13}C or - even more specific - coupled to ^{13}C in a certain chemical group. The indirect monitoring of ^{13}C nuclei overcomes the sensitivity problem by selectively referring to ^{1}H nuclei coupled to ^{13}C. In principle, these "edited" signals are recorded with proton sensitivity as concerns chemical groups containing ^{13}C.

The selective detection of hydrocarbons via ^{13}C NMR is to be compared with magnetic resonance spectroscopic imaging (see Chap. 28 and [252]). MRSI is fast and easy to implement. However, the selectivity to certain hydrocarbon compounds is somewhat unspecific. The two indirect ^{13}C imaging techniques to be described in the following distinguish themselves by a strong spectroscopic selectivity. Important features also are how susceptible a method is to motions, and how much RF power is deposited in tissue while recording an image. The first editing method to be outlined in the following is multiple-quantum filtering [267].

41.1.1
Multiple-Quantum Edited Hydrocarbon Maps

The "hydrogen/carbon tomography" (HYCAT) method is based on multiple-quantum filtering of heteronuclear ^{1}H/^{13}C coherences in analogy to the heteronuclear MQF-VOSING technique described in Sect. 40.1. Spectral editing and imaging are performed in an integrated way within one pulse train. The number of RF pulses and RF-power deposition is therefore particularly low. However, a drawback for in vivo applications is that the method is prone to motion artifacts.

Figure 41.1 shows the double-resonance pulse sequence employed for the HYCAT method. The initial excitation occurs on the proton side with the aid of a slice-selective 90° pulse. The proton coherences evolve during an interval $\tau_1 = 1/(2J)$ where J is the proton/^{13}C spin-spin coupling constant. The isochromats of the proton doublets arising from spin-spin coupling to ^{13}C are then antiphase to each other.

The antiphase proton coherences are converted to multiple-quantum ^{1}H/^{13}C coherences by the second pulse which is applied in the ^{13}C channel in the absence of any magnetic-field gradient. This pulse may be "soft", i.e., chemical-shift selective. The coherences of interest are of zeroth and second order. The slice-selective 180° proton pulse in the middle of the subsequent τ_2 interval converts zero- into double-quantum coherences and vice versa.

The τ_2 interval is terminated by a 90° pulse in the ^{13}C channel. The multiple-quantum coherences are partly transferred into single-quantum proton coherences which are refocused eventually as a coherence-transfer echo. This can be acquired as a phase and frequency encoded signal suitable for 2D-image reconstruction.

The coherence pathway described so far must be discriminated from any other pathway, in particular from a homonuclear one potentially leading to an echo superimposed on the coherence-transfer echo of interest. This is performed by dephas-

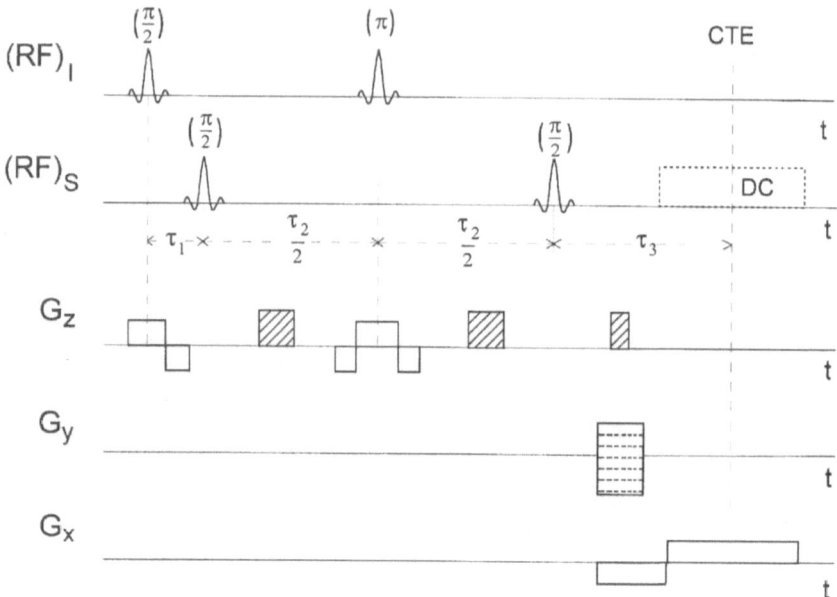

Fig. 41.1. HYCAT method for proton-detected ^{13}C imaging. The signal to be recorded in the presence of an optional ^{13}C decoupling pulse (DC) is a multiple-quantum coherence transfer echo (CTE). The hatched gradient pulses serve for editing of the coherence pathway of the heteronuclear spin system of interest.

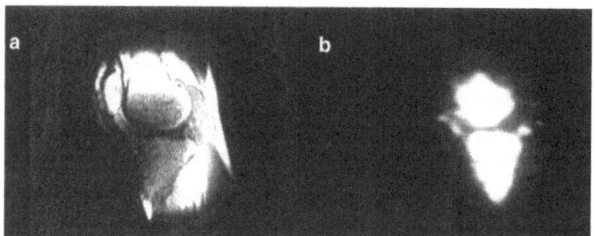

Fig. 41.2. a) Conventional proton image of a human knee recorded in a 2 T whole-body scanner. The slice thickness is 5 mm, the pixel matrix amounts 256 × 256. b) HYCAT map adjusted to methylene group signals of the same knee. In this case the slice thickness is 1.5 cm. The image consists of a pixel matrix of 128 × 128. The total acquisition time was 60 min. (Reproduced by permission from ref. [269])

ing the multiple-quantum coherences in the τ_2 interval and rephasing the single-quantum coherences into which they have been transferred in the τ_3 interval. The corresponding field-gradient pulses (hatched in Fig. 41.1) must be appropriately weighted as described in Sect. 40.1. The τ_3 interval is given in Eq. 40.3. The coherence pathway marked in this manner in particular excludes the formation of Hahn echoes of protons uncoupled to ^{13}C.

A HYCAT picture of a human knee is shown in Fig. 41.2 in comparison to an ordinary proton image. It demonstrates that there is a considerable portion of substances containing methylene groups within the bone near the joint.

41.1.2
Cross-Polarization Edited Hydrocarbon Maps

"Cyclic cross-polarization" (CYCLCROP) tomography also combines spectral editing and imaging [294]. The principle of this technique may be illustrated by the following scheme:

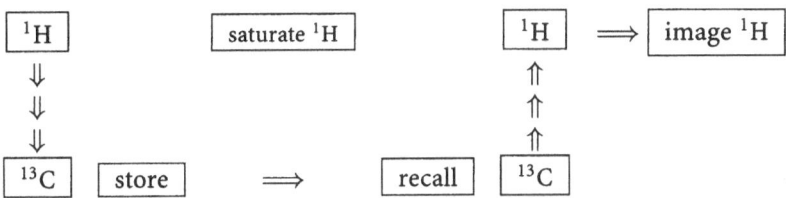

Figure 41.3 shows the pulse sequence for CYCLCROP imaging. As the name says, the editing principle is cyclic J cross polarization [292]. The polarization of a selected chemical species of coupled protons is transferred to the ^{13}C coupling partners and stored in z direction. All proton spins, i.e., all coupled and all uncou-

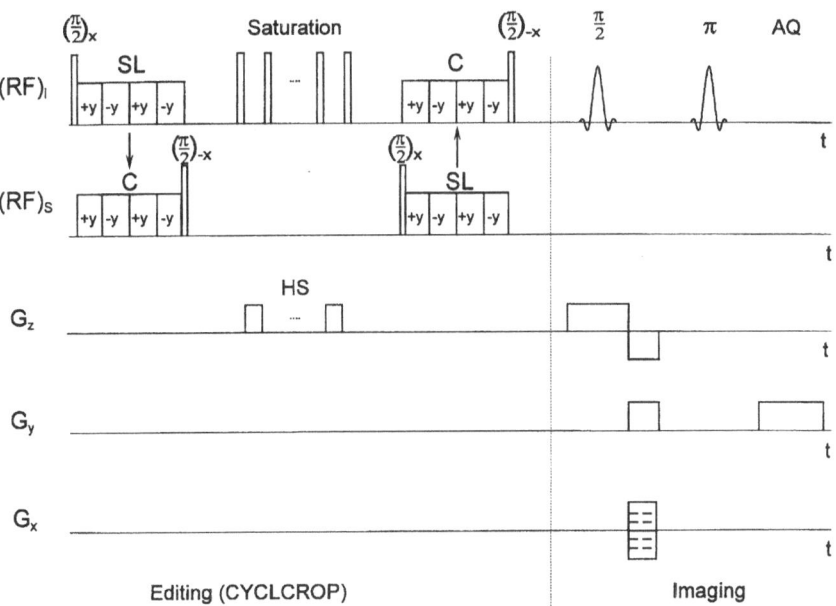

Fig. 41.3. CYCLCROP tomography pulse sequence. SL, spin-lock RF pulse; C, contact RF pulse; HS, homospoil gradient pulse; AQ, acquisition.

pled spins, are then entirely saturated by a comb of 90° pulses in combination with spoil gradients.

In the next step, the selected species of the coupled proton spins are repolarized by J cross polarization from the ^{13}C nuclei to their coupling partners on the proton side. There are no free-evolution intervals up to now so that this method is much less prone to motional artifacts compared with HYCAT at the expense of a somewhat higher RF-power deposition. The subsequent imaging part of the pulse sequence is conventional. It selectively renders maps of the spatial distribution of the protons coupled to ^{13}C nuclei of interest.

The cross-polarization pulses in the editing section are composed of a spin-locking pulse on the primary side and a contact pulse on the secondary side. In order to obtain effective polarization transfer of a selected chemical group the JCP interval has to be optimized not only with respect to the Hartmann/Hahn condition (Eq. 38.1) but also to the frequencies (chemical shifts) and to the pulse length. The optimal contact interval for maximal polarization transfer $I \rightarrow S$ depends on the spin-spin coupling constant J, the size of the spin system and the potential influence of Hartmann/Hahn mismatch. For AX, AX_2 and AX_3 systems and under perfect Hartmann/Hahn matching conditions the optimal contact times are given by $\tau_{op} = J^{-1}$, $\tau_{op} = (\sqrt{2}J)^{-1}$ and $\tau_{op} = 0.61J^{-1}$, respectively [357]. (The full treatment for AX spin systems is given in Chap. 39.)

The efficiency of J cross polarization theoretically is 100% for an AX system if Hartmann/Hahn matching is perfect. For larger spin systems the relative efficiency is less in principle. Nevertheless, the signals are enhanced relative to AX systems, for example, by a factor of 1.5 for AX_3 systems [459, 460].

The Hartmann/Hahn pulse sequence is susceptible to inaccurate adjustments of the RF amplitudes. As a remedy, such mismatch effects can be compensated to a certain degree by the MOIST modification of the pulse sequence [298]. With this variant, the phases of the spin-locking and contact pulses are simultaneously reversed in intervals short compared with the reciprocal coupling constant. The contact time for maximal polarization transfer is then the same as for the conventional Hartmann/Hahn experiment. The flip angles of the MOIST pulse segments are uncritical. The prerequisite is merely that the RF power of the Hartmann/Hahn pulses exceeds the minimum amplitude for cross polarization as determined by J coupling and the desired chemical shift range.

The sensitivity of indirect detection of ^{13}C nuclei is theoretically enhanced relative to conventional ^{13}C spectroscopy by a factor of $(\gamma_I/\gamma_S)^3 \approx 64$. A further signal improvement results from the higher signal accumulation rates permitted because all detected signals are derived from the initial proton magnetization. Spin-lattice relaxation of protons, however, is normally much faster than that of ^{13}C nuclei. The CYCLCROP method thus promises the maximal sensitivity reachable in heteronuclear NMR in principle. For comparison, the sensitivity of direct ^{13}C detection can be improved by any polarization transfer $^1H \rightarrow \ ^{13}C$ or as a consequence of the nuclear Overhauser effect by a factor of $\gamma_I/\gamma_S \approx 4$ at most.

Note that signal enhancements by spin decoupling are likewise possible with direct and indirect detection. If ^1H is detected under ^{13}C decoupling, the two doublet

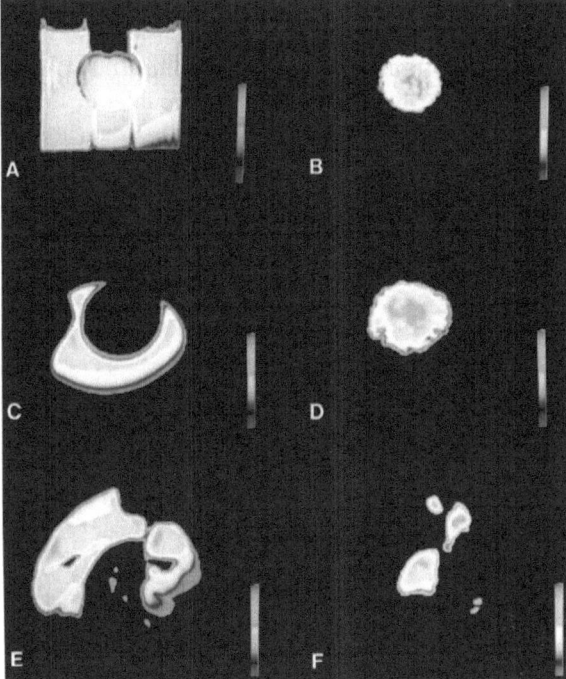

Fig. 41.4. ^{13}C maps recorded with the CYCLCROP tomography sequence Fig. 41.3) in comparison to conventional proton images. The proton-detected ^{13}C images refer to natural isotope abundance. The spin-echo time typically was $T_E = 13$ ms. The field of view is 8 cm.

A Conventional proton image of a phantom sample (plastic ball filled with methanol and positioned in water) recorded without the editing part of the pulse scheme in Fig. 41.3. Water as well as methanol signals contribute. (Specific image data: slice thickness, 1 cm; recycle delay, 2.5 s; pixel matrix, 256 x 256).

B CYCLCROP image of the same phantom recorded with the complete pulse sequence in Fig. 41.3. The contrasts represent the spatial distribution of the ^{13}C containing methyl groups. No water signal is visible any longer. (Specific image data: same as with (A); number of accumulated transients, 8; total acquisition time, 35 min).

C Conventional proton image of a hen's egg recorded without the editing part of the pulse scheme in Fig. 41.3. Signals from the yolk are strongly suppressed because of the short transverse relaxation time. (Specific image data: slice thickness, 1 cm; recycle delay, 2 s; pixel matrix, 128 x 128).

D CYCLCROP image of the same hen's egg recorded with the complete pulse sequence in Fig. 41.3. The contrasts represent the spatial distribution of ^{13}C containing methylene groups. Water signals and the very low fat content of egg-white remarkably are no longer visible. (Specific image data: same as with (C); number of accumulated transients, 128).

E Conventional proton image of a fresh porcine shank recorded without the editing part of the pulse scheme in Fig. 41.3. Muscle tissue is visible but bone marrow and fat are strongly suppressed due to fast transverse relaxation. (Specific image data: same as with (C)).

F CYCLCROP image of the same porcine shank recorded with the complete pulse sequence in Fig. 41.3. The contrasts represent the spatial distribution of ^{13}C containing methylene groups in fatty acid residues of bone marrow and fat. Muscle tissue is no longer visible. (Specific image data: same as with (D)).

(Reproduced by permission from ref. [294])

lines are added up, whereas in the reverse case the ^{13}C multiplet lines are combined to a single line. For simplicity or for the sake of low RF power deposition, it may be desirable to avoid spin decoupling. Indirect ^{13}C detection then has the advantage that each resonance line is only divided into two doublet lines of half intensity.

The second part of the pulse scheme (Fig. 41.3) consists of a conventional Fourier transform NMR imaging pulse sequence. That is, the z magnetization of the protons coupled to the ^{13}C nuclei of interest is slice-selectively excited again and encoded by the spatial phase and frequency distribution. The proton images generated in this way render the local ^{13}C number density with respect to a predetermined chemical group.

Figure 41.4 shows CYCLCROP indirect ^{13}C images of different objects in comparison to conventional proton images. In the case of the hen egg, the methylene-group signal is restricted to the yolk region whereas the conventional image is dominated by the egg-white.

IV

Analytical NMR Toolbox

CHAPTER 42

Miscellaneous Formulae and Rules

42.1
Some Algebraic Symbols

The Kronecker symbol takes the values

$$\delta_{s,t} = \begin{cases} 1 & \text{for } s = t \\ 0 & \text{otherwise} \end{cases} \qquad (s = x, y, z; \ t = x, y, z) \tag{42.1}$$

The ϵ symbol is defined by the cyclic permutation relations

$$
\begin{aligned}
\epsilon_{sss} &= 0 \qquad (s = x, y, z) \\
\epsilon_{sst} = \epsilon_{tss} = \epsilon_{sts} &= 0 \qquad (s, t = x, y, z) \\
\epsilon_{xyz} = \epsilon_{yzx} = \epsilon_{zxy} &= 1 \\
\epsilon_{xzy} = \epsilon_{yxz} = \epsilon_{zyx} &= -1
\end{aligned}
\tag{42.2}
$$

42.2
The Delta Function

Dirac's δ function can be defined by the following properties:

$$\int f(x)\, \delta(x - x_0)\, dx = f(x_0) \tag{42.3}$$

$$\delta\left(\frac{x}{a} - x_0\right) = |a|\, \delta(x - ax_0) \tag{42.4}$$

$$\delta(r - r_0) = \delta(x - x_0)\, \delta(y - y_0)\, \delta(z - z_0) \tag{42.5}$$

It can also be considered as the limit of diverse ordinary functions:

$$\delta(x) = \lim_{a \to \infty} \sqrt{\frac{a}{\pi}}\, e^{-ax^2} \tag{42.6}$$

$$= \lim_{a \to \infty} \frac{a}{2}\, e^{-a|x|} \tag{42.7}$$

$$= \lim_{a \to 0} \frac{a}{\pi(a^2 + x^2)} \tag{42.8}$$

$$= \lim_{a \to \infty} \frac{\sin(ax)}{\pi x} \tag{42.9}$$

$$= \lim_{a \to \infty} \frac{1}{2\pi} \int_{-a}^{a} e^{\pm ixy}\, dy \tag{42.10}$$

$$= \lim_{a \to \infty} \frac{\sin^2(ax)}{\pi a x^2} \qquad (42.11)$$

A selection of useful Fourier transform pairs including δ functions is listed in Table 42.1.

42.3
Fourier Transforms

The Fourier transformation links functions of conjugated variables. Relevant examples are time vs (angular) frequency, wave vector component vs space coordinate, and velocity wave vector component vs velocity space coordinate. Referring to the corresponding functions, one speaks of the "time" and "frequency domains" or "reciprocal-" and "real-space domains", respectively. If different time scales or more than one spatial vector components are considered, the Fourier transformation becomes "multidimensional."

The most common definition of the transform pairs $f(t)$ and $F(\omega)$, for the time and (angular) frequency domains, respectively, are

$$F(\omega) = \int_{-\infty}^{\infty} f(t)e^{-i\omega t}\, dt \qquad (42.12)$$

$$f(t) = \frac{1}{2\pi} \int_{-\infty}^{\infty} F(\omega)e^{i\omega t}\, d\omega \qquad (42.13)$$

Analogously the transform pairs $\rho(r)$ and $S(k)$, for the position vector and wave vector domains respectively, obey the relations

$$S(k) = \int \rho(r)e^{-i k \cdot r}\, d^\varsigma r \qquad (42.14)$$

$$\rho(r) = \frac{1}{(2\pi)^\varsigma} \int S(k)e^{i k \cdot r}\, d^\varsigma k \qquad (42.15)$$

where the integrals refer to the whole z dimensional r and k spaces, respectively. For details see [84]. Analytical Fourier transform pairs of interest in NMR are listed in Tables 42.1 to 42.3 in a handy form for the convenience of the reader.

There are several excellent books on numerical Fourier transformation procedures for NMR purposes [62, 328, 444]. The parameters which limit numerical Fourier transforms of spectroscopic (imaging) data sets are the spectral width (field of view) and the spectral (spatial) resolution. A discussion and comparison of these quantities can be found in Chap. 32.

A relation crucial for digital-acquisition mode methods is the **Nyquist sampling theorem**. For a time/frequency Fourier transform pair, for instance, it states that the maximum frequency that can be detected is

$$v_m = \frac{1}{2\tau} \qquad (42.16)$$

where τ is the "dwell time", i.e., the sampling time per data point in the time domain. At least two data points are required per period in order to identify a frequency component. Otherwise the so-called "aliasing" will occur resulting in mirror lines. The total frequency range unambiguously probed by a dwell time τ is thus $-\nu_m \leq \nu \leq \nu_m$. On the other hand, the numerical resolution in the frequency domain is determined by the total acquisition time T_a,

$$\Delta \nu = \frac{1}{T_a} \qquad (42.17)$$

Table 42.1. Fourier transform pairs including δ functions.

$f(t)$	$\rightarrow$	$F(\omega) = \int\limits_{-\infty}^{\infty} f(t) e^{-i\omega t} \, dt$
$A\delta(t)$	$\rightarrow$	A
$A\delta(t - t_0)$	$\rightarrow$	$A e^{-i\omega t_0}$
$\delta(t - t_0) + \delta(t + t_0)$	$\rightarrow$	$2\cos(\omega t_0)$
$\sum\limits_{n=-\infty}^{\infty} \delta(t - nt_0)$	$\rightarrow$	$\frac{2\pi}{t_0} \sum\limits_{n=-\infty}^{\infty} \delta(\omega - n\frac{2\pi}{t_0})$
$\sum\limits_{n=-\infty}^{\infty} \delta(t - [n + \frac{1}{2}]t_0)$	$\rightarrow$	$\frac{2\pi}{t_0} \sum\limits_{n=-\infty}^{\infty} (-1)^n \delta(\omega - n\frac{2\pi}{t_0})$
A	$\rightarrow$	$2\pi\delta(\omega)$
$\cos(\omega_0 t)$	$\rightarrow$	$\pi[\delta(\omega - \omega_0) + \delta(\omega + \omega_0)]$
$\sin(\omega_0 t)$	$\rightarrow$	$i\pi[\delta(\omega + \omega_0) - \delta(\omega - \omega_0)]$
$e^{\pm i\omega_0 t}$	$\rightarrow$	$2\pi\delta(\omega \mp \omega_0)$
$\cos^2(\omega_0 t)$	$\rightarrow$	$\pi[\frac{1}{2}\delta(\omega + 2\omega_0) + \delta(\omega) + \frac{1}{2}\delta(\omega - 2\omega_0)]$
$\sin^2(\omega_0 t)$	$\rightarrow$	$\pi[-\frac{1}{2}\delta(\omega + 2\omega_0) + \delta(\omega) - \frac{1}{2}\delta(\omega - 2\omega_0)]$
$\cos(\omega_0 t)\,[A + a\cos(\omega_m t)]$	$\rightarrow$	$\pi A[\delta(\omega - \omega_0) + \delta(\omega + \omega_0)]$ $+ \frac{a\pi}{2}[\delta(\omega - \omega_0 - \omega_m)$ $+ \delta(\omega - \omega_0 + \omega_m)$ $+ \delta(\omega + \omega_0 - \omega_m) + \delta(\omega + \omega_0 + \omega_m)]$

Table 42.2. Some typical NMR Fourier transform pairs linking the time and (angular) frequency domains. The symbols have the usual meanings in context with NMR. That is, $R_2^* = 1/T_2^*$ is the effective decay rate of a FID, and M_2 is the second moment of the lineshape $F(\omega)$.

$f(t)$	$\rightarrow$	$F(\omega) = \int\limits_{-\infty}^{\infty} f(t)e^{-i\omega t}\, dt$
$\cos(\Omega t)\, e^{-R_2^* t}$ for $t \geq 0$ 0 otherwise	$\rightarrow$	$\frac{1}{2}\left[\frac{R_2^*}{R_2^{*2}+(\Omega+\omega)^2} + \frac{R_2^*}{R_2^{*2}+(\Omega-\omega)^2}\right]$ $\qquad +\frac{i}{2}\left[\frac{\Omega-\omega}{R_2^{*2}+(\Omega-\omega)^2} - \frac{\Omega+\omega}{R_2^{*2}+(\Omega+\omega)^2}\right]$ $\frac{R_2^*}{R_2^{*2}+\omega^2} - i\frac{\omega}{R_2^{*2}+\omega^2}$ (if $\Omega = 0$)
$\sin(\Omega t)\, e^{-R_2^* t}$ for $t \geq 0$ 0 otherwise	$\rightarrow$	$\frac{1}{2}\left[\frac{\Omega-\omega}{R_2^{*2}+(\Omega-\omega)^2} + \frac{\Omega+\omega}{R_2^{*2}+(\Omega+\omega)^2}\right]$ $\qquad +\frac{i}{2}\left[\frac{R_2^*}{R_2^{*2}+(\Omega+\omega)^2} - \frac{R_2^*}{R_2^{*2}+(\Omega-\omega)^2}\right]$ 0 (if $\Omega = 0$)
$\cos(\Omega t)\, e^{-R_2^*|t|}$	$\rightarrow$	$\frac{R_2^*}{R_2^{*2}+(\Omega-\omega)^2} + \frac{R_2^*}{R_2^{*2}+(\Omega+\omega)^2}$ $\frac{2R_2^*}{R_2^{*2}+\omega^2}$ (if $\Omega = 0$)
$\sin(\Omega t)\, e^{-R_2^*|t|}$	$\rightarrow$	$i\left[\frac{R_2^*}{R_2^{*2}+(\Omega+\omega)^2} - \frac{R_2^*}{R_2^{*2}+(\Omega-\omega)^2}\right]$ 0 (if $\Omega = 0$)
$e^{i\Omega t - R_2^* t}$ for $t \geq 0$ 0 otherwise	$\rightarrow$	$\frac{R_2^*}{R_2^{*2}+(\Omega-\omega)^2} + i\frac{\Omega-\omega}{R_2^{*2}+(\Omega-\omega)^2}$ $\frac{R_2^*}{R_2^{*2}+\omega^2} - i\frac{\omega}{R_2^{*2}+\omega^2}$ (if $\Omega = 0$)
$\exp\left\{\frac{1}{2}M_2 t^2\right\}$	$\rightarrow$	$\sqrt{\frac{2\pi}{M_2}}\, \exp\left\{-\frac{\omega^2}{2M_2}\right\}$
$\frac{\Omega}{\pi}\, \mathrm{sinc}\left(\frac{\Omega t}{\pi}\right) = \frac{\sin(\Omega t)}{\pi t}$	$\rightarrow$	1 for $|\omega| \leq \Omega$ 0 otherwise

Table 42.3. Some typical NMR Fourier transform pairs linking reciprocal- and real-space domains, i.e., for example, functions of the x components of the wave vector and the real-space position vector, respectively. In magnetic-resonance imaging applications, the wave vector component is determined by the corresponding field gradient component and the interval in which the gradient is effective: $k_x = \gamma_n G_x t_x$. The symbol ξ_2 then refers to the free-induction decay during signal acquisition. It is defined by $\xi_2 \equiv 1/(T_2^* \gamma_n G_x)$.

$S(k_x)$	$\rightarrow$	$\rho(x) = \frac{1}{2\pi} \int\limits_{-\infty}^{\infty} F(k_x) e^{ik_x x} \, dk_x$

$$
\left.\begin{array}{ll} \cos(k_x x_a)\, e^{-\xi_2 k_x} & \text{for } k_x \geq 0 \\ \\ 0 & \text{otherwise} \end{array}\right\} \rightarrow \begin{cases} \frac{1}{4\pi}\left[\frac{\xi_2}{\xi_2^2+(x_a-x)^2} + \frac{\xi_2}{\xi_2^2+(x_a+x)^2}\right] \\ \quad + \frac{i}{4\pi}\left[\frac{x_a+x}{\xi_2^2+(x_a+x)^2} - \frac{x_a-x}{\xi_2^2+(x_a-x)^2}\right] \\ \frac{1}{2\pi}\left[\frac{\xi_2}{\xi_2^2+x^2} + i\frac{x}{\xi_2^2+x^2}\right] \quad (\text{if } x_a = 0) \end{cases}
$$

$$
\left.\begin{array}{ll} \sin(k_x x_a)\, e^{-\xi_2 k_x} & \text{for } k_x \geq 0 \\ \\ 0 & \text{otherwise} \end{array}\right\} \rightarrow \begin{cases} \frac{1}{4\pi}\left[\frac{x_a+x}{\xi_2^2+(x_a+x)^2} + \frac{x_a-x}{\xi_2^2+(x_a-x)^2}\right] \\ \quad + \frac{i}{4\pi}\left[\frac{\xi_2}{\xi_2^2+(x_a-x)^2} - \frac{\xi_2}{\xi_2^2+(x_a+x)^2}\right] \\ 0 \quad (\text{if } x_a = 0) \end{cases}
$$

$$
\cos(k_x x_a)\, e^{-\xi_2 |k_x|} \rightarrow \begin{cases} \frac{1}{2\pi}\frac{\xi_2}{\xi_2^2+(x_a+x)^2} + \frac{\xi_2}{\xi_2^2+(x_a-x)^2} \\ \frac{1}{\pi}\frac{\xi_2}{\xi_2^2+x^2} \quad (\text{if } x_a = 0) \end{cases}
$$

$$
\sin(k_x x_a)\, e^{-\xi_2 |k_x|} \rightarrow \begin{cases} \frac{i}{2\pi}\left[\frac{\xi_2}{\xi_2^2+(x_a-x)^2} - \frac{\xi_2}{\xi_2^2+(x_a+x)^2}\right] \\ 0 \quad (\text{if } x_a = 0) \end{cases}
$$

$$
\left.\begin{array}{ll} e^{ix_a k_x - \xi_2 k_x} & \text{for } k_x \geq 0 \\ \\ 0 & \text{otherwise} \end{array}\right\} \rightarrow \begin{cases} \frac{1}{2\pi}\left[\frac{\xi_2}{\xi_2^2+(x_a+x)^2} + i\frac{x_a+x}{\xi_2^2+(x_a+x)^2}\right] \\ \frac{1}{2\pi}\left[\frac{\xi_2}{\xi_2^2+x^2} + i\frac{x}{\xi_2^2+x^2}\right] \quad (\text{if } x_a = 0) \end{cases}
$$

$$
e^{-ak_x^2} \rightarrow \frac{1}{2\sqrt{\pi a}}e^{-x^2/(4a)}
$$

42.4
Spherical Harmonics

Spherical harmonics, $Y_{l,m}(\vartheta, \varphi) \equiv Y_{l,m}(u)$, are functions of the orientation of a unit vector u defined by the polar and azimuthal angles, ϑ and φ, respectively. The analytical form is identical to that of the **eigenfunctions of the angular momentum operator**[1] L,

$$(L)^2 Y_{l,m}(\vartheta, \varphi) = l(l+1)Y_{l,m}(\vartheta, \varphi), \qquad L_z Y_{l,m}(\vartheta, \varphi) = m Y_{l,m}(\vartheta, \varphi) \quad (42.18)$$

which are characterized by the quantum numbers $l = 0, 1, 2, \ldots$ and $m = -l, -l + 1, \ldots, +l$. Another reading of the left-hand eigenvalue equation is

$$\nabla^2 Y_{l,m} = \frac{1}{r}\frac{\partial^2}{\partial r^2}(rY_{l,m}) + \frac{1}{r^2 \sin\vartheta}\frac{\partial}{\partial\vartheta}\left(\sin\vartheta\frac{\partial Y_{l,m}}{\partial\vartheta}\right) + \frac{1}{r^2 \sin^2\vartheta}\frac{\partial^2 Y_{l,m}}{\partial\varphi^2}$$

$$= -\frac{l(l+1)}{r^2}Y_{l,m}(\vartheta, \varphi) \qquad (42.19)$$

where the Laplace operator has been expressed in polar coordinates. Referring to the orientation of the unit vector, the coordinate r takes the fixed value 1, of course. The spherical harmonics which are of particular importance in NMR are listed in Table 42.4.

The **complex conjugate** of spherical harmonics obeys

$$Y_{l,m}^*(\vartheta, \varphi) = (-1)^m Y_{l,-m}(\vartheta, \varphi) \qquad (42.20)$$

The **parity** of spherical harmonics with even l is positive, that of functions with odd l is negative:

$$Y_{l,m}(\pi - \vartheta, \pi + \varphi) = (-1)^l Y_{l,m}(\vartheta, \varphi) \qquad (42.21)$$

The spherical harmonics form a **complete orthonormal set** of functions in the ranges $0 \leq \vartheta \leq \pi$ and $0 \leq \varphi \leq 2\pi$. That is,

$$\int_0^{2\pi} d\varphi \int_0^{\pi} d\vartheta \, \sin\vartheta \, Y_{l,m}^*(\vartheta, \varphi)Y_{l',m'}(\vartheta, \varphi) = \delta_{ll'}\delta_{mm'} \qquad (42.22)$$

The $(2l + 1)$ functions $Y_{l,m}(\vartheta, \varphi)$ of rank l for the $(2l + 1)$ allowed values of m are linearly independent. Any **linear combination** $\sum_{m=-l}^{+l} a_{l,m}Y_{l,m}(\vartheta, \varphi)$ again forms a spherical harmonic of the same rank.

Functions of polar coordinates, $f(r, \vartheta, \varphi)$, can be expanded according to

$$f(r, \vartheta, \varphi) = \sum_{l=0}^{\infty} \sum_{m=-l}^{+l} a_{l,m}(r)Y_{l,m}(\vartheta, \varphi) \qquad (42.23)$$

---

[1]In this book, *angular momenta* are throughout used as dimensionless quantities, i.e., in units $\hbar$. If the real angular momentum is meant, the quantity is multiplied by $\hbar$.

Table 42.4. Spherical harmonics up to rank 2 expressed in polar and orthogonal Cartesian coordinates (see Sect. 48.2).

$$Y_{0,0} = \sqrt{\frac{1}{4\pi}}$$

$$Y_{1,0} = \sqrt{\frac{3}{4\pi}} \cos\vartheta \qquad\qquad = \sqrt{\frac{3}{4\pi}} \frac{z}{r}$$

$$Y_{1,\pm1} = \mp\sqrt{\frac{3}{8\pi}} \sin\vartheta\, e^{\pm i\varphi} \qquad = \mp\sqrt{\frac{3}{8\pi}} \frac{x\pm iy}{r}$$

$$Y_{2,0} = \sqrt{\frac{5}{16\pi}} (3\cos^2\vartheta - 1) \qquad = \sqrt{\frac{5}{16\pi}} \frac{2z^2-x^2-y^2}{r^2}$$

$$Y_{2,\pm1} = \mp\sqrt{\frac{15}{8\pi}} \cos\vartheta \sin\vartheta\, e^{\pm i\varphi} = \mp\sqrt{\frac{15}{8\pi}} \frac{(x\pm iy)z}{r^2}$$

$$Y_{2,\pm2} = \sqrt{\frac{15}{32\pi}} \sin^2\vartheta\, e^{\pm 2i\varphi} \qquad = \sqrt{\frac{15}{32\pi}} \frac{(x\pm iy)^2}{r^2}$$

where the coefficients $a_{l,m}(r)$ are given by

$$a_{l,m}(r) = \int_0^{2\pi}\int_0^{\pi} f(r,\vartheta,\varphi) Y_{l,m}^*(\vartheta,\varphi) \sin\vartheta\, d\vartheta\, d\varphi \qquad (42.24)$$

Expansions of particular interest are

$$\delta(u - u') \equiv \frac{\delta(\vartheta - \vartheta')\delta(\varphi - \varphi')}{\sin\vartheta} = \sum_{l=0}^{\infty}\sum_{m=-l}^{+l} Y_{l,m}^*(\vartheta,\varphi) Y_{l,m}(\vartheta',\varphi') \qquad (42.25)$$

$$\delta(r - r') \equiv \frac{\delta(\vartheta - \vartheta')\delta(\varphi - \varphi')\delta(r - r')}{r^2 \sin\vartheta} = \frac{\delta(r - r')}{r^2}\sum_{l=0}^{\infty}\sum_{m=-l}^{+l} Y_{l,m}^*(\vartheta,\varphi) Y_{l,m}(\vartheta',\varphi')$$

$$(42.26)$$

so that

$$\int_V \delta(r - r')r^2 \sin\vartheta\, d\vartheta\, d\varphi = \begin{cases} 1 & \text{for } r' \text{ in } V \\ 0 & \text{for } r' \text{ not in } V \end{cases} \qquad (42.27)$$

A frequently employed formula is

$$\frac{1}{|r - r'|} = 4\pi\sum_{l=0}^{\infty}\sum_{m=-l}^{+l} \frac{1}{2l+1}\frac{r^l}{(r')^{l+1}} Y_{l,m}(\vartheta,\varphi) Y_{l,m}^*(\vartheta',\varphi') \qquad (r' > 0) \quad (42.28)$$

Exponential functions, as they occur in context with spatial Fourier transforms (see Eqs. 42.14 and 42.15), can be expanded according to

$$e^{i\mathbf{k}\cdot\mathbf{r}} = 4\pi \sum_{l=0}^{\infty} \sum_{m=-l}^{+l} i^l \sqrt{\frac{\pi}{2kr}} J_{l+\frac{1}{2}}(kr) Y_{l,m}^*(\vartheta_k, \varphi_k) Y_{l,m}(\vartheta_r, \varphi_r) \qquad (42.29)$$

where $J_{l+\frac{1}{2}}$ is the Bessel function of order $l + \frac{1}{2}$ and the subscripts k and r refer to the vectors $\mathbf{k}$ and $\mathbf{r}$, respectively.

Spherical harmonics **transform under rotations** of the reference frame as

$$Y_{l,m}(\vartheta, \varphi) \rightarrow R(\tilde{\alpha}, \tilde{\beta}, \tilde{\gamma}) Y_{l,m}(\vartheta, \varphi) R^{-1}(\tilde{\alpha}, \tilde{\beta}, \tilde{\gamma}) = \sum_{m'=-l}^{+l} Y_{l,m'}(\vartheta, \varphi) D_{m',m}^{(l)}(\tilde{\alpha}, \tilde{\beta}, \tilde{\gamma})$$

$$(42.30)$$

where $D_{m',m}^{(l)}(\tilde{\alpha}, \tilde{\beta}, \tilde{\gamma})$ represents the Wigner rotation matrix elements

$$D_{m',m}^{(l)}(\tilde{\alpha}, \tilde{\beta}, \tilde{\gamma}) = \langle l, m' | R(\tilde{\alpha}, \tilde{\beta}, \tilde{\gamma}) | l, m \rangle \qquad (42.31)$$

of the rotation operator $R(\tilde{\alpha}, \tilde{\beta}, \tilde{\gamma})$ for rotations about the Euler angles $\tilde{\alpha}, \tilde{\beta}, \tilde{\gamma}$ (see Chaps. 48 and 49).

The **addition theorem** for spherical harmonics is of particular interest in NMR. Consider two vectors r_1 and r_2 with the polar coordinates φ_1, ϑ_1 and φ_2, ϑ_2, respectively. The angle spanned by these vectors is denoted γ (Fig. 42.1). The addition theorem relates the Legendre polynomial of order l of the angle γ to the spherical

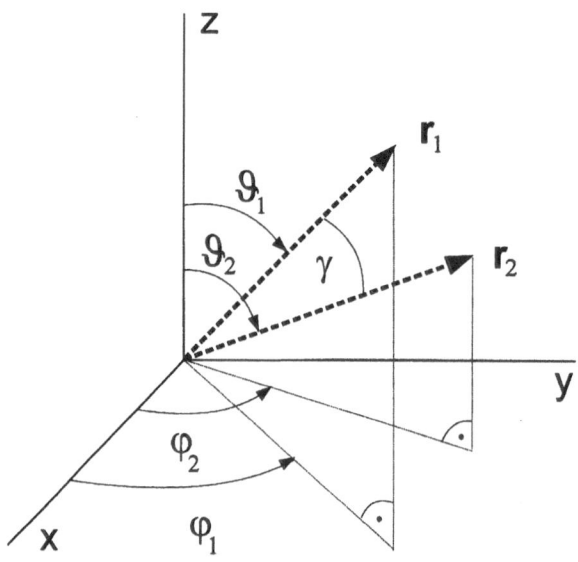

Fig. 42.1. Polar $(\vartheta_1, \vartheta_2)$ and azimuthal (φ_1, φ_2) angles of two vectors r_1 and r_2 relative to an orthogonal Cartesian reference frame.

harmonics of the same order of the polar angles of the two vectors:

$$P_l(\cos \gamma) = \frac{4\pi}{2l+1} \sum_{m=-l}^{l} Y_{l,m}^*(\vartheta_1, \varphi_1) Y_{l,m}(\vartheta_2, \varphi_2) \tag{42.32}$$

where

$$\cos \gamma = \cos \vartheta_1 \cos \vartheta_2 + \sin \vartheta_1 \sin \vartheta_2 \cos(\varphi_2 - \varphi_1) \tag{42.33}$$

The first few Legendre polynomials are

$$P_0(x) = 1 \tag{42.34}$$

$$P_1(x) = x \tag{42.35}$$

$$P_2(x) = \frac{1}{2}(3x^2 - 1) \tag{42.36}$$

$$P_3(x) = \frac{1}{2}(5x^3 - 3x) \tag{42.37}$$

$$P_4(x) = \frac{1}{8}(35x^4 - 30x^2 + 3) \tag{42.38}$$

Based on the addition theorem one finds for the average second-order Legendre polynomial of the vector r_1 rotating (or jumping in steps of at least threefold symmetry) about the fixed vector r_2 with a fixed angle γ in between:

$$\langle 3 \cos^2 \vartheta_1 - 1 \rangle = \frac{1}{2}(3 \cos^2 \gamma - 1)(3 \cos^2 \vartheta_2 - 1) \tag{42.39}$$

42.5
Classification of Operators

Consider an operator $\mathcal{O}$ and two wavefunctions $|\varphi\rangle$ and $|\psi\rangle$. The operator $\mathcal{O}^\dagger$ adjoint to $\mathcal{O}$ is defined by

$$\langle \varphi | \mathcal{O} \psi \rangle = \langle \mathcal{O}^\dagger \varphi | \psi \rangle = \langle \psi | \mathcal{O}^\dagger \varphi \rangle^* \tag{42.40}$$

where the star indicates the conjugate complex value. A Hermitian (or self-adjoint) operator complies with

$$\mathcal{O}^\dagger = \mathcal{O} \tag{42.41}$$

An operator is unitary if

$$\mathcal{O}^\dagger = \mathcal{O}^{-1} \tag{42.42}$$

where $\mathcal{O}^{-1}$ is the inverse operator specified by $\mathcal{O}^{-1}\mathcal{O} = 1$. That is, unitary operators obey $\mathcal{O}^\dagger\mathcal{O} = \mathcal{O}\mathcal{O}^\dagger = 1$, and have eigenvalues of modulus unity.

42.6
Spin-Operator Relations

42.6.1
Spin-Operator Representations

Spin is a particle-inherent angular momentum of a quantized nature. In the present context, it is an intrinsic property of certain nuclei and electrons. A spin system is a selection of spin-bearing particles coupled with each other by spin interactions. Examples are the nuclei with finite spin in isolated molecules or in molecules under motional-averaging conditions. In solids, where the spin interactions normally cannot be demarcated, the spin system tends to comprise the whole sample.

The spin vector operator

$$I = \begin{pmatrix} I_x \\ I_y \\ I_z \end{pmatrix} \tag{42.43}$$

is given in units $\hbar$, i.e., dimensionless. The magnitude, $\mathcal{I} = \sqrt{I(I+1)}$, is defined by the spin quantum number I. The z component, I_z, has $2I+1$ eigenkets $|\Psi_{I,m}\rangle$ and eigenvalues (the magnetic quantum numbers m) with the values $m = -I, -I+1, \ldots, +I$. The corresponding eigenvalue equations are

$$I_z|\Psi_{I,m}\rangle = m|\Psi_{I,m}\rangle, \qquad I^2|\Psi_{I,m}\rangle = I(I+1)|\Psi_{I,m}\rangle \tag{42.44}$$

On this basis, the spin-1/2 operators may be represented in the form of the "Pauli spin matrices"

$$I_x = \frac{1}{2}\begin{pmatrix} 0 & 1 \\ 1 & 0 \end{pmatrix}, \qquad I_y = \frac{1}{2}\begin{pmatrix} 0 & -i \\ i & 0 \end{pmatrix}, \qquad I_z = \frac{1}{2}\begin{pmatrix} 1 & 0 \\ 0 & -1 \end{pmatrix} \tag{42.45}$$

The eigenkets and eigenbras of spin-1/2 operators are

$$|\alpha\rangle = \begin{pmatrix} 1 \\ 0 \end{pmatrix}, \qquad\qquad |\beta\rangle = \begin{pmatrix} 0 \\ 1 \end{pmatrix} \tag{42.46}$$

and

$$\langle\alpha| = \begin{pmatrix} 1 & 0 \end{pmatrix}, \qquad\qquad \langle\beta| = \begin{pmatrix} 0 & 1 \end{pmatrix} \tag{42.47}$$

respectively.

The matrix elements $\langle\Psi_{I,m_i}|I_l|\Psi_{I,m_j}\rangle$ ($m_i, m_j = \pm 1/2$ and $l = x, y, z$) are formed using the eigenvalue at Eq. 42.44, expressing I_x and I_y by raising and lowering operators as described below, and by taking into account the orthonormal property of the eigenfunctions

$$\langle\Psi_{I,m}|\Psi_{I,m'}\rangle = \delta_{mm'} \tag{42.48}$$

The matrix representations of spin-1 operators are

$$I_x = \frac{1}{\sqrt{2}}\begin{pmatrix} 0 & 1 & 0 \\ 1 & 0 & 1 \\ 0 & 1 & 0 \end{pmatrix}, \qquad I_y = \frac{i}{\sqrt{2}}\begin{pmatrix} 0 & -1 & 0 \\ 1 & 0 & -1 \\ 0 & 1 & 0 \end{pmatrix},$$

$$I_z = \begin{pmatrix} 1 & 0 & 0 \\ 0 & 0 & 0 \\ 0 & 0 & -1 \end{pmatrix} \tag{42.49}$$

The eigenkets and eigenbras of spin-1 operators are

$$|+1\rangle = \begin{pmatrix} 1 \\ 0 \\ 0 \end{pmatrix}, \qquad |0\rangle = \begin{pmatrix} 0 \\ 1 \\ 0 \end{pmatrix}, \qquad |-1\rangle = \begin{pmatrix} 0 \\ 0 \\ 1 \end{pmatrix} \tag{42.50}$$

and

$$\langle +1| = \begin{pmatrix} 1 & 0 & 0 \end{pmatrix}, \qquad \langle 0| = \begin{pmatrix} 0 & 1 & 0 \end{pmatrix}, \qquad \langle -1| = \begin{pmatrix} 0 & 0 & 1 \end{pmatrix}$$
$$(42.51)$$

respectively.

The commutators among the components (Eq. 44.10) do not vanish, so that only one component can be determined in an experiment at a time. In other words, there is no eigenfunction common to any two components. This is in contrast to pairs of one component and the square of the magnitude of the spin vector (Eq. 44.11).

42.6.2
Ladder Operators

Raising and lowering operators can be expressed by transverse spin-operator components and vice versa:

$$I^+ = I_x + iI_y, \qquad\qquad I^- = I_x - iI_y \qquad\qquad (42.52)$$

$$I_x = \tfrac{1}{2}(I^+ + I^-), \qquad\qquad I_y = \tfrac{1}{2i}(I^+ - I^-) \qquad\qquad (42.53)$$

I^+ and I^- are adjoint (but not inverse) to each other. Using the commutators of the spin operators (see Chap. 44), one readily finds relations such as

$$(I^+)^2 + (I^-)^2 = 2(I_x^2 - I_y^2) \qquad\qquad (42.54)$$

$$I_x^2 + I_y^2 = \frac{1}{2}(I^+ I^- + I^- I^+) \qquad\qquad (42.55)$$

$$I^+ I^- = I_x^2 + I_y^2 + I_z \qquad\qquad (42.56)$$

$$I^- I^+ = I_x^2 + I_y^2 - I_z \qquad\qquad (42.57)$$

$$I^+ I^- + I^- I^+ = 2(I_x^2 + I_y^2) = 2(I - I_z^2) \qquad\qquad (42.58)$$

$$I = I^- I^+ + I_z + I_z^2 = I^+ I^- - I_z + I_z^2 \qquad\qquad (42.59)$$

Raising and lowering operators act on eigenkets of a spin I according to

$$I^+ |\Psi_{I,m}\rangle = \sqrt{I(I+1) - m(m+1)}\,|\Psi_{I,m+1}\rangle \qquad\qquad (42.60)$$
$$I^- |\Psi_{I,m}\rangle = \sqrt{I(I+1) - m(m-1)}\,|\Psi_{I,m-1}\rangle \qquad\qquad (42.61)$$

For example, spins $I = 1/2$ obey

$$I^+ |\Psi_{1/2,-1/2}\rangle = |\Psi_{1/2,+1/2}\rangle \qquad\qquad (42.62)$$
$$I^+ |\Psi_{1/2,+1/2}\rangle = 0 \qquad\qquad (42.63)$$
$$I^- |\Psi_{1/2,+1/2}\rangle = |\Psi_{1/2,-1/2}\rangle \qquad\qquad (42.64)$$
$$I^- |\Psi_{1/2,-1/2}\rangle = 0 \qquad\qquad (42.65)$$

Equations 42.53 thus yield the relations

$$I_x \left| \Psi_{1/2,-1/2} \right\rangle = \frac{1}{2} \left| \Psi_{1/2,+1/2} \right\rangle \tag{42.66}$$

$$I_x \left| \Psi_{1/2,+1/2} \right\rangle = \frac{1}{2} \left| \Psi_{1/2,-1/2} \right\rangle \tag{42.67}$$

$$I_y \left| \Psi_{1/2,-1/2} \right\rangle = \frac{1}{2i} \left| \Psi_{1/2,+1/2} \right\rangle \tag{42.68}$$

$$I_y \left| \Psi_{1/2,+1/2} \right\rangle = -\frac{1}{2i} \left| \Psi_{1/2,-1/2} \right\rangle \tag{42.69}$$

(For completeness, these equations may be supplemented by the first of Eqs. 42.44.)

42.6.3
Spherical Spin Operators

All spin interactions can be expressed on the basis of the set of the unnormalized "spherical" operators I^+, I^-, and I_z. Apart from this, a set of normalized spherical operators defined by

$$I^{(+1)} = -\tfrac{1}{\sqrt{2}} I^+ , \qquad\qquad I^{(-1)} = \tfrac{1}{\sqrt{2}} I^- , \qquad\qquad I^{(0)} = I_z \tag{42.70}$$

is in use. Cartesian spin operators are related to the normalized spherical operators by

$$I_x = -\tfrac{1}{\sqrt{2}} \left(I^{(+1)} - I^{(-1)} \right) , \qquad I_y = \tfrac{i}{\sqrt{2}} \left(I^{(+1)} + I^{(-1)} \right) , \qquad I_z = I^{(0)} \tag{42.71}$$

42.6.4
Relations for Spin-1/2 Operators

Useful relations for the Cartesian components of the vector operator of a spin $I = 1/2$ are

$$
\begin{aligned}
&I_\nu^2 = \tfrac{1}{4}, && I_\nu^k = \tfrac{1}{4} I_\nu^{k-2} && \text{(for } \nu = x, y, z \text{ and } k \geq 2) \\
&I_x I_y = \tfrac{i}{2} I_z , && I_y I_z = \tfrac{i}{2} I_x , && I_z I_x = \tfrac{i}{2} I_y \\
&I_y I_x = -\tfrac{i}{2} I_z , && I_z I_y = -\tfrac{i}{2} I_x , && I_x I_z = -\tfrac{i}{2} I_y
\end{aligned}
\tag{42.72}
$$

42.6.5
Relations for Spin-1 Operators

For spins $I = 1$, we have the relations

$$
\begin{aligned}
&(I^+)^3 = 0, && (I^-)^3 = 0 \\
&I_\nu^3 = I_\nu , && I_\nu I_\mu I_\nu = 0, \\
&I_\nu I_\mu^2 + I_\mu^2 I_\nu = I_\nu && \text{(for } \nu, \mu = x, y, z \text{ and } \nu \neq \mu)
\end{aligned}
\tag{42.73}
$$

42.6.6
Two-Spin Systems

For a system of two spins with the vector operators

$$I = \begin{pmatrix} I_x \\ I_y \\ I_z \end{pmatrix} \quad \text{and} \quad S = \begin{pmatrix} S_x \\ S_y \\ S_z \end{pmatrix} \tag{42.74}$$

the following relations are often useful:

$$
\begin{aligned}
I \cdot S &= \tfrac{1}{2}(I^+S^- + I^-S^+) + I_zS_z \\
I^+S^+ &= I_xS_x - I_yS_y + i(I_xS_y + I_yS_x) \\
I^-S^- &= I_xS_x - I_yS_y - i(I_xS_y + I_yS_x) \\
I^-S^+ &= I_xS_x + I_yS_y - i(I_yS_x - I_xS_y) \\
I^+S^- &= I_xS_x + I_yS_y + i(I_yS_x - I_xS_y) \\
I_xS_x &= \tfrac{1}{4}(I^+S^+ + I^-S^- + I^+S^- + I^-S^+) \\
I_yS_y &= \tfrac{1}{4}(I^+S^- + I^-S^+ - I^+S^+ - I^-S^-) \\
I_xS_y &= \tfrac{1}{4i}(I^+S^+ - I^-S^- - I^+S^- + I^-S^+) \\
I_yS_x &= \tfrac{1}{4i}(I^+S^+ - I^-S^- + I^+S^- - I^-S^+)
\end{aligned} \tag{42.75}
$$

42.6.7
Single-Transition Operators

Zero- and double-quantum transitions of a system consisting of spins I and S can be represented by single-transition operators spanning the zero-quantum Liouville subspace with the coordinates [497, 522]

$$
\begin{aligned}
I_x^{(0qt)} &= \tfrac{1}{2}(I^+S^- + I^-S^+) &= (I_xS_x + I_yS_y) \\
I_y^{(0qt)} &= \tfrac{1}{2i}(I^+S^- - I^-S^+) &= (I_yS_x - I_xS_y) \\
I_z^{(0qt)} &= \tfrac{1}{2}(I_z - S_z)
\end{aligned} \tag{42.76}
$$

and the double-quantum Liouville subspace with the coordinates

$$
\begin{aligned}
I_x^{(2qt)} &= \tfrac{1}{2}(I^+S^+ + I^-S^-) &= (I_xS_x - I_yS_y) \\
I_y^{(2qt)} &= \tfrac{1}{2i}(I^+S^+ - I^-S^-) &= (I_xS_y + I_yS_x) \\
I_z^{(2qt)} &= \tfrac{1}{2}(I_z + S_z)
\end{aligned} \tag{42.77}
$$

The single-transition operators obey the same rules as ordinary spin operators. This in particular refers to Eqs. 42.72 and to the commutator relations (see Chap. 44)

$$
\begin{aligned}
\left[I_\alpha^{(2qt)}, I_\beta^{(0qt)} \right] &= 0 \\
\left[I_\alpha^{(0qt)}, I_\beta^{(0qt)} \right] &= iI_\gamma^{(0qt)} \\
\left[I_\alpha^{(2qt)}, I_\beta^{(2qt)} \right] &= iI_\gamma^{(2qt)}
\end{aligned} \tag{42.78}
$$

where the subscript set (α, β, γ) is a cyclic permutation of the set (x, y, z).

42.6.8
An Instructive Example

Consider the zero-quantum coherence of a weakly J-coupled two-spin-1/2 system represented by the spherical product operator

$$\sigma(0) = 2I^{(+1)}S^{(-1)} \tag{42.79}$$

(see Table 51.4 on page 490). From the physical standpoint it is easily perceived that weak J coupling among the active spins cannot affect this coherence. Formally this is proven by applying the unitary transformation

$$\sigma(t) = e^{-i\pi Jt2I^{(0)}S^{(0)}} \, 2I^{(+1)}S^{(-1)} \, e^{i\pi Jt2I^{(0)}S^{(0)}} \tag{42.80}$$

The unitary transformation can be applied subsequently to each of the two spin operators. Following the rules listed in Table 51.5 on page 491, we find

$$\begin{aligned}
\sigma(t) &= 2e^{-i\pi Jt2S^{(0)}I^{(0)}} \, I^{(+1)} \, e^{i\pi Jt2I^{(0)}S^{(0)}} e^{-i\pi Jt2I^{(0)}S^{(0)}} \, S^{(-1)} \, e^{i\pi Jt2S^{(0)}I^{(0)}} \\
&= 2[I^{(+1)}\cos(\pi Jt) - 2iI^{(+1)}S^{(0)}\sin(\pi Jt)][S^{(-1)}\cos(\pi Jt) + 2iS^{(-1)}I^{(0)}\sin(\pi Jt)] \\
&= 2[I^{(+1)}S^{(-1)}\cos^2(\pi Jt) + 4I^{(+1)}S^{(0)}S^{(-1)}I^{(0)}\sin^2(\pi Jt) \\
&\quad -iI^{(+1)}S^{(0)}S^{(-1)}\sin(2\pi Jt) + iI^{(+1)}S^{(-1)}I^{(0)}\sin(2\pi Jt)]
\end{aligned} \tag{42.81}$$

One must remember, with expressions like this, that spin operators of distinguishable particles commute, whereas different spin operators of the same particle do not. Therefore, any desirable rearrangement of the operators within a product must not violate the sequence of the operators of the same particle. According to Eqs. 42.62 to 42.65, the considered zero-quantum coherence can only exist if the initial state of the system is

$$|m_I, m_S\rangle = \left|-\frac{1}{2}, +\frac{1}{2}\right\rangle \tag{42.82}$$

Applying Eq. 42.81 fictitiously to this ket, where we comply strictly to the sequence of the operators counted from the right, and taking into account the first equation at Eq. 42.44, shows that the last two terms cancel. The combined effect of the first two terms, on the other hand, is the same as that of the original product operator. Therefore

$$e^{-i\pi Jt2I^{(0)}S^{(0)}} \, 2I^{(+1)}S^{(-1)} \, e^{i\pi Jt2I^{(0)}S^{(0)}} = 2I^{(+1)}S^{(-1)} \tag{42.83}$$

as it must be.

CHAPTER 43

Rules for Traces

Traces occur particularly in calculations of average expectation values using density operators. The average expectation value of the spin energy with the Hamiltonian $\mathcal{H}$, for instance, is given by

$$\overline{<\mathcal{H}>} = \text{Tr}\{\rho\mathcal{H}\} \tag{43.1}$$

With NMR or NQR problems one is mainly dealing with traces of products and sums of spin operators. Some of the following rules apply also to arbitrary operators (designated here as $\mathcal{O}$ and $\mathcal{P}$).

(a) Traces of **operator sums:**

$$\text{Tr}\{\mathcal{O}\pm\mathcal{P}\} = \text{Tr}\{\pm\mathcal{P}+\mathcal{O}\} = \text{Tr}\{\mathcal{O}\}\pm\text{Tr}\{\mathcal{P}\} = \pm\text{Tr}\{\mathcal{P}\}+\text{Tr}\{\mathcal{O}\} \tag{43.2}$$

(b) Traces of **operator products:**

$$\text{Tr}\{\mathcal{O}\mathcal{P}\} = \text{Tr}\{\mathcal{P}\mathcal{O}\} \tag{43.3}$$

Consequently

$$\text{Tr}\{[\mathcal{O},\mathcal{P}]\} = 0 \tag{43.4}$$

(c) Traces are invariant with any **unitary transformation.** That is,

$$\text{Tr}\{\mathcal{U}\,\mathcal{O}\,\mathcal{U}^{-1}\} = \text{Tr}\{\mathcal{O}\} \tag{43.5}$$

(d) As a consequence, **traces of operator expressions changing the sign under a unitary transformation vanish.** An example for spin operators is

$$\text{Tr}\{I_{kz}I_{lz}\} = \text{Tr}\{e^{i\pi I_{kx}}I_{kz}I_{lz}e^{-i\pi I_{kx}}\} = \text{Tr}\{(-I_{kz})I_{lz}\} = 0 \tag{43.6}$$

where I_{kz} and I_{lz} are the spin operator components of two (like or unlike) nuclei numbered by k and l. Analogously one finds

$$\text{Tr}\{I_{kx}I_{lx}\} = 0, \qquad \text{Tr}\{I_{ky}I_{ly}\} = 0 \tag{43.7}$$

and

$$\text{Tr}\{I_k \cdot I_l\} = \text{Tr}\{I_{kx}I_{lx} + I_{ky}I_{ly} + I_{kz}I_{lz}\} = 0 \tag{43.8}$$

(e) Traces of **squares of spin operators** for spins I obey

$$\text{Tr}\{I_x^2\} = \text{Tr}\left\{e^{iI_z\pi/2}I_x^2e^{-iI_z\pi/2}\right\} = \text{Tr}\{I_y^2\}$$

$$= \text{Tr}\left\{e^{iI_x\pi/2}I_y^2e^{-iI_x\pi/2}\right\} = \text{Tr}\{I_z^2\} = \frac{1}{3}I(I+1)\text{Tr}\{\mathcal{E}\} \tag{43.9}$$

$$\text{Tr}\{I^2\} = \text{Tr}\{I_x^2 + I_y^2 + I_z^2\} = I(I+1)\text{Tr}\{\mathcal{E}\} \tag{43.10}$$

Thus single spins $I = 1/2$ follow

$$\text{Tr}\{I_x^2\} = \text{Tr}\{I_y^2\} = \text{Tr}\{I_z^2\} = \frac{1}{2} \tag{43.11}$$

(f) Traces of some **products of Cartesian operator components of arbitrary spins** I:

$$\text{Tr}\{I_s\} = 0$$

$$\text{Tr}\{I_sI_t\} = \frac{I(I+1)\text{Tr}\{\mathcal{E}\}}{3}\delta_{st}$$

$$\text{Tr}\{I_sI_tI_u\} = i\frac{I(I+1)\text{Tr}\{\mathcal{E}\}}{6}\epsilon_{stu} \tag{43.12}$$

$$\text{Tr}\{I_sI_tI_uI_v\} = \frac{I(I+1)\text{Tr}\{\mathcal{E}\}}{15}$$

$$\left\{\left[I(I+1)+\frac{1}{2}\right](\delta_{st}\delta_{uv}+\delta_{sv}\delta_{tu}) + [I(I+1)-2]\delta_{su}\delta_{tv}\right\}$$

s, t, u, v may be equal to x, y, z. The unit operator is designated by $\mathcal{E}$.

(g) Traces of **spin operators of a system of N spins** I:
I_x, I_y, I_z are the total spin operator components

$$I_s = \sum_{k=1}^{N} I_{ks} \qquad (s = x, y, z) \tag{43.13}$$

Then

$$\text{Tr}\{I_s^2\} = \sum_{k=1}^{N} \text{Tr}\{I_{ks}^2\} = \frac{1}{3}NI(I+1)(2I+1)^N \tag{43.14}$$

$$\text{Tr}\{I_{js}I_{kt}\} = \frac{1}{3}I(I+1)(2N+1)^N\delta_{jk}\delta_{st} \qquad (s, t = x, y, z) \tag{43.15}$$

(h) Traces of some **commutator expressions**:

$$\text{Tr}\{[I_s, I_t]\} = 0 \tag{43.16}$$

$$\text{Tr}\{[I_s, I_t]I_u\} = \text{Tr}\{[I_t, I_u]I_s\} = \text{Tr}\{[I_u, I_s]I_t\} \quad \text{(cyclic permutation)} \tag{43.17}$$

(i) Traces of the **unity operator:**
For a single spin I we have

$$\text{Tr}\{\mathcal{E}\} = 2I + 1 \qquad (43.18)$$

If I_x, I_y, I_z refer to the total spin operator components

$$I_s = \sum_{k=1}^{N} I_{ks} \qquad (s = x, y, z) \qquad (43.19)$$

of a system of N spins I, the trace of $\mathcal{E}$ is

$$\text{Tr}\{\mathcal{E}\} = N(2I + 1)^N \qquad (43.20)$$

If I_x, I_y, I_z represent the spin-operator components of an individual spin in a system of N spins I, we have

$$\text{Tr}\{\mathcal{E}\} = (2I + 1)^N \qquad (43.21)$$

(j) Traces of **direct matrix products** (see Chap. 51) obey

$$\text{Tr}\{\mathcal{O} \otimes \mathcal{P}\} = \text{Tr}\{\mathcal{O}\}\text{Tr}\{\mathcal{P}\} \qquad (43.22)$$

For further relations of traces see [475, 492] and the appendix of [523].

CHAPTER 44

Commutator Algebra

44.1
General Operators

The commutator of two arbitrary operators $\mathcal{O}_1$ and $\mathcal{O}_2$ is

$$[\mathcal{O}_1, \mathcal{O}_2] = \mathcal{O}_1\mathcal{O}_2 - \mathcal{O}_2\mathcal{O}_1 \qquad (44.1)$$

From this definition the following properties can be derived directly:

$$[\mathcal{O}_1, (\mathcal{O}_2 + \mathcal{O}_3)] = [\mathcal{O}_1, \mathcal{O}_2] + [\mathcal{O}_1, \mathcal{O}_3] \qquad (44.2)$$

$$[\mathcal{O}_1, \mathcal{O}_2] = -[\mathcal{O}_2, \mathcal{O}_1] \qquad (44.3)$$

$$[\mathcal{O}_1\mathcal{O}_2, \mathcal{O}_3] = \mathcal{O}_1[\mathcal{O}_2, \mathcal{O}_3] + [\mathcal{O}_1, \mathcal{O}_3]\mathcal{O}_2 \qquad (44.4)$$

$$[\mathcal{O}_1, \mathcal{O}_2\mathcal{O}_3] = [\mathcal{O}_1, \mathcal{O}_2]\mathcal{O}_3 + \mathcal{O}_2[\mathcal{O}_1, \mathcal{O}_3] \qquad (44.5)$$

$$[\mathcal{O}_1\mathcal{O}_2, \mathcal{O}_3\mathcal{O}_4] = \mathcal{O}_1[\mathcal{O}_2, \mathcal{O}_3]\mathcal{O}_4 + \mathcal{O}_1\mathcal{O}_3[\mathcal{O}_2, \mathcal{O}_4] + [\mathcal{O}_1, \mathcal{O}_3]\mathcal{O}_4\mathcal{O}_2 + \mathcal{O}_3[\mathcal{O}_1, \mathcal{O}_4]\mathcal{O}_2 \qquad (44.6)$$

$$[\mathcal{O}_1, [\mathcal{O}_2, \mathcal{O}_3]] + [\mathcal{O}_2, [\mathcal{O}_3, \mathcal{O}_1]] + [\mathcal{O}_3, [\mathcal{O}_1, \mathcal{O}_2]] = 0 \qquad (44.7)$$

$$[\mathcal{O}_1^m, \mathcal{O}_2] = \sum_{j-0}^{m-1} \mathcal{O}_1^j[\mathcal{O}_1, \mathcal{O}_2]\mathcal{O}_1^{m-j-1} \quad \text{where} \quad m = 1, 2, 3 \ldots \qquad (44.8)$$

$$[\mathcal{O}_1, \mathcal{O}_2]^\dagger = [\mathcal{O}_2^\dagger, \mathcal{O}_1^\dagger] \qquad (44.9)$$

44.2
Spin Operators

The basic **commutators of spin operators** of a certain particle are

$$[I_x, I_y] = iI_z, \qquad [I_y, I_z] = iI_x, \qquad [I_z, I_x] = iI_y \qquad (44.10)$$

Other frequently encountered commutators of spin operators are

$$[I_x, I^2] = 0, \qquad [I_y, I^2] = 0, \qquad [I_z, I^2] = 0 \qquad (44.11)$$

$$[I^+, I^2] = 0, \qquad [I^-, I^2] = 0, \qquad [I_z, I^+I^-] = 0, \qquad [I_z, I^-I^+] = 0 \qquad (44.12)$$

$$[I^+, I_z] = -I^+, \qquad [I^-, I_z] = I^-, \qquad [I^+, I^-] = 2I_z \qquad (44.13)$$

Consequences of these relations are

$$[I^+S^-, I_z + S_z] = 0, \qquad [I^-S^+, I_z + S_z] = 0 \qquad (44.14)$$

where I and S are spins of different particles. Generally, the spin operators of different particles commute:

$$[I_j, S_k] = 0 \qquad (j, k = x, y, z) \qquad (44.15)$$

$$[I_j S_k, I_l S_m] = 0 \qquad (j, k, l, m = x, y, z; \ j = l \text{ and } k = m \text{ or } j \neq l \text{ and } k \neq m) \qquad (44.16)$$

The **anti-commutator** of spin-1/2 operators is

$$[I_j, I_k]_+ \equiv I_j I_k + I_k I_j = \frac{1}{2}\delta_{jk} \qquad (j, k = x, y, z) \qquad (44.17)$$

A comprehensive representation of the spin-1 commutator algebra can be found in [87].

The rules for commutations of spin operators and **irreducible spherical tensor operators** are given in Sect. 49.2. Commutators of **single-transition operators** correspond to those of ordinary spin operators (Eq. 44.10) and are listed in Sect. 42.6.7.

CHAPTER 45

Exponential and Trigonometric Operators

Consider a Hermitian operator $\mathcal{O}$. The **exponential operator,**

$$e^{i\mathcal{O}} = 1 + i\mathcal{O} + \frac{(i\mathcal{O})^2}{2!} + \frac{(i\mathcal{O})^3}{3!} + \cdots \qquad (45.1)$$

forms a unitary (but non-Hermitian) operator. If the operator $\mathcal{P}$ commutes with the operator $\mathcal{O}$, that is $[\mathcal{O}, \mathcal{P}] = 0$, then it also commutes with the exponential operator, so that

$$\mathcal{P}e^{i\mathcal{O}} = e^{i\mathcal{O}}\mathcal{P} \quad \text{or} \quad [\mathcal{P}, e^{i\mathcal{O}}] = 0 \qquad (45.2)$$
$$e^{i(\mathcal{P}+\mathcal{O})} = e^{i\mathcal{O}}e^{i\mathcal{P}} = e^{i\mathcal{P}}e^{i\mathcal{O}} \quad \text{or} \quad [e^{i\mathcal{O}}, e^{i\mathcal{O}}] = 0 \qquad (45.3)$$

In unitary transformations and in sandwich propagator expressions for the evolution of density operators under the action of stationary Hamiltonians, terms occur of the form (see Eq. 47.33)

$$\mathcal{P}' = e^{-i\lambda\mathcal{O}}\mathcal{P}e^{i\lambda\mathcal{O}} \qquad (45.4)$$

where $\mathcal{O}$ is a Hermitian operator and λ is a real parameter. A suitable expansion can be derived employing the **Baker/Hausdorff lemma:**

$$e^{i\lambda\mathcal{O}}\mathcal{P}e^{-i\lambda\mathcal{O}} = \mathcal{P} + i\lambda[\mathcal{O}, \mathcal{P}] + \frac{(i\lambda)^2}{2!}[\mathcal{O}, [\mathcal{O}, \mathcal{P}]] + \frac{(i\lambda)^3}{3!}[\mathcal{O}, [\mathcal{O}, [\mathcal{O}, \mathcal{P}]]] + \cdots$$
$$+ \frac{(i\lambda)^n}{n!}[\mathcal{O}, [\mathcal{O}, [\mathcal{O}, \ldots [\mathcal{O}, \mathcal{P}]]]\ldots] + \cdots \qquad (45.5)$$

The calculation of sandwich propagators is thus reduced to the evaluation of commutators which are often quite elementary. Numerous applications of this expansion can be found in [86].

Trigonometric operators are again interpreted by series expansions. That is

$$\cos\mathcal{O} = 1 - \frac{\mathcal{O}^2}{2!} + \frac{\mathcal{O}^4}{4!} \mp \cdots \qquad (45.6)$$

$$\sin\mathcal{O} = \mathcal{O} - \frac{\mathcal{O}^3}{3!} + \frac{\mathcal{O}^5}{5!} \mp \cdots \qquad (45.7)$$

If $\mathcal{O}$ is proportional to the z component of a spin-1/2 operator, $\mathcal{O} = aS_z$, one can make use of the relation $S_z^2|\Psi\rangle = (1/4)|\Psi\rangle$ or simply $S_z^2 = 1/4$. Thus for

$n = 0, 1, 2, 3, \ldots$

$$S_z^{2n} = \left(\frac{1}{4}\right)^n \quad \text{and} \quad S_z^{2n+1} = \left(\frac{1}{4}\right)^n S_z \tag{45.8}$$

That is,

$$\cos(aS_z) = \cos(a/2) \tag{45.9}$$

$$\sin(aS_z) = 2S_z \sin(a/2) \tag{45.10}$$

CHAPTER 46

Spin Hamiltonians

The following list of Hamilton operators refers to the **laboratory** or the **molecular frames of reference**. In practical treatments, the use of rotated reference frames is ubiquitous. Corresponding transforms are presented below in Chap. 48 after introduction of unitary transformations.

46.1
Zeeman Interaction

Consider a particle having a magnetic dipole moment[1]

$$\mu = \gamma_n \hbar I \tag{46.3}$$

The magnitude is $\mu = |\gamma_n|\hbar I$, the z component $\mu_z = \gamma_n \hbar m$. Depending on the particle properties the gyromagnetic ratio $\gamma_n = g\mu_n/\hbar$ can be positive or negative. The nuclear magneton is $\mu_n = e\hbar/(2m_p)$, where e is the positive elementary charge and m_p is the proton rest mass. The "g factor," g, is to be taken as negative if γ_n is negative.

The Zeeman Hamiltonian follows from the dipole term of the multipole expansion, i.e., the Taylor series about the origin, of the magnetic energy of a classical current density distribution $j_n(r)$ in a vector potential $A_0(r)$ (see Eq. 49.40), and the classical definition of the magnetic dipole moment, Eq. 46.1. For a magnetic flux density $B_0 = B_0 u_z$, the standard versions of the Zeeman Hamiltonian are

$$\mathcal{H}_0 = -\mu \cdot B_0 = -\gamma_n \hbar I \cdot B_0 = -\gamma_n B_0 \hbar I_z = -\omega_0 \hbar I_z \tag{46.4}$$

---

[1]The classical precursor of the nuclear dipole moment is defined in connection with the corresponding term of the multipole-expansion of the magnetic energy of a current density j exposed to an external vector potential (see Eq. 49.40) as

$$\mu_{class} = \frac{1}{2} \int [r \times j(r)]\, d^3r \tag{46.1}$$

The current density formed by a discrete particle of charge q, position r_q, and orbital velocity v_q is $j(r) = qv_q\delta(r - r_q)$ where $\delta(r - r_q)$ is Dirac's delta function. As the orbital angular momentum of a particle of mass m is given by $L = m(r_q \times v_q)$, the classical magnetic dipole moment is readily found to be

$$\mu_{class} = \frac{q}{2m}L \tag{46.2}$$

where $\omega_0 = \gamma_n B_0$.

46.2
RF Irradiation

The RF flux density is assumed to oscillate along the x axis (unit vector u_x) of the laboratory frame. That is

$$B_{rf} = 2B_1 \cos(\omega_c t)\, u_x = B_1 \left(e^{i\omega_c t} + e^{-i\omega_c t} \right) u_x \qquad (46.5)$$

where $2B_1$ is the amplitude and ω_c the "carrier frequency." The Hamiltonian of a magnetic dipole in this field is

$$\mathcal{H}_{rf} = -\boldsymbol{\mu} \cdot \boldsymbol{B}_{rf} = -2\hbar\omega_1 I_x \cos(\omega_c t) = -\frac{1}{2}\hbar\omega_1 [I^+ + I^-]\left(e^{i\omega_c t} + e^{-i\omega_c t} \right) \quad (46.6)$$

where $\omega_1 = \gamma_n B_1$. Note that Eq. 46.6 refers to both counterrotating circular polarized components of the RF field.

46.3
Internal Spin Interactions

"Internal" means that the spin interactions reflect phenomena taking place inside the samples. These interactions can be intra- or inter-molecular in nature. Chemical shift interaction, dipole-dipole coupling, indirect spin coupling, quadrupolar interaction, and - by circumstance - spin-rotation interaction are of particular interest.

The Hamiltonians of internal spin interactions generally consist of spin-dependent and coordinate-dependent parts referring to the "spin space" and the "real space," respectively. The common Cartesian tensor form of spin interaction Hamiltonians is

$$\mathcal{H}_\lambda = \boldsymbol{I} \cdot \boldsymbol{C}^{(\lambda)} \cdot \boldsymbol{V}^{(\lambda)} \qquad (46.7)$$

where $\boldsymbol{I}$ is a (row) vector operator of a spin, $\boldsymbol{C}^{(\lambda)}$ is the second-rank Cartesian tensor of the spin coupling depending on real-space coordinates, and $\boldsymbol{V}^{(\lambda)}$ is a (column) vector depending on the type of the spin interaction, λ, as listed in Table 46.1.

The second-rank coupling tensors of dipole-dipole and quadrupole interactions are symmetric and traceless so that no isotropic contributions exist. Chemical shift, spin-rotation, and indirect spin-spin couplings, on the other hand, provide isotropic contributions. These reveal themselves under motional averaging conditions as scalar interaction constants

$$C^{(\lambda)} = \frac{1}{3}\mathrm{Tr}\{\boldsymbol{C}^{(\lambda)}\} \qquad (46.8)$$

There are also antisymmetric contributions which, however, are negligible in NMR spectroscopy.

Rotations in spin and real space can conveniently be described on a spherical basis. This means that eigenfunctions of angular momentum are used as a basis, i.e., spherical harmonics. The spin-interaction Hamiltonians can correspondingly

Table 46.1. Nomenclature of the internal spin interaction Hamiltonians (Eq. 46.7).

coupling	λ	$\mathcal{H}_\lambda = I \cdot C^{(\lambda)} \cdot V^{(\lambda)}$	$C^{(\lambda)}$	$V^{(\lambda)}$
chemical shift	cs	$\mathcal{H}_{cs} = \hbar\gamma_n I \cdot \sigma \cdot B_0$	shielding tens.	static magn.-flux density
dipole-dipole	d	$\mathcal{H}_d = \frac{\mu_0}{4\pi}\gamma_I\gamma_S\hbar^2 I \cdot D \cdot S$	dip.-int. tens.	spin vector op. S coupled to I
indirect	J	$\mathcal{H}_J = h I \cdot J \cdot S$	J-coupling tens.	spin vector op. S coupled to I
quadrupole	Q	$\mathcal{H}_q = \frac{eQ}{2I(2I-1)} I \cdot \Gamma \cdot I$	el.-field grad. tens.	spin vector op. I
spin rotation	sr	$\mathcal{H}_{sr} = h I \cdot C^{(sr)} \cdot J$	spin-rotation tens.	molecular angular mom. op. J

be represented as "irreducible spherical tensor (IST) operators." The general theory of IST operators can be found in [63, 422], for instance, and is outlined in Chap. 49 and [86, 179, 336].

The internal spin interaction Hamiltonians consist of secular and non-secular terms. The term "non-secular" in this context indicates that spin transitions are induced which do not conserve the Zeeman energy of the spins. This sort of transition is relevant for relaxation but not for the evolution of spin coherences in the high-field limit. For spectroscopic purposes one can therefore restrict oneself to the consideration of the secular parts, generating either no spin transitions or only transitions conserving the spin Zeeman energy. Typical expressions of internal spin interaction Hamiltonians are listed in Tables 46.2 to 46.5.

Table 46.2. Representations of the chemical shift Hamiltonian of a spin I [433, 465] (IST, irreducible spherical tensor; LAS, laboratory axes system (B_0 z-axis); PAS, principal axes system).

specification	Hamiltonian
basic form:	$\mathcal{H}_{cs} = \hbar\gamma_n \, I \cdot \sigma \cdot B_0 \quad$ where $\quad B_0 = \begin{pmatrix} 0 \\ 0 \\ B_0 \end{pmatrix}$ $= \hbar\gamma_n(\sigma_{xz}I_x + \sigma_{yz}I_y + \sigma_{zz}I_z)B_0$
IST operators:	$\mathcal{H}_{cs} = a_{cs} \sum\limits_{l=0,2} \sum\limits_{m=-l}^{l} (-1)^m A_{l,-m} T_{l,m}$ $a_{cs} = \hbar\gamma_n B_0$ $A_{0,0} = \sigma_i$ $A_{2,0} = \sqrt{\frac{4\pi}{5}}(\sigma_{33}-\sigma_i)\left[\sqrt{\frac{3}{2}}Y_{2,0}(\tilde{\beta},\tilde{\gamma})\right.$ $\qquad \left. +\frac{\eta}{2}\left(Y_{2,+2}(\tilde{\beta},\tilde{\gamma}) + Y_{2,-2}(\tilde{\beta},\tilde{\gamma})\right)\right]$ $A_{2,\pm1} = \frac{1}{2}(\sigma_{33}-\sigma_i)\sin\tilde{\beta}\,e^{\mp i\tilde{\alpha}}\left\{\pm3\cos\tilde{\beta}\right.$ $\qquad \left. +\eta\left[\sinh(i2\tilde{\gamma}) \mp \cos\tilde{\beta}\cosh(i2\tilde{\gamma})\right]\right\}$ $T_{0,0} = I^{(0)} = I_z$ $T_{2,0} = \sqrt{\frac{2}{3}}\,I^{(0)} = \sqrt{\frac{2}{3}}\,I_z$ $T_{2,\pm1} = \sqrt{\frac{1}{2}}\,I^{(\pm1)} = \mp\frac{1}{2}\,I^{\pm}$ $T_{2,\pm2} = 0$
secular part:	$\mathcal{H}_{cs}^{(0)} = \hbar\gamma_n I_z B_0 \sigma_{zz}$ $\mathcal{H}_{cs}^{(0)} = \hbar\gamma_n I_z B_0(\sigma_{11}\cos^2\varphi\sin^2\vartheta + \sigma_{22}\sin^2\varphi\sin^2\vartheta + \sigma_{33}\cos^2\vartheta)$ $= \hbar\gamma_n I_z B_0\left\{\sigma_i + \sqrt{\frac{4\pi}{5}}(\sigma_{33}-\sigma_i)[Y_{2,0}(\vartheta,\varphi)\right.$ $\qquad \left. +\frac{1}{2}\eta\sqrt{\frac{2}{3}}(Y_{2,2}(\vartheta,\varphi) + Y_{2,-2}(\vartheta,\varphi))]\right\}$ $= \hbar\gamma_n I_z B_0\left\{\sigma_i + (\sigma_{33}-\sigma_i)\frac{1}{2}[3\cos^2\vartheta - 1 - \eta\sin^2\vartheta\cos(2\varphi)]\right\}$
non-secular part:	$\mathcal{H}_{cs}^{ns} = \hbar\gamma_n(\sigma_{33}-\sigma_i)\left[I_x B_{0x} + I_y B_{0y} + \frac{1}{2}\eta(B^+I^+ + B^-I^-)\right]$
definitions:	$\sigma_{11}, \sigma_{22}, \sigma_{33}$ elements of σ in the PAS σ_{ij} ($i,j = x, y, z$) elements of σ in the LAS $\tilde{\alpha}, \tilde{\beta}, \tilde{\gamma}$ Euler angles bringing the LAS $\qquad$ into coincidence with the PAS $\varphi \equiv \tilde{\gamma}$ azimuthal angle $\left.\right\}$ of B_0 $\vartheta \equiv \tilde{\beta}$ polar angle $\left.\right\}$ rel. to the PAS $\sigma_i = \frac{1}{3}\text{Tr}\{\sigma\} = \frac{1}{3}(\sigma_{11} + \sigma_{22} + \sigma_{33})$ isotropic shielding const. $\eta = (\sigma_{22}-\sigma_{11})/(\sigma_{33}-\sigma_i)$ anisotropy constant $B^{\pm} = B_{0x} \pm iB_{0y}$

46.3.1
Chemical-Shift Interaction

The background of the chemical-shift Hamiltonian (Table 46.2) is the local field at the nucleus

$$B_l = B_0 - \sigma B_0 \tag{46.9}$$

That is, the externally applied flux density B_0 is shielded[2] by the electron shell which in this respect is characterized by the second-rank shielding tensor σ. The Zeeman Hamiltonian,

$$\mathcal{H}_l = -\boldsymbol{\mu} \cdot \boldsymbol{B}_l \tag{46.10}$$

can thus be analyzed into

$$\mathcal{H}_l = \mathcal{H}_0 - \mathcal{H}_{cs} \tag{46.11}$$

where the chemical-shift Hamiltonian is defined as

$$\mathcal{H}_{cs} = \hbar\gamma_n \boldsymbol{I} \cdot \sigma \cdot \boldsymbol{B}_0 \tag{46.12}$$

The shielding tensor can be related to the states of the electron shell in an external magnetic field with the aid of first-order perturbation theory [388, 393].

Table 46.3. Representations of the indirect spin-spin coupling (or J coupling) Hamiltonian for a spin pair I and S located in the same molecule. Under most practical circumstances merely the isotropic part is relevant. Note that the same isotropic coupling form arises for unpaired electrons with a finite residence probability in a spin-bearing nucleus ("scalar interaction"). The scalar nature of this interaction makes an IST representation superfluous, because rotational-transformation properties do not matter here.

specification	Hamiltonian
basic form	$\mathcal{H}_J = h\boldsymbol{I} \cdot \boldsymbol{J} \cdot \boldsymbol{S}$
isotropic part	$\mathcal{H}_J^{iso} = hJ\boldsymbol{I} \cdot \boldsymbol{S}$
	$\quad = hJ\left[I_z S_z + \frac{1}{2}(I^+S^- + I^-S^+)\right]$
	$\quad$ where $J = \frac{1}{3}\mathrm{Tr}\{\boldsymbol{J}\}$
form effective in the weak-coupl. limit	$\mathcal{H}_J^{wc} = hJI_z S_z \quad (J \ll \lvert\nu_I - \nu_S\rvert)$

---

[2]This term should not be taken literally in general, because paramagnetic contributions possibly dominate. In that case, the local flux density exceeds that of the externally applied field. Anyway, under normal conditions the chemical shift is of a diamagnetic nature. The local field is then reduced relative to B_0.

46.3.2
Dipolar, Scalar, and Indirect Couplings

The Hamiltonian of the **dipole-dipole interaction** (Table 46.5) can be traced back to the classical energy of a magnetic dipole in a magnetic field (compare Eq. 46.4)

$$W_m = -\mu_1 \cdot B_d \tag{46.13}$$

of a magnetic dipole μ_1 in the magnetic field B_d of a current loop. Using Biot/Savart's law, one finds in the far-field limit:

$$B_d(r) = \frac{\mu_0}{4\pi} \left(\frac{3(\mu_2 \cdot r)r}{r^5} - \frac{\mu_2}{r^3} \right) \tag{46.14}$$

The magnetic dipole moment of the current loop is defined by the current I and the loop area vector A as $\mu_2 = IA$ (Eq. 46.1).

The **indirect spin-spin coupling** between the nuclei of a molecule (Table 46.3) is mediated by the common electron shell. The electron states are perturbed by the presence of nuclear spins in principally two different ways. First, the orbital contribution to the electron Hamiltonian is perturbed by the vector potential originating from the nuclear dipoles. Second, nuclear and electronic magnetic dipoles with the dipole moments μ_n and μ_e, respectively, interact with each other, so that an additional energy contribution arises. This turns out to be the dominating effect.

The Hamiltonian of an electronic magnetic dipole in the magnetic field of a nucleus is given by

$$\mathcal{H}_{IS} = -\mu_e \cdot B_n \tag{46.15}$$

where the dipolar magnetic flux density of the nucleus, B_n, is related to the vector potential A_n by

$$B_n = \nabla \times A_n \tag{46.16}$$

The vector potential of the nuclear dipole field is given by

$$A_n = \frac{\mu_0}{4\pi} \frac{\mu_n \times r}{r^3} = \frac{\mu_0}{4\pi} \nabla \times \frac{\mu_n}{r} \tag{46.17}$$

where the origin, $r = 0$, is in the nucleus. The interaction Hamiltonian hence is

$$\mathcal{H}_{IS} = -\mu_e \cdot \frac{\mu_0}{4\pi} \left[\nabla \times \left(\nabla \times \frac{\mu_n}{r} \right) \right] \tag{46.18}$$

Using the relation

$$\nabla \times (\nabla \times a) = \nabla(\nabla \cdot a) - \nabla^2 a \tag{46.19}$$

for a vector a, the above Hamiltonian can be rewritten as

$$\mathcal{H}_{IS} = \underbrace{\frac{\mu_0}{4\pi} \frac{2}{3} (\mu_e \cdot \mu_n) \nabla^2 \left(\frac{1}{r} \right)}_{\text{term (a)}} - \underbrace{\frac{\mu_0}{4\pi} \left[(\mu_e \cdot \nabla)(\mu_n \cdot \nabla) - \frac{1}{3} (\mu_e \cdot \mu_n) \nabla^2 \right] \left(\frac{1}{r} \right)}_{\text{term (b)}}$$

$$\tag{46.20}$$

Table 46.4. Representations of the Hamiltonian of the quadrupole interaction of a spin $I > 1/2$. For a derivation see Sect. 49.6.

specification	Hamiltonian
basic form	$\mathcal{H}_q = \frac{eQ}{2I(2I-1)} \, \boldsymbol{I} \cdot \boldsymbol{\Gamma} \cdot \boldsymbol{I}$ $\boldsymbol{\Gamma} = \left\{ \left. \frac{\partial^2 \Phi_0}{\partial x_\mu \partial x_\nu} \right\vert_{r=0} \right\}$ $(\mu, \nu = 1, 2, 3)$ EFG tensor
PAS of EFG tensor: (low magn. fields, i.e., $\langle \mathcal{H}_q \rangle \ll \langle \mathcal{H}_0 \rangle$)	$\mathcal{H}_q = \frac{e^2 qQ}{4I(2I-1)} \left[3I_z^2 - I^2 + \frac{\eta}{2}\{(I^+)^2 + (I^-)^2\} \right]$
definitions:	$\Gamma_{\mu\mu} = \left. \frac{\partial^2 \Phi_0}{\partial x_\mu^2} \right\vert_{r=0} = \frac{1}{4\pi\epsilon_0} \int \frac{3x_\mu^2 - r^2}{r^5} \rho_0(r)\, \mathrm{d}^3 r$ (principal axes values of the EFG tensor) $\Gamma_{11} + \Gamma_{22} + \Gamma_{33} = 0$ $q = \Gamma_{33}/e$ $\eta = (\Gamma_{11} - \Gamma_{22})/\Gamma_{33}$ where $\lvert\Gamma_{33}\rvert \geq \lvert\Gamma_{22}\rvert \geq \lvert\Gamma_{11}\rvert$ and $0 \leq \eta \leq 1$ (anisotropy constant) $Q = \langle m = I \vert Q_{33} \vert m = I \rangle$ (quadrupole moment (PAS)) $Q_{\alpha\beta} = \frac{1}{e} \int \rho_n(r) \left(3x_\alpha x_\beta - \delta_{\alpha\beta} r^2 \right) \mathrm{d}^3 r$ (quadrupole moment tensor)
IST operators (PAS)	$\mathcal{H}_q = a_q \sum\limits_{m=-2}^{2} (-1)^m A_{2,-m} T_{2,m}$ $a_q = \sqrt{\frac{3}{2}} \frac{eQ}{I(2I-1)}$ $T_{2,0} = \frac{1}{\sqrt{6}} \left[3I^{(0)2} - I(I+1) \right]$ $\quad\;\; = \frac{1}{\sqrt{6}} \left[3I_z^2 - I(I+1) \right]$ $T_{2,\pm1} = \frac{1}{\sqrt{2}} \left[I^{(0)} I^{(\pm1)} + I^{(\pm1)} I^{(0)} \right]$ $\quad\;\; = \mp\frac{1}{2} \left[I_z I^\pm + I^\pm I_z \right]$ $T_{2,\pm2} = \left(I^{(\pm1)} \right)^2$ $\quad\;\; = \frac{1}{2} \left(I^\pm \right)^2$ $A_{2,0} = \frac{1}{2} eq$ $A_{2,\pm1} = 0$ $A_{2,\pm2} = \frac{1}{2\sqrt{6}} eq\eta$
laboratory system: (high magn. fields, i.e., $\langle \mathcal{H}_q \rangle \gg \langle \mathcal{H}_0 \rangle$), secular part ($\eta$ arb.)	$\mathcal{H}_q^{(0)} = \frac{e^2 qQ}{8I(2I-1)} \left[3I_z^2 - I(I+1) \right] \left[(3\cos^2 \vartheta - 1) + \eta \sin^2 \vartheta \cos(2\varphi) \right]$
tot. Ham. ($\eta = 0$)	$\mathcal{H}_q = \frac{e^2 qQ}{8I(2I-1)} \big\{ (3\cos^2 \vartheta - 1) \left[3I_z^2 - I(I+1) \right]$ $\qquad\qquad + 3\sin\vartheta \cos\vartheta \left[I_z(I^+ + I^-) + (I^+ + I^-)I_z \right]$ $\qquad\qquad + \frac{3}{2} \sin^2 \vartheta [(I^+)^2 + (I^-)^2] \big\}$ (φ, ϑ azim. + pol. angles of $\boldsymbol{B}_0$ in the PAS of the EFG tensor)

Table 46.5. Representations of the dipole-dipole interaction Hamiltonian for a spin pair I and S separated by the dipole-dipole distance vector r with the Cartesian components x_1, x_2, and x_3. The polar and azimuthal angles are ϑ and φ, respectively (compare Fig. 42.1).

specification	Hamiltonian		
basic form	$\mathcal{H}_d$	$=$	$\frac{\mu_0}{4\pi}\frac{\gamma_I\gamma_S\hbar^2}{r^3}\left\{I\cdot S - \frac{3(I\cdot r)(S\cdot r)}{r^2}\right\}$ $\quad (r>0)$
		$=$	$f_d\sum_{k=-2}^{2}F^{(k)}\mathcal{O}^{(k)}$
		f_d $=$	$\frac{\mu_0}{4\pi}\gamma_I\gamma_S\hbar^2$
		$\mathcal{O}^{(-2)}$ $=$	$-\frac{3}{4}I^-S^-$
		$\mathcal{O}^{(-1)}$ $=$	$-\frac{3}{2}(I^-S_z + I_zS^-)$
		$\mathcal{O}^{(0)}$ $=$	$I_zS_z - \frac{1}{4}(I^+S^- + I^-S^+)$
		$\mathcal{O}^{(1)}$ $=$	$-\frac{3}{2}(I^+S_z + I_zS^+)$
		$\mathcal{O}^{(2)}$ $=$	$-\frac{3}{4}I^+S^+$
		$F^{(0)}$ $=$	$F^{(0)*} = \frac{1}{r^3}(1-3\cos^2\vartheta)$
		$=$	$-\frac{1}{r^3}\sqrt{\frac{16\pi}{5}}\,Y_{2,0}(\vartheta,\varphi)$
		$F^{(1)}$ $=$	$F^{(-1)*} = \frac{1}{r^3}\sin\vartheta\,\cos\vartheta\,e^{-i\varphi}$
		$=$	$\frac{1}{r^3}\sqrt{\frac{8\pi}{15}}\,Y_{2,-1}(\vartheta,\varphi)$
		$F^{(2)}$ $=$	$F^{(-2)*} = \frac{1}{r^3}\sin^2\vartheta\,e^{-2i\varphi}$
		$=$	$\frac{1}{r^3}\sqrt{\frac{32\pi}{15}}\,Y_{2,-2}(\vartheta,\varphi)$
truncated form	$\mathcal{H}_d^{tr}$	$=$	$\frac{\mu_0}{4\pi}\frac{\gamma_I\gamma_S\hbar^2}{r^3}\frac{1-3\cos^2\vartheta}{2}[3I_zS_z - I\cdot S]$
form effective in the weak-coupl. limit	$\mathcal{H}_d^{tr,wc}$	$=$	$\frac{\mu_0}{4\pi}\frac{\gamma_I\gamma_S\hbar^2}{r^3}(1-3\cos^2\vartheta)I_zS_z$ $\left(\frac{\mu_0}{4\pi}\frac{\gamma_I\gamma_S\hbar^2}{r^3}\ll |\omega_I-\omega_S|\right)$
Cartesian tensor	$\mathcal{H}_d$	$=$	$\frac{\mu_0}{4\pi}\gamma_I\gamma_S\hbar^2\,I\cdot D\cdot S$ $D_{kl} = -\frac{3x_kx_l-r^2\delta_{kl}}{r^5}$ $\quad (k,l=1,2,3)$
IST operators	$\mathcal{H}_d$	$=$	$a_d\sum_{m=-2}^{2}(-1)^m A_{2,-m}T_{2,m}$
		a_d $=$	$-\frac{\mu_0}{4\pi}\frac{\gamma_I\gamma_S\hbar^2}{r^3}$
		$T_{2,0}$ $=$	$\frac{1}{\sqrt{6}}[3I^{(0)}S^{(0)} - I\cdot S]$
		$=$	$\frac{1}{\sqrt{6}}[3I_zS_z - I\cdot S]$
		$T_{2,\pm1}$ $=$	$\frac{1}{\sqrt{2}}[I^{(\pm1)}S^{(0)} + I^{(0)}S^{(\pm1)}]$
		$=$	$\mp\frac{1}{2}(I_zS^\pm + I^\pm S_z)$
		$T_{2,\pm2}$ $=$	$-I^{(\pm1)}S^{(\pm1)}$
		$=$	$\frac{1}{2}I^\pm S^\pm$
		$A_{2,-m}$ $=$	$\sqrt{\frac{24\pi}{5}}\,Y_{2,-m}(\vartheta,\varphi)$

Two cases can be distinguished: $r = 0$, i.e., the electron is at the position of the nucleus, and $r > 0$. It can be shown that term (b) vanishes in the limit $r = 0$ whereas term (a) may be reformulated with the aid of the relation $\nabla^2(1/r) = -4\pi\delta(r)$ as

$$\mathcal{H}_F = -\frac{2}{3}\mu_0\,(\boldsymbol{\mu}_e \cdot \boldsymbol{\mu}_n)\delta(r) \tag{46.21}$$

This result represents the **Fermi contact interaction** between nuclear and electronic magnetic dipoles. In context with nuclear relaxation, it is also called **scalar interaction** because of its orientation independent character (by contrast to dipolar coupling and in principle also to indirect interaction of two nuclei via the electrons within a molecule).

On the other hand, in the case $r > 0$, the Fermi contact interaction obviously vanishes whereas the term (b) after carrying out the derivatives takes the form

$$\mathcal{H}_d = \frac{\mu_0}{4\pi}\,\frac{(\boldsymbol{\mu}_e \cdot \boldsymbol{\mu}_n)r^2 - 3(\boldsymbol{\mu}_n \cdot \boldsymbol{r})(\boldsymbol{\mu}_e \cdot \boldsymbol{r})}{r^5} \tag{46.22}$$

which we recognize as the (electron-nuclear) dipole-dipole interaction Hamiltonian. The total Hamiltonian is thus represented by

$$\mathcal{H}_{IS} = \mathcal{H}_F + \mathcal{H}_d \tag{46.23}$$

With diamagnetic molecules, first-order perturbation theory does not yield any effect on the basis of the above Hamiltonians because all electron spins are paired in the ground state. The first non-vanishing contribution is found with the aid of second-order perturbation theory [388, 394]. It turns out that the energy shifts provided by the Fermi contact interaction dominate the indirect coupling of nuclei via the electron shell by far. This so-called J coupling is mainly an effect of the finite probability density of electrons at the positions of the coupled nuclei.

46.3.3
Quadrupolar and Spin-Rotation Couplings

The derivation of the **quadrupolar Hamiltonian** requires somewhat more extended considerations. This is delineated in a separate section based on the multipole expansion of the electrostatic energy (Sect. 49.6). Various versions of the result are summarized in Table 46.4.

The **spin-rotation interaction** is mentioned here for completeness. It refers to couplings of nuclear dipoles with magnetic dipoles of the rotating charge distribution of the molecule, i.e., of the nuclear spin with the molecular angular momentum. It is clear that this sort of interaction only becomes perceptible in gases or liquids with rather low viscosity.

CHAPTER 47

The Density Operator

47.1
Definition of the Density Operator

Density operator treatments [140] are appropriate for all cases dealing with state mixtures. A **"state mixture"** is understood to characterize an ensemble of independent systems, each of them having a pure but potentially incoherent state. A **"pure state"** is characterized by a certain linear combination of the eigenstate wavefunctions (a certain vector in Hilbert space). As a special case, the state of a system may be one of its eigenfunctions themselves, so that an eigenfunction may also represent a pure state. The ensemble described by a state mixture (a distribution of vectors in Hilbert space) therefore implies a distribution of the coefficients of the linear combination forming the pure states of the ensemble members.

The crucial point is now that in the case of a state mixture the expectation value of an observable is not determined by a pure state function. Rather, one must form the average over the ensemble of independent systems. This is most conveniently done with the aid of the density operator.

Let us consider pure states characterized by the ket

$$|\Psi_n\rangle = \sum_{m=1}^{z} c_m^{(n)} |\varphi_m\rangle = \begin{pmatrix} c_1^{(n)} \\ \cdot \\ \cdot \\ \cdot \\ c_z^{(n)} \end{pmatrix} \tag{47.1}$$

and the bra

$$\langle\Psi_n| = \sum_{m'=1}^{z} c_{m'}^{(n)*} \langle\varphi_n| = (c_1^{(n)*}, \ldots, c_z^{(n)*}) \tag{47.2}$$

The functions φ_i represent eigenstates. The coefficients $c_j^{(n)}$ are complex constants with normalized squared magnitudes. The sub- or superscripts n indicate the state vectors occurring in the ensemble of identical quantum-mechanical systems. Within this ensemble, the state vectors of the systems may vary with respect to their eigenstate compositions (magnitudes of the coefficients $c_j^{(n)}$) as well as regarding the phases of the eigenfunction contributions (complex phase factors of the coefficients $c_j^{(n)}$).

With the statistical weight w_n of the nth pure state within the ensemble, the density operator is defined by

$$\rho = \sum_n w_n |\Psi_n\rangle\langle\Psi_n|$$

$$= \sum_{n,m',m} w_n c_m^{(n)} c_{m'}^{(n)*} |\varphi_m\rangle\langle\varphi_{m'}| \qquad (47.3)$$

The density-matrix elements are then

$$\rho_{ij} = \langle\varphi_i|\rho|\varphi_j\rangle = \sum_n w_n c_i^{(n)} c_j^{(n)*} = \overline{c_i^{(n)} c_j^{(n)*}} \qquad (47.4)$$

A direct consequence of this definition of the density operator is that the (ensemble average) expectation value of an operator $\mathcal{O}$ is given by the relation

$$\boxed{\overline{\langle\mathcal{O}\rangle} = \text{Tr}\{\rho\mathcal{O}\} = \text{Tr}\{\mathcal{O}\rho\}} \qquad (47.5)$$

From this equation, it follows in particular that the **magnetization vector** is related to the spin vector operator by

$$M = n\gamma_n\hbar\text{Tr}\{\rho I\} = n\gamma_n\hbar b\text{Tr}\{\sigma I\} \qquad (47.6)$$

where n is the number density of spins. σ is the **reduced density operator** (see footnote 2 on page 431). The factor b is defined in Eq. 47.27 for thermal-equilibrium conditions.[1] With the rules of Chap. 43, it can easily be perceived that the three components of M are determined by density operator terms implying the spin operators I_x, I_y, and I_z respectively. The **complex transverse magnetization** is given by

$$\boxed{m = M_x + iM_y = n\gamma_n\hbar b\text{Tr}\{(I_x + iI_y)\sigma\}} \qquad (47.7)$$

The density operator implies the information of an ensemble of spin systems as far as it is accessible by measurements at all. It is therefore the quantity which is - explicitly or implicitly - employed in all treatments concerning the evolution of spin coherences, longitudinal spin polarizations, and spin order. For a general introduction into the theory of density operators the reader is referred to [50, 140].

The diagonal elements of the density matrix represent the populations of the eigenstates,

$$\rho_{ii} = \overline{|c_i^{(n)}|^2} \qquad (47.8)$$

---

[1]The quantity b is also valid for non-equilibrium orientations of the local magnetization as long as the magnitude is not affected. That is, the spin states may be arbitrarily manipulated by RF pulses, but the magnitude of the local magnetization must not be subject to irreversible processes such as relaxation or diffusion.

obeying

$$\text{Tr}\{\rho\} = 1 \tag{47.9}$$

The difference of the populations of spin-up and spin-down states is characterized as the "longitudinal spin polarization," P_z, which is proportional to the longitudinal (or simply z) magnetization, M_z (see the following section). In the case of uncoupled spins 1/2, these quantities are given by

$$P_z = \overline{|c_1^{(n)}|^2} - \overline{|c_2^{(n)}|^2} \tag{47.10}$$

$$M_z = \frac{1}{2}P_z n \gamma_n \hbar \tag{47.11}$$

"Longitudinal spin order" or simply "spin order" indicates that populations of the spin states are correlated with the interaction state. For instance, a spin-up state of the coupling partner may be accompanied by a high population whereas the spin-down state means the opposite. A typical procedure for producing spin order during a certain interval is the Jeener/Broekaert pulse sequence [208] originally proposed for dipolar coupling in solids. It is, however, of a quite general nature and leads to "longitudinal scalar order" or "J order" if indirect coupling dominates, "dipolar order" if dipole-dipole interaction prevails, and "quadrupolar order" for quadrupole nuclei. It should be noted that spin-order terms in density-operator expressions do not include longitudinal polarization of the spins. The spin-up and spin-down partial z magnetizations cancel. Superimposed longitudinal polarizations are represented by separate terms.

Finite values of the off-diagonal elements,

$$\rho_{ij} = \rho_{ji}^* = \overline{c_i^{(n)} c_j^{(n)*}} \qquad (i \neq j) \tag{47.12}$$

indicate "spin coherences" within the ensemble. This includes transverse magnetization if the states "i" and "j" are linked by a single-quantum transition, i.e., $i - j = \pm 1$. It is important to note that zero- and multiple-quantum transitions (i.e., $i - j = 0$ and $i - j = \pm n$, $n = 2, 3, 4, \ldots$, respectively) are addressed alike. In this case, however, the transitions are not connected with detectable electromagnetic radiation. Dipole radiation of this sort is forbidden because it violates the conservation of angular momentum. As radiationless phenomena, **multiple-quantum coherences** nevertheless exist and can be detected indirectly by conversion into single-quantum coherences after a certain evolution interval. It is also evident that in thermal equilibrium there are no coherences, i.e., the off-diagonal elements of the density matrix vanish.

47.2
Thermal Equilibrium

Consider an ensemble of independent spin systems with a Hamiltonian $\mathcal{H}$. "Independent" means that the wavefunctions form non-overlapping entities. In this case, the population or statistical weight of a state i of the systems is determined by the

Boltzmann factor, i.e.,

$$w_i = \frac{e^{-E_i/(k_B T)}}{\sum_m e^{-E_m/(k_B T)}} \tag{47.13}$$

where

$$E_i = \langle \varphi_i | \mathcal{H} | \varphi_i \rangle \tag{47.14}$$

is the eigenenergy corresponding to the eigenket $|\varphi_i\rangle$, k_B is Boltzmann's constant, and T the absolute temperature. The denominator is the partition function. With Eqs. 45.1 and 47.14 the Boltzmann factor becomes

$$e^{-E_i/(k_B T)} = e^{-\langle \varphi_i | \mathcal{H} | \varphi_i \rangle/(k_B T)} = \langle \varphi_i | e^{-\mathcal{H}/(k_B T)} | \varphi_i \rangle \tag{47.15}$$

so that

$$w_i = \frac{\langle \varphi_i | e^{-\mathcal{H}/(k_B T)} | \varphi_i \rangle}{\sum_m \langle \varphi_m | e^{-\mathcal{H}/(k_B T)} | \varphi_m \rangle} = \frac{\langle \varphi_i | e^{-\mathcal{H}/(k_B T)} | \varphi_i \rangle}{\mathrm{Tr}\left\{ e^{-\mathcal{H}/(k_B T)} \right\}} \tag{47.16}$$

The diagonal matrix elements of the equilibrium density operator are defined by the equilibrium populations,

$$\langle \varphi_i | \rho_0 | \varphi_i \rangle = w_i \tag{47.17}$$

It follows by comparison that

$$\rho_0 = \frac{e^{-\mathcal{H}/(k_B T)}}{\mathrm{Tr}\left\{ e^{-\mathcal{H}/(k_B T)} \right\}} \tag{47.18}$$

The equilibrium density matrix elements are consequently

$$\{\rho_0\}_{ij} = \frac{e^{-E_i/(k_B T)}}{\sum_m e^{-E_m/(k_B T)}} \delta_{ij} \tag{47.19}$$

Off-diagonal elements are zero as it must be.

In the high-temperature limit, $E_i/(k_B T) \ll 1$, the **equilibrium density operator** can be written in linear approximation as

$$\rho_0 \approx \frac{1 - \mathcal{H}/(k_B T)}{\mathrm{Tr}\left\{ 1 - \mathcal{H}/(k_B T) \right\}} \tag{47.20}$$

For spins I in a field $B_0 = B_0 u_z$, the Zeeman Hamilton operator is

$$\mathcal{H}_0 = -\mu \cdot B_0 = -\gamma_n \hbar B_0 I_z \tag{47.21}$$

With

$$\mathrm{Tr}\left\{ 1 + \frac{\gamma_n \hbar B_0 I_z}{k_B T} \right\} = \mathrm{Tr}\{\mathcal{E}\} \tag{47.22}$$

we obtain[2]

$$\rho_0 \approx \frac{1+\gamma_n \hbar B_0 I_z/(k_B T)}{\text{Tr}\{\mathcal{E}\}}$$

$$= a + b\,\sigma_0 \qquad (47.24)$$

$$= a + b\,I_z$$

where

$$\sigma_0 = I_z \qquad (47.25)$$

$$a \equiv \frac{1}{\text{Tr}\{\mathcal{E}\}} \qquad (47.26)$$

$$b \equiv \frac{\gamma_n \hbar B_0}{k_B T \text{Tr}\{\mathcal{E}\}} \qquad (47.27)$$

For uncoupled spins the denominator in Eq. 47.24 is

$$\text{Tr}\{\mathcal{E}\} = 2I + 1 \qquad (47.28)$$

With the equilibrium density operator in the high-temperature approximation, the equilibrium magnetization M_0 is given by

$$M_0 = n\text{Tr}\{\rho_0 \mu\} \qquad (47.29)$$

where μ is the dipole moment, and n is the number density of the spins. This leads to $M_{0x} = M_{0y} = 0$ and

$$M_{0z} = M_0 = \frac{n\gamma_n \hbar}{2I + 1} \text{Tr}\left\{I_z + \frac{\gamma_n \hbar B_0}{k_B T} I_z^2\right\} \qquad (47.30)$$

After evaluation of the trace (see Chap. 43), we find the **Curie magnetization**

$$M_0 = \frac{n\gamma_n^2 \hbar^2 I(I + 1)}{3k_B T} B_0 \qquad (47.31)$$

which is valid for paramagnetic particles in the high-temperature approximation.

---

[2] The constants a and b defined by Eq. 47.24 are irrelevant for the evolution of spin coherences. This is demonstrated with Eq. 2.12, for instance, for the density operator description of the Hahn spin echo. Therefore it is often more convenient to treat coherence evolution pathways by the **reduced density operator** σ defined by

$$\rho(t) = a + b\sigma(t) \qquad (47.23)$$

For this definition the high-temperature limit must be assumed, of course. Note also that the reduced density operator σ does not comply to the normalization rule (Eq. 47.9).

47.3
Evolution of the Density Operator

Starting from the thermal equilibrium, for instance, the density operator evolves in the course of time as a consequence of RF field irradiation and the influence of spin interactions. The general equation of motion is the **Liouville/von Neumann equation**

$$\frac{\partial \rho(t)}{\partial t} = \frac{i}{\hbar} \left[\rho(t), \mathcal{H} \right]$$

(47.32)

Provided that there is no explicit time dependence of the Hamilton operator, the solution is given by the "sandwich" expression

$$\rho(t) = e^{-(i/\hbar)\mathcal{H}t} \rho(0) e^{(i/\hbar)\mathcal{H}t}$$

(47.33)

This solution for time-independent Hamiltonians can easily be verified by inserting it into Eq. 47.32. The time evolution operator,

$$U(t) \equiv e^{-(i/\hbar)\mathcal{H}t}$$

(47.34)

is also called "propagator." Note that the solution (Eq. 47.33) is of the unitary-transformation type which will be considered in more detail in Chap. 48. If the Hamiltonian refers to the Zeeman interaction this expression is equivalent to the precession of the magnetization about the z direction (see Eq. 48.58). Generalizing this issue to Hamiltonians of the same analytical form, we speak of "precession-type solutions."

In practical situations, the total Hamilton operator often is a time-dependent function at least because in pulsed NMR experiments the RF is switched on and off. The formal solution of the Liouville/von Neumann equation is then obtained with the aid of the propagator for time dependent Hamiltonians,

$$U(t) = \mathcal{T} \exp \left\{ -\frac{i}{\hbar} \int_0^t \mathcal{H}(t') \, dt' \right\}$$

(47.35)

where $\mathcal{T}$ is the Dyson time ordering operator. It ensures that the propagator acts in the proper order of the infinitesimal intervals dt' in case the Hamilton operators, at different times t' and t'', are not independent from each other, i.e., are not commuting: $[\mathcal{H}(t'), \mathcal{H}(t'')] \neq 0$.

The propagator (Eq. 47.35) is often expanded in analogy to the Baker/Campbell/Hausdorff formula (Eq. 45.5). The result is the Magnus expansion [315] which reads

$$U(t) = \exp\{-(i/\hbar)[\overline{\mathcal{H}}^{(0)} + \overline{\mathcal{H}}^{(1)} + \overline{\mathcal{H}}^{(2)} + \ldots]t\}$$

(47.36)

where

$$\overline{\mathcal{H}}^{(0)} = \frac{1}{t} \int_0^t \mathcal{H}(t') \, dt'$$

(47.37)

$$\overline{\mathcal{H}}^{(1)} = -\frac{i}{2t} \int_0^t dt_2 \int_0^{t_2} dt_1 \, [\mathcal{H}(t_2), \mathcal{H}(t_1)] \tag{47.38}$$

$$\overline{\mathcal{H}}^{(2)} = -\frac{1}{6t} \int_0^t dt_3 \int_0^{t_3} dt_2 \int_0^{t_2} dt_1 \, \{[\mathcal{H}(t_3), [\mathcal{H}(t_2), \mathcal{H}(t_1)]]$$
$$+ [\mathcal{H}(t_1), [\mathcal{H}(t_2), \mathcal{H}(t_3)]]\} \tag{47.39}$$

The two principal means for achieving a situation where the precession-type version of the solution (Eq. 47.33) applies (see Eq. 48.58) are

- segmented treatments of pulse sequences as delineated in the following section
- unitary transformations to a reference frame in which the Hamiltonian merely contains z components of the spin operators, that is, components along the direction of quantization.[3]

47.3.1
Segmented Treatments

Frequently, it is possible to treat pulse sequences on the basis of Eq. 47.33 by splitting the pulse train into several intervals τ_j in which the Hamilton operator takes stationary values $\mathcal{H}_j$. Such intervals used to be demarcated by RF pulses and free-evolution periods. In this case, the solution (Eq. 47.33) can be written as intercalated sandwich operators. The result of the propagator action in each time step forms the initial value of the density operator in the subsequent interval. In a period t segmented into n such intervals, the density operator is propagated according to

$$\rho\left(t = \sum_{j=1}^n \tau_j\right) =$$

$$\underbrace{e^{-(i/\hbar)\mathcal{H}^{(n)}\tau_n} \dots e^{-(i/\hbar)\mathcal{H}^{(2)}\tau_2} \overbrace{e^{-(i/\hbar)\mathcal{H}^{(1)}\tau_1} \underbrace{\rho(0)}_{} e^{(i/\hbar)\mathcal{H}^{(1)}\tau_1}}^{\rho(\tau_1+\tau_2)} e^{(i/\hbar)\mathcal{H}^{(2)}\tau_2} \dots e^{(i/\hbar)\mathcal{H}^{(n)}\tau_n}}_{\rho(\tau_1+\dots+\tau_n)} \tag{47.40}$$

The evaluation of Eq. 47.40 for a given pulse sequence and secular Hamiltonians is convenient if the effective field is aligned along the z axis of the rotating reference frame. The task is therefore reduced to the transformation of the Hamiltonian to the appropriate coordinate system. This is relevant in the presence of offsets by stationary spin interactions, RF irradiation, or magnetic field gradients leading to effective fields tilted with respect to the rotating frame axes. Evolution is then treated by transforming the Hamiltonian to the tilted rotating frame (see Sect. 48.5). Non-

---

[3]This operation is sometimes called "diagonalization" of the Hamiltonian in allusion to its matrix representation.

secular contributions arising as a consequence of these transformations must be truncated again, of course. Examples are given in Sects. 48.9 to 48.11.

Pulse sequences can produce complicated superpositions of spin coherences, longitudinal spin polarizations, and spin order following different coherence pathways. These become more transparent by the application of product operator treatments which are feasible for small spin systems (see Chap. 51). Note that these convenient formalisms are entirely equivalent variants of the more general propagation rule at Eq. 47.40.

47.3.2
Average Hamiltonians

A second source of time dependence is molecular dynamics leading to fluctuating spin interactions. This is the origin of relaxation processes and, under appropriate conditions, of motional averaging of anisotropic contributions to the spin interactions. In the treatment of coherence evolution during pulse sequences, relaxation can normally be ignored or considered separately as an independent phenomenon. Motional averaging, on the other hand, is crucial for any sort of high-resolution NMR.

Let τ_c be a time constant characteristic for the time variation of the Hamiltonian. In context with stochastic fluctuations it is identical with the correlation time.[4] Motional averaging is then effective in a coherence-evolution period τ if the condition

$$\tau \gg \tau_c \tag{47.41}$$

is fulfilled. The interval τ is considered to be shorter than the mean coherence lifetime, i.e., the transverse relaxation time. Then the evolution is effectively governed by the average Hamiltonian

$$\overline{\mathcal{H}} = \frac{1}{\tau} \int\limits_0^\tau \mathcal{H}(\tau') \, d\tau' \tag{47.42}$$

In context with low-viscous isotropic liquids, the condition at Eq. 47.41 is normally uncritical for the segmented treatment of pulse sequences as sketched above (Eq. 47.40). Solid-like and anisotropic liquids such as mesomorphic systems, however, can have correlation times reaching or even exceeding the usual time scale in which intervals are incremented in spin-echo or phase-encoding pulse experiments. Incrementing a pulse delay in this case means that the average Hamiltonian of this interval is also changed, and - hence - is not a constant any more. The evolution of spin coherences is then superimposed by correlation effects of the Hamiltonian between the free evolution intervals.

---

[4]Note that motional averaging does not only occur in context with fast Brownian motions but can also be produced artificially either by rapidly spinning the sample [430] or - in a somewhat generalized sense - by applying multipulse sequences. The conditions for the applicability of average Hamiltonians for periodic time dependences of the Hamiltonian (and "stroboscopic" signal acquisition) have been examined in detail in the literature [139, 178, 179].

CHAPTER 48

Unitary Transformations in NMR

Unitary transformations are ubiquitous in treatments of NMR problems. They are based on unitary operators or matrices (see Eq. 42.42) which are represented in the form of exponential operators (Chap. 45). Examples of applications are the evolution of spin coherences in the presence of time-independent Hamilton operators, the transformation between coordinate systems rotated against each other, or the rotational diffusion of molecules.

A major part of the following sections formally refers to rotations of the reference frame leaving the physical object untouched. Such transformations are formally equivalent to the opposite rotation of the physical object leaving the reference frame untouched. Therefore, one often refers to "rotation" as such, leaving it a matter of the context whether the rotation refers to the reference frame or to the physical object. Note, however, that the sign of the rotation angles is reversed with these two views.

48.1
Diagonalization of a Matrix

48.1.1
The Eigenvalue Problem

A matrix is diagonalized, i.e., transformed to the principal-axes system, by solving the eigenvalue problem. This is best demonstrated by considering a given 2×2 matrix as a simple example,

$$A \equiv \begin{pmatrix} A_{11} & A_{12} \\ A_{21} & A_{22} \end{pmatrix} \tag{48.1}$$

The eigenvalue equation reads

$$\begin{pmatrix} A_{11} & A_{12} \\ A_{21} & A_{22} \end{pmatrix} \begin{pmatrix} c_1 \\ c_2 \end{pmatrix} = \lambda \begin{pmatrix} c_1 \\ c_2 \end{pmatrix} \tag{48.2}$$

or

$$\begin{pmatrix} A_{11} - \lambda & A_{12} \\ A_{21} & A_{22} - \lambda \end{pmatrix} \begin{pmatrix} c_1 \\ c_2 \end{pmatrix} = 0 \tag{48.3}$$

where λ stands for the eigenvalues, and c_1 and c_2 for the components of the eigenvectors. Equation 48.3 is equivalent to a set of two linear homogeneous equations which possess nontrivial solutions only if the **secular determinant** vanishes:

$$\begin{vmatrix} A_{11} - \lambda & A_{12} \\ A_{21} & A_{22} - \lambda \end{vmatrix} = 0 \tag{48.4}$$

This is the so-called **characteristic equation**. In the present case, it is equivalent to an equation of second order in λ:

$$(A_{11} - \lambda)(A_{22} - \lambda) - A_{12}A_{21} = 0 \tag{48.5}$$

The solutions are the two eigenvalues λ_1 and λ_2. The eigenvector components can then be determined for each eigenvalue by inserting the eigenvalues in Eq. 48.3. For the first eigenvalue, for instance, this gives

$$\begin{pmatrix} A_{11} - \lambda_1 & A_{12} \\ A_{21} & A_{22} - \lambda_1 \end{pmatrix} \begin{pmatrix} c_1^{(1)} \\ c_2^{(1)} \end{pmatrix} = 0 \tag{48.6}$$

or

$$(A_{11} - \lambda_1)c_1^{(1)} + A_{12}c_2^{(1)} = 0 \tag{48.7}$$

$$(A_{22} - \lambda_1)c_1^{(1)} + A_{21}c_2^{(1)} = 0 \tag{48.8}$$

That is, we have a system of two linear equations determining the components $c_1^{(1)}$ and $c_2^{(1)}$ of the first eigenvector. Likewise, the system of linear equations for the components of the second eigenvector reads

$$(A_{11} - \lambda_2)c_1^{(2)} + A_{12}c_2^{(2)} = 0 \tag{48.9}$$

$$(A_{22} - \lambda_2)c_1^{(2)} + A_{21}c_2^{(2)} = 0 \tag{48.10}$$

With the two eigenvectors

$$c^{(1)} = \begin{pmatrix} c_1^{(1)} \\ c_2^{(1)} \end{pmatrix} ; \qquad c^{(2)} = \begin{pmatrix} c_1^{(2)} \\ c_2^{(2)} \end{pmatrix} \tag{48.11}$$

and the two complex eigenvalues λ_1 and λ_2 we have solved Eq. 48.2. If the matrix A is symmetric, i.e., $A_{12} = A_{21}$, the eigenvalues are real, and the eigenvectors are orthogonal to each other.

48.1.2
Transformation Matrices

On the same basis, the matrices U with the eigenvectors as columns,

$$U \equiv \begin{pmatrix} c_1^{(1)} & c_1^{(2)} \\ c_2^{(1)} & c_2^{(2)} \end{pmatrix} \tag{48.12}$$

and Λ with the eigenvalues as diagonal elements,

$$\Lambda \equiv \begin{pmatrix} \lambda_1 & 0 \\ 0 & \lambda_2 \end{pmatrix} \tag{48.13}$$

may be defined. It is then easy to show that the eigenvalue equation at Eq. 48.2 can be rewritten

$$\begin{pmatrix} A_{11} & A_{12} \\ A_{21} & A_{22} \end{pmatrix} \begin{pmatrix} c_1^{(1)} & c_1^{(2)} \\ c_2^{(1)} & c_2^{(2)} \end{pmatrix} = \begin{pmatrix} c_1^{(1)} & c_1^{(2)} \\ c_2^{(1)} & c_2^{(2)} \end{pmatrix} \begin{pmatrix} \lambda_1 & 0 \\ 0 & \lambda_2 \end{pmatrix} \tag{48.14}$$

or in a symbolized form,

$$AU = U\Lambda \tag{48.15}$$

Multiplying this equation by the inverse matrix U^{-1} either from the right-hand or from the left-hand sides, and using the property $UU^{-1} = \mathcal{E}$ (unity matrix), gives the matrix transformation equations

$$A = U\Lambda U^{-1} \tag{48.16}$$

$$\Lambda = U^{-1}AU \tag{48.17}$$

These equations obviously link the matrix A with its diagonal matrix Λ and vice versa.

48.1.3
The Inverse Matrix

If the matrix A is symmetric, the eigenvectors are orthogonal on each other, and the inverse eigenvector matrix, U^{-1}, is equal to the transposed matrix, i.e., $U^{-1} = \tilde{U}$ with the elements $(\tilde{U})_{ij} = (U)_{ji}$. Otherwise the inverse matrix of

$$U \equiv \begin{pmatrix} U_{11} & U_{12} \\ U_{21} & V_{22} \end{pmatrix} \tag{48.18}$$

is formed according to

$$U^{-1} = \frac{1}{||U||} \begin{pmatrix} V_{11} & V_{21} \\ V_{12} & V_{22} \end{pmatrix} \tag{48.19}$$

where the determinant of U is given by

$$||U|| = U_{11}U_{22} - U_{12}U_{21} \tag{48.20}$$

The algebraic complement V_{ik} of U is generally equal to $(-1)^{i+k}$ times the subdeterminant obtained from the residual matrix of U after crossing off the i-th row and the k-th column. In the case of 2×2 matrices considered here, we have

$$\begin{pmatrix} V_{11} & V_{21} \\ V_{12} & V_{22} \end{pmatrix} = \begin{pmatrix} U_{22} & -U_{12} \\ -U_{21} & U_{11} \end{pmatrix} \tag{48.21}$$

so that the inverse matrix reads

$$U^{-1} = \frac{1}{U_{11}U_{22} - U_{12}U_{21}} \begin{pmatrix} U_{22} & -U_{12} \\ -U_{21} & U_{11} \end{pmatrix} \tag{48.22}$$

48.2
Coordinate Systems

In NMR, three different coordinate systems are in use depending on the problem to be described. The coordinates of a vector

$$r = \begin{pmatrix} x \\ y \\ z \end{pmatrix}$$

in the standard *right-handed rectangular Cartesian system* are related to *polar coordinates*[1], r (magnitude), ϑ ("polar angle" spanned between the z-axis and r), and φ ("azimuthal angle" spanned between the x-axis and the projection of r on the x, y plane), by (compare Fig. 42.1)

$$
\begin{array}{lll}
x = r \sin \vartheta \cos \varphi, & r = \sqrt{x^2 + y^2 + z^2} & (0 \leq r < \infty) \\
y = r \sin \vartheta \sin \varphi, & \vartheta = \arccos \frac{z}{\sqrt{x^2+y^2+z^2}} & (0 \leq \vartheta \leq \pi) \\
z = r \cos \vartheta, & \varphi = \arccos \frac{x}{\sqrt{x^2+y^2}} & (0 \leq \varphi \leq 2\pi)
\end{array}
$$

$$(48.23)$$

The coordinates in the *spherical system* (in the proper sense), x_{+1}, x_0, x_{-1}, are related to those in the rectangular Cartesian and polar systems by

$$
\begin{array}{rclcl}
x_{+1} & = & -\frac{1}{\sqrt{2}}(x + iy) & = & -\frac{1}{\sqrt{2}}r \sin \vartheta \, e^{i\varphi}, \\
x_0 & = & z & = & r \cos \vartheta, \\
x_{-1} & = & \frac{1}{\sqrt{2}}(x - iy) & = & \frac{1}{\sqrt{2}}r \sin \vartheta \, e^{-i\varphi},
\end{array}
\qquad
\begin{array}{l}
x = -\frac{1}{\sqrt{2}}(x_{+1} - x_{-1}) \\
y = \frac{i}{\sqrt{2}}(x_{+1} + x_{-1}) \\
z = x_0
\end{array}
$$

$$(48.24)$$

48.3
Euler Angles

The Euler angles, $\tilde{\alpha}, \tilde{\beta}, \tilde{\gamma}$, specify an arbitrary rotation of an orthogonal Cartesian coordinate system $S\{x, y, z\}$ about the origin. The rotation is performed in three steps,

$$S\{x, y, z\} \xrightarrow{\tilde{\alpha}} S'\{x', y', z'\} \xrightarrow{\tilde{\beta}} S''\{x'', y'', z''\} \xrightarrow{\tilde{\gamma}} S'''\{x''', y''', z'''\}$$

defined by[2]

        a) a rotation about the original z axis through an angle $\tilde{\alpha}$
        b) a rotation about the new y' axis through an angle $\tilde{\beta}$
        c) a rotation about the new z'' axis through an angle $\tilde{\gamma}$

---

[1]In the literature the latter system is often called "spherical." Note that the third type of coordinate system is also termed "spherical" so that one must be somewhat careful to avoid confusion.

[2]We are applying the convention used by Brink and Satchler [63] and Rose [422], for instance. Note that other definitions are also common in the literature.

Positive angles are defined by the right-hand screw sense. Cartesian (column) vectors, r, and Cartesian tensors, σ, are transformed to a correspondingly rotated frame by

$$r''' = R r \quad \text{and} \quad \sigma''' = R \sigma R^{-1},\tag{48.25}$$

respectively. The Cartesian rotation matrix is [63, 422]

$$R = \begin{pmatrix} \cos\tilde\alpha \cos\tilde\beta \cos\tilde\gamma - \sin\tilde\alpha \sin\tilde\gamma & \sin\tilde\alpha \cos\tilde\beta \cos\tilde\gamma + \cos\tilde\alpha \sin\tilde\gamma & -\sin\tilde\beta \cos\tilde\gamma \\ -\cos\tilde\alpha \cos\tilde\beta \sin\tilde\gamma - \sin\tilde\alpha \cos\tilde\gamma & -\sin\tilde\alpha \cos\tilde\beta \sin\tilde\gamma + \cos\tilde\alpha \cos\tilde\gamma & \sin\tilde\beta \sin\tilde\gamma \\ \cos\tilde\alpha \sin\tilde\beta & \sin\tilde\alpha \sin\tilde\beta & \cos\tilde\beta \end{pmatrix}\tag{48.26}$$

The inverse matrix R^{-1} can in this case be formed as the transposed matrix, i.e., $R^{-1} = \tilde R$ with the elements $(\tilde R)_{ij} = (R)_{ji}$.

For example, a rotation of the reference frame through an angle δ about the z axis is generated by $\delta = \tilde\alpha$; $\tilde\beta = \tilde\gamma = 0$, or, equivalently, by $\tilde\alpha = \tilde\beta = 0$; $\tilde\gamma = \delta$. That is, the unit vectors u_1, u_2, u_3 along the x, y and z axes in S, respectively,

$$u_1 = \begin{pmatrix} 1 \\ 0 \\ 0 \end{pmatrix}; \quad u_2 = \begin{pmatrix} 0 \\ 1 \\ 0 \end{pmatrix}; \quad u_3 = \begin{pmatrix} 0 \\ 0 \\ 1 \end{pmatrix}\tag{48.27}$$

appear in S''' as

$$\begin{aligned} u_1''' &= R(\tilde\alpha = \delta, \tilde\beta = 0, \tilde\gamma = 0)u_1 = R(\tilde\alpha = 0, \tilde\beta = 0, \tilde\gamma = \delta)u_1 \\ &= \begin{pmatrix} \cos\delta \\ -\sin\delta \\ 0 \end{pmatrix} \\ u_2''' &= R(\tilde\alpha = \delta, \tilde\beta = 0, \tilde\gamma = 0)u_2 = R(\tilde\alpha = 0, \tilde\beta = 0, \tilde\gamma = \delta)u_2 \\ &= \begin{pmatrix} \sin\delta \\ \cos\delta \\ 0 \end{pmatrix} \\ u_3''' &= R(\tilde\alpha = \delta, \tilde\beta = 0, \tilde\gamma = 0)u_3 = R(\tilde\alpha = 0, \tilde\beta = 0, \tilde\gamma = \delta)u_3 \\ &= \begin{pmatrix} 0 \\ 0 \\ 1 \end{pmatrix} \end{aligned}\tag{48.28}$$

respectively.

48.4
Transformation of the Chemical-Shift Tensor

The chemical-shift screening tensor in the principal-axes system $(S\{x_p, y_p, z_p\})$ fixed on the molecule is given by

$$\sigma_{cs} = \begin{pmatrix} \sigma_{11} & 0 & 0 \\ 0 & \sigma_{22} & 0 \\ 0 & 0 & \sigma_{33} \end{pmatrix}\tag{48.29}$$

where σ_{ii} are the principal-axes values. The transformation equations 48.25 bringing the principal axes system into coincidence with the laboratory frame[3] yields the third diagonal element of the tensor relative to the new reference frame as a function of the original diagonal elements,

$$\sigma_{zz} = \sigma_{11} \cos^2 \tilde{\alpha} \sin^2 \tilde{\beta} + \sigma_{22} \sin^2 \tilde{\alpha} \sin^2 \tilde{\beta} + \sigma_{33} \cos^2 \tilde{\beta} \qquad (48.30)$$

As the external magnetic field is aligned along the z axis of the laboratory frame, the tensor element (Eq. 48.30) determines the secular part of the chemical-shift Hamiltonian. It is independent of the Euler angle $\tilde{\gamma}$ because this would describe rotations about the magnetic field direction which are irrelevant for NMR.[4]

48.5
Transformations by Single-Spin Operators

In the following we consider rotations either of the reference frame or of the object (the particle or the vectors and tensors linked to the particle orientation). Let the rotation angle be φ and the rotation axis coincide with the k axis of the (orthogonal Cartesian) reference frame.

The unitary operator generating

either

a **counterclockwise** rotation of the **reference frame** through an angle φ when looking along the positive k axis,

or

a **clockwise** rotation of the **object** through an angle φ when looking along the positive k axis

is given by

$$R(\varphi) = e^{-i\varphi I_k} \qquad (k = x, y, z) \qquad (48.31)$$

Changing the sign of φ changes the rotation sense.

Any operator $\mathcal{O}$ complying with the cyclic commutation rules at Eq. 44.10 is transformed to an operator $\mathcal{O}'$ according to[5]

$$\mathcal{O}' = R\,\mathcal{O}\,R^{-1} = e^{-i\varphi I_k}\,\mathcal{O}\,e^{i\varphi I_k} \qquad (k = x, y, z) \qquad (48.32)$$

The operator $\mathcal{O}'$ acting in the rotated state is expressed by the operator $\mathcal{O}$ acting in the original situation. The inverse operation is

$$\mathcal{O} = R^{-1}\,\mathcal{O}'\,R = e^{i\varphi I_k}\,\mathcal{O}'\,e^{-i\varphi I_k} \qquad (k = x, y, z) \qquad (48.33)$$

---

[3]Note that this transformation refers to the coordinate system while the object (the molecule) is left fixed in space.

[4]Note that the Euler angles here refer to the molecular frame which is brought into coincidence with the laboratory frame by the three Euler angle rotations. The angle $\tilde{\gamma}$ therefore describes a rotation about the field direction.

[5]Quantities and operators referring to the rotated state are marked by primes or corresponding subscripts only if they are to represent *functions* of time, rotation angles, or spin operators. This particularly means that spin operator symbols are normally written without reference to the frame to which they apply. There is no danger of confusion because of this convention.

meaning that the operator $\mathcal{O}$ acting in the original state is expressed by the operator $\mathcal{O}'$ after the rotation. Wave functions are transformed to their counterparts after the rotation as

$$|\Psi'\rangle = R\,|\Psi\rangle = e^{-i\varphi I_k}\,|\Psi\rangle \qquad (k = x, y, z) \qquad (48.34)$$

and vice versa,

$$|\Psi\rangle = R^{-1}\,|\Psi'\rangle = e^{i\varphi I_k}\,|\Psi'\rangle \qquad (k = x, y, z) \qquad (48.35)$$

The "sandwich" operations defined in Eqs. 48.32 or 48.33) can easily be carried out by representing all operators by their matrices. For instance,

$$I'_x = e^{-i\varphi I_z}\,I_x\,e^{i\varphi I_z} \qquad (48.36)$$

can be rewritten for spins 1/2 with the aid of the relations

$$e^{-i\varphi I_z} = \begin{pmatrix} \exp\{-i\varphi/2\} & 0 \\ 0 & \exp\{+i\varphi/2\} \end{pmatrix} \qquad (48.37)$$

$$I_x = \frac{1}{2}\begin{pmatrix} 0 & 1 \\ 1 & 0 \end{pmatrix} \qquad (48.38)$$

$$e^{i\varphi I_z} = \begin{pmatrix} \exp\{i\varphi/2\} & 0 \\ 0 & \exp\{-i\varphi/2\} \end{pmatrix} \qquad (48.39)$$

Multiplying the matrices leads to

$$I'_x = \frac{1}{2}\begin{pmatrix} 0 & \exp\{-i\varphi\} \\ \exp\{+i\varphi\} & 0 \end{pmatrix} \qquad (48.40)$$

$$= \frac{1}{2}\begin{pmatrix} 0 & \cos\varphi - i\sin\varphi \\ \cos\varphi + i\sin\varphi & 0 \end{pmatrix}$$

$$= \frac{1}{2}\begin{pmatrix} 0 & 1 \\ 1 & 0 \end{pmatrix}\cos\varphi + \frac{1}{2}\begin{pmatrix} 0 & -i \\ i & 0 \end{pmatrix}\sin\varphi \qquad (48.41)$$

$$= I_x\cos\varphi + I_y\sin\varphi \qquad (48.42)$$

The result can be interpreted as a transformation equation to a reference frame rotated by an angle φ about the z axis while keeping the object fixed. The x component in the rotated frame is expressed by the x and y components in the original system. The rotation sense is counterclockwise when looking along the positive z axis. Another interpretation justified as well is the rotation of the object by φ in the opposite rotation sense while keeping the reference frame fixed.

A more general derivation for arbitrary spins is to formally introduce functions

$$f(\varphi) \equiv e^{-i\varphi I_j}\,I_k\,e^{i\varphi I_j} \qquad (j, k = x, y, z) \qquad (48.43)$$

The first and second derivatives are

$$\frac{df}{d\varphi} = e^{-i\varphi I_j}\,i[I_k, I_j]\,e^{i\varphi I_j} \qquad (48.44)$$

$$\frac{d^2 f}{d\varphi^2} = e^{-i\varphi I_j}\,[I_j, [I_k, I_j]]\,e^{i\varphi I_j} \qquad (48.45)$$

Table 48.1. Unitary transformations of spin operators corresponding to rotations by angles φ about the coordinate axes with the unit vectors u_k where $k = x, y, z$, respectively.

$\mathcal{O}$	$\mathcal{O}' = e^{-i\varphi I_x} \mathcal{O} e^{i\varphi I_x}$	$\mathcal{O}' = e^{-i\varphi I_y} \mathcal{O} e^{i\varphi I_y}$	$\mathcal{O}' = e^{-i\varphi I_z} \mathcal{O} e^{i\varphi I_z}$
I_x	$I_x' = I_x$	$I_x' = I_x \cos\varphi - I_z \sin\varphi$	$I_x' = I_x \cos\varphi + I_y \sin\varphi$
I_y	$I_y' = I_y \cos\varphi + I_z \sin\varphi$	$I_y' = I_y$	$I_y' = I_y \cos\varphi - I_x \sin\varphi$
I_z	$I_z' = I_z \cos\varphi - I_y \sin\varphi$	$I_z' = I_z \cos\varphi + I_x \sin\varphi$	$I_z' = I_z$
I^+	$I^{+'} = iI_z \sin\varphi + I^+ \cos^2\frac{\varphi}{2}$ $+ I^- \sin^2\frac{\varphi}{2}$	$I^{+'} = -I_z \sin\varphi + I^+ \cos^2\frac{\varphi}{2}$ $- I^- \sin^2\frac{\varphi}{2}$	$I^{+'} = I^+ e^{-i\varphi}$
I^-	$I^{-'} = -iI_z \sin\varphi + I^+ \sin^2\frac{\varphi}{2}$ $+ I^- \cos^2\frac{\varphi}{2}$	$I^{-'} = -I_z \sin\varphi - I^+ \sin^2\frac{\varphi}{2}$ $+ I^- \cos^2\frac{\varphi}{2}$	$I^{-'} = I^- e^{i\varphi}$

Using the commutators, Eqs. 44.10, leads to the linear differential equation

$$\frac{d^2 f}{d\varphi^2} = -f \tag{48.46}$$

The solution is of the form

$$f(\varphi) = f(0) \cos\varphi + \left.\frac{df}{d\varphi}\right|_{\varphi=0} \sin\varphi \tag{48.47}$$

If $k = x$ and $j = z$, for instance, this gives

$$e^{-i\varphi I_z} I_x e^{i\varphi I_z} = I_x \cos\varphi + I_y \sin\varphi \tag{48.48}$$

reproducing the above result.

Finally the Baker/Hausdorff lemma, Eq. 45.5, may conveniently be employed for the derivation of the unitary transformation equations for spin operators. The resulting expressions are summarized in Table 48.1 for the convenience of the reader.

Unitary transformations of the form described above form a most useful tool for the treatment of continuous or intermittent precession of spins about stationary (effective) fields, whatever the origin or nature is. This certainly is one of the most important concepts of magnetic resonance.

48.5.1
Transformation to the Rotating Frame

The unitary operator for the transformation to a frame rotating with an angular frequency ω about the z axis[6] has the form

$$R = e^{-i\omega t I_z} \tag{48.49}$$

---

[6]The rotation sense is assumed to be equal to that of the Larmor precession defined by $\omega_L = -\gamma_n B_0$ where $B_0 = B_0 u_z$. If not stated otherwise, the gyromagnetic ratio γ_n is assumed to be positive, so that the rotating frame rotates counterclockwise when looking along the positive z axis.

The time-dependent laboratory frame density operator, $\rho(t)$, is converted into the rotating-frame density operator

$$\rho'(t) = e^{-i\omega t I_z}\rho(t)e^{i\omega t I_z} \tag{48.50}$$

$$= e^{-i(\omega-\omega_0)t I_z}\rho(0)e^{i(\omega-\omega_0)t I_z} \tag{48.51}$$

where we have inserted Eq. 47.33 assuming that the total Hamiltonian acting on the spins refers to the Zeeman interaction, i.e., $\mathcal{H} = \mathcal{H}_0 = -\omega_0\hbar I_z$.

If the new reference frame rotates with the resonance frequency, $\omega = \omega_0$, all time dependences are removed, and the rotating-frame density operator becomes stationary, i.e.,

$$\rho'(t) = \rho(0) \tag{48.52}$$

This corresponds to a stationary expectation value of the magnetization in the rotating frame. However, if the Hamiltonian contains a term with a transverse spin vector operator component, the rotating-frame density operator becomes time dependent as outlined in the following section.

48.5.2
RF Pulses and Unitary Transformations

The rotating-frame Hamiltonian during an RF pulse with B_1 aligned along the x' axis, for instance, is given by

$$\mathcal{H}'_{rf} = -\hbar\gamma_n B_1 I_x = -\hbar\omega_1 I_x \tag{48.53}$$

where $\omega_1 = \gamma_n B_1$, and γ_n is again assumed to be positive. The evolution of the reduced density operator under the action of this Hamiltonian is determined by

$$\sigma(t) = e^{-(i/\hbar)\mathcal{H}'_{rf}t}\sigma(0)e^{(i/\hbar)\mathcal{H}'_{rf}t}$$

$$= e^{i\omega_1 t I_x}\sigma(0)e^{-i\omega_1 t I_x} \tag{48.54}$$

For example, if $\sigma(0) = I_z$, we obtain according to Table 48.1

$$\sigma(t) = I_z\cos(\omega_1 t) + I_y\sin(\omega_1 t) \tag{48.55}$$

For a 90° pulse, for instance, that is, $\omega_1 t_{90} = \pi/2$, the reduced density operator is converted into $\sigma(t_{90}) = I_y$. The magnetization $M(0) = M_0 u_z$ is correspondingly reoriented along the y' axis, i.e., $M(t_{90}) = M_0 u'_y$.

48.5.3
Precession

The solution of the Liouville/von Neumann equation

$$\frac{\partial\rho(t)}{\partial t} = \frac{i}{\hbar}\left[\rho(t),\mathcal{H}\right] \tag{48.56}$$

under the action of the stationary Zeeman Hamilton operator

$$\mathcal{H}_0 = -\gamma_n \hbar B_0 I_z = -\hbar \omega_0 I_z \tag{48.57}$$

is of the unitary-transformation type (see Eq. 47.33)

$$\begin{aligned} \rho(t) &= e^{-(i/\hbar)\mathcal{H}_0 t} \, \rho(0) \, e^{(i/\hbar)\mathcal{H}_0 t} \\ &= e^{i\omega_0 t I_z} \, \rho(0) \, e^{-i\omega_0 t I_z} \end{aligned} \tag{48.58}$$

where $\omega_0 = \gamma_n B_0$. The precession of the complex transverse magnetization readily follows from

$$m(t) = M_x(t) + iM_y(t) = n\gamma_n \hbar \, \mathrm{Tr} \left\{ \rho(t) \left(I_x + iI_y \right) \right\} \tag{48.59}$$

if the initial density operator, $\rho(0) = a + b\sigma(0)$, implies transverse spin operators corresponding to single-quantum coherences.[7] For example, a $(90°)_{-y}$ pulse leads for positive γ_n to

$$\begin{aligned} \rho(0) &= a + b I_x \\ \sigma(0) &= I_x \\ m(0) &= M_0 \end{aligned} \tag{48.60}$$

The transverse magnetization is purely real at this instant, i.e., it is aligned along the x axis.

Using Table 48.1, we find the laboratory frame quantities

$$\begin{aligned} \rho(t) &= a + b\,[I_x \cos(\omega_0 t) - I_y \sin(\omega_0 t)] \\ \sigma(t) &= I_x \cos(\omega_0 t) - I_y \sin(\omega_0 t) \\ m(t) &= M_0\,[\cos(\omega_0 t) - i \sin(\omega_0 t)] \end{aligned} \tag{48.61}$$

The transverse magnetization, $m(t)$, precesses with the angular frequency

$$\omega_L = -\gamma_n B_0 = -\omega_0 u_z \tag{48.62}$$

about the z axis (unit vector u_z). For positive γ_n, the precession sense is counter-clockwise when looking along the B_0 direction.[8]

48.6
Transformations by Bilinear Spin-1/2 Operators

The secular parts of the indirect spin-spin coupling and of the dipole-dipole interaction Hamiltonians (see Table 46.5) contain products of the form $I_z S_z$. These

---

[7]Note that the only time dependence in this expression arises from $\rho(t) = a + b\sigma(t)$.

[8]The comparison of the exponent signs in Eqs. 48.50 and 48.58 verifies the formal equivalence of the rotation of the reference frame on the one hand, and, on the other hand, the rotation of the object, i.e., the spin vector operator, in the *opposite* sense.

Table 48.2. Unitary transformations of spin-1/2 operators by bilinear spin operator terms.

$e^{-iaS_zI_zt}I_xe^{iaS_zI_zt}$	$e^{-iaS_zI_zt}I_ye^{iaS_zI_zt}$	$e^{-iaS_zI_zt}I_ze^{iaS_zI_zt}$
$I_x\cos\left(\frac{a}{2}t\right)+2I_yS_z\sin\left(\frac{a}{2}t\right)$	$I_y\cos\left(\frac{a}{2}t\right)-2I_xS_z\sin\left(\frac{a}{2}t\right)$	I_z

spin operator terms are crucial for NMR spectroscopy. A typical example of unitary transforms mediated by single spin operators was (see Table 48.1)

$$e^{-i\varphi I_z}I_xe^{i\varphi I_z} = I_x\cos\varphi + I_y\sin\varphi \tag{48.63}$$

The angle φ can now be formally interpreted as $\varphi = atS_z$, where a is a constant. Thus

$$e^{-iaS_zI_zt}I_xe^{iaS_zI_zt} = I_x\cos(aS_zt) + I_y\sin(aS_zt) \tag{48.64}$$

Inserting Eqs. 45.9 and 45.10, which are valid for spins 1/2, results in

$$e^{-iaS_zI_zt}I_xe^{iaS_zI_zt} = I_x\cos\left(\frac{a}{2}t\right) + 2I_yS_z\sin\left(\frac{a}{2}t\right) \tag{48.65}$$

The transforms of the other spin components are derived analogously (see Table 48.2).

48.7
Equations of Motion in the Rotating Frame

48.7.1
Classical Precession Equation

A magnetic dipole μ in a magnetic field B obeys the laboratory-frame equation of motion

$$\frac{d\mu}{dt} = \gamma\mu \times B \tag{48.66}$$

where γ is the (positive) gyromagnetic ratio. The magnetic flux density is assumed to consist of the stationary part $B_0 = B_0u_z$ and a component oscillating along the x axis, $2B_1 = (B_1\exp\{i\omega t\} + B_1\exp\{-i\omega t\})u_x$. Here we have analyzed the linearly polarized oscillation into two counterrotating, circularly polarized components represented in complex notation.

The transformation of this equation of motion to a frame rotating with the angular frequency ω assumed to be equal to the oscillation frequency of the time-varying part of the magnetic flux density is performed by applying the following rule known from classical mechanics to the left-hand side:

$$\frac{d\mu}{dt} = \left(\frac{d\mu}{dt}\right)' + \omega \times \mu \tag{48.67}$$

where the prime refers to the rotating frame. The right-hand side is transformed by replacing

$$B = B_0 u'_z + B_1 u'_x + B_1[\cos(2\omega t)u'_x + \sin(2\omega t)u'_y] \tag{48.68}$$

The magnetic flux density in the rotating frame consists of a stationary part including the originally oscillating component which was circularly polarized in the same sense as the rotation angular frequency of the reference frame. This is in contrast to the originally circularly polarized component rotating in the opposite sense, which has doubled its rotation angular frequency relative to the new frame. As only secular terms are able to affect the vector direction of μ, and, hence, contribute the time derivative, the rapidly rotating field can safely be neglected.

The transformed equation of motion thus becomes

$$\left(\frac{d\mu}{dt}\right)' = \gamma\mu \times B_e \tag{48.69}$$

where the magnetic flux density effective in the rotating frame is

$$B_e = B_1 u'_x + \left(B_0 - \frac{\omega}{\gamma}\right)u'_z \tag{48.70}$$

48.7.2
Liouville/von Neumann Equation

The general equation of motion in the laboratory frame is given by the Liouville/von Neumann equation

$$\frac{\partial\rho}{\partial t} = -\frac{i}{\hbar}[\mathcal{H}, \rho] \tag{48.71}$$

The transformation operator to the frame rotating at the angular frequency ω about the z axis is given by

$$R = e^{-i\omega t I_z} \tag{48.72}$$

The laboratory frame density operator is then related to its rotating frame counterpart by

$$\rho = e^{i\omega t I_z}\rho' e^{-i\omega t I_z} \tag{48.73}$$

The time derivative is

$$\frac{\partial\rho}{\partial t} = i\omega I_z e^{i\omega t I_z}\rho' e^{-i\omega t I_z} + e^{i\omega t I_z}\left(\frac{\partial\rho'}{\partial t}e^{-i\omega t I_z} - i\omega\rho' I_z e^{-i\omega t I_z}\right) \tag{48.74}$$

Inserting Eqs. 48.73 and 48.74 into Eq. 48.71 and multiplying by $\exp\{-i\omega t I_z\}$ from the left and $\exp\{i\omega t I_z\}$ from the right gives

$$\frac{\partial\rho'}{\partial t} = -\frac{i}{\hbar}[\mathcal{H}_e, \rho'] \tag{48.75}$$

where we have introduced the Hamiltonian effective in the rotating frame,

$$\mathcal{H}_e = R\mathcal{H}R^{-1} + \hbar\omega I_z$$
$$= \mathcal{H}' + i\hbar\dot{R}R^{-1} \tag{48.76}$$

The transformation of the laboratory frame expression, Eq. 48.71, to the rotating frame thus leads to the same analytical form. Note the difference between the effective Hamiltonian, $\mathcal{H}_e$, and the unitary transform, $\mathcal{H}' = R\mathcal{H}R^{-1}$.

48.7.3
Time Dependent Schrödinger Equation

The time dependent Schrödinger equation in the laboratory frame,

$$\frac{\partial|\Psi\rangle}{\partial t} = -\frac{i}{\hbar}\mathcal{H}|\Psi\rangle \tag{48.77}$$

can likewise be transformed to the frame rotating at the angular frequency ω about the z axis. The laboratory-frame wavefunction, $|\Psi\rangle$, is related to its rotating-frame counterpart, $|\Psi'\rangle$, by

$$|\Psi\rangle = R^{-1}|\Psi'\rangle = e^{i\omega t I_z}|\Psi'\rangle \tag{48.78}$$

The derivative is

$$\frac{\partial|\Psi\rangle}{\partial t} = i\omega I_z|\Psi'\rangle + e^{i\omega t I_z}\frac{\partial|\Psi'\rangle}{\partial t} \tag{48.79}$$

Inserting Eqs. 48.78 and 48.79 into Eq. 48.77 and multiplying by $e^{-i\omega t I_z}$ from the left leads to the rotating-frame Schrödinger equation

$$\frac{\partial|\Psi'\rangle}{\partial t} = -\frac{i}{\hbar}\mathcal{H}_e|\Psi'\rangle \tag{48.80}$$

where the Hamiltonian effective in the rotating frame, $\mathcal{H}_e$ is given by Eq. 48.76.

48.7.4
Heisenberg Equation

Another frequently used equation of motion is the Heisenberg equation

$$\frac{\partial\mathcal{O}}{\partial t} = \frac{i}{\hbar}[\mathcal{H}, \mathcal{O}] \tag{48.81}$$

where $\mathcal{O}$ is a time dependent operator. In contrast to the Schrödinger representation, the wavefunctions are now assumed to be time independent. The transformation to the frame rotating at the angular frequency ω about the z axis is entirely analogous to the treatment of the Liouville/von Neumann equation in Sect. 48.7.2. The result is

$$\frac{\partial\mathcal{O}'}{\partial t} = \frac{i}{\hbar}[\mathcal{H}_e, \mathcal{O}'] \tag{48.82}$$

where

$$\mathcal{O}' = e^{-i\omega t I_z} \mathcal{O} e^{i\omega t I_z} \tag{48.83}$$

The different equations of motion are, in principle, equivalent and can be derived from each other: The choice is more a question of which representation is more adapted to the problem. The most important result of the above considerations is that the same quantum mechanics is valid in the rotating frame provided that the Hamiltonian is replaced by the effective Hamilton operator at Eq. 48.76.

48.8
Rotating-Frame Hamilton Operators

The laboratory frame Hamiltonian is assumed to consist of the Zeeman, the radio frequency, and the spin interaction terms. That is

$$\mathcal{H} = \mathcal{H}_0 + \mathcal{H}_{rf} + \mathcal{H}_i \tag{48.84}$$

where

$$\mathcal{H}_0 = -\hbar\omega_0 I_z \tag{48.85}$$

$$\mathcal{H}_{rf} = -\hbar\frac{\omega_1}{2}\left(I^+ + I^-\right)\left(e^{i\omega_c t} + e^{-i\omega_c t}\right) \tag{48.86}$$

and $\omega_0 = \gamma_n B_0$ and $\omega_1 = \gamma_n B_1$. The RF field is assumed to oscillate along the x axis with an amplitude $2B_1$ and the carrier angular frequency ω_c.

The unitary transformation to a frame rotating at an angular frequency $\omega = -\omega u_z$ about the z axis with the unit vector u_z is mediated by the operator

$$R = e^{-i\omega t I_z} \tag{48.87}$$

According to Eq. 48.76 the Hamiltonian effective in the rotating frame is then[9]

$$\mathcal{H}_e = \mathcal{H}_0' + i\hbar\dot{R}R^{-1} + \mathcal{H}_{rf}' + \mathcal{H}_i' \tag{48.88}$$

where

$$\mathcal{H}_0' = R\,\mathcal{H}_0\,R^{-1} \tag{48.89}$$

$$\mathcal{H}_i' = R\,\mathcal{H}_i\,R^{-1} \tag{48.90}$$

$$\mathcal{H}_{rf}' = R\,\mathcal{H}_{rf}\,R^{-1} \tag{48.91}$$

Carrying out the unitary transformation of Eqs. 48.85 and 48.86 gives

$$\mathcal{H}_0' = -\hbar\omega_0 I_z \tag{48.92}$$

$$\mathcal{H}_{rf}' = -\hbar\frac{\omega_1}{2}\left[I^+\left(e^{-i(\omega-\omega_c)t} + e^{-i(\omega+\omega_c)t}\right)\right.$$
$$\left. + I^-\left(e^{i(\omega+\omega_c)t} + e^{i(\omega-\omega_c)t}\right)\right] \tag{48.93}$$

$$i\hbar\dot{R}R^{-1} = \hbar\omega I_z \tag{48.94}$$

---

[9]Rotating-frame operators are indicated by primes. Spin operators in the rotating frame are, however, written without primes for simplicity with the exception of a few ambiguous cases.

If the frame rotates with the carrier angular frequency, i.e., $\omega = \omega_c$, and after neglect of the non-secular terms oscillating with $\exp(\pm i2\omega_c t)$, Eq. 48.93 becomes

$$\mathcal{H}'_{rf} = -\hbar\frac{\omega_1}{2}\left[I^+ + I^-\right] = -\hbar\omega_1 I_x \tag{48.95}$$

This expression corresponds to a stationary RF flux density aligned along the x' axis of the rotating frame. Combining Eqs. 48.88, 48.92, and 48.95, and summarizing the spin operator components in vector form leads to to the effective Hamiltonian

$$\mathcal{H}_e = -\gamma_n\hbar\mathbf{I}\cdot\mathbf{B}_e + \mathcal{H}'_i \tag{48.96}$$

where the effective magnetic-flux density is

$$\mathbf{B}_e = \mathbf{B}'_1 + \mathbf{B}_0 + \frac{\omega_c}{\gamma_n} \tag{48.97}$$

where $\mathbf{B}_0 = B_0\mathbf{u}'_z$, $\mathbf{B}'_1 = B_1\mathbf{u}'_x$, and $\boldsymbol{\omega}_c = -\omega_c\mathbf{u}'_z$ (if γ_n positive). The unit vectors along the x' and z' axes of the rotating frame are designated by $\mathbf{u}'_x$ and $\mathbf{u}'_z$, respectively.

The modulus of the effective flux density is[10]

$$B_e = \sqrt{\left(B_0 - \frac{\omega_c}{\gamma_n}\right)^2 + B_1^2} \tag{48.98}$$

where we have again assumed a positive γ_n. The polar tilt angle of the effective flux density against the z direction is

$$\Theta = \arctan\left(\frac{\omega_1}{\omega_0 - \omega_c}\right) = \arccos\left(\frac{\omega_0 - \omega_c}{\omega_e}\right) \tag{48.99}$$

where $\omega_e = \gamma_n B_e$.

If the carrier angular frequency equals the resonance angular frequency, i.e., $\omega_c = \omega_0 = -\gamma_n B_0$, the effective Hamltonian becomes

$$\mathcal{H}_e = -\gamma_n\hbar\mathbf{I}\cdot\mathbf{B}'_1 + \mathcal{H}'_i \tag{48.100}$$

i.e., $B_e = B_1$.

48.9
Dipolar Hamiltonian in the Tilted Rotating Frame

Consider a system of like spins, i.e., $\gamma_I = \gamma_S = \gamma_n$. The secular dipolar Hamiltonian of this system is (see Chap. 46)

$$\mathcal{H}_d^{(0)} = \sum_{k<l} d_{kl} \tag{48.101}$$

---

[10]The RF flux density vector synchronously rotating with the rotating frame is marked by a prime, whereas its modulus is symbolized by B_1.

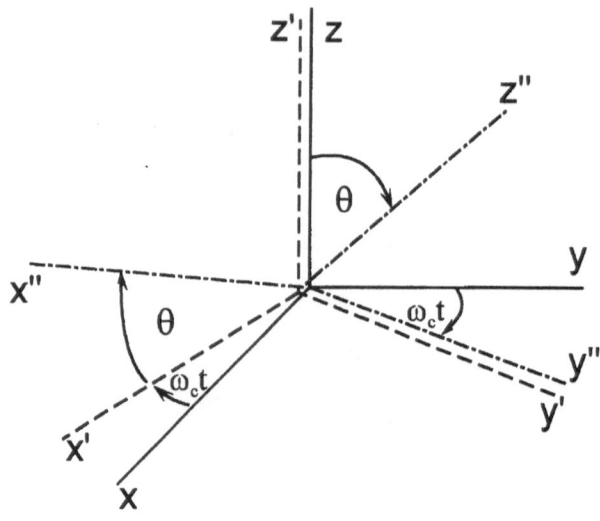

Fig. 48.1. Laboratory (x, y, z), rotating (x', y', z') and tilted rotating (x'', y'', z'') frames of reference.

where

$$d_{kl} = a_{kl}[2I_{kz}I_{lz} - \frac{1}{2}(I_k^+ I_l^- + I_k^- I_l^+)]$$
$$= a_{kl}[3I_{kz}I_{lz} - (I_k \cdot I_l)] \tag{48.102}$$

and

$$a_{kl} = \frac{\mu_0}{4\pi} \frac{\gamma_n^2 \hbar^2}{r_{kl}^3} \frac{1}{2}(1 - 3\cos^2 \vartheta_{kl}) \tag{48.103}$$

ϑ_{kl} is the polar angle between the z axis and the internuclear vector between the spins labeled with the subscripts k and l.

The objective is now to transform the secular dipolar Hamiltonian valid in the laboratory frame to the corresponding Hamiltonian acting in the tilted rotating frame (Fig. 48.1). In other words, the spin operator components in the Hamiltonian of the tilted rotating frame are to be expressed as a function of the spin operator components relative to the laboratory or the rotating frame.

Two unitary transformation steps are required according to

$$\mathcal{H}_{d,TR} = T \, R \, \mathcal{H}_d^{(0)} \, R^{-1} \, T^{-1} \tag{48.104}$$

where

$$R = e^{-i\omega_c t(I_{kz}+I_{lz})} \tag{48.105}$$

provides the transformation from the laboratory frame to the frame rotating at an angular frequency ω_c about the z axis. The operator

$$T = e^{-i\Theta(I_{ky}+I_{ly})} \tag{48.106}$$

transforms the result to the frame tilted by an angle Θ about the y' axis.

The scalar product in Eq. 48.102 is invariant with any coordinate transformation and need not be considered for the moment. The spin operator expression $I_{kz}I_{lz}$ is transformed by R into itself, i.e.,

$$I'_{kz}I'_{lz} = R \ I_{kz}I_{lz} \ R^{-1} = I_{kz}I_{lz} \tag{48.107}$$

The secular dipolar Hamiltonians in the laboratory and in the rotating frame are identical. The transformation to the tilted rotating frame leads to

$$
\begin{aligned}
I''_{kz}I''_{lz} &= T \ I_{kz}I_{lz} \ T^{-1} \\
&= (I_{kz}\cos\Theta + I_{kx}\sin\Theta)(I_{lz}\cos\Theta + I_{lx}\sin\Theta) \\
&= I_{kz}I_{lz}\cos^2\Theta + I_{kx}I_{lx}\sin^2\Theta \\
&\quad + (I_{kz}I_{lx} + I_{kx}I_{lz})\cos\Theta\sin\Theta
\end{aligned} \tag{48.108}
$$

Thus we have

$$
\begin{aligned}
d_{kl}^{TR} &= T \ R \ d_{kl} \ R^{-1} \ T^{-1} \\
&= a_{kl}\left[\frac{1}{2}(3\cos^2\Theta - 1)(3I_{kz}I_{lz} - I_k \cdot I_l)\right. \\
&\quad + \frac{3}{4}\sin^2\Theta \ (I_k^+ I_l^+ + I_k^- I_l^-) \\
&\quad \left. + \frac{3}{2}\sin\Theta\cos\Theta \ \{I_{kz}(I_l^+ + I_l^-) + I_{lz}(I_k^+ + I_k^-)\}\right]
\end{aligned} \tag{48.109}
$$

This expression consists of a secular and two nonsecular terms. Note that the latter are the result of the transformations because all nonsecular terms of the Hamiltonian in the original frame were truncated. The total Hamiltonian of dipolar interaction in the tilted rotating frame is

$$\mathcal{H}_{d,TR} = P_2(\cos\Theta) \ \mathcal{H}_d^{(0)} + \sin^2\Theta \ \mathcal{P} + \sin\Theta \ \cos\Theta \ \mathcal{Q} \tag{48.110}$$

The secular part of the Hamiltonian acting in the tilted rotating frame is thus related to the secular Hamiltonian in the laboratory frame with the second-order Legendre polynomial as a proportionality factor. That is

$$\mathcal{H}_{d,TR}^{(0)} = P_2(\cos\Theta) \ \mathcal{H}_d^{(0)} \tag{48.111}$$

where

$$P_2(\cos\Theta) = \frac{1}{2}(3\cos^2\Theta - 1) \tag{48.112}$$

$\mathcal{H}_{d,TR}^{(0)}$ is the secular dipolar Hamiltonian acting in the tilted rotating frame. It is expressed as a function of the secular dipolar Hamiltonian in the laboratory frame, $\mathcal{H}_d^{(0)}$. Note that $\mathcal{H}_{d,TR}^{(0)}$ vanishes for the magic angle $\Theta = \arccos(1/\sqrt{3}) \approx 54°44'$.

The nonsecular operators are given by

$$P = \frac{3}{4} \sum_{k<l} \{I_k^+ I_l^+ + I_k^- I_l^-\}$$

(48.113)

$$Q = \frac{3}{2} \sum_{k<l} \{I_{kz}(I_l^+ + I_l^-) + I_{lz}(I_k^+ + I_k^-)\}$$

(48.114)

48.10
Quadrupolar Hamiltonian in the Tilted Rotating Frame

In the high-field case, the axis of quantization is the external magnetic field rather than the electric field gradient. We therefore designate the laboratory reference frame as "Zeeman frame." The secular quadrupolar Hamiltonian of a spin $I > 1/2$ is in this frame (see Chap. 46 and [106, 455])

$$\mathcal{H}_q^{(0)} = \frac{e^2 qQ}{4I(2I-1)} \frac{3\cos^2 \vartheta - 1}{2} [3I_z^2 - I^2]$$

(48.115)

where ϑ is the polar angle between the z axis of the Zeeman frame and the z principal axis of an axially symmetric electric field-gradient tensor.

Let us now transform this expression from the laboratory to the tilted rotating frame (Fig. 48.1). I^2 is invariant with coordinate transformations. The operator I_z^2 is transformed in two steps as with dipolar interaction. The first transformation refers to a frame rotating with a frequency ω_c about the z axis. It follows

$$I_z'^2 = R\, I_z^2\, R^{-1} = I_z^2$$

(48.116)

The transformation from the rotating to the tilted rotating frame leads to an operator expression

$$\begin{aligned}
I_z''^2 &= T\, I_z^2\, T^{-1}\\
&= T\, I_z\, T^{-1}T\, I_z\, T^{-1}\\
&= (I_z \cos\Theta + I_x \sin\Theta)(I_z \cos\Theta + I_x \sin\Theta)\\
&= I_z^2 \cos^2\Theta + I_x^2 \sin^2\Theta\\
&\quad + (I_z I_x + I_x I_z)\cos\Theta\sin\Theta
\end{aligned}$$

(48.117)

Thus we have

$$\begin{aligned}
\mathcal{H}_{q,TR} &= T\, R\, \mathcal{H}_q^{(0)}\, R^{-1}\, T^{-1}\\
&= \frac{e^2 qQ}{4I(2I-1)} \frac{3\cos^2 \vartheta - 1}{2} \Big[\frac{1}{2}(3\cos^2\Theta - 1)(3I_z^2 - I^2)\\
&\quad + \frac{3}{4}\sin^2\Theta\,[(I^+)^2 + (I^-)^2]\\
&\quad + \frac{3}{2}\sin\Theta\cos\Theta\,\{I_z(I^+ + I^-) + (I^+ + I^-)I_z\}\Big]
\end{aligned}$$

(48.118)

where we have used the relation

$$I_x^2 = \frac{1}{2}(I^2 - I_z^2) + \frac{1}{4}[(I^+)^2 + (I^-)^2] \tag{48.119}$$

Equation 48.118 consists of a secular and two nonsecular terms. The latter arise as a consequence of the transformations. The secular part of the Hamiltonian acting in the tilted rotating frame is related to the secular Hamiltonian acting in the laboratory frame by

$$\mathcal{H}_{q,TR}^{(0)} = P_2(\cos\Theta)\mathcal{H}_q^{(0)} \tag{48.120}$$

48.11
Spin-Spin Coupling in the Doubly-Rotating Frame

Indirect spin-spin coupling between different nuclear species reduces under motional averaging conditions to the scalar form[11]

$$\mathcal{H}_J = hJI \cdot S \tag{48.121}$$

where J is the spin-spin coupling constant (expressed in frequency units) and I and S are the dimensionless spin vectors. In the homonuclear case, the scalar product is invariant to any transformation to a rotating frame, and the interaction Hamiltonian remains of a secular nature. With heteronuclear spin systems, spin-spin coupling implies nonsecular terms. This is of interest in context with double-resonance experiments such as J cross-polarization [189]. Figure 38.1a shows a corresponding pulse scheme.

During the spin-locking/contact pulse pair the total laboratory-frame Hamiltonian of the spin system is

$$\mathcal{H} = -\hbar\omega_{0,I}I_z - \hbar\omega_{0,S}S_z + hJI \cdot S - 2\hbar\omega_{1,I}I_y\cos(\omega_I t) - 2\hbar\omega_{1,S}S_y\cos(\omega_S t) \tag{48.122}$$

where $\omega_{0,I} = \gamma_I B_{0,I}$, $\omega_{0,S} = \gamma_S B_{0,S}$, $\omega_{1,I} = \gamma_I B_{1,I}$, $\omega_{1,S} = \gamma_S B_{1,S}$. ω_I, and ω_S are the angular carrier frequencies of the two RF channels.

Performing the unitary transformation to the doubly-rotating frame,

$$\mathcal{H}_e = R\,\mathcal{H}\,R^{-1} + i\hbar\dot{R}R^{-1} \tag{48.123}$$

where

$$R = e^{-i(\omega_I I_z + \omega_S S_z)t} \tag{48.124}$$

leads to

$$\begin{aligned}
\mathcal{H}_e = {}& -\hbar(\omega_{0,I} - \omega_I)I_z - \hbar(\omega_{0,S} - \omega_S)S_z - \hbar\omega_{1,I}I_y - \hbar\omega_{1,S}S_y \\
& - \hbar\omega_{1,I}[I_y\cos(2\omega_I t) - I_x\sin(2\omega_I t)] \\
& - \hbar\omega_{1,S}[S_y\cos(2\omega_S t) - S_x\sin(2\omega_S t)] \\
& + hJ\{I_x S_x\cos[(\omega_I - \omega_S)t] + I_y S_y\cos[(\omega_I - \omega_S)t] \\
& - I_x S_y\sin[(\omega_I - \omega_S)t] + I_y S_x\sin[(\omega_I - \omega_S)t] + I_z S_z\}
\end{aligned} \tag{48.125}$$

---

[11]Note that the subsequent theory contains elements also valid for weakly dipolar-coupled spin pairs as outlined in Sect. 51.4.

The terms oscillating with frequencies $(\omega_I - \omega_S)$, $2\omega_I$, or $2\omega_S$ have zero time averages and practically do not influence the spin evolution. They are therefore discarded, so that

$$\mathcal{H}_e = -\hbar(\omega_{0,I} - \omega_I)I_z - \hbar(\omega_{0,S} - \omega_S)S_z - \hbar\omega_{1,I}I_y - \hbar\omega_{1,S}S_y + hJI_zS_z \quad (48.126)$$

If the carrier frequencies are resonant, Eq. 48.126 simplifies to

$$\mathcal{H}_e = -\hbar\omega_{1,I}I_y - \hbar\omega_{1,S}S_y + hJI_zS_z \quad (48.127)$$

Let us now define the average frequency in the doubly-rotating frame, ω_1 and the deviations from it, $\pm\Delta\omega_1/2$, by the equations

$$\omega_{1,I} = \omega_1 + \frac{\Delta\omega_1}{2} \quad (48.128)$$

$$\omega_{1,S} = \omega_1 - \frac{\Delta\omega_1}{2} \quad (48.129)$$

Inserting these expressions into Eq. 48.127 gives

$$\mathcal{H}_e = -\hbar\left(\omega_1 + \frac{\Delta\omega_1}{2}\right)I_y - \hbar\left(\omega_1 - \frac{\Delta\omega_1}{2}\right)S_y + hJI_zS_z \quad (48.130)$$

Finally, we perform a transformation to a reference frame rotating at the angular frequency ω_1 about the y' axis of the doubly-rotating frame according to

$$\mathcal{H}'_e = R' \mathcal{H}_e R'^{-1} + i\hbar\dot{R}'R'^{-1} \quad (48.131)$$

where

$$R' = e^{-i\omega_1(I_y + S_y)t} \quad (48.132)$$

The result is

$$\mathcal{H}'_e = -\hbar\frac{\Delta\omega_1}{2}I_y + \hbar\frac{\Delta\omega_1}{2}S_y + hJ\left[I_zS_z \cos^2(\omega_1 t) + I_xS_x \sin^2(\omega_1 t)\right.$$
$$\left. - I_zS_x \cos(\omega_1 t)\sin(\omega_1 t) - I_xS_z \cos(\omega_1 t)\sin(\omega_1 t)\right] \quad (48.133)$$

Rapid oscillations are again irrelevant for the evolution of the spin system. Therefore replacing the corresponding terms by their time averages, we obtain

$$\mathcal{H}'_e = -\hbar\frac{\Delta\omega_1}{2}I_y + \hbar\frac{\Delta\omega_1}{2}S_y + \frac{hJ}{2}[I_zS_z + I_xS_x] \quad (48.134)$$

Under Hartmann/Hahn matching conditions, i.e.,

$$\omega_{1,I} = \omega_{1,S} = \omega_1, \quad (48.135)$$

or

$$\Delta\omega_1 = 0 \quad (48.136)$$

the Hamiltonian reduces to

$$\mathcal{H}'_e = \frac{hJ}{2}[I_zS_z + I_xS_x] \quad (48.137)$$

This is the Hamiltonian effective in the "rotating doubly-rotating frame" for J cross polarization under Hartmann/Hahn matching conditions.

CHAPTER 49

Irreducible Spherical Tensor Operators

All Hamiltonians of spin interactions, which are based on second-rank tensors (see Tables 46.1 to 46.5), can be expressed in the general form

$$\mathcal{H}_i = a_i \sum_{l=0,2} \sum_{m=-l}^{l} (-1)^m A_{l,-m} T_{l,m} \tag{49.1}$$

where a_i is a constant specific for the interaction. The quantities $A_{l,-m}$ are functions of the spatial orientation and distances (usually given in polar coordinates (r, φ, ϑ) as defined in Chap. 48), whereas $T_{l,m}$ represents functions of spin operators. The analytical form of these expressions are chosen in such a way that the Hamiltonians transform under rotations like **spherical harmonics of rank** l. They are therefore referred to as **irreducible spherical tensor (IST) operators** of the spatial and the spin coordinates, respectively. "Irreducible" means that no decomposition in terms transforming under rotations other than spherical harmonics of the same rank l is possible.[1] The subscript m indicates the **order of multiple-quantum transition** induced by $T_{l,m}$.

---

[1] This is to be seen in contrast to Cartesian tensors $\tilde{T}$. For example, consider a Cartesian tensor of rank 2 formed by the dyadic product of two vectors u and v

$$\{\tilde{T}_{ij}\} = \{u_i v_j\} \tag{49.2}$$

This is a *reducible* tensor as can be seen by (the straightforward) decomposition in terms that have different rotational transformation properties:

$$u_i v_j = \underbrace{\frac{u \cdot v}{3} \delta_{ij}}_{\text{scalar product}} + \underbrace{\frac{u_i v_j - u_j v_i}{2}}_{\text{antisymm. tensor}} + \underbrace{\left(\frac{u_i v_j + u_j v_i}{2} - \frac{u \cdot v}{3} \delta_{ij} \right)}_{\text{symm. traceless tensor}} \tag{49.3}$$

The scalar product is invariant under rotations and is specified by a single component. The second term is an element of an antisymmetric tensor and has a form $\epsilon_{ijk}(u \times v)_k$. This tensor is determined by three independent components. The last term is an element of a symmetric traceless tensor which is characterized by five independent components.

Altogether we have nine independent components corresponding to the 3×3 elements of the second-rank Cartesian tensor assumed. We note that the numbers of independent components of the three objects, into which we have decomposed the Cartesian tensor, are related as the multiplicities for angular momentum quantum numbers $l = 0, 1, 2$. This suggests that we have decomposed the reducible Cartesian tensor into three different irreducible tensors.

49.1
Rotational Transformation of IST Operators

The rotational transformation properties defining the analytical form of spherical tensor components are the same as those of spherical harmonics (see Sect. 42.4). The component $T_{l,m}$ of an irreducible spherical tensor $T^{(l)}$ of rank l is transformed into

$$T'_{l,m} = RT_{l,m}R^{-1} = \sum_{m'=-l}^{l} T_{l,m'} D^{(l)}_{m',m}(\tilde{\alpha}, \tilde{\beta}, \tilde{\gamma}) \qquad (49.4)$$

This in particular means

$$e^{iI^{(0)}\varphi} T_{lm} e^{-iI^{(0)}\varphi} = T_{lm} e^{im\varphi} \qquad (49.5)$$

The rotation operators, R and R^{-1}, referring to subsequent rotations through the Euler angles, $\tilde{\alpha}, \tilde{\beta}, \tilde{\gamma}$ (Sect. 48.3) are given by (Sect. 48.5)

$$R = R(\tilde{\alpha}, \tilde{\beta}, \tilde{\gamma}) = e^{-i\tilde{\gamma}l_{z''}} e^{-i\tilde{\beta}l_{y'}} e^{-i\tilde{\alpha}l_z} \qquad (49.6)$$

$$R^{-1} = R^{-1}(\tilde{\alpha}, \tilde{\beta}, \tilde{\gamma}) = e^{i\tilde{\alpha}l_z} e^{i\tilde{\beta}l_{y'}} e^{i\tilde{\gamma}l_{z''}} \qquad (49.7)$$

where the rotations refer to the z, y' and z'' axes of the original, $S\{x, y, z\}$, and the two intermediate reference systems, $S'\{x', y', z'\}$ and $S''\{x'', y'', z''\}$, which are defined by the Euler angles.

The system $S''\{x'', y'', z''\}$ is the result of a rotation of $S'\{x', y', z'\}$ about the y' axis through an angle $\tilde{\beta}$, so that the exponential operator $\exp(-i\tilde{\gamma}l_{z''})$ complies with the unitary transformation

$$e^{-i\tilde{\gamma}l_{z''}} = e^{-i\tilde{\beta}l_{y'}} e^{-i\tilde{\gamma}l_{z'}} e^{i\tilde{\beta}l_{y'}} \qquad (49.8)$$

Likewise, the $S'\{x', y', z'\}$ system results from a rotation of $S\{x, y, z\}$ about the z axis through an angle $\tilde{\alpha}$, that is, the exponential operator $\exp(-i\tilde{\beta}l_{y'})$ is generated by

$$e^{-i\tilde{\beta}l_{y'}} = e^{-i\tilde{\alpha}l_z} e^{-i\tilde{\beta}l_y} e^{i\tilde{\alpha}l_z} \qquad (49.9)$$

Finally, we may formally write

$$e^{-i\tilde{\gamma}l_{z'}} = e^{-i\tilde{\alpha}l_z} e^{-i\tilde{\gamma}l_z} e^{i\tilde{\alpha}l_z} \qquad (49.10)$$

Inserting the transformation relations, Eqs. 49.8 to 49.10, into Eq. 49.6 (in that order) readily gives the total transformation operator expressed by operators for rotations about the axes of the original reference frame, $S\{x, y, z\}$ through angles $\tilde{\gamma}, \tilde{\beta}, \tilde{\alpha}$ (in that inverse sequence),

Table 49.1. Reduced Wigner rotation matrix elements for the transformation of irreducible spherical tensors of rank $l = 2$ [63].

$$
\begin{aligned}
d_{0,0}^{(2)} &= \tfrac{1}{2}(3\cos^2\tilde{\beta} - 1) \\[4pt]
d_{1,0}^{(2)} &= -\sqrt{\tfrac{3}{2}}\sin\tilde{\beta}\cos\tilde{\beta} \\[4pt]
d_{1,\pm 1}^{(2)} &= \pm\tfrac{1}{2}(2\cos^2\tilde{\beta} \pm \cos\tilde{\beta} - 1) \\[4pt]
d_{2,0}^{(2)} &= \sqrt{\tfrac{3}{8}}\sin^2\tilde{\beta} \\[4pt]
d_{2,\pm 1}^{(2)} &= \tfrac{1}{2}\sin\tilde{\beta}\,(1 \pm \cos\tilde{\beta}) \\[4pt]
d_{2,\pm 2}^{(2)} &= \tfrac{1}{4}(1 \pm \cos\tilde{\beta})^2
\end{aligned}
$$

$$
R = R(\tilde{\alpha}, \tilde{\beta}, \tilde{\gamma}) = e^{-i\tilde{\alpha}l_z}e^{-i\tilde{\beta}l_y}e^{-i\tilde{\gamma}l_z} \tag{49.11}
$$

The **Wigner rotation matrix elements**, $D_{m',m}^{(l)}$, are then defined as the expectation values of the total rotation operator for all $2l + 1$ values of the magnetic quantum numbers m and m' of the angular momentum l generating the rotations. That is

$$
D_{m',m}^{(l)} = \left\langle l, m' \left| e^{-i\tilde{\alpha}l_z}e^{-i\tilde{\beta}l_y}e^{-i\tilde{\gamma}l_z} \right| l, m \right\rangle \tag{49.12}
$$

The exponential operators with l_z in the exponent can be evaluated in terms of the eigenvalues m, m' so that the Wigner matrix elements may be factored in the form

$$
D_{m',m}^{(l)} = e^{-i\tilde{\alpha}m}d_{m',m}^{(l)}(\tilde{\beta})e^{-i\tilde{\gamma}m'} \tag{49.13}
$$

where the reduced Wigner rotation matrix elements are

$$
d_{m',m}^{(l)}(\tilde{\beta}) = \left\langle l, m' \left| e^{-i\tilde{\beta}l_y} \right| l, m \right\rangle \tag{49.14}
$$

The only elements of interest in NMR are those for $l = 2$. They are listed in Table 49.1.

49.2
Commutation of IST and Spin Operators

The characteristic transformation behavior of irreducible spherical tensor operators under rotations also reveals itself in the commutation properties with spin operators. The corresponding rules can therefore be used for a check of the irreducible spherical character of an operator expression. A component $T_{l,m}$ of an irreducible spherical tensor operator obeys

$$
[I_z, T_{l,m}] = [I^{(0)}, T_{l,m}] = m\,T_{l,m} \tag{49.15}
$$

$$[I^+, T_{l,m}] = -\sqrt{2}\,[I^{(+1)}, T_{l,m}] = \sqrt{l(l+1) - m(m+1)}\,T_{l,m+1} \tag{49.16}$$

$$[I^-, T_{l,m}] = +\sqrt{2}\,[I^{(-1)}, T_{l,m}] = \sqrt{l(l+1) - m(m-1)}\,T_{l,m-1} \tag{49.17}$$

49.3
Analytical Form of IST Operators

The analytical form of irreducible spherical tensor operators readily follows from that of spherical harmonics depending on the orientation of any vectorial quantity (Table 42.4 on page 403). In context with magnetic resonance, one refers either to the orientation of a position vector $r = x u_x + y u_y + z u_z$ or to that of an angular momentum vector, e.g., a spin vector $I = I_x u_x + I_y u_y + I_z u_z$. An irreducible spherical tensor operator of rank 1 consequently has - apart from a constant numerical factor - an analytical form identical to that of spherical harmonics of rank 1. For the vector r, we thus formally define the set of spherical tensor components

$$\begin{aligned}
T^{(1)} &= (T_{1,-1},\ T_{1,0},\ T_{1,+1}) \\
&= \left(\frac{1}{\sqrt{2}}(x - iy),\ z,\ -\frac{1}{\sqrt{2}}(x + iy)\right)
\end{aligned} \tag{49.18}$$

The corresponding set of spherical-tensor components for the spin vector operator I are

$$\begin{aligned}
T^{(1)} &= \left(\frac{1}{\sqrt{2}}(I_x - iI_y),\ I_z,\ -\frac{1}{\sqrt{2}}(I_x + iI_y)\right) \\
&= \left(\frac{1}{\sqrt{2}}I^-,\ I_z,\ -\frac{1}{\sqrt{2}}I^+\right) \\
&= \left(I^{(-1)},\ I^{(0)},\ I^{(+1)}\right)
\end{aligned} \tag{49.19}$$

Analogously, the analytical form of second-rank irreducible spherical tensors is the same as that of spherical harmonics of rank 2 apart from a numerical factor. That is, formally we are dealing with a set of spherical-tensor components

$$T^{(2)} = (T_{2,-2},\ T_{2,-1},\ T_{2,0},\ T_{2,+1},\ T_{2,+2}) \tag{49.20}$$

The second-rank IST components expressed in Cartesian components of the vector r are defined as

$$\begin{aligned}
T_{2,2} &= \tfrac{1}{2}(x + iy)^2 \\
T_{2,+1} &= -z(x + iy) \\
T_{2,0} &= \tfrac{1}{\sqrt{6}}(3z^2 - r^2) \\
T_{2,-1} &= z(x - iy) \\
T_{2,-2} &= \tfrac{1}{2}(x - iy)^2
\end{aligned} \tag{49.21}$$

In terms of the spin vector operator I we have

$$
\begin{array}{llll}
T_{2,2} & = & (I^{(+1)})^2 & = & \frac{1}{2}(I^+)^2 \\
T_{2,+1} & = & \frac{1}{\sqrt{2}}(I^{(0)}I^{(+1)} + I^{(+1)}I^{(0)}) & = & -\frac{1}{2}(I_z I^+ + I^+ I_z) \\
T_{2,0} & = & \frac{1}{\sqrt{6}}(3(I^{(0)})^2 - I^2) & = & \frac{1}{\sqrt{6}}(3I_z^2 - I^2) \\
T_{2,-1} & = & \frac{1}{\sqrt{2}}(I^{(0)}I^{(-1)} + I^{(-1)}I^{(0)}) & = & \frac{1}{2}(I_z I^- + I^- I_z) \\
T_{2,-2} & = & (I^{(-1)})^2 & = & \frac{1}{2}(I^-)^2
\end{array}
\tag{49.22}
$$

From the spin-operator expressions, the meaning of the IST subscripts becomes obvious: m is the **coherence order**; the rank l is equal to the **multipolar order** of the spin system. That is, for N spins 1/2 we have the restrictions $0 \le l \le N$ and $-l \le m \le l$.

49.4
Wigner/Eckart Theorem

The Wigner/Eckart theorem[2] serves factoring of the matrix elements of irreducible spherical tensor components into Clebsch/Gordan (alias Wigner) coefficients[3], $(Ikmq|I'm')$, on the one hand, and reduced matrix elements, $\langle I'||T^{(k)}||I\rangle$, on the other:

$$\langle I'm'|T_{k,q}|Im\rangle = (Ikmq|I'm')\,\langle I'||T^{(k)}||I\rangle \tag{49.23}$$

The actual analytical form of the factors on the right-hand side is of minor importance in this context. We rather concentrate on the properties that can be attributed to these quantities.

The **Clebsch/Gordan coefficients**[4] link three quantities of angular momentum character here represented by the angular-momentum quantum numbers I, k, I' and

---

[2] A derivation can be found in Slichter's book [455], for instance.
[3] Instead of the Clebsch/Gordan coefficients, Wigner's $3j$ symbols are also in common use [63, 422].
[4] The Clebsch/Gordan coefficients play two important roles in quantum physics. Apart from the subject of this section, they were introduced in the first instance for the treatment of coupled pairs of angular momenta. Consider two angular momenta with the quantum numbers j_1, m_{j1} and j_2, m_{j2} which are coupled by any interaction so that a coupled system is formed with the quantum numbers of the total angular momentum, J, m_J. The coupling leads to a set of commuting operators for the observables $j_1^2, j_2^2, J = j_1 + j_2, J_z = j_{1z} + j_{2z}$. That is, we have four "good quantum numbers", j_1, j_2, J, m_J, characterizing the eigenfunctions of the coupled system. These are given as a linear combination of the eigenfunctions of the uncoupled system:

$$\underbrace{|j_1 j_2 J m_J\rangle}_{\text{coupled}} = \sum_{m_{j1}, m_{j2}} \underbrace{(j_1 j_2 m_{j1} m_{j2}|J m_J)}_{\text{Clebsch/Gordan}} \underbrace{|j_1 j_2 m_{j1} m_{j2}\rangle}_{\text{uncoupled}} \tag{49.24}$$

or vice versa:

$$\underbrace{|j_1 j_2 m_{j1} m_{j2}\rangle}_{\text{uncoupled}} = \sum_{J=|j_1-j_2|}^{j_1+j_2} \underbrace{(j_1 j_2 m_{j1} m_{j2}|J m_J)}_{\text{Clebsch/Gordan}} \underbrace{|j_1 j_2 J m_J\rangle}_{\text{coupled}} \tag{49.25}$$

Thus, the Clebsch/Gordan coefficients mediate a unitary transformation between the coupled and uncoupled systems. They can be found in explicit and tabulated form in [98], for instance.

the magnetic quantum numbers m, q, m'. By contrast, the **reduced matrix elements** solely depend on the quantum numbers I, k, I', but not on the "projection" quantum numbers m, q, m'. That is, the Clebsch/Gordan coefficients refer to the magnitude as well as the orientation of the angular momenta involved, whereas the reduced matrix elements are determined by the magnitudes alone. We are therefore mainly interested in the Clebsch/Gordan coefficients, the selection rules of which determine the angular momentum transitions addressed by the irreducible spherical tensor operators (see footnote 4).

For example, second-rank irreducible spherical tensor operators ($k = 2$) link only angular momenta with quantum numbers: $m' - m = 0, \pm 1, \pm 2$, $I' - I = 0, \pm 1, \pm 2$, where $I = I' = 0$ is excluded. As $m' = q + m$ or $q = m - m'$, irreducible spherical tensors of rank $k = 2$ have five independent components for $q = 0, \pm 1, \pm 2$ corresponding to a traceless symmetric Cartesian tensor.

Likewise, rank $k = 0$ tensors connect angular momenta according to the rules $m' = m$ and $I' = I$. That is, there is only one component ($q = 0$): we are dealing with a scalar.

From the Wigner/Eckart theorem an important relation between matrix elements of any pair of irreducible spherical tensor operators of the same rank follows. For instance, consider rank-k irreducible spherical tensor operators as functions of the position vector and the spin vector, i.e., $T^{(k)}(\mathbf{r})$ and $T^{(k)}(\mathbf{I})$, respectively. Factoring the matrix elements of the components of the two tensors with the aid of the Wigner/Eckart theorem, Eq. 49.23, and eliminating the (common) Clebsch/Gordan coefficients, leads to

$$\langle Im|T_{k,q}(\mathbf{r})|I'm'\rangle = C\langle Im|T_{k,q}(\mathbf{I})|I'm'\rangle \tag{49.27}$$

where

$$C = \frac{\langle I'||T^{(k)}(\mathbf{r})||I\rangle}{\langle I'||T^{(k)}(\mathbf{I})||I\rangle} \tag{49.28}$$

is constant with respect to m', m. On this basis, matrix elements of any like-rank pair of irreducible spherical tensor operators can be converted into each other. The same applies to linear combinations of like-rank irreducible spherical tensor components.

Physically the squared Clebsch/Gordan coefficients may be interpreted as the probability for the occurrence of the single-spin magnetic quantum numbers m_{j1} and m_{j2} when the coupled system is in the state $|j_1 j_2 J m_J\rangle$.

The eigenfunctions of the uncoupled system are given as product functions of the eigenfunctions of the two (independent) angular momenta considered. The selection rules for finite Clebsch/Gordan coefficients are

$$m_J = m_{j1} + m_{j2} \quad \text{and} \quad |j_1 - j_2| \le J \le j_1 + j_2 \tag{49.26}$$

(The latter is known as the "triangle rule.")

49.5
Selection Rules for Stationary Nuclear Moments

A fundamental application of the Wigner/Eckart theorem is the determination of the allowed orders of stationary electric and magnetic nuclear multipoles for a given nuclear spin.[5] The electrostatic energy of the nuclear charge density distribution, ρ_n, in the potential Φ_0 of an external charge density distribution, ρ_0,

$$W_E = \int \rho_n \Phi_0 \, d^3 r = \frac{1}{4\pi\epsilon_0} \int \int \frac{\rho_0(r_0)\rho_n(r_n)}{|r_0 - r_n|} \, d^3 r_0 \, d^3 r_n \qquad (49.29)$$

can be expanded in terms of spherical harmonics, i.e.,

$$\frac{1}{|r_0 - r_n|} = 4\pi \sum_{k=0}^{\infty} \sum_{q=-k}^{+k} \frac{1}{2k+1} \frac{r_n^k}{r_0^{k+1}} Y_{k,q}(\vartheta_n, \varphi_n) \, Y_{k,q}^*(\vartheta_0, \varphi_0) \qquad (r_0 > 0) \quad (49.30)$$

where $r_n, \vartheta_n, \varphi_n$ and $r_0, \vartheta_0, \varphi_0$ are the polar coordinates of r_n (nuclear charge distribution) and r_0 (external charge distribution), respectively. The origin is assumed in the center of the nucleus. Furthermore, the continuous charge densities $\rho^{(n)}$ and $\rho^{(e)}$ may be replaced by

$$\rho^{(n)}(r) = e\delta(r - r_\nu) \qquad (49.31)$$
$$\rho^{(e)}(r) = q_\kappa^{(e)}\delta(r - r_\kappa) \qquad (49.32)$$

referring to the proton charges e in the nucleus at $r = r_\nu$ and to the external point charges $q_\kappa^{(e)}$ at $r = r_\kappa$.

Interpreting the terms of the above expansion as (irreducible spherical tensor) operators and replacing the complex conjugate spherical harmonics according to Eq. 42.20 readily leads to the Hamiltonian in the scalar-product form

$$\mathcal{H}_E = \sum_{k} \sum_{q=-k}^{k} (-1)^q A_{k,-q} B_{k,q} \qquad (49.33)$$

---

[5]Here we are referring to stationary nuclear moments. However, in context with "electromagnetic multipole radiation," nuclear multipoles of an oscillating nature also arise and entail characteristic angular distributions of the radiation. Note that - apart from the triangle rule (see footnote 4) for the angular momenta involved - the selection rules for the transient occurrence of oscillating multipoles are different from those for the stationary case, because the nuclear wavefunctions may change their parity in the course of a radiative transition. Thus, in context with γ radiation, even and odd orders of multipole radiation may occur for electric as well as magnetic multipoles. For example, electric dipole radiation ("E1") or magnetic quadrupole radiation ("M2") are permitted if the parity of the nuclear wavefunctions is changed ("yes" transition) and the triangle rule is fulfilled. On the other hand, atomic spectra are practically dominated by electric dipole (E1, yes) transitions, whereas magnetic resonance and nuclear quadrupole resonance are connected with magnetic dipole (M1, no) transitions.

where[6]

$$A_{k,q} = \frac{1}{4\pi\epsilon_0} \sqrt{\frac{4\pi}{2k+1}} \sum_\kappa q_\kappa^{(e)} r_\kappa^{-(k+1)} Y_{k,q}(\vartheta_\kappa, \varphi_\kappa) \tag{49.34}$$

refers to the external point charges $q_\kappa^{(e)}$ at $r_\kappa, \vartheta_\kappa, \varphi_\kappa$, and

$$B_{k,q} = \sqrt{\frac{4\pi}{2k+1}} \, e \sum_\nu r_\nu^k Y_{k,q}(\vartheta_\nu, \varphi_\nu) \tag{49.35}$$

represents the proton point charges e at $r_\nu, \vartheta_\nu, \varphi_\nu$ within the nucleus. The nuclear multipole moment tensors are defined as the expectation values of the nuclear operators $B_{k,q}$.

For $k = 0$, this results in the **electric monopole moment** (or total charge of the nucleus of atomic number Z):

$$\langle B_{0,0} \rangle = eZ \tag{49.36}$$

For $k = 1$, we find the (vanishing) **electric dipole moment tensor**:

$$\langle B_{1,0} \rangle = \langle B_{1,\pm 1} \rangle = 0 \tag{49.37}$$

There is no stationary electric dipole moment contribution because the parity of spherical harmonics of rank 1 is odd, and the nuclear wavefunctions have a well-defined (either even or odd) parity, so that the expectation value integral vanishes. The same applies to all terms of higher odd ranks k.

The rank $k = 2$ terms lead to the **electric quadrupole moment tensor**. The expectation value of the operator $B_{2,0}$ for the highest magnetic quantum number, i.e., $m = I$, is of particular interest as the definition of the proper nuclear quadrupole moment:

$$Q \equiv \frac{2}{e} \langle II|B_{2,0}|II \rangle \tag{49.38}$$

Note that the second-rank spherical harmonics have even parity so that the electric quadrupole moment tensor (and those of all higher even orders) may be finite. However, there is a further selection rule owing to the Wigner/Eckart theorem applied to the expectation values of the operators $B_{k,q}$:

$$\langle Im|B_{k,q}|Im \rangle = (Ikmq|Im) \, \langle I||B^{(k)}||I \rangle \tag{49.39}$$

The Clebsch/Gordan coefficients in this expression vanish unless $0 \leq k \leq 2I$. That is, for a given spin I, the allowed electric multipole orders are restricted to even values of $k \leq 2I$.

The non-vanishing stationary magnetic multipole moments can be derived in an analogous manner starting from the magnetic energy of the nuclear current distribution j_n in the vector potential A_0 of an external current distribution j_0

$$W_B = -\int j_n \cdot A_0 \, d^3r = -\frac{\mu_0}{4\pi} \int \int \frac{j_n \cdot j_0}{|r_0 - r_n|} \, d^3r_0 \, d^3r_n \tag{49.40}$$

---

[6]Linear combinations of spherical harmonics referring to orientations of independent vectors are also forms of spherical tensor operators, because spherical tensor operators are solely defined by the corresponding transformation behavior under rotations (see Eq. 49.4).

Table 49.2. Allowed stationary nuclear multipole moments up to the order 3 (Z, atomic number; μ, magnetic dipole moment; Q, electric quadrupole moment; O, magnetic octopole moment).

spin	orders	electric	magnetic
	k	2^k-poles (k even)	2^k-poles (k odd)
$I = 0$	$k = 0$	$Ze \neq 0$	0
$I = \frac{1}{2}$	$k = 0, 1$	$Ze \neq 0$	$\mu \neq 0$
$I = 1$	$k = 0, 1, 2$	$Ze \neq 0; Q \neq 0$	$\mu \neq 0$
$I = \frac{3}{2}$	$k = 0, 1, 2, 3$	$Ze \neq 0; Q \neq 0$	$\mu \neq 0; O \neq 0$

It turns out that only odd multipole orders k are allowed in this case, where the order must again comply to $k \leq 2I$ for a given nuclear spin I. Table 49.2 summarizes the nuclear moments relevant for most experiments.[7]

49.6
IST Representation of the Quadrupolar Hamiltonian

In this section we express the Hamiltonian of the electric quadrupole interaction in terms of spin operators. The electric quadrupole contribution, i.e., the rank 2 term in the expansion, Eq. 49.33, is

$$\mathcal{H}_q = \sum_{q=-2}^{2} (-1)^q A_{2,-q} B_{2,q} \tag{49.41}$$

where the spherical harmonics in the IST terms $A_{2,q}$ and $B_{2,q}$ can be expressed by the Cartesian coordinates of the external point charges at the positions $r_\kappa, x_\kappa, y_\kappa, z_\kappa$, and by those of the proton position in the nucleus $r_\nu, x_\nu, y_\nu, z_\nu$, (see Table 42.4 on page 403). The external-charge terms are then

$$
\begin{aligned}
A_{2,\pm 2} &= \frac{1}{4\pi\epsilon_0} \frac{\sqrt{6}}{4} \sum_\kappa q_\kappa^{(e)} \frac{(x_\kappa \pm iy_\kappa)^2}{r_\kappa^5} \\
A_{2,\pm 1} &= \frac{1}{4\pi\epsilon_0} \frac{\sqrt{6}}{2} \sum_\kappa q_\kappa^{(e)} \frac{z_\kappa(x_\kappa \pm iy_\kappa)}{r_\kappa^5} \\
A_{2,0} &= \frac{1}{4\pi\epsilon_0} \frac{1}{2} \sum_\kappa q_\kappa^{(e)} \frac{3z_\kappa^2 - r_\kappa^2}{r_\kappa^5}
\end{aligned}
\tag{49.42}
$$

---

[7]The interaction of the stationary nuclear moments with fields originating from outside the nucleus leads to energy splittings between which electromagnetic resonance transitions can be induced as everybody knows. Thus, magnetic dipoles in external magnetic fields can be the subject of nuclear magnetic resonance (NMR). Electric quadrupoles in electric field gradients make nuclear quadrupole resonance (NQR) experiments possible if the magnetic dipole interaction is suppressed by keeping magnetic fields low or absent. The next experiment in this series would be magnetic octopole resonance (OMR) [1]. This experiment has not been carried out up to now. The difficulty is to suppress sufficiently the interactions of the lower nuclear moments with external fields that otherwise dominate.

whereas the nuclear charge distribution is represented by

$$B_{2,\pm 2} = \tfrac{\sqrt{6}}{4} e \sum_{\nu}(x_\nu \pm iy_\nu)^2$$
$$B_{2,\pm 1} = \tfrac{\sqrt{6}}{2} e \sum_{\nu} z_\nu(x_\nu \pm iy_\nu) \qquad (49.43)$$
$$B_{2,0} = \tfrac{1}{2} e \sum_{\nu}(3z_\nu^2 - r_\nu^2)$$

The irreducible spherical tensor components referring to the nuclear properties, $B_{2,q}$, are now to be converted into functions of the spin operators using the theorem at Eq. 49.27 and the corresponding spin operator expressions given in Eq. 49.22:

$$\langle Im| B_{2,q}(r_\nu)|Im\rangle = C \langle Im| T_{2,q}(I)|Im\rangle \qquad (49.44)$$

The constant C readily follows from the definition of the electric quadrupole moment, Q, given in Eq. 49.38 and this theorem. That is

$$\frac{eQ}{2} = \langle II|B_{2,0}|II\rangle \overset{!}{=} C \langle II|T_{2,0}|II\rangle = C \frac{I(2I-1)}{\sqrt{6}} \qquad (49.45)$$

so that

$$C = \sqrt{\frac{3}{2}} \frac{eQ}{I(2I-1)} \qquad (49.46)$$

The operators $B_{2,q}$ can then be replaced by

$$B_{2,q} = \sqrt{\frac{3}{2}} \frac{eQ}{I(2I-1)} T_{2,q} \qquad (49.47)$$

where the standard irreducible spherical tensor operators in spin coordinates are listed in Eq. 49.22.

The quantities $A_{2,q}$ characterize the electric environment of the nucleus. In particular, they can be related to the electric field gradient tensor taken at the origin in the center of the nucleus. The (Cartesian) components of the electric field gradient tensor, Γ, are given as the second derivatives of the electrostatic potential of the external charge distribution:

$$\{\Gamma_{ij}\} = \left\{\frac{\partial^2 \Phi_0}{\partial x_i \partial x_j}\right\}_{r=0} \qquad (i,j=1,2,3) \qquad (49.48)$$

where

$$\Phi_0(r) = \frac{1}{4\pi\epsilon_0} \sum_{\kappa} \frac{q_\kappa^{(e)}}{|r_\kappa - r|} \qquad (49.49)$$

and $r = (x_1, x_2, x_3)$, $r_\kappa = (x_{\kappa 1}, x_{\kappa 2}, x_{\kappa 3})$. Using

$$\frac{1}{|r_\kappa - r|} = [r_\kappa^2 + r^2 - 2(r_\kappa \cdot r)]^{-1/2} = \left(r_\kappa^2 + r^2 - 2\sum_{i=1}^{3} x_{\kappa i} x_i\right)^{-1/2} \qquad (49.50)$$

we readily find for the second derivatives of the electrostatic potential

$$\frac{\partial^2}{\partial x_i \partial x_j} \frac{1}{|r_\kappa - r|} = \frac{3(x_{\kappa i} - x_i)(x_{\kappa j} - x_j)}{|r_\kappa - r|^5} - \frac{\delta_{ij}}{|r_\kappa - r|^3} \tag{49.51}$$

The elements of the field gradient tensor at $r = 0$ are then

$$\Gamma_{ij} = \frac{1}{4\pi\epsilon_0} \sum_\kappa \frac{3x_{\kappa i} x_{\kappa j} - \delta_{ij} r_\kappa^2}{r_\kappa^5} q_\kappa^{(e)} \tag{49.52}$$

In the principal-axes representation, all off-diagonal elements vanish, i.e., $\Gamma_{ij} = 0$ for $i \neq j$. The convention is to choose the principal axes in such a way that the diagonal elements obey

$$|\Gamma_{33}| \geq |\Gamma_{22}| \geq |\Gamma_{11}| \tag{49.53}$$

The field gradient tensor is then determined by two independent parameters[8]

$$\begin{aligned} q &\equiv \Gamma_{33}/e & (e \text{ positive}) \\ \eta &= (\Gamma_{11} - \Gamma_{22})/\Gamma_{33} & (0 \leq \eta \leq 1) \end{aligned} \tag{49.54}$$

i.e., by the largest principal-axes field gradient value and the asymmetry parameter, respectively. On the basis of these parameters the field gradient tensor reads

$$\Gamma = \begin{pmatrix} \Gamma_{11} & 0 & 0 \\ 0 & \Gamma_{22} & 0 \\ 0 & 0 & \Gamma_{33} \end{pmatrix} = eq \begin{pmatrix} -\frac{1}{2}(1-\eta) & 0 & 0 \\ 0 & -\frac{1}{2}(1+\eta) & 0 \\ 0 & 0 & 1 \end{pmatrix} \tag{49.55}$$

Two cases are of particular interest. For rotational symmetry about the "3" direction of the pricipal-axes system, i.e., about a corresponding molecular symmetry axis, one expects $\Gamma_{11} = \Gamma_{22}$ so that $\eta = 0$. Second, in case of cubic symmetry, the three principal-axes values must be equal, $\Gamma_{11} = \Gamma_{22} = \Gamma_{33}$. This in turn means that all field gradient tensor elements are zero, $\Gamma_{ii} = 0$, owing to Laplace's equation. In particular we have then $q = 0$.

Combining Eqs. 49.42, 49.52, and 49.54, the components of the irreducible spherical tensor $A^{(2)}$ can thus be expressed in terms of the (Cartesian) field gradient tensor elements:

$$\begin{aligned} A_{2,\pm 2} &= \frac{1}{2\sqrt{6}}(\Gamma_{11} - \Gamma_{22} \pm 2i\Gamma_{12}) &= \frac{1}{2\sqrt{6}} eq\eta \\ A_{2,\pm 1} &= \frac{1}{\sqrt{6}}(\Gamma_{13} \pm i\Gamma_{23}) &= 0 \\ A_{2,0} &= \frac{1}{2}\Gamma_{33} &= \frac{1}{2} eq \end{aligned} \tag{49.56}$$

where the fact was used that all off-diagonal elements of Γ vanish in the principal-axes system.

The resulting quadrupolar Hamiltonian thus adopts the form

$$\mathcal{H}_q = \sqrt{\frac{3}{2}} \frac{eQ}{I(2I-1)} \sum_{m=-2}^{2} (-1)^m A_{2,-m} T_{2,m} \tag{49.57}$$

---

[8]The third parameter needed for the characterization of a diagonal second-rank tensor follows from Laplace's equation, i.e., $\text{Tr}\{\Gamma\} = 0$.

where

$$T_{2,0} = \frac{1}{\sqrt{6}}[3I^{(0)2} - I(I+1)] = \frac{1}{\sqrt{6}}[3I_z^2 - I(I+1)]; \quad A_{2,0} = \frac{1}{2}eq$$

$$T_{2,\pm1} = \frac{1}{\sqrt{2}}[I^{(0)}I^{(\pm1)} + I^{(\pm1)}I^{(0)}] = \mp\frac{1}{2}[I_zI^\pm + I^\pm I_z]; \quad A_{2,\pm1} = 0$$

$$T_{2,\pm2} = (I^{(\pm1)})^2 = \frac{1}{2}(I^\pm)^2; \quad A_{2,\pm2} = \frac{1}{2\sqrt{6}}eq\eta$$

(49.58)

Further formulations of the total quadrupolar Hamiltonian, $\mathcal{H}_q$, derived as outlined above, are listed in Table 46.4 on page 424. Note that the principal-axes systems to which the above irreducible sperical tensor operators $T^{(2)}$ and $A^{(2)}$ refer are identical. In the absence of strong magnetic fields the axis of quantization evidently coincides with the "3" direction of the principal-axes system of Γ, so that the definition of the nuclear quadrupole moment, Eq. 49.38, is consistent with the use of the principal-axes system of Γ.

CHAPTER 50

Derivation of Basic NMR Spectra

50.1
Pake Spectrum

50.1.1
Dipolar Coupling

Consider isolated pairs of like spins 1/2, I_1 and I_2, in a solid environment. Chemical shift and indirect spin interactions are assumed to be negligible. The spin Hamiltonian is then dominated by the Zeeman interaction and dipolar coupling:

$$\mathcal{H} = \mathcal{H}_0 + \mathcal{H}_d \tag{50.1}$$

For spectroscopy the secular part of the dipolar Hamiltonian is relevant, i.e., the truncated form (for like spins) given in Table 46.5 on page 425:

$$\mathcal{H}_d^{(0)} = \mathcal{H}_d^{tr} = \frac{\mu_0}{4\pi} \frac{\gamma_n^2 \hbar^2}{r^3} \frac{1 - 3\cos^2 \vartheta}{2} [3I_{1z}I_{2z} - I_1 \cdot I_2] \tag{50.2}$$

The single-spin wavefunctions for the magnetic quantum numbers $m = +1/2$ and $m = -1/2$ are denoted by α and β, respectively. The eigenkets of the two-spin system are composed of the basis product function kets,[1] $|\alpha\alpha\rangle$, $|\alpha\beta\rangle$, $|\beta\alpha\rangle$, and $|\beta\beta\rangle$, which are eigenkets to $\mathcal{H}_0$. As a general rule of quantum mechanics, systems of identical particles may only have eigenfunctions which are either symmetric or antisymmetric against particle interchange. These are the symmetric "triplet states"

$$|\psi_1\rangle = |\alpha\alpha\rangle \tag{50.3}$$

$$|\psi_2\rangle = \frac{1}{\sqrt{2}} (|\alpha\beta\rangle + |\beta\alpha\rangle) \tag{50.4}$$

$$|\psi_3\rangle = |\beta\beta\rangle \tag{50.5}$$

and the antisymmetric "singlet state"

$$|\psi_4\rangle = \frac{1}{\sqrt{2}} (|\alpha\beta\rangle - |\beta\alpha\rangle) \tag{50.6}$$

---

[1] The first factor in the product ket is to refer to spin I_1 and the second to spin I_2, for instance.

That is, the time-independent Schrödinger equation,

$$\mathcal{H}|\psi_i\rangle = E|\psi_i\rangle \tag{50.7}$$

is fulfilled by these states. The energy eigenvalues are

$$E_1 = -\gamma_n \hbar B_0 - \frac{\mu_0}{4\pi} \frac{\gamma_n^2 \hbar^2}{4r^3} \left(3\cos^2\vartheta - 1\right) \tag{50.8}$$

$$E_2 = \frac{\mu_0}{4\pi} \frac{\gamma_n^2 \hbar^2}{2r^3} \left(3\cos^2\vartheta - 1\right) \tag{50.9}$$

$$E_3 = \gamma_n \hbar B_0 - \frac{\mu_0}{4\pi} \frac{\gamma_n^2 \hbar^2}{4r^3} \left(3\cos^2\vartheta - 1\right) \tag{50.10}$$

$$E_4 = 0 \tag{50.11}$$

The selection rules for transitions induced by electromagnetic radiation imply $\Delta m = \pm 1$, and the conservation of the symmetry. Consequently there are two allowed transitions of the triplet system, i.e., $|\psi_1\rangle \leftrightarrow |\psi_2\rangle$ and $|\psi_2\rangle \leftrightarrow |\psi_3\rangle$. The resonance frequencies are

$$\begin{aligned}
\omega_{1,2} &= \gamma_n B_0 \pm \frac{3\mu_0}{16\pi} \frac{\gamma_n^2 \hbar}{r^3} \left(3\cos^2\vartheta - 1\right) \\
&= \omega_0 \pm \tilde{\omega}
\end{aligned} \tag{50.12}$$

where $\omega_0 = \gamma_n B_0$, $\tilde{\omega} = \delta\left(3\cos^2\vartheta - 1\right)$, $\delta = 3\mu_0\gamma_n^2\hbar/(16\pi r^3)$. The line intensities are equal. In the following, the resonances corresponding to the plus and minus signs in Eq. 50.12 will be called "+ transition" and "− transition", respectively.

In a powder, all orientations of the internuclear vector r are equally probable. Therefore the spectrum consists of a superposition of doublets weighted according to the probability, $p(\vartheta)\,d\vartheta$, that the polar angle is found in the range $\vartheta \ldots \vartheta + d\vartheta$ where $0 \le \vartheta \le \pi$. This probability is given by the quotient of the solid angle defined by this polar-angle range, and the full solid angle. That is

$$p(\vartheta)\,d\vartheta = \frac{2\pi \sin\vartheta\,d\vartheta}{4\pi} = \frac{1}{2}\sin\vartheta\,d\vartheta = \frac{1}{2}|d(\cos\vartheta)| \tag{50.13}$$

We express $\cos\vartheta$ as a function of $\tilde{\omega}$

$$\cos\vartheta = \left[\frac{1}{3}\left(\frac{\tilde{\omega}}{\delta} + 1\right)\right]^{1/2} \tag{50.14}$$

The probability, $g_+(\tilde{\omega})\,d\tilde{\omega}$, that the "+ transition" takes place in a range $\tilde{\omega} \ldots \tilde{\omega} + d\tilde{\omega}$, is

$$g_+(\tilde{\omega})\,d\tilde{\omega} = p(\vartheta)\,d\vartheta \tag{50.15}$$

The lineshape function of the "+ transition" thus is

$$g_+(\tilde{\omega}) = \frac{1}{2}\left|\frac{d(\cos\vartheta)}{d\tilde{\omega}}\right| = \frac{1}{12\delta}\left[\frac{1}{3}\left(\frac{\tilde{\omega}}{\delta} + 1\right)\right]^{-1/2} \tag{50.16}$$

The relevant frequency range is $-\delta \leq \tilde{\omega} \leq 2\delta$ corresponding to polar angles $\vartheta = \pi/2 \ldots 0(\pi)$.

The lineshape function for the "− transition" is readily found by replacing $\tilde{\omega}$ by $-\tilde{\omega}$ in the above expression. That is,

$$g_-(\tilde{\omega}) = \frac{1}{12\delta} \left[\frac{1}{3} \left(-\frac{\tilde{\omega}}{\delta} + 1 \right) \right]^{-1/2} \tag{50.17}$$

In this case, the relevant frequency range is $-2\delta \leq \tilde{\omega} \leq \delta$ corresponding to polar angles $\vartheta = 0(\pi) \ldots \pi/2$.

The total lineshape function,

$$g(\tilde{\omega}) = \frac{1}{2} \left[g_-(\tilde{\omega}) + g_+(\tilde{\omega}) \right] \tag{50.18}$$

is referred to as "**Pake spectrum**" [379]. It is plotted in Fig. 50.1. The relevant frequency range is $-2\delta \leq \tilde{\omega} \leq 2\delta$. There are singularities at $\tilde{\omega} = \pm\delta$ corresponding to the polar angle $\vartheta = \pi/2$, whereas $\vartheta = 0$ (or π) is attributed to the outer edges of the total spectrum.

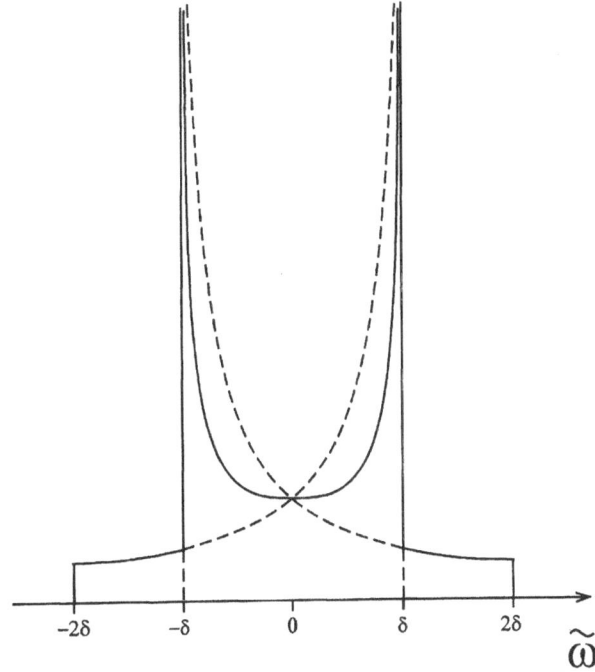

Fig. 50.1. The Pake spectrum of powdery solids. It refers to dipolar-coupled pairs of spins 1/2 as well as to quadrupolar-coupled spins 1 in axially symmetric electric-field gradients. Lifetime broadening has not been taken into account so that the singularities at $\tilde{\omega} = \pm\delta$ and the edges are not smeared out. The dashed lines represent the partial spectra corresponding to the + and − transitions. The solid line is the total spectrum.

50.1.2
Quadrupolar Coupling

The same sort of (high-field) solid-state spectrum is obtained for quadrupole coupling of spins 1 in axially symmetric electric-field gradients ($\eta = 0$). Chemical
shift, indirect or direct spin interactions may be considered to be negligible. The
spin Hamiltonian is then dominated by the Zeeman interaction and quadrupolar
coupling:

$$\mathcal{H} = \mathcal{H}_0 + \mathcal{H}_q \tag{50.19}$$

For spectroscopy purposes, only the secular part of the dipolar Hamiltonian is
relevant (Table 46.4 on page 424):

$$\mathcal{H}_q^{(0)} = \frac{e^2 qQ}{8} (3I_z^2 - 2)(3\cos^2 \vartheta - 1) \tag{50.20}$$

where ϑ is the polar angle of B_0 in the principal-axes system of the electric-field
gradient tensor.

$\mathcal{H}_q^{(0)}$ commutes with $\mathcal{H}_0$ so that there are common eigenfunctions. The eigen-
energies are

$$E_m = -\gamma_n \hbar B_0 m + \left(m^2 - \frac{2}{3} \right) \frac{3e^2 qQ}{8}(3\cos^2 \vartheta - 1) \tag{50.21}$$

where the magnetic quantum number takes the values $m = -1, 0, +1$. With the
selection rule for electromagnetic excitation, $\Delta m = \pm 1$, we obtain two resonance
lines at

$$\omega_{1,2} = \omega_0 \pm \tilde{\omega}' \tag{50.22}$$

where $\omega_0 = \gamma_n B_0$, $\tilde{\omega}' = \delta' \left(3\cos^2 \vartheta - 1\right)$, and $\delta' = 3e^2 qQ/(8\hbar)$. The line intensities
are equal. Formally we thus have the same situation as with Eq. 50.12. The powder
spectrum is derived as in the previous section. The result again is a "Pake spectrum"
as plotted in Fig. 50.1.

50.2
Chemical-Shift Anisotropy Spectrum

Chemical-shift anisotropy becomes relevant if spin couplings are weak, and if motional averaging is absent. Such a situation arises with isolated, i.e., well separated,
spins 1/2 in solids. The only anisotropy is then due to the chemical-shift interaction.
The spin Hamiltonian can be written as

$$\mathcal{H} = \mathcal{H}_0 + \mathcal{H}_{cs} \tag{50.23}$$

The shielding tensor which determines the chemical-shift distribution of a nucleus
in a certain compound in the solid state is denominated by σ. The secular part
of the chemical-shift Hamiltonian, which is relevant for spectral properties, is (see
Table 46.2 on page 421)

$$\mathcal{H}_{cs}^{(0)} = \hbar \gamma_n I_z B_0 \sigma_{zz} \tag{50.24}$$

where σ_{zz} is the "secular" diagonal component of the chemical-shift tensor in the laboratory frame. The orientation of the chemical-shift principal axes relative to the external magnetic field can be defined by the Euler angles $\tilde{\alpha}, \tilde{\beta}, \tilde{\gamma}$ as described in Chap. 48. On this basis, the "secular" chemical-shift tensor component is given by Eq. 48.30, i.e.,

$$\sigma_{zz} = \sigma_{11} \cos^2 \tilde{\alpha} \sin^2 \tilde{\beta} + \sigma_{22} \sin^2 \tilde{\alpha} \sin^2 \tilde{\beta} + \sigma_{33} \cos^2 \tilde{\beta} \tag{50.25}$$

where σ_{ii} are the principal-axes values of the chemical-shift tensor.

In a powder, the probability that the third principal axis points in a solid-angle range $\Omega \ldots \Omega + d\Omega$ defined by corresponding polar and azimuthal-angles ranges is

$$p(\Omega) \, d\Omega = \frac{1}{4\pi} d\Omega \tag{50.26}$$

where $p(\Omega)$ is a probability density with respect to the solid angle. Equation 50.25 shows that there is an unambiguous relation between the tensor orientation and the chemical shift, so that this probability can be equated to the probability $g(\sigma_{zz}) \, d\sigma_{zz}$ that the chemical shift is found in a range $\sigma_{zz} \ldots \sigma_{zz} + d\sigma_{zz}$. That is,

$$\frac{1}{4\pi} d\Omega = g(\sigma_{zz}) \, d\sigma_{zz} \tag{50.27}$$

or

$$g(\sigma_{zz}) = \frac{1}{4\pi} \left(\frac{d\sigma_{zz}}{d\Omega} \right)^{-1} \tag{50.28}$$

For rotational symmetry, i.e., $\sigma_{11} = \sigma_{22}$, which applies to linear molecules, Eq. 50.25 becomes

$$\sigma_{zz} = \sigma_{11} + (\sigma_{33} - \sigma_{11}) \cos^2 \tilde{\beta} \tag{50.29}$$

Using the explicit expression for the solid angle element for rotational symmetry,

$$d\Omega = 2\pi \sin \tilde{\beta} \, d\tilde{\beta} \tag{50.30}$$

and

$$\cos \tilde{\beta} = \left(\frac{\sigma_{zz} - \sigma_{11}}{\sigma_{33} - \sigma_{11}} \right)^{1/2} \tag{50.31}$$

we find

$$\begin{aligned}
g(\sigma_{zz}) &= \frac{1}{2} \left(\frac{d\sigma_{zz}}{\sin \tilde{\beta} \, d\tilde{\beta}} \right)^{-1} \\
&= \frac{1}{2} \sin \tilde{\beta} \left(2[\sigma_{33} - \sigma_{11}] \cos \tilde{\beta} \sin \tilde{\beta} \right)^{-1} \\
&= \frac{1}{4(\sigma_{33} - \sigma_{11})^{1/2} (\sigma_{zz} - \sigma_{11})^{1/2}}
\end{aligned} \tag{50.32}$$

where $\sigma_{11} \le \sigma_{zz} \le \sigma_{33}$. Otherwise $g(\sigma_{zz}) = 0$. This chemical-shift anisotropy lineshape function has a singularity at $\sigma_{zz} = \sigma_{11}$. A graphical representation is

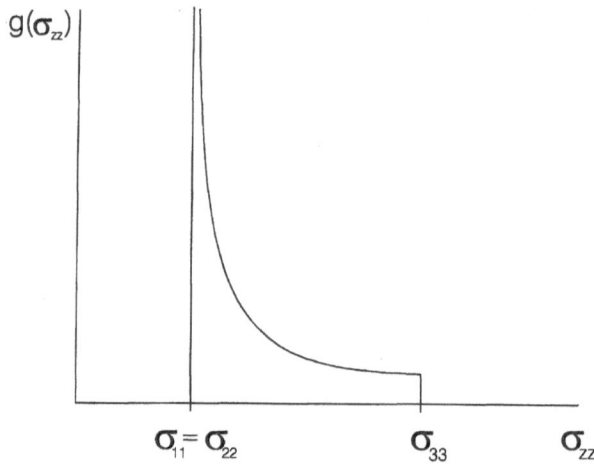

Fig. 50.2. Chemical-shift anisotropy lineshape function for rotational symmetry. Lifetime broadening has not been taken into account.

shown in Fig. 50.2. The general case without rotational symmetry was treated in [44, 336, 455]. Under motional-narrowing conditions, the spectrum coalesces to a narrow line at the isotropic chemical shift given by the shielding constant

$$\sigma^{(i)} = \frac{1}{3}\text{Tr}\{\sigma\} = \frac{1}{3}(\sigma_{11} + \sigma_{22} + \sigma_{33}) \tag{50.33}$$

50.3
A₂, AB, and AX Spectra

We consider a spin system consisting of two J-coupled spins $I_a = 1/2$ and $I_b = 1/2$. We further assume motional-narrowing conditions That is, the spin system is in a low-viscous, isotropic, liquid environment, so that all dipolar couplings and chemical-shift anisotropies are averaged out. The chemical-shift difference (in frequency units) is denominated by $\Delta = \gamma_n[B_a - B_b]/(2\pi) = \gamma_n B_0[\sigma_b^{(i)} - \sigma_a^{(i)}]/(2\pi)$, where B_a, B_b denote the local magnetic flux densities at the positions of the nuclei. Depending on whether $\Delta \gg J$ or $\Delta \overset{<}{\approx} J$, we speak of AX or AB spectra, respectively.[2] If $\Delta = 0$, the nuclei are equivalent, and the spin system is of the type A₂.
The Hamiltonian is given by[3]

$$\mathcal{H} = \mathcal{H}_0 + \mathcal{H}_J = -\gamma_n \hbar(B_a I_{az} + B_b I_{bz}) + hJ\left[\frac{1}{2}(I_a^+ I_b^- + I_a^- I_b^+) + I_{az}I_{bz}\right] \tag{50.34}$$

---

[2] A typical compound providing proton spectra of the AX type is uracil (compare the two-dimensional spectra of this substance in Figs. 7.1, 7.2, and 7.4.

[3] The exact treatment in the following demonstrates that the spin-spin coupling Hamiltonian in the weak-coupling limit, $J \ll \Delta$, can be approximated by its secular form $\mathcal{H}_J \approx hJI_{az}I_{bz}$.

The single-spin wavefunctions for the magnetic quantum numbers $m = +1/2$ and $m = -1/2$ are denoted by α and β, respectively. The (orthonormal) basis functions of the two-spin system are formed as products of the single-spin functions.[4] The kets $|\psi_1\rangle = |\alpha\alpha\rangle$ and $|\psi_4\rangle = |\beta\beta\rangle$, are eigenkets to $\mathcal{H}_0$ as well as to $\mathcal{H}_J$, so that the corresponding energy eigenvalues readily follow from the time-independent Schrödinger equation:

$$E_1 = \hbar\gamma_n B_0 \left(-1 + \frac{1}{2}\sigma_a^{(i)} + \frac{1}{2}\sigma_b^{(i)}\right) + \frac{1}{4}hJ \tag{50.35}$$

$$E_4 = \hbar\gamma_n B_0 \left(1 - \frac{1}{2}\sigma_a^{(i)} - \frac{1}{2}\sigma_b^{(i)}\right) + \frac{1}{4}hJ \tag{50.36}$$

The other two eigenkets can be represented by linear combinations[5]

$$|\psi_{2,3}\rangle = c_1|\alpha\beta\rangle + c_2|\beta\alpha\rangle \tag{50.37}$$

where the coefficients c_1 and c_2 must be determined in such a way that $|\psi_{2,3}\rangle$ forms eigenkets to $\mathcal{H}_0$ as well as to $\mathcal{H}_J$.

We start with the time-independent Schrödinger equation

$$\mathcal{H}|\psi\rangle = E|\psi\rangle. \tag{50.38}$$

Multiplying this equation with the basis functions $\langle\alpha\beta|$ and $\langle\beta\alpha|$ from the left produces a set of two linear equations for the coefficients c_1 and c_2, which can be represented in matrix form as

$$\begin{pmatrix} \langle\alpha\beta|\mathcal{H}|\alpha\beta\rangle - E & \langle\alpha\beta|\mathcal{H}|\beta\alpha\rangle \\ \langle\beta\alpha|\mathcal{H}|\alpha\beta\rangle & \langle\beta\alpha|\mathcal{H}|\beta\alpha\rangle - E \end{pmatrix} \begin{pmatrix} c_1 \\ c_2 \end{pmatrix} = \begin{pmatrix} 0 \\ 0 \end{pmatrix} \tag{50.39}$$

Equation 50.39 is soluble if the secular determinant vanishes, i.e.,

$$\begin{vmatrix} -\frac{h}{2}\Delta - \frac{hJ}{4} - E & \frac{hJ}{2} \\ \frac{hJ}{2} & \frac{h}{2}\Delta - \frac{hJ}{4} - E \end{vmatrix} = 0 \tag{50.40}$$

The solutions are the eigenvalues

$$E_{2,3} = -\frac{hJ}{4} \mp \frac{h}{2}\left(\Delta^2 + J^2\right)^{1/2} \tag{50.41}$$

From Eq. 50.39 we obtain the coefficients for the corresponding eigenfunctions. The result is

$$|\psi_2\rangle = \frac{1}{\sqrt{2}}\left(1 + \frac{\Delta}{\sqrt{J^2 + \Delta^2}}\right)^{1/2}|\alpha\beta\rangle - \frac{1}{\sqrt{2}}\left(1 - \frac{\Delta}{\sqrt{J^2 + \Delta^2}}\right)^{1/2}|\beta\alpha\rangle \tag{50.42}$$

$$|\psi_3\rangle = \frac{1}{\sqrt{2}}\left(1 - \frac{\Delta}{\sqrt{J^2 + \Delta^2}}\right)^{1/2}|\alpha\beta\rangle + \frac{1}{\sqrt{2}}\left(1 + \frac{\Delta}{\sqrt{J^2 + \Delta^2}}\right)^{1/2}|\beta\alpha\rangle \tag{50.43}$$

---

[4] The first factor in the product ket is to refer to spin I_a and the second to spin I_b, for instance.

[5] Recall that the spin-bearing particles considered here are not necessarily identical as was assumed in the case of a dipolar-coupled two-spin system treated in Sect. 50.1.1!

The transition matrix elements lead to the relative line intensities S_i of the four allowed resonances v_i:

$$\left.\begin{array}{l} v_1 = \bar{v} + \frac{J}{2} + \frac{1}{2}\sqrt{\Delta^2 + J^2} \\ v_2 = \bar{v} - \frac{J}{2} - \frac{1}{2}\sqrt{\Delta^2 + J^2} \end{array}\right\} \quad S_{1,2} = 1 - \frac{J}{\sqrt{\Delta^2 + J^2}} \tag{50.44}$$

$$\left.\begin{array}{l} v_3 = \bar{v} + \frac{J}{2} - \frac{1}{2}\sqrt{\Delta^2 + J^2} \\ v_4 = \bar{v} - \frac{J}{2} + \frac{1}{2}\sqrt{\Delta^2 + J^2} \end{array}\right\} \quad S_{3,4} = 1 + \frac{J}{\sqrt{\Delta^2 + J^2}} \tag{50.45}$$

where $\bar{v} = (v_a + v_b)/2 = \gamma_n(B_a + B_b)/(4\pi)$. Different limits of this result are schematically shown in Fig. 50.3.

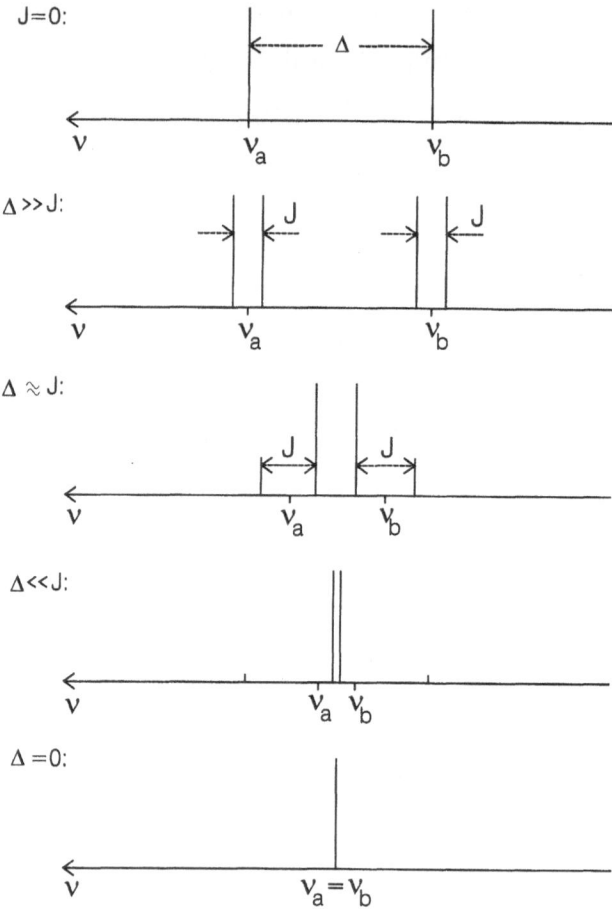

Fig. 50.3. Schematic representation of A_2 ($\Delta = 0$), AB ($\Delta \approx J$), and AX ($\Delta \gg J$) high-resolution NMR spectra. Lifetime broadening has not been taken into account.

CHAPTER 51

Product Operators for Spins 1/2

The evolution of spin coherences, longitudinal spin polarizations, and spin order in the course of pulse sequences can generally be treated on the basis of the density operator formalism (see Chap. 47). After several pulses and evolution periods, however, the result tends to be complex, and it is difficult to pursue the coherence pathways leading to the final signal contributions. Depending on the spin system and the pulse sequence, all sorts of multiple-quantum coherences, longitudinal spin polarization, and spin order states may appear together at a time. What is of interest for the experimenter is that only the single-quantum coherences can be detected as a signal. It is therefore often more favorable to trace the individual coherence pathways from the beginning. All contributions which do not lead to single-quantum coherences in the detection interval can then be omitted from further treatment as soon as this becomes obvious.

For this purpose, a method is required permitting the decomposition of the density operator in terms which directly represent the possible appearances of spin states. The different physical phenomena are to be represented by specific terms, which can easily be recognized as such. This objective is approached by product operator formalisms which permit a convenient treatment of the propagation of density operators in the course of coherence evolution in RF pulse intervals, and coherence transfer by RF pulses. So to speak, product operators keep track with the different states of the spin system generated and converted into each other by sequences of RF pulses.

The strategy of product operator formalisms is to express the reduced density operator by a linear combination of the elements of a so-called "complete set of basis operators," B_j,

$$\sigma(t) = \sum_{j=1}^{d^2} c_j(t) B_j \tag{51.1}$$

where d is the dimensionality of the Hilbert space. The basis operators are constituted in the form of spin operator products. In the following, a plausible example of a product operator basis set, the Cartesian two-spin-1/2 product operators, is described first. After outlining the orthogonality properties required for such operators to be eligible as basis operators, we present an interpretation revealing the physical phenomena represented by these operator products. In a further section, the Hamiltonians suitable for the description of the evolution of spin states and co-

herences in this approach are classified and discussed. Finally, the evolution rules
for Cartesian as well as for spherical product operators are specified and tabulated.

51.1
Simple Example of a Product Operator Basis Set

The term "product operator" refers to direct products[1] of spin operators of different
particles. A further factor in such products can be the unity operator. Let us first
formally define a typical set of basis operators, then discuss the physical meaning
and the evolution properties afterwards. Depending on the chosen representation,
the spin operators can be Cartesian or spherical components (see Sect. 42.6). As
a simple example, we consider the Cartesian components of two spin-1/2 vector
operators I and S. Together with the unity operator, we have four single operators
per spin, with which we can form the following 16 product operators:[2]

$$\begin{array}{cccccc}
\tfrac{1}{2}\mathcal{E}\mathcal{E}, & I_x\mathcal{E}, & S_x\mathcal{E}, & 2\,I_xS_x, & 2\,I_yS_x, & 2\,I_zS_x, \\
& I_y\mathcal{E}, & S_y\mathcal{E}, & 2\,I_xS_y, & 2\,I_yS_y, & 2\,I_zS_y, \\
& I_z\mathcal{E}, & S_z\mathcal{E}, & 2\,I_xS_z, & 2\,I_yS_z, & 2\,I_zS_z
\end{array} \tag{51.2}$$

These products form a **"complete set of basis operators"** of a 2-spin-1/2 system.
The number of 16 different operators is equal to the square of the Hilbert space
dimensionality corresponding to the four product eigenfunctions $\alpha\alpha$, $\alpha\beta$, $\beta\alpha$, and
$\beta\beta$.

Note that the basis operator set includes the ordinary Cartesian spin-vector
operator components and the terms of the reduced equilibrium density operator
(see Eq. 47.24 and footnote 2 on page 431),

$$\sigma_0 = I_z + S_z \tag{51.3}$$

which defines the starting point of all evolution processes.[3]

Larger spin systems stipulate correspondingly larger sets of basis operators. If
d designates the dimensionality of Hilbert space, i.e., the total number of spin
eigenstates, then d^2 product operators are needed for completeness. That is a set of
4^N operators for a N-spin-1/2 system.

---

[1] For a definition of a "direct product" see Sect. 51.3.

[2] As will become obvious below, it is somewhat more convenient to use half the unity operator
and to multiply all product operators by a factor of 2 for formal reasons. This of course does
not touch the principle of the product operator representation and has no physical meaning.
Besides, the unity operator is not explicitly written in practical treatments, but is tacitly assumed
with each single operator occurring in expressions for a two-spin system, for instance.

[3] Another simplification is the disregard of relaxation which must be distinguished from
the evolution processes considered here. The latter refer to spin coherences, longitudinal spin
polarizations, and spin order which are affected by rf pulses, chemical shifts, offset fields, and
spin couplings.

51.2
Orthogonality

An arbitrary reduced density operator can always be expressed as a linear combination of the basis operators $\mathcal{B}_j$ formed by a complete set of product operators[4] [140], as specified in Eq. 51.1. Product operators are orthogonal (but not necessarily normalized), i.e., they obey

$$\text{Tr}\{\mathcal{B}_l\mathcal{B}_m\} \propto \delta_{lm} \tag{51.4}$$

This property readily permits one to identify the density operator terms leading to finite expectation values of the observables. The transverse magnetization components, for instance, are proportional to the expectation values of the transverse spin operators. It is thus clear that the expected signal is determined only by those product operator terms in the density operator which are **not** orthogonal to the transverse spin operators. The x magnetization in the rotating frame at time t is, for instance,

$$M'_x(t) \propto \text{Tr}\{\sigma(t) I_x\} = \text{Tr}\left\{\sum_{j=1}^{d^2} c_j(t)\mathcal{B}_j I_x\right\} \propto \frac{1}{2}c_{(I_x)}(t) \tag{51.5}$$

where $c_{(I_x)}$ is the coefficient of the $\mathcal{B}_j \propto I_x$ term in the linear combination at Eq. 51.1. Therefore, the I_x and I_y terms in density operator terms are sometimes used synonymously with the x and y magnetization components, respectively.

51.3
Matrix Interpretation

The meaning of product operators such as those given in the above list (Eq. 51.2) becomes obvious if the operators are represented in matrix form and the products are carried out in the form of direct products:

$$
\begin{array}{cccccc}
\frac{1}{2}\mathcal{E} \otimes \mathcal{E}, & I_x \otimes \mathcal{E}, & S_x \otimes \mathcal{E}, & 2I_x \otimes S_x, & 2I_y \otimes S_x, & 2I_z \otimes S_x, \\
& I_y \otimes \mathcal{E}, & S_y \otimes \mathcal{E}, & 2I_x \otimes S_y, & 2I_y \otimes S_y, & 2I_z \otimes S_y, \\
& I_z \otimes \mathcal{E}, & S_z \otimes \mathcal{E}, & 2I_x \otimes S_z, & 2I_y \otimes S_z, & 2I_z \otimes S_z
\end{array} \tag{51.6}
$$

The direct product of two 2×2 matrices is defined by

$$
\mathcal{O} \otimes \mathcal{P} = \begin{pmatrix} a & b \\ c & d \end{pmatrix} \otimes \begin{pmatrix} a' & b' \\ c' & d' \end{pmatrix} = \begin{pmatrix} aa' & ab' & ba' & bb' \\ ac' & ad' & bc' & bd' \\ & & & \\ ca' & cb' & da' & db' \\ cc' & cd' & dc' & dd' \end{pmatrix} \tag{51.7}
$$

---

[4]There is an analogy to wave mechanics where it is known that arbitrary wavefunctions can be expressed by linear combinations of orthogonal basis functions provided that these form a complete set (e.g., all eigenfunctions of the system). The counterpart to the Hilbert space in wave mechanics is the Liouville space considered in density operator formalisms. The dimensionality of the Liouville space is, however, equal to the square of that of the Hilbert space for a given system.

More generally, the direct product of two matrices, $\mathcal{O} \otimes \mathcal{P}$, of arbitrary and potentially different dimensions are formed by replacing each element o_{nm} of $\mathcal{O}$ by the matrix $o_{nm}\mathcal{P}$. The dimension of the resulting matrix is then given by the product of the dimensions of the original matrices.

The matrices of the spin operators of spins 1/2, the Pauli spin matrices (Eqs. 42.45), are supplemented by half the unity matrix, so that we have the matrix set

$$
\frac{1}{2}\mathcal{E} = \frac{1}{2}\begin{pmatrix} 1 & 0 \\ 0 & 1 \end{pmatrix}; \qquad\qquad I_x, S_x = \frac{1}{2}\begin{pmatrix} 0 & 1 \\ 1 & 0 \end{pmatrix}
$$
$$
I_y, S_y = \frac{1}{2}\begin{pmatrix} 0 & -i \\ i & 0 \end{pmatrix}; \qquad\qquad I_z, S_z = \frac{1}{2}\begin{pmatrix} 1 & 0 \\ 0 & -1 \end{pmatrix}
\tag{51.8}
$$

Based on the Pauli spin matrices, the above direct products can readily be evaluated for two-spin 1/2 systems. Let us discuss three typical cases.

51.3.1
Example 1: In-phase Single-quantum Coherences

As a first example, we consider a density operator term given by the product operator $S_x \otimes \mathcal{E}$. Its explicit matrix form is

$$
\sigma_1 = S_x \otimes \mathcal{E} = \frac{1}{2}\begin{pmatrix} 0 & 1 & 0 & 0 \\ 1 & 0 & 0 & 0 \\ 0 & 0 & 0 & 1 \\ 0 & 0 & 1 & 0 \end{pmatrix}
\tag{51.9}
$$

The single-spin eigenfunctions are denoted by α and β for the spin-up and spin-down states, respectively. The matrix column numbers correspond then to a basis given in the form of product function kets $|\alpha\alpha\rangle$, $|\alpha\beta\rangle$, $|\beta\alpha\rangle$, $|\beta\beta\rangle$ provided that spin couplings are weak so that the magnetic quantum numbers of the individual spins are good quantum numbers (see below). The first position in the kets refers to the I spin, the second to S. The row numbers analogously indicate the bras in the sequence $\langle\alpha\alpha|$, $\langle\alpha\beta|$, $\langle\beta\alpha|$, $\langle\beta\beta|$. In the above example the finite matrix elements are

$$
\langle\alpha\alpha|\sigma_1|\alpha\beta\rangle = \langle\alpha\beta|\sigma_1|\alpha\alpha\rangle = \langle\beta\alpha|\sigma_1|\beta\beta\rangle = \langle\beta\beta|\sigma_1|\beta\alpha\rangle = \frac{1}{2}
\tag{51.10}
$$

These matrix elements symbolize only transitions of the second spin, i.e., the S spin, whereas the I spin remains passive. The product operator (Eq. 51.9) therefore represents solely single-spin coherences of the S spins generating the transverse component of the S spin magnetization in the rotating frame according to

$$
M'_{S,x} \propto \mathrm{Tr}\{(S_x \otimes \mathcal{E})\sigma_1\} = 1
\tag{51.11}
$$

where we understand that the operators of the magnetization components in the matrix representation of this treatment are analogously given in the form of direct products. Contributions to other magnetization components do not arise, i.e.,

$$M'_{S,y} \propto \text{Tr}\{(S_y \otimes \mathcal{E})\sigma_1\} = 0$$
$$M'_{S,z} \propto \text{Tr}\{(S_z \otimes \mathcal{E})\sigma_1\} = 0$$
$$M'_{I,x} \propto \text{Tr}\{(I_x \otimes \mathcal{E})\sigma_1\} = 0$$
$$M'_{I,y} \propto \text{Tr}\{(I_y \otimes \mathcal{E})\sigma_1\} = 0$$
$$M'_{I,z} \propto \text{Tr}\{(I_z \otimes \mathcal{E})\sigma_1\} = 0 \tag{51.12}$$

This is not the only information this operator manifests. The matrix elements $\langle \alpha\alpha|\sigma_1|\alpha\beta \rangle$ and $\langle \alpha\beta|\sigma_1|\alpha\alpha \rangle$ reflect S spin transitions while the I spin remains in state α. The other two finite matrix elements, $\langle \beta\alpha|\sigma_1|\beta\beta \rangle$ and $\langle \beta\beta|\sigma_1|\beta\alpha \rangle$, address S spin transitions with the I spin being in the β state. That is, the product operator $S_x \otimes \mathcal{E}$ symbolizes both doublet isochromats at a time. The x-magnetizations of these isochromats add constructively, i.e., both contributions are in-phase. The full meaning of the product operator $S_x \otimes \mathcal{E}$ thus is **in-phase single-quantum coherences of the S spins in the x direction** of the rotating frame.

51.3.2
Example 2: Multiple-quantum Coherences

A second example is a reduced density operator term assumed to be represented by $2\,I_x \otimes S_x$, i.e.,

$$\sigma_2 = 2\,I_x \otimes S_x = \frac{i}{2} \begin{pmatrix} 0 & 0 & 0 & +1 \\ 0 & 0 & +1 & 0 \\ 0 & +1 & 0 & 0 \\ +1 & 0 & 0 & 0 \end{pmatrix} \tag{51.13}$$

This matrix indicates the presence of transitions of the type $\alpha\alpha \leftrightarrow \beta\beta$ and $\alpha\beta \leftrightarrow \beta\alpha$, i.e., **double-quantum and zero-quantum coherences**. Similarly, all other product operators combining two transverse spin-vector operator components, i.e., $2\,I_x \otimes S_y$, $2\,I_y \otimes S_x$, and $2\,I_y \otimes S_y$ reflect superpositions of double-quantum and zero-quantum coherences. All magnetization components connected with this sort of product operators vanish. That is, in the present case,

$$M'_{S,x} \propto \text{Tr}\{(S_x \otimes \mathcal{E})\sigma_2\} = 0$$
$$M'_{S,y} \propto \text{Tr}\{(S_y \otimes \mathcal{E})\sigma_2\} = 0$$
$$M'_{S,z} \propto \text{Tr}\{(S_z \otimes \mathcal{E})\sigma_2\} = 0$$
$$M'_{I,x} \propto \text{Tr}\{(I_x \otimes \mathcal{E})\sigma_2\} = 0$$
$$M'_{I,y} \propto \text{Tr}\{(I_y \otimes \mathcal{E})\sigma_2\} = 0$$
$$M'_{I,z} \propto \text{Tr}\{(I_z \otimes \mathcal{E})\sigma_2\} = 0 \tag{51.14}$$

This result directly manifests the orthogonality property (Eq. 51.4). A more physical reading is, n-quantum coherences represent transitions which are forbidden for electromagnetic radiation. The transitions are radiationless. On these grounds no finite transverse magnetization can arise, whereas the z components are not represented by coherence terms anyway.

The matrix representation of reduced density operator terms of the form $2\,I_x \otimes S_z$, $2\,I_y \otimes S_z$, $2\,S_x \otimes I_z$, or $2\,S_y \otimes I_z$ reveal **antiphase single-quantum coherences** of the I and S spins, respectively, as the physical background. The expectation values of all magnetization components vanish as with the n-quantum coherences. However, the explanations of these findings are of a different category. Antiphase single-quantum coherences represent transitions which are certainly connected with electromagnetic radiation. The reason why the expectation values of the transverse magnetization components vanish is that the Cartesian product operator formalism does not keep the isochromats of a multiplet explicitly distinct.[5] Therefore the antiphase isochromat partial magnetizations formally cancel each other.

However, in high-resolution Fourier-transform spectroscopy, these partial magnetizations may be resolved and differentiated from each other owing to the different resonance frequencies of the isochromats. The two partial magnetizations then reveal themselves as distinct signals. Single-quantum coherences which are antiphase at the beginning of the acquisition interval appear as a pair of lines with positive and negative intensities.

The interpretation of the product operators is also directly possible without detour to matrix representations. Omitting the direct-product symbol, the operator $2\,I_x \otimes S_x$, for instance, can be expressed by raising and lowering operators:

$$\sigma_2 = 2I_x S_x = \frac{1}{2}(\underbrace{I^+S^- + I^-S^+}_{\text{0Q-coh.}} + \underbrace{I^+S^+ + I^-S^-}_{\text{2Q-coh.}}) \tag{51.15}$$

(Compare Eqs. 42.76 and 42.77.) The physical meaning is obvious: as above with the matrix representation, it is discernible that this operator represents zero-quantum and double-quantum coherences.

This direct consideration of the operators is the usual, less tutorial but more general way of interpreting product operators since operator algebra does not refer to a particular basis of wavefunctions as was assumed above. The independence of any matrix representations in particular means that one need not care whether the magnetic quantum numbers of the individual spins are good or not.

51.3.3
Example 3: Longitudinal Scalar Order

Finally, consider a density operator term given by the direct product $2I_z \otimes S_z$. The explicit matrix form is

$$\sigma_3 = 2I_z \otimes S_z = \frac{1}{2}\begin{pmatrix} 1 & 0 & 0 & 0 \\ 0 & -1 & 0 & 0 \\ 0 & 0 & -1 & 0 \\ 0 & 0 & 0 & 1 \end{pmatrix} \tag{51.16}$$

---

[5]This is in contrast to the spherical product operators to be described below.

The expectation values of all magnetization components vanish for this density operator. However, this does not mean that all isochromat magnetization components are zero. The situation is similar to the antiphase single-spin coherences above. That is, the z components of the partial magnetizations of the isochromats have opposite signs for each spin, and, therefore, cancel each other.

Closer analysis of the situation reveals the following isochromat magnetization polarities:

S in state α $\rightarrow$ magnetization of I positive
S in state β $\rightarrow$ magnetization of I negative
I in state α $\rightarrow$ magnetization of S positive
I in state β $\rightarrow$ magnetization of S negative

The polarity of the isochromat partial magnetization obviously is correlated to the spin state of the coupling partner. Therefore one speaks of "**longitudinal scalar order**" or simply of "**J order.**" A further term in use is "**zz order.**"

51.3.4
Summary of the Interpretations of Cartesian Product Operators

The interpretations of all product operators of the set at Eq. 51.2 are summarized in Table 51.1.[6] Products referring to up to three spins in a n-spin 1/2 system are considered.

51.4
Hamiltonians and Applicability Limits

The product operator formalisms in particular refer to spin systems, i.e., to coupled spins. The strengths of the spin interactions are characterized by the coupling constants J in the case of indirect spin-spin interactions and

$$c_d = \frac{\mu_0}{4\pi} \frac{3}{2} \frac{\gamma_I \gamma_S \hbar}{2\pi r^3} (1 - 3\cos^2 \vartheta) \qquad (51.17)$$

for secular dipole-dipole interaction (see Table 46.5 on page 425). The Hamiltonians governing the evolution of density operators in NMR refer to the Zeeman interaction with the local magnetic field or RF fields, and to the spin interactions. The latter determine to what extent product operator formalisms are feasible. Table 51.2 specifies the different cases for which product operator formalisms have been suggested.

The criteria for the applicability of product operator formalisms are based on whether we have

* indirect (J) coupling or dipole-dipole interaction

---

[6]The unity operator $\mathcal{E}$ and the direct-product symbol $\otimes$ have been introduced for completeness and with respect to the didactic matrix representation. In practical product operator treatments these symbols are usually omitted. Note also that the spin-operator products are multiplied by a factor of 2 as usual.

Table 51.1. Interpretation of typical Cartesian product operator terms arising in treatments of the evolution of the density operator. Each product term globally represents all isochromats arising from spin-spin coupling. A "coherence" is understood to comprise "emissive" as well as "absorptive" transitions. This ambiguity may be avoided by the use of the spherical-product-operator formalism (see Table 51.4).

---

	terms addressing spin "k" of a ($n \geq 1$)-spin 1/2 system
I_{kx}	single-quantum coherences of spin "k"; isochromats in-phase in $+x$ direction
I_{ky}	single-quantum coherences of spin "k"; isochromats in-phase in $+y$ direction
I_{kz}	longitudinal polarization of spin "k"; all isochromats in $+z$ direction

	additional terms addressing spins "k", "l" of a ($n \geq 2$)-spin 1/2 system
$2I_{kx}I_{lz}$	single-quantum coherences of spin "k"; spin "k" isochromats caused by spin "l" up/down are aligned antiphase ($+x$ and $-x$ directions); spin "k" isochromats with respect to other spins are in-phase
$2I_{ky}I_{lz}$	single-quantum coherences of spin "k"; spin "k" isochromats caused by spin "l" up/down are aligned antiphase ($+y$ and $-y$ directions); spin "k" isochromats with respect to other spins are in-phase
$2I_{kx}I_{lx}$ $2I_{kx}I_{ly}$ $2I_{ky}I_{lx}$ $2I_{ky}I_{ly}$	superimposed 0Q and 2Q coherences of spins "k", "l"; all other spins are passive; isochromats of spins "k", "l" with respect to passive spins are in-phase
$2I_{kz}I_{lz}$	longitudinal order of spins "k", "l"; spin "k" isochromats caused by spin "l" up/down are antiparallel and vice versa ($+z$ and $-z$ directions); isochromats with respect to other spins are parallel

	additional terms addressing spins "k", "l", "m" of a ($n \geq 3$)-spin 1/2 system
$4I_{kx}I_{lz}I_{mz}$	single-quantum coherences of spin "k"; spin "k" isochromats caused by spin "l" up/down or spin "m" up/down are antiphase in each case ($+x$ and $-x$ directions); all other isochromats are in-phase
$4I_{kx}I_{lx}I_{mz}$	superimposed 0Q and 2Q coherences of spins "k", "l"; all other spins are passive; the 0Q and 2Q isochromats caused by spin "m" up/down are antiphase; isochromats with respect to other passive spins are in-phase
$4I_{kx}I_{lx}I_{mx}$	superimposed 1Q and 3Q three-spin coherences of spins "k", "l", "m"; all other spins are passive; isochromats with respect to passive spins are in-phase
$4I_{kz}I_{lz}I_{mz}$	longitudinal order of spins "k", "l", "m"; isochromats of each of the spins "k", "l", "m" caused by spin up/down of each of the coupling partners within the triple are antiparallel ($+z$ and $-z$ directions); isochromats with respect to other spins are parallel

Table 51.2. Survey on strong (strg) and weak (wk) interactions in a system of two like or unlike spins 1/2, I and S. In the last column, references of suitable product operator formalisms (POF) are listed. α and β are the single-spin eigenfunctions for spin "up" and "down." The difference between the resonance frequencies $\Delta = \nu_I - \nu_S$ and the spin-spin coupling constant J are assumed to be positive. The coefficients in the strong-J-coupling wavefunctions are $c_+ = \sqrt{1 + \Delta/\sqrt{J^2 + \Delta^2}}$ and $c_- = \sqrt{1 - \Delta/\sqrt{J^2 + \Delta^2}}$. The dipolar coupling constant is expressed as $c_d = \frac{\mu_0}{4\pi}\frac{3}{2}\frac{\gamma_I \gamma_S \hbar}{2\pi r^3}(1 - 3\cos^2 \vartheta)$ (see Table 46.5).

case	γ_I, γ_S	chem. sh.	Hamiltonian	eigenfunctions	POF
$\mathcal{H}_J$ strg	like $\gamma_I = \gamma_S$	$\Delta \lessapprox J$	$-\hbar(\omega_I I_z + \omega_s S_z) + hJI \cdot S$	$\lvert\alpha\alpha\rangle$ $\frac{1}{\sqrt{2}}\left[c_+\lvert\alpha\beta\rangle - c_-\lvert\beta\alpha\rangle\right]$ $\frac{1}{\sqrt{2}}\left[c_+\lvert\beta\alpha\rangle + c_-\lvert\alpha\beta\rangle\right]$ $\lvert\beta\beta\rangle$	[220, 254]
$\mathcal{H}_J$ wk	like $\gamma_I = \gamma_S$	$\Delta \gg J$	$\approx -\hbar(\omega_I I_z + \omega_s S_z) + hJI_z S_z$	$\lvert\alpha\alpha\rangle$ $\approx \lvert\alpha\beta\rangle$ $\approx \lvert\beta\alpha\rangle$ $\lvert\beta\beta\rangle$	[360, 378] [457, 489]
	unlike $\gamma_I \neq \gamma_S$	arb.	$-\hbar(\omega_I I_z + \omega_s S_z) + hJI_z S_z$	$\lvert\alpha\alpha\rangle$ $\lvert\alpha\beta\rangle$ $\lvert\beta\alpha\rangle$ $\lvert\beta\beta\rangle$	[360, 378] [457, 489]
$\mathcal{H}_d$ strg	like $\gamma_I = \gamma_S$	$\Delta = 0$	$-\hbar\omega_0(I_z + S_z) + hc_d(I_z S_z - \frac{1}{3}I \cdot S)$	$\lvert\alpha\alpha\rangle$ $\frac{1}{\sqrt{2}}\left[\lvert\alpha\beta\rangle + \lvert\beta\alpha\rangle\right]$ $\lvert\beta\beta\rangle$ $\frac{1}{\sqrt{2}}\left[\lvert\alpha\beta\rangle - \lvert\beta\alpha\rangle\right]$	[502]
$\mathcal{H}_d$ wk	like $\gamma_I = \gamma_S$	$\Delta \gg c_d$	$\approx -\hbar(\omega_I I_z + \omega_s S_z) + hc_d I_z S_z$	$\lvert\alpha\alpha\rangle$ $\approx \lvert\alpha\beta\rangle$ $\approx \lvert\beta\alpha\rangle$ $\lvert\beta\beta\rangle$	[360, 378] [457, 489]
	unlike $\gamma_I \neq \gamma_S$	arb.	$-\hbar(\omega_I I_z + \omega_s S_z) + \frac{2}{3}hc_d I_z S_z$	$\lvert\alpha\alpha\rangle$ $\lvert\alpha\beta\rangle$ $\lvert\beta\alpha\rangle$ $\lvert\beta\beta\rangle$	[360, 378] [457, 489]

- **homonuclear** ("like spins", $\gamma_I = \gamma_S$) or **heteronuclear** ("unlike spins", $\gamma_I \neq \gamma_S$) interactions
- **weak coupling** ($J \ll \Delta$; $c_d \ll \Delta$) or **strong coupling** ($J \overset{>}{\approx} \Delta$; $c_d \gg \Delta$ or even $\Delta = 0$)

The product operator formalisms available for the treatment of cases defined in these terms refer to **Cartesian** [200, 378, 457, 458, 489], **spherical** [220, 360] or **c3po** [254] product operator bases.

Interestingly, the cases for weak homo- or heteronuclear J or dipolar coupling and for dipolar interaction of equivalent nuclei can commonly be treated using an effective form of the spin-interaction Hamiltonian.

- The **J interaction Hamiltonian of like spins in the weak-coupling limit** can be approached by its secular part, i.e.,

$$\mathcal{H}_J^{(0)} = hJI_zS_z \tag{51.18}$$

  (see Table 46.3 on page 422 and the general treatment in Sect. 50.3).
- The **J interaction Hamiltonian of unlike spins at usual magnetic flux densities** always complies with the weak-coupling limit so that $\mathcal{H}_J^{(0)}$ is again applicable.
- The **secular dipole-dipole interaction Hamiltonian of like spins in the weak-coupling limit** can be approached by the corresponding secular part, i.e.,

$$\mathcal{H}_d^{(0)} = \frac{2}{3}hc_dI_zS_z \tag{51.19}$$

  (see Table 46.5 on page 425).
- The **secular dipole-dipole interaction Hamiltonian of unlike spins** at usual magnetic flux densities likewise fulfill the weak-coupling limit, so that $\mathcal{H}_d^{(0)}$ applies again.
- The **secular dipole-dipole interaction Hamiltonian of like and moreover equivalent spin-1/2 pairs** ($\Delta = \nu_I - \nu_S = 0$) may be described by a Hamiltonian effective for the evolution of the spin coherences[7]

$$\mathcal{H}_d^{(ev)} = hc_dI_zS_z \tag{51.23}$$

---

[7]The dipole-dipole interaction Hamiltonian of like and moreover equivalent nuclei ($\Delta = \nu_I - \nu_S = 0$) is first approached by its secular part,

$$\mathcal{H}_d^{(0)} = hc_d\left(I_zS_z - \frac{1}{3}I \cdot S\right) \tag{51.20}$$

With systems of equivalent spins (indistinguishable particles) one considers the total spin

$$F = I + S \tag{51.21}$$

The scalar product can then be replaced according to

$$I \cdot S = \frac{1}{2}(F^2 - I^2 - S^2) = \frac{1}{2}[F(F+1) - I(I+1) - S(S+1)] \tag{51.22}$$

In the case of a 2-spin-1/2 system ($I = 1/2$, $S = 1/2$), the possible spin quantum numbers are $F = \frac{1}{2} - \frac{1}{2} = 0$ ("singlet state") and $F = \frac{1}{2} + \frac{1}{2} = 1$ ("triplet state"). The singlet state does

Thus, a common form of the secular spin-interaction Hamilton operator,

the **standard $I_z S_z$ interaction Hamiltonian**

for coherence evolution can be formulated:

$$\mathcal{H}_{J,d}^{(ev)} = h c_{IS} I_z S_z \tag{51.24}$$

where the following cases for the coupling constant are to be distinguished:

weak J coupling of like or unlike nuclei: $\quad c_{IS} = J$	
weak dipole-dipole int. of	
like or unlike nuclei: $\qquad\qquad c_{IS} = \frac{2}{3} c_d$	
$\qquad\qquad\qquad\qquad\qquad = \frac{\mu_0}{4\pi} \frac{\gamma_I \gamma_S \hbar}{2\pi r^3} (1 - 3\cos^2 \vartheta)$	
dipole-dipole int. of	
equivalent spin-1/2 pairs: $\qquad c_{IS} = c_d$	
$\qquad\qquad\qquad\qquad\qquad = \frac{\mu_0}{4\pi} \frac{3}{2} \frac{\gamma_n^2 \hbar}{2\pi r^3} (1 - 3\cos^2 \vartheta)$	

We restrict ourselves to density-operator evolutions for spin interaction Hamiltonians of the form at Eq. 51.24 for "standard $I_z S_z$" or - referring to a selected spin pair k and l - "standard $I_{kz} I_{lz}$ interactions." The treatment is performed either on a Cartesian or on a spherical product operator basis. There are alternative choices of basis operators which are equivalent and which by circumstance provide certain advantages compared with the standard basis sets. An example is described in [254]. It should also be mentioned that the product operator formalisms are particularly suited for computer algebra treatments [215, 446].

51.5
Evolution Rules for Cartesian Product Operators

The Cartesian product operators listed in Table 51.1 evolve under the action of RF pulses, chemical shifts, field/frequency offsets, and standard $I_{kz} I_{lz}$ interactions [200, 457, 502]. We note that the initial reduced density operator of an N-spin system at equilibrium (see Eq. 47.24) itself is composed of Cartesian product operators:

$$\sigma_0 = \sum_{k=1}^{N} I_{kz} \tag{51.25}$$

A pulse sequence is treated in a segmented way (compare Sect. 47.3.1). The evolution steps then consist of unitary transformations analogous to Eq. 47.33

$$\sigma(t) = e^{-(i/\hbar)\mathcal{H}'t} \sigma(0) e^{(i/\hbar)\mathcal{H}'t} \tag{51.26}$$

---

not show up in NMR experiments. For the triplet state the scalar product is equal to 3/4 (see Eqs. 42.72) and can be dropped from the Hamilton operator, because it does not influence the evolution of the density operator. It merely reduces all triplet energy eigenvalues by a constant amount $-\frac{1}{12} h c_d$. The Hamiltonian effective for the evolution of two-spin 1/2 coherences is hence $\mathcal{H}_d^{(ev)} = h c_d I_z S_z$. Note that this Hamilton operator is solely introduced for the treatment of the evolution of spin coherences, but not for the calculation of dipolar spectra. In that case the full truncated Hamiltonian $\mathcal{H}_d^{(0)}$ must be used (see Sect. 50.1.1).

Table 51.3. Evolution of the reduced density operator in terms of Cartesian product operators for spins 1/2 in the standard $I_{kz}I_{lz}$ spin coupling case. The coupling constant between spins k and l, c_{kl}, can refer to indirect spin-spin or dipolar interactions (see Eq. 51.24). α is the pulse flip angle (always has a positive value); ν indicates the rotating frame direction of the RF field B_1; $\Omega_k = \gamma_k \Delta B_k$ is the frequency offset of spin k in the rotating frame. The gyromagnetic ratios γ_k and γ_l are assumed to be positive. The operator expressions $\sigma_{0qc,x}^{(k,l)}$, $\sigma_{0qc,y}^{(k,l)}$, $\sigma_{2qc,x}^{(k,l)}$, $\sigma_{2qc,y}^{(k,l)}$ represent zero-quantum and double-quantum coherences of (only) two active spins, respectively. For definitions see Eqs. 42.76 and 42.77.

RF pulses: $(\alpha)_{\pm\nu}$

"hard"	$\sigma(t+)$	=	$\exp\left\{\pm i\alpha \sum_k I_{k\nu}\right\} \sigma(t-) \exp\left\{\mp i\alpha \sum_k I_{k\nu}\right\}$
"selective"	$\sigma(t+)$	=	$\exp\{\pm i\alpha I_{k\nu}\}\, \sigma(t-)\, \exp\{\mp i\alpha I_{k\nu}\}$

$\nu = x$
$(\alpha)_{\pm x}$
$B_1 \uparrow\uparrow \pm u_x$

$$I_{kx} \rightarrow I_{kx}$$
$$I_{ky} \rightarrow I_{ky}\cos\alpha \mp I_{kz}\sin\alpha$$
$$I_{kz} \rightarrow I_{kz}\cos\alpha \pm I_{ky}\sin\alpha$$

$\nu = y$
$(\alpha)_{\pm y}$
$B_1 \uparrow\uparrow \pm u_y$

$$I_{kx} \rightarrow I_{kx}\cos\alpha \pm I_{kz}\sin\alpha$$
$$I_{ky} \rightarrow I_{ky}$$
$$I_{kz} \rightarrow I_{kz}\cos\alpha \mp I_{kx}\sin\alpha$$

(Continued next page)

Table 51.3. (cont.)

chem. shift: (frequ. offset Ω_k)

$$\sigma(t) = \prod_k e^{i\Omega_k t I_{kz}}\,\sigma(0)\,\prod_k e^{-i\Omega_k t I_{kz}}$$

I_{kx}	$\rightarrow$ $I_{kx}\cos(\Omega_k t) - I_{ky}\sin(\Omega_k t)$
I_{ky}	$\rightarrow$ $I_{ky}\cos(\Omega_k t) + I_{kx}\sin(\Omega_k t)$
I_{kz}	$\rightarrow$ I_{kz}
$\sigma_{0qc,x}^{(k,l)}$	$\rightarrow$ $\sigma_{0qc,x}^{(k,l)}\cos[(\Omega_k-\Omega_l)t] - \sigma_{0qc,y}^{(k,l)}\sin[(\Omega_k-\Omega_l)t]$
$\sigma_{0qc,y}^{(k,l)}$	$\rightarrow$ $\sigma_{0qc,y}^{(k,l)}\cos[(\Omega_k-\Omega_l)t] + \sigma_{0qc,x}^{(k,l)}\sin[(\Omega_k-\Omega_l)t]$
$\sigma_{2qc,x}^{(k,l)}$	$\rightarrow$ $\sigma_{2qc,x}^{(k,l)}\cos[(\Omega_k+\Omega_l)t] - \sigma_{2qc,y}^{(k,l)}\sin[(\Omega_k+\Omega_l)t]$
$\sigma_{2qc,y}^{(k,l)}$	$\rightarrow$ $\sigma_{2qc,y}^{(k,l)}\cos[(\Omega_k+\Omega_l)t] + \sigma_{2qc,x}^{(k,l)}\sin[(\Omega_k+\Omega_l)t]$

k, l are the only active spins

$I_{kz}I_{lz}$ coupling: (coupl. const. c_{kl})

$$\sigma(t) = \prod_{k<l} e^{-i\pi c_{kl}t\,2I_{kz}I_{lz}}\,\sigma(0)\,\prod_{k<l} e^{i\pi c_{kl}t\,2I_{kz}I_{lz}}$$

I_{kx}	$\rightarrow$ $I_{kx}\cos(\pi c_{kl}t) + 2I_{ky}I_{lz}\sin(\pi c_{kl}t)$
I_{ky}	$\rightarrow$ $I_{ky}\cos(\pi c_{kl}t) - 2I_{kx}I_{lz}\sin(\pi c_{kl}t)$
I_{kz}	$\rightarrow$ I_{kz}
$2I_{kx}I_{lz}$	$\rightarrow$ $2I_{kx}I_{lz}\cos(\pi c_{kl}t) + I_{ky}\sin(\pi c_{kl}t)$
$2I_{ky}I_{lz}$	$\rightarrow$ $2I_{ky}I_{lz}\cos(\pi c_{kl}t) - I_{kx}\sin(\pi c_{kl}t)$
$2I_{kx,y}I_{lx,y}$	$\rightarrow$ $2I_{kx,y}I_{lx,y}$
$\sigma_{0qc,x}^{(k,l)}$	$\rightarrow$ $\sigma_{0qc,x}^{(k,l)}\cos[\pi(c_{km}-c_{lm})t] + 2I_{mz}\sigma_{0qc,y}^{(k,l)}\sin[\pi(c_{km}-c_{lm})t]$
$\sigma_{0qc,y}^{(k,l)}$	$\rightarrow$ $\sigma_{0qc,y}^{(k,l)}\cos[\pi(c_{km}-c_{lm})t] - 2I_{mz}\sigma_{0qc,x}^{(k,l)}\sin[\pi(c_{km}-c_{lm})t]$
$\sigma_{2qc,x}^{(k,l)}$	$\rightarrow$ $\sigma_{2qc,x}^{(k,l)}\cos[\pi(c_{km}+c_{lm})t] + 2I_{mz}\sigma_{2qc,y}^{(k,l)}\sin[\pi(c_{km}+c_{lm})t]$
$\sigma_{2qc,y}^{(k,l)}$	$\rightarrow$ $\sigma_{2qc,y}^{(k,l)}\cos[\pi(c_{km}+c_{lm})t] - 2I_{mz}\sigma_{2qc,x}^{(k,l)}\sin[\pi(c_{km}+c_{lm})t]$

k, l are the only active spins, m is passive

where $\mathcal{H}'$ represents the rotating-frame Hamiltonians (see Sect. 48.8) dominating in the corresponding intervals.

During "hard" RF pulses,[8] the relevant Hamiltonian is

$$\mathcal{H}'_{rf} = -\hbar\omega_1 \sum_k I_{kv} \tag{51.27}$$

where v indicates the rotating-frame phase direction of the RF field and $\omega_1 = \gamma_n B_1$. A "selective" RF pulse is correspondingly characterized by

$$\mathcal{H}'_{rf,k} = -\hbar\omega_1 I_{kv} \tag{51.28}$$

In the free-evolution intervals, frequency offsets (particularly by chemical shifts) and the standard $I_{kz}I_{lz}$ interactions are assumed to dominate. The rotating-frame Hamiltonian is then

$$\mathcal{H}' = -\hbar \sum_k \Omega_k I_{kz} + h \sum_{k<l} c_{kl} I_{kz} I_{lz} \tag{51.29}$$

where $\Omega_k = \gamma_k \Delta B_k$ is the resonance offset of the kth spin. The coupling constant, c_{kl}, between the kth and the lth spin is due to J or dipolar couplings (including different numerical factors; see Eq. 51.24).

Both terms in Eq. 51.29 commute so that the unitary transformations at Eq. 51.26 can be performed in arbitrary order, interaction by interaction, and, as concerns the standard $I_{kz}I_{lz}$ interaction, spin pair by spin pair of a given (multi) spin system. The basic relations are listed in Table 51.3.[9]

Note that **RF pulses can change the coherence order**, whereas it is **conserved during free evolution** at high fields. If zero- or multiple-quantum coherences occur, one distinguishes "passive" and "active" interaction partners (with respect to the participation in the n-quantum coherence). Only the spin-spin coupling to passive spins has an influence. Spin-spin coupling among active spins does not affect the evolution of multiple-quantum coherences.

51.6
Evolution Rules for Spherical Product Operators

The advantage of product operators on a spherical basis (Eqs. 42.70) is that physical interpretation is even more direct than with the Cartesian basis. This refers in particular to multiple-quantum coherences. With spherical product operators, each

---

[8]That is, the bandwidth of the pulses which are to be applied in the middle of the spectrum is much broader than the spectral range.

[9]We define the rotating-frame direction corresponding to the RF phase by the direction of the vector B_1. Note that in the literature the RF phase is often related to the vector of the precession frequency about B_1 instead of to the opposite direction. The signs of the operator terms resulting from the action of a RF pulse are then partly reversed. Both conventions lead to equivalent descriptions, of course.

Table 51.4. Interpretation of spherical product operator terms (for the definition of spherical spin operators see Eqs. 42.70), typically appearing in treatments of the evolution of the density operator of spin 1/2 systems. Each product term globally represents all isochromats arising from spin-spin coupling. The attributes "emissive" (em.) and "absorptive" (abs.) indicate the direction of the transition corresponding to the respective coherences (assuming positive gyromagnetic ratios). Note that the transitions characterized in this way do not necessarily involve electromagnetic radiation. This applies only in the case of single-quantum coherences.

	terms addressing spin "k" of an ($n \geq 1$)-spin system
$I_k^{(0)}$	longitudinal polarization of spin "k"; all isochromats in $+z$ direction
$I_k^{(+1)}$	emissive single-quantum coherences of spin "k"; all isochromats in-phase
$I_k^{(-1)}$	absorptive single-quantum coherences of spin "k"; all isochromats in-phase

	terms addressing spins "k", "l" of an ($n \geq 2$)-spin system
$2I_k^{(0)}I_l^{(0)}$	longitudinal order of spins "k", "l"; spin "k" isochr. caused by spin "l" up/down are antiparallel and vice versa ($+z$ and $-z$ dir.); all other isochr. are parallel
$2I_k^{(+1)}I_l^{(0)}$	emissive single-quantum coherences of spin "k"; spin "k" isochr. caused by spin "l" up/down are antiphase; spin "k" isochr. due to all other spins are in-phase
$2I_k^{(-1)}I_l^{(0)}$	absorp. single-quantum coherences of spin "k"; spin "k" isochromats caused by spin "l" up/down are antiphase; spin "k" isochromats due to all other spins are in-phase
$2I_k^{(+1)}I_l^{(+1)}$	"emissive" 2Q coherences of spins "k", "l"; all other spins are passive; 2Q isochromats due to passive spins are in-phase
$2I_k^{(-1)}I_l^{(-1)}$	"absorptive" 2Q coherences of spins "k", "l"; all other spins are passive; 2Q isochromats due to passive spins are in-phase
$2I_k^{(\pm 1)}I_l^{(\mp 1)}$	"emissive" or "absorptive" 0Q coherences of spins "k", "l"; all other spins are passive; 0Q isochromats due to passive spins are in-phase

	additional terms addressing spins "k", "l", "m" of an ($n \geq 3$)-spin system
$4I_k^{(\pm 1)}I_l^{(0)}I_m^{(0)}$	em. or abs. single-quantum coherences of spin "k"; spin "k" isochr. caused by spin "l" up/down or spin "k" up/down are antiphase; all other isochr. are in-phase
$4I_k^{(\pm 1)}I_l^{(\pm 1)}I_m^{(0)}$	"em." or "abs." 2Q coherences of spins "k", "l"; all other spins are passive; 2Q isochr. caused by spin "m" up/down are antiphase; all other isochromats are in-phase
$4I_k^{(\pm 1)}I_l^{(\mp 1)}I_m^{(0)}$	"em." or "abs." 0Q coherences of spins "k", "l"; all other spins are passive; 0Q isochr. caused by spin "m" up/down are antiphase; all other isochromats are in-phase
$4I_k^{(\pm 1)}I_l^{(\pm 1)}I_m^{(\pm 1)}$	"em." or "abs." 3Q coherences of spins "k", "l", "m"; all other spins are passive; 3Q isochromats due to passive spins are in-phase
$4I_k^{(\pm 1)}I_l^{(\pm 1)}I_m^{(\mp 1)}$	em. or abs. 3-spin-1Q coherences of spins "k", "l", "m"; all other spins are passive; 3-spin-1Q isochromats due to passive spins are in-phase

Table 51.5. Evolution of the reduced density operator in terms of spherical product operators for spins 1/2 coupled by standard $I_k^{(0)}I_l^{(0)}$ interactions. The coupling constant between spins k and l, c_{kl}, can represent J or dipolar couplings constant with different numerical factors (see Eq. 51.24). The spherical spin operators are defined in Eqs. 42.70. α is the pulse flip angle (which is specified by a positive value irrespective of the pulse phase); ν indicates the rotating frame direction of the RF field B_1; $\Omega_k = \gamma_k \Delta B_k$ is the resonance offset of spin k in the rotating frame (e.g., by chemical shift). The gyromagnetic ratios γ_k and γ_l are assumed to be positive.

RF pulses: $(\alpha)_{\pm\nu}$

"hard"

$$\sigma(t+) = \exp\left\{\pm i\alpha \sum_k I_{k\nu}\right\} \sigma(t-) \exp\left\{\mp i\alpha \sum_k I_{k\nu}\right\}$$

chem. shift. sel.

$$\sigma(t+) = \exp\{\pm i\alpha I_{k\nu}\}\, \sigma(t-)\, \exp\{\mp i\alpha I_{k\nu}\}$$

$\nu = x$
$(\alpha)_{\pm x}$
$B_1 \uparrow\uparrow \pm u_x$

$$I_k^{(0)} \rightarrow I_k^{(0)}\cos\alpha \pm \frac{i}{\sqrt{2}}\left(I_k^{(+1)}+I_k^{(-1)}\right)\sin\alpha$$
$$I_k^{(+1)} \rightarrow \pm\frac{i}{\sqrt{2}}I_k^{(0)}\sin\alpha + \frac{1}{2}I_k^{(+1)}(1+\cos\alpha) + \frac{1}{2}I_k^{(-1)}(-1+\cos\alpha)$$
$$I_k^{(-1)} \rightarrow \pm\frac{i}{\sqrt{2}}I_k^{(0)}\sin\alpha + \frac{1}{2}I_k^{(+1)}(-1+\cos\alpha) + \frac{1}{2}I_k^{(-1)}(1+\cos\alpha)$$

$\nu = y$
$(\alpha)_{\pm y}$
$B_1 \uparrow\uparrow \pm u_y$

$$I_k^{(0)} \rightarrow I_k^{(0)}\cos\alpha \pm \frac{1}{\sqrt{2}}\left(I_k^{(+1)}-I_k^{(-1)}\right)\sin\alpha$$
$$I_k^{(+1)} \rightarrow \mp\frac{1}{\sqrt{2}}I_k^{(0)}\sin\alpha + \frac{1}{2}I_k^{(+1)}(1+\cos\alpha) - \frac{1}{2}I_k^{(-1)}(-1+\cos\alpha)$$
$$I_k^{(-1)} \rightarrow \pm\frac{1}{\sqrt{2}}I_k^{(0)}\sin\alpha - \frac{1}{2}I_k^{(+1)}(-1+\cos\alpha) + \frac{1}{2}I_k^{(-1)}(1+\cos\alpha)$$

(Continued next page)

Table 51.5. (cont.)

chem. shift: (frequ. offset Ω_k)	$\sigma(t) \;=\; \prod_k e^{i\Omega_k t I_{kz}}\, \sigma(0) \prod_k e^{-i\Omega_k t I_{kz}}$	
	$I_k^{(0)} \rightarrow I_k^{(0)}$	
	$I_k^{(+1)} \rightarrow I_k^{(+1)} e^{i\Omega_k t}$	
	$I_k^{(-1)} \rightarrow I_k^{(-1)} e^{-i\Omega_k t}$	
$I_k^{(0)} I_l^{(0)}$ coupling: ($\equiv I_{kz}I_{lz}$ coupling)	$\sigma(t) \;=\; \prod_{k<l} e^{-i\pi c_{kl} t\, 2 I_k^{(0)} I_l^{(0)}}\, \sigma(0) \prod_{k<l} e^{i\pi c_{kl} t\, 2 I_k^{(0)} I_l^{(0)}}$	
	$I_k^{(0)} \rightarrow I_k^{(0)}$	
	$I_k^{(\pm1)} \rightarrow I_k^{(\pm1)} \cos(\pi c_{kl} t)\; 2 i I_k^{(\pm1)} I_l^{(0)} \sin(\pi c_{kl} t)$	
	$2 I_k^{(\pm1)} I_l^{(0)} \rightarrow 2 I_x^{(\pm1)} I_l^{(0)} \cos(\pi c_{kl} t) \mp i I_k^{(\pm1)} \sin(\pi c_{kl} t)$	
	$2 I_k^{(\pm1)} I_l^{(\pm1)} \rightarrow 2 I_k^{(\pm1)} I_l^{(\pm1)}$	
	$2 I_k^{(\pm1)} I_l^{(\mp1)} \rightarrow 2 I_k^{(\pm1)} I_l^{(\mp1)}$	

multiple-quantum coherence is represented by a separate term in contrast to the Cartesian basis where the corresponding terms refer to superimposed orders of multiple-quantum coherences. Moreover, the terms for "emissive" and "absorptive" transitions are distinguished. A series of typical product operator terms arising in treatments of the evolution of the reduced density operator of spin 1/2 systems is listed in Table 51.4.

The initial reduced density operator of an N-spin system is given by Eq. 51.25 again, which now reads

$$\sigma_0 = \sum_{m=1}^{N} I_m^{(0)} \tag{51.30}$$

The rules for the evolution of σ_0 in the course of a segmented treatment of a pulse sequence can easily be derived from those for a Cartesian basis (Table 51.3) by substituting the Cartesian components according to Eqs. 42.71. The results are listed in Table 51.5.

As concerns the interactions of spins, the same range of applicability as outlined in Sect. 51.4 applies. The rules are again given for the standard $I_{kz}I_{lz}$ type of spin-spin interaction (see Eq. 51.24), which now should be spelt "$I_k^{(0)}I_l^{(0)}$ interaction." Strong coupling of inequivalent nuclei was treated on a spherical basis in [220].

The inspection of the rules in Table 51.5 shows that the order of multiple-quantum coherences obviously can only be changed by RF pulses, whereas in the free-evolution intervals the coherence level is conserved. It is also clear that the evolution of multiple-quantum coherences is not affected by couplings among the spins actively participating in these coherences. Merely passive spins within the spin system can influence the evolution.

CHAPTER 52

Spin Operators for $I = 1$ Quadrupole Nuclei

The most frequent NMR application to nuclei with quadrupole interaction refers to deuterons ($I = 1$) in the high-field case. The situation is particularly simple if the electric field gradient is **axially symmetric** (asymmetry parameter $\eta = 0$) and if the quadrupole interaction is the dominating coupling. We restrict ourselves to this standard situation. A further assumption is that chemical-shift and susceptibility offsets are negligible, so that the RF can be assumed to be resonant.

The principle of the treatment of coherence evolution is to expand the density operator, as shown previously (Eq. 51.1), in terms of a suitable operator basis set. The basis operators are chosen in a way that the propagators under the action of the relevant Hamiltonians can be applied conveniently. Different basis sets have been suggested in the literature. This includes the fictitious spin 1/2 operators [496], single-transition operators [522], and irreducible spherical tensor operators [56, 170, 466].[1]

The basis set to be described in the following [20, 48, 495] is closely related to the product operator formalisms of the previous chapter. Although the physical meaning of the operators cannot be expressed in such simple terms as with the product operators, it can be elucidated and rationalized in a certain analogy. Formally speaking the basis operators are linear combinations of product operators so that they become Hermitian and represent observables in principle.

A complete basis set consists of $(2I + 1)^2$ operators $\mathcal{O}_j$, i.e., in the present case 9 operators including the unity operator. All Hamiltonians determining the evolution of the coherences must be expressed in terms of these operators. Moreover, the basis operators must be orthogonal so that they satisfy

$$\text{Tr}\{\mathcal{O}_j\mathcal{O}_k\} \propto \delta_{jk} \qquad (52.1)$$

A suitable basis set is

$$
\begin{array}{llll}
\mathcal{O}_0 & = & \mathcal{E} & \qquad \mathcal{O}_1 & = & I_x \\
\mathcal{O}_2 & = & I_y & \qquad \mathcal{O}_3 & = & I_z \\
\mathcal{O}_4 & = & I_z^2 - \frac{2}{3} & \qquad \mathcal{O}_5 & = & I_xI_z + I_zI_x \\
\mathcal{O}_6 & = & I_yI_x + I_xI_y & \qquad \mathcal{O}_7 & = & I_zI_y + I_yI_z \\
\mathcal{O}_8 & = & I_x^2 - I_y^2 &
\end{array}
\qquad (52.2)
$$

---

[1]The latter formalism was also extended to spins 3/2, 2, and 5/2 [57].

With this complete set of operators one can generally express the reduced density operator as a linear combination

$$\sigma = \sum_{j=0}^{8} c_j \mathcal{O}_j \tag{52.3}$$

The Hamiltonians of interest are also related to these basis operators. That is the secular part of the high-field quadrupole interaction Hamiltonian (Table 46.4 on page 424),

$$\mathcal{H}_q = \hbar\omega_q \left[I_z^2 - \frac{1}{3}I^2 \right] = \hbar\omega_q \left[I_z^2 - \frac{2}{3} \right] = \hbar\omega_q \mathcal{O}_4 \tag{52.4}$$

the rotating frame RF Hamiltonians (Eq. 48.95),

$$\begin{aligned}
\mathcal{H}'_{rf} &= -\hbar\omega_1 I_\nu & (\nu = x, y) \\
&= -\hbar\omega_q \mathcal{O}_j & (j = 1, 2)
\end{aligned} \tag{52.5}$$

and the rotating-frame frequency offset Hamiltonian,

$$\mathcal{H}'_0 = -\hbar\Omega I_z = -\hbar\omega_q \mathcal{O}_3 \tag{52.6}$$

where $\omega_q = 3e^2 qQ(3\cos^2 \vartheta - 1)/[8I(2I - 1)] = 3e^2 qQ(3\cos^2 \vartheta - 1)/8$.

The physical meaning of the basis operators can be visualized by considering their matrix representation (compare Sect. 51.3 and Eq. 42.49). For example, the operators $\mathcal{O}_1$ and $\mathcal{O}_2$ represent single-quantum coherences,

$$\sigma = \mathcal{O}_1 = I_x = \sqrt{\frac{1}{2}} \begin{pmatrix} 0 & 1 & 0 \\ 1 & 0 & 1 \\ 0 & 1 & 0 \end{pmatrix} \tag{52.7}$$

and

$$\sigma = \mathcal{O}_2 = I_y = \sqrt{\frac{1}{2}} \begin{pmatrix} 0 & -i & 0 \\ i & 0 & -i \\ 0 & i & 0 \end{pmatrix} \tag{52.8}$$

Double-quantum coherences are described by

$$\sigma = \mathcal{O}_6 = I_y I_x + I_x I_y = \begin{pmatrix} 0 & 0 & -i \\ 0 & 0 & 0 \\ i & 0 & 0 \end{pmatrix} \tag{52.9}$$

and

$$\sigma = \mathcal{O}_8 = I_x^2 - I_y^2 = \begin{pmatrix} 0 & 0 & 1 \\ 0 & 0 & 0 \\ 1 & 0 & 0 \end{pmatrix} \tag{52.10}$$

Finally, the formal representation of quadrupolar order is

$$\sigma = \mathcal{O}_4 = I_z^2 - \frac{2}{3} = \frac{1}{3} \begin{pmatrix} 1 & 0 & 0 \\ 0 & -2 & 0 \\ 0 & 0 & 1 \end{pmatrix} \tag{52.11}$$

The evolution of the reduced density operator in a time t under the action of the constant Hamiltonian $\mathcal{H}$ is described by the propagator expression (see Eq. 47.33)

$$\sigma(t) = e^{-(i/\hbar)\mathcal{H}t}\sigma(0)e^{(i/\hbar)\mathcal{H}t} \tag{52.12}$$

Since the Hamiltonians can also be expressed by the basis operators, we are confronted with the evaluation of propagator expressions of the type

$$f \equiv e^{-i\varphi\mathcal{O}_j}\mathcal{O}_k e^{i\varphi\mathcal{O}_j} \tag{52.13}$$

Generalizing Eqs. 48.44 and 48.45 to operators of our basis set gives

$$\frac{df}{d\varphi} = e^{-i\varphi\mathcal{O}_j}i[\mathcal{O}_k,\mathcal{O}_j]e^{i\varphi\mathcal{O}_j} \tag{52.14}$$

$$\frac{d^2f}{df^2} = e^{-i\varphi\mathcal{O}_j}[\mathcal{O}_j,[\mathcal{O}_k,\mathcal{O}_j]]e^{i\varphi\mathcal{O}_j} \tag{52.15}$$

The operators of the above basis set are selected to fulfill the cyclic permutation relations

$$[\mathcal{O}_j,\mathcal{O}_k] = ia\mathcal{O}_l \quad \text{and} \quad [\mathcal{O}_l,\mathcal{O}_j] = ia\mathcal{O}_k \tag{52.16}$$

where $\mathcal{O}_l$ is a basis operator or a linear combination of basis operators. The quantity a is a numerical constant. The second derivative can then be rewritten as

$$\frac{d^2f}{df^2} = -a^2 e^{-i\varphi\mathcal{O}_j}\mathcal{O}_k e^{i\varphi\mathcal{O}_j} = -f \tag{52.17}$$

in analogy to Eq. 48.46. The solution of this linear differential equation is

$$f(\varphi) = f(0)\cos(a\varphi) + \frac{df}{d\varphi}\bigg|_{\varphi=0}\sin(a\varphi) \tag{52.18}$$

That is

$$e^{-i\varphi\mathcal{O}_j}\mathcal{O}_k e^{i\varphi\mathcal{O}_j} = \mathcal{O}_k\cos(a\varphi) + \mathcal{O}_l\sin(a\varphi) \tag{52.19}$$
$$e^{i\varphi\mathcal{O}_j}\mathcal{O}_k e^{-i\varphi\mathcal{O}_j} = \mathcal{O}_k\cos(a\varphi) - \mathcal{O}_l\sin(a\varphi) \tag{52.20}$$

where

$$\mathcal{O}_l = \frac{1}{ia}[\mathcal{O}_j,\mathcal{O}_k] \tag{52.21}$$

Thus, knowing the commutators of the basis operators (or of linear combinations of them) permits one to calculate the propagator sandwich expression of any operator pair. The quantities entering into these expressions, i.e., the constants a and the operators $\mathcal{O}_l$, are implicitly listed in Table 52.1 in context with the commutators of the basis operators.

Table 52.1. Commutators of the basis operators for quadrupole nuclei. In the derivation of these expressions, Eq. 44.5 was employed in particular.

$[\mathcal{O}_1, \mathcal{O}_2]$	$=$	$[I_x, I_y] = iI_z$	$=$	$i\mathcal{O}_3$
$[\mathcal{O}_1, \mathcal{O}_3]$	$=$	$[I_x, I_z] = -iI_y$	$=$	$-i\mathcal{O}_2$
$[\mathcal{O}_1, \mathcal{O}_4]$	$=$	$[I_x, I_z^2 - 2/3] = -i(I_yI_z + I_zI_y)$	$=$	$-i\mathcal{O}_7$
$[\mathcal{O}_1, \mathcal{O}_5]$	$=$	$[I_x, I_xI_z + I_zI_x] = -i(I_xI_y + I_yI_x)$	$=$	$-i\mathcal{O}_6$
$[\mathcal{O}_1, \mathcal{O}_6]$	$=$	$[I_x, I_yI_x + I_xI_y] = i(I_xI_z + I_zI_x)$	$=$	$i\mathcal{O}_5$
$[\mathcal{O}_1, \mathcal{O}_7]$	$=$	$[I_x, I_zI_y + I_yI_z] = 2i(I_z^2 - I_y^2)$	$=$	$i(3\mathcal{O}_4 + \mathcal{O}_8)$
$[\mathcal{O}_1, \mathcal{O}_8]$	$=$	$[I_x, I_x^2 - I_y^2] = -i(I_yI_z + I_zI_y)$	$=$	$-i\mathcal{O}_7$
$[\mathcal{O}_2, \mathcal{O}_3]$	$=$	$[I_y, I_z] = iI_x$	$=$	$i\mathcal{O}_1$
$[\mathcal{O}_2, \mathcal{O}_4]$	$=$	$[I_y, I_z^2 - 2/3] = i(I_zI_x + I_xI_z)$	$=$	$i\mathcal{O}_5$
$[\mathcal{O}_2, \mathcal{O}_5]$	$=$	$[I_y, I_xI_z + I_zI_x] = -2i(I_z^2 - I_x^2)$	$=$	$-i(3\mathcal{O}_4 - \mathcal{O}_8)$
$[\mathcal{O}_2, \mathcal{O}_6]$	$=$	$[I_y, I_yI_x + I_xI_y] = -i(I_yI_z + I_zI_y)$	$=$	$-i\mathcal{O}_7$
$[\mathcal{O}_2, \mathcal{O}_7]$	$=$	$[I_y, I_zI_y + I_yI_z] = i(I_xI_y + I_yI_x)$	$=$	$i\mathcal{O}_6$
$[\mathcal{O}_2, \mathcal{O}_8]$	$=$	$[I_y, I_x^2 - I_y^2] = -i(I_zI_x + I_xI_z)$	$=$	$-i\mathcal{O}_5$
$[\mathcal{O}_3, \mathcal{O}_4]$	$=$	$[I_z, I_z^2 - 2/3]$	$=$	0
$[\mathcal{O}_3, \mathcal{O}_5]$	$=$	$[I_z, I_xI_z + I_zI_x] = i(I_yI_z + I_zI_y)$	$=$	$i\mathcal{O}_7$
$[\mathcal{O}_3, \mathcal{O}_6]$	$=$	$[I_z, I_yI_x + I_xI_y] = -2i(I_x^2 - I_y^2)$	$=$	$-2i\mathcal{O}_8$
$[\mathcal{O}_3, \mathcal{O}_7]$	$=$	$[I_z, I_zI_y + I_yI_z] = -i(I_zI_x + I_xI_z)$	$=$	$-i\mathcal{O}_5$
$[\mathcal{O}_3, \mathcal{O}_8]$	$=$	$[I_z, I_x^2 - I_y^2] = 2i(I_xI_y + I_yI_x)$	$=$	$2i\mathcal{O}_6$
$[\mathcal{O}_4, \mathcal{O}_5]$	$=$	$[I_z^2 - 2/3, I_xI_z + I_zI_x] \stackrel{(I=1)}{=} iI_y$	$=$	$i\mathcal{O}_2$
$[\mathcal{O}_4, \mathcal{O}_6]$	$=$	$[I_z^2 - 2/3, I_yI_x + I_xI_y]$	$\stackrel{(I=1)}{=}$	0
$[\mathcal{O}_4, \mathcal{O}_7]$	$=$	$[I_z^2 - 2/3, I_zI_y + I_yI_z] \stackrel{(I=1)}{=} -iI_x$	$=$	$-i\mathcal{O}_1$
$[\mathcal{O}_4, \mathcal{O}_8]$	$=$	$[I_z^2 - 2/3, I_x^2 - I_y^2]$	$\stackrel{(I=1)}{=}$	0
$[\mathcal{O}_5, \mathcal{O}_6]$	$=$	$[I_xI_z + I_zI_x, I_yI_x + I_xI_y] \stackrel{(I=1)}{=} -iI_x$	$=$	$-i\mathcal{O}_1$
$[\mathcal{O}_5, \mathcal{O}_7]$	$=$	$[I_xI_z + I_zI_x, I_zI_y + I_yI_z] \stackrel{(I=1)}{=} iI_z$	$=$	$i\mathcal{O}_3$
$[\mathcal{O}_5, \mathcal{O}_8]$	$=$	$[I_xI_z + I_zI_x, I_x^2 - I_y^2] \stackrel{(I=1)}{=} iI_y$	$=$	$i\mathcal{O}_2$
$[\mathcal{O}_6, \mathcal{O}_7]$	$=$	$[I_yI_x + I_xI_y, I_zI_y + I_yI_z] \stackrel{(I=1)}{=} -iI_y$	$=$	$-i\mathcal{O}_2$
$[\mathcal{O}_6, \mathcal{O}_8]$	$=$	$[I_yI_x + I_xI_y, I_x^2 - I_y^2] \stackrel{(I=1)}{=} -iI_z$	$=$	$-i\mathcal{O}_3$
$[\mathcal{O}_7, \mathcal{O}_8]$	$=$	$[I_zI_y + I_yI_z, I_x^2 - I_y^2] \stackrel{(I=1)}{=} iI_x$	$=$	$i\mathcal{O}_1$

References*

[1] Abragam A (1961) The principles of nuclear magnetism. Clarendon Press, Oxford — 7, 97, 113, 125, 463

[2] Abragam A, Goldman M (1982) Nuclear magnetism: order and disorder. Clarendon Press, Oxford — 42

[3] Ackerman JJH, Grove TH, Wong GG, Gadian DG, Radda GK (1980) Nature 283: 167 — 21, 313, 340

[4] Ailion DC (1971) Advan Magn Reson 5: 177 — 84

[5] Aleksandrov IV (1966) The theory of nuclear magnetic resonance. Academic Press, New York — 97

[6] Allen PS, Cowking A (1968) J Chem Phys 49: 789 — 86

[7] Allen PS (1974) J Phys C: Solid State Phys 7: L22 — 86

[8] Allerhand A, Gutowsky HS (1964) J Chem Phys 41: 2115 — 221

[9] Anderson PW, Weiss PR (1953) Rev Mod Physics 25: 269 — 118, 121

[10] Anderson PW (1954) J Phys Soc Japan 9: 316 — 121

[11] Andrew ER (1971) Progr NMR Spectr 8: 1 — 320

[12] Ardelean J, Stapf S, Demco DE, Kimmich R (1997) J Magn Reson 124: 506 — 16

[13] Atherton NM (1993) Principles of electron spin resonance. Ellis Horwood, Chichester — 222

[14] Aue WP, Bartholdi E, Ernst RR (1976) J Chem Phys 64: 2229 — 55

[15] Aue WP, Müller S, Cross TA, Seelig J (1984) J Magn Reson 56: 350 — 325, 341

[16] Aue WP, Müller S, Seelig J (1985) J Magn Reson 61: 392 — 362

[17] Axel L, Dougherty L (1989) Radiology 172: 349 — 303, 306

[18] Bagguley DMS (1992) In: Bagguley DMS (ed) Pulsed magnetic resonance: NMR, ESR, and optics. Clarendon Press, Oxford, p 5 — 4

[19] Bailes DR, Bryant DJ (1984) Contemp Phys 25: 441 — 240

[20] Barbara TM (1986) J Magn Reson 67: 491 — 493

---

* The numbers on the right-hand margins are page numbers of this book indicating where the references are cited.

[21] Barish A, Bradley MS, Johnson CS (1986) Rev Sci Instrum 57: 904   194

[22] Barrall GA, Lee YK, Chingas GC (1994) J Magn Reson A 106: 132   148

[23] Barth P, Hafner S, Kuhn W (1994) J Magn Reson A 110: 198   46, 318

[24] Basser PJ, Mattiello J, LeBihan D (1994) Biophys J 66: 259   290

[25] Bax A (1982) Two-dimensional nuclear magnetic resonance in liquids.   72
      Delft University Press, Dordrecht

[26] Beckmann N, Müller S (1991) J Magn Reson 93: 186   388

[27] Bénard H (1900) Rev Gen Sci Pure Appl 11: 1261   306

[28] Berendsen HJC (1962) J Chem Phys 36: 3297   152

[29] Bielecki A, Murdoch JB, Weitekamp DP, Zax DB, Zilm KW, Zimmer-   5
      mann H, Pines A (1984) J Chem Phys 80: 2232

[30] Binnig G, Rohrer H, Gerber C, Weibel E (1983) Phys Rev Lett 50: 120   323

[31] Bito Y, Hirata S, Nabeshima T, Yamamoto E (1995) Magn Reson Med   290
      33: 69

[32] Bittoun J, Bourrel E, Jolivet O, Idy-Peretti I, Mousseaux E, Tardivon A,   283
      Pernneau P (1993) Magn Reson Med 29: 674

[33] Bittoun J, Gonord P, Ruaud J-P, Kan S, Sauzade M (1995) J Trace &   298
      Microprobe Techn 13: 267

[34] Blanz M, Rayner TJ, Smith JAS (1993) Meas Sci Technol 4: 48   140

[35] Blechta V, Schraml J (1990) J Magn Reson 87: 601   207

[36] Blicharski JS (1972) Acta Physica Polonica A 41: 223   115

[37] Blicharski JS (1972) Z Naturforsch 27a: 1355   147

[38] Blinc R, Hogenboom DL, O'Reilly DE, Peterson EM (1969) Phys Rev   126
      Lett 23: 969

[39] Blinc R, Mali M, Osredkar R, Prelesnik A, Seliger J, Zupančič I, Ehren-   138
      berg L (1972) J Chem Phys 57: 5087

[40] Bloch F (1946) Phys Rev 70: 460   90

[41] Bloch F (1956) Phys Rev 102: 104   97

[42] Bloch F (1957) Phys Rev 1206: 105   97

[43] Bloembergen N, Purcell EM, Pound RV (1948) Phys Rev 73: 679   97, 110,
                                                                         125

[44] Bloembergen N, Rowland JA (1953) Acta Met 1: 731   472

[45] Bloembergen N, Pound RV (1954) Phys Rev 95: 8   7

[46] Bloom M, Hahn EL, Herzog B (1955) Phys Rev 97: 1699   228

[47] Bloom M, Legros MS (1986) Can J Phys 64: 1522   34

[48] Bloom M, Morrison C, Sternin E, Thewalt JL (1992). In: Bagguley DMS   39, 493
      (ed) Pulsed magnetic resonance: NMR, ESR, and optics. Clarendon
      Press, Oxford, p 274

[49] Blümler P, Blümich B (1994) In: Diehl P, Fluck E, Kosfeld R (eds) NMR basic principles and progress, vol 30. Springer, Berlin Heidelberg New York, p 209
317

[50] Blum K (1981) Density matrix theory and applications. Plenum Press, New York
428

[51] Boden N, Levine YK, Lightowlers D, Squires RT (1975) Chem Phys Lett 31: 511
26

[52] Bodenhausen G (1981) Progr NMR Spectr 14: 137
51

[53] Bodurka J, Gutsze A, Buntkowsky G, Limbach HH (1995) Z Phys Chem 190: 99
149

[54] Borgia GC, Brown RJS, Fantazzini P (1995) Phys Rev E 51: 2104
88

[55] Bourgeois D, Decorps M (1991) J Magn Reson 91: 128
21, 200

[56] Bowden GJ, Hutchison WD (1986) J Magn Reson 67: 403
493

[57] Bowden GJ, Hutchison WD, Khachan J (1986) J Magn Reson 67: 415
493

[58] Bowtell R, Bowley RM, Glover P (1990) J Magn Reson 88: 643
16

[59] Bowtell R (1992) J Magn Reson 100: 1
71

[60] Braunschweiler L, Ernst RR (1983) J Magn Reson 53: 521
368

[61] Brereton MG (1990) Macromolecules 23: 1119
89, 118

[62] Brigham EO (1988) The fast fourier transform. Prentice Hall, Englewood Cliffs
398

[63] Brink DM, Satchler GR (1968) Angular momentum. Clarendon Press, Oxford
420, 438, 439, 457, 459

[64] Broekaert P, Jeener J (1995) J Magn Reson A 113: 60
7, 71

[65] Broekaert P, Vlassenbroek A, Jeener J, Lippens G, Wieruszeski J-M (1996) J Magn Reson A 120: 97
71

[66] Bruce CR, Norberg RE, Pake GE (1956) Phys Rev 104: 419
71

[67] Brunner P, Reinhold M, Ernst RR (1980) J Chem Phys 73: 1086
362

[68] Bunde A, Havlin S (eds) (1991) Fractals and disordered systems. Springer, Berlin Heidelberg New York
178

[69] Bychuk OV, O'Shaughnessy B (1994) J Chem Phys 101: 772
126, 136, 178

[70] Callaghan PT (1984) Aust J Phys 37: 359
175, 186

[71] Callaghan PT, Eccles CD (1987) J Magn Reson 71: 426
301

[72] Callaghan PT, Eccles CD (1988) J Magn Reson 78: 1
300

[73] Callaghan PT (1990) J Magn Reson 87: 304
266, 301

[74] Callaghan PT (1990) J Magn Reson 88: 493
186

[75] Callaghan PT, MacGowan D, Packer KJ, Zelaya FO (1990) J Magn Reson 90: 177
191, 193

[76] Callaghan PT (1991) Principles of nuclear magnetic resonance microscopy. Clarendon Press, Oxford   192, 193, 244, 245, 249

[77] Callaghan PT, Coy A, MacGowan D, Packer KJ, Zelaya FO (1991) Nature 351: 467   191

[78] Callaghan PT, Coy A, Forde LC, Rofe CJ (1993) J Magn Reson A 101: 347   260

[79] Callaghan PT, Clark CJ, Forde LC (1995) J Biophys Chem 50: 225   258, 290

[80] Callaghan PT, Manz B (1994) J Magn Reson A 106: 260   216, 223

[81] Canet D, Diter B, Belmajdoub A, Brondeau J, Boubel JC, Elbayed K (1989) J Magn Reson 81: 1   21, 200, 204

[82] Caprihan A, Fukushima E (1990) Physics Reports 198: 195   283

[83] Carr HY, Purcell EM (1954) Phys Rev 94: 630   186, 252

[84] Champeney DC (1973) Fourier transforms and their physical applications. Academic Press, London   134, 398

[85] Chandrakumar N, Visalakshi GV, Ramaswamy D, Subramanian S (1986) J Magn Reson 67: 307   381

[86] Chandrakumar N, Subramanian S (1987) Modern techniques in high-resolution FT-NMR. Springer, Berlin Heidelberg New York   72, 368, 416, 420

[87] Chandrakumar N (1996) Spin-1 NMR. In: Diehl P, Fluck E, Günther H, Kosfeld R, Seelig J (eds) NMR basic principles and progress, vol. 34. Springer, Berlin Heidelberg New York   415

[88] Chandrasekhar S (1943) Rev Mod Physics 15: 1   175, 177

[89] Chandrasekhar S (1977) Liquid crystals. Cambridge University Press, Cambridge   169

[90] Chingas GC, Garroway AN, Bertrand RD, Moniz WB (1979) J Magn Reson 35: 283   362, 378

[91] Chingas GC, Garroway AN, Moniz WB, Bertrand RD (1980) J Am Chem Soc 102: 2526   362

[92] Chingas GC, Garroway AN, Bertrand RD, Moniz WB (1981) J Chem Phys 74: 127   362, 381

[93] Chingas GC, Garroway AN, Moniz WB (1984). In: Levy GC (ed) Topics in carbon-13 NMR spectroscopy. Wiley, New York, p 160   362

[94] Chingas GC, Miller JB, Garroway AN (1986) J Magn Reson 66: 530   320

[95] Cho CH, Ahn CB, Juh SC, Lee HK, Jacobs RE, Lee S, Yi JH, Jo JM (1988) Med Phys 15: 815   266, 298

[96] Clague ADH (1985) Helv Phys Acta 58: 121   175

[97] Cohen-Addad JP, Dupeyre R (1985) Macromolecules 18: 1101   89, 118

[98] Condon EU, Shortley GH (1935) Theory of atomic spectra. Cambridge University Press, Cambridge   459

[99] Cooley JW, Tukey JW (1965) Math Comput 19: 297   293

[100] Cory DG, Garroway AN (1990) J Magn Reson 14: 435   191

[101] Cory DG, Miller JB, Garroway AN (1990) J Magn Reson 90: 544   325

[102] Cory DG, Miller JB, Turner B, Garroway AN (1990) Mol Phys 70: 331   320

[103] Cotts RM, Hoch MJR, Sun T, Markert JT (1983) J Magn Reson 83: 252   207

[104] Crank J (1975) The mathematics of diffusion. Clarendon Press, Oxford   175

[105] Creel RB, von Meerwall ED, Barnes RG (1977) Chem Phys Lett 49: 501   34

[106] Cunningham AC, Day SM (1966) Phys Rev 152: 287   452

[107] Damnion RA, Vennart W, Summers IR, Ellis RE (1994) Magn Reson   290
       Imag 12: 873

[108] Das TP, Hahn EL (1958) Nuclear quadrupole resonance. In: Solid state   5
       physics, Suppl 1. Academic Press, New York, p 71

[109] Daszkiewicz OK, Hennel JW, Lubas B (1963) Nature 200: 1006.   152, 154

[110] Davis JH, Jeffrey KR, Bloom M, Valic MI, Higgs TP (1976) Chem Phys   37
       Lett 42: 390

[111] de Gennes PG (1971) J Chem Phys 55: 572   211

[112] de Gennes PG (1974). The physics of liquid crystals. Clarendon Press,   169
       Oxford

[113] Demco DE (1973) Phys Lett A 45: 113   42

[114] Demco DE, Tegenfeldt J, Waugh JS (1975) Phys Rev 11: 4133   82, 381

[115] Demco DE, Kimmich R, Hafner S, Weber H-W (1991) J Magn Reson   325, 328
       94: 317

[116] Demco DE, Hafner S, Kimmich R (1991) J Magn Reson 94: 333   319

[117] Demco DE, Hafner S, Kimmich R (1992) J Magn Reson 96: 307   47, 50,
       319, 321

[118] Demco DE, Johansson A, Tegenfeldt J (1994) J Magn Reson A 110: 183   187

[119] Demco DE, Köstler H, Kimmich R (1994) J Magn Reson A 110: 136   325, 367

[120] Demco DE, Hafner S, Kimmich R (1995) Appl Magn Reson 9: 267   325, 362,
       382

[121] Demco DE, Hafner S, Ardelean I, Kimmich R (1995) Appl Magn Reson   325, 362,
       9: 491   382, 388

[122] Deville G, Bernier M, Delrieux JM (1979) Phys Rev B 19: 5666   16, 18

[123] Doane JW, Visintainer JJ (1969) Phys Rev Lett 23: 1421   126

[124] Doddrell DM, Pegg DT, Bendall MR (1982) J Magn Reson 48: 323   362

[125] Doddrell DM, Brooks WM, Bulsing JM, Field J, Irving M, Baddeley H   325, 341
       (1986) J Magn Reson 68: 367

[126] Doi M, Edwards SF (1986) The theory of polymer dynamics. Clarendon   208, 210,
       Press, Oxford   212

[127] Dong RY (1994) Nuclear magnetic resonance of liquid crystals. Springer,   169
       Berlin Heidelberg New York

[128] Douglass DC, McCall DW (1958) J Phys Chem 62: 1102   181

[129] Drazin PG, Reed WH (1981) Hydrodynamic stability. Cambridge University Press, Cambridge — 306

[130] Dumoulin CL, Hart HR (1986) Radiology 161: 717 — 283

[131] Dupeyre R, Devoulon P, Bourgeois D, Decorps M (1991) J Magn Reson 95: 589 — 22, 200, 207

[132] Duyn JH, Mattay VS, Sexton RH, Sobering GS, Barrios FA, Liu G, Frank JA, Weinberger DR, Moonen CTW (1994) Magn Reson Med 32: 150 — 258

[133] Ebinger HD, Jänsch HJ, Polenz C, Polivka B, Preyss W, Saier V, Veith R, Fick D (1996) Phys Rev Lett 76: 656 — 302

[134] Edelstein WA, Hutchison JMS, Johnson G, Redpath T (1980) Phys Med Biol 25: 751 — 238

[135] Edmonds DT (1977) Phys Rep 29: 233 — 138

[136] Edzes HT, Samulski ET (1978) J Magn Reson 31: 207 — 149

[137] Eichler H, Salje G, Stahl H (1973) J Appl Phys 44: 5383 — 194

[138] Einzel D, Eska G, Hirayoshi Y, Kopp T, Wölfle P (1984) Phys Rev Lett 53: 2312 — 16

[139] Ernst RR, Bodenhausen G, Wokaun A (1987) Principles of nuclear magnetic resonance in one and two dimensions. Clarendon Press, Oxford — 51, 55, 60, 72, 72, 75, 222, 434

[140] Fano U (1957) Rev Mod Phys 29: 74 — 427, 428, 477

[141] Fatkullin NF (1991) Sov Phys JETP 72: 563 — 210

[142] Fatkullin N, Kimmich R, Weber HW (1993) Phys Rev E 47: 4600 — 126

[143] Fatkullin N (1994) Magnetic resonance and related phenomena. Extended abstracts of the XXVIIth Congress Ampere, Kazan, p 235 — 97, 98

[144] Fatkullin N, Kimmich R (1994) J Chem Phys 101: 822 — 126

[145] Fatkullin N, Kimmich R (1995) Phys Rev E 52: 3273 — 211

[146] Feiner LF, Locher PR (1980) Appl Phys 22: 257 — 316

[147] Feller W (1968) An introduction to probability theory and its applications. Wiley, New York — 181

[148] Fischer HW, van Haverbeke Y, Rinck PA, Schmitz-Feuerhake I, Muller RN (1989) Magn Reson Med 9: 315 — 157

[149] Fischer E, Kimmich R, Fatkullin N (1996) J Chem Phys 104: 9174 — 214

[150] Frahm J, Hänicke W (1984) J Phys E: Sci Instrum 17: 612 — 316

[151] Frahm J, Merboldt KD, Hänicke W, Haase A (1985) J Magn Reson 64: 177 — 343

[152] Frahm J, Merboldt K, Hänicke W (1993) Magn Reson Med 29: 139 — 258

[153] Fujara F, Ilyina E, Nienstaedt H, Sillescu H, Spohr R, Trautmann C (1994) Magn Reson Imaging 12: 245 — 186

[154] Fukushima E, Roeder SBW (1981) Experimental pulse NMR, Addison-Wesley, Reading, MA — 3

[155] Gadian DG (1982) Nuclear magnetic resonance and its applications to living systems. Clarendon Press, Oxford   357

[156] Garroway AN, Grannell PK, Mansfield P (1974) J Phys C 7: L457   238

[157] Glover P, Bowtell RW, Mansfield P (1994) Magn Reson Med 31: 423   302

[158] Goldman M (1970) Spin temperature and nuclear magnetic resonance in solids. Clarendon Press, Oxford   84, 336

[159] Görke U, Kimmich R, Weis J (1996) J Magn Reson B 111: 236   281, 289, 306

[160] Golovchenko JA (1986) Science 232: 48   323

[161] Gordon RE, Hanley PE, Shaw D (1982) Progr NMR Spectr 15: 1   340

[162] Graf V, Noack F, Béné GJ (1980) J Chem Phys 72: 861   112, 263

[163] Grant DM, Harris RK (eds) (1996) Encyclopedia of nuclear magnetic resonance. Wiley, Chichester   vi

[164] Gravina S, Cory DG (1994) J Magn Reson B 104: 53   316

[165] Grinberg F, Kimmich R (1995) J Chem Phys 103: 365   170, 173

[166] Grinberg F, Kimmich R (1996) J Chem Phys 105: 3301   174

[167] Grinberg F, Kimmich R, Möller M, Molenberg A (1996) J Chem Phys 105: 9657   174

[168] Grössl G, Winter F, Kimmich R (1985) J Phys E: Sci Instrum 18: 358   140

[169] Gründer W (1974) Wiss Zeitschrift Karl-Marx-Univ Leipzig Math Naturwiss R 23: 466   117

[170] Günther E, Blümich B, Spiess HW (1990) Mol Phys 71: 477   388, 493

[171] Gupta RK, Gupta P (1982) J Magn Reson 47: 344   357

[172] Guyer RA (1985) Phys Rev A 32: 2324   178

[173] Haase A, Malloy C, Radda GK (1983) J Magn Reson 55: 164   314

[174] Haase A, Hänicke W, Frahm J (1984) J Magn Reson 56: 401   204, 313

[175] Haase A, Frahm J, Matthaei D, Hänicke W, Merboldt K-D (1986) J Magn Reson 67: 258   251

[176] Haase A, Matthaei D (1987) J Magn Reson 71: 550   267

[177] Haase A (1990) Magn Reson Med 13: 77   251

[178] Haeberlen U, Waugh JS (1968) Phys Rev 175: 453   434

[179] Haeberlen U (1976) High resolution NMR in solids. Academic Press, New York   316, 420, 434

[180] Hafner S, Rommel E, Kimmich R (1990) J Magn Reson 88: 449   325, 326, 328, 359

[181] Hafner S, Demco DE, Kimmich R (1991) Meas Sci Technol 2: 882   319, 321

[182] Hafner S, Demco DE, Kimmich R (1991) Chem Phys Lett 187: 53   325, 362, 382

[183] Hafner S, Barth P, Kuhn W (1994) J Magn Reson A 108: 21   324

[184] Hafner S (1994) Magn Reson Imaging 12: 1047   316

[185] Hafner S, Demco D, Kimmich R (1996) Solid State Nucl Magn Reson 6: 275   44, 321, 322

[186] Hahn E (1950) Phys Rev 80: 580   4, 7, 13, 186

[187] Hahn EL, Maxwell DE (1952) Physical Review 88: 1070   217

[188] Halse MR, Rahman HJ, Strange JH (1994) Physica B 203: 169   175

[189] Hartmann SR, Hahn EL (1962) Phys Rev 128: 2042   360, 362, 453

[190] Hashi T, Fukuda Y, Tanigawa M (1992) In: Bagguley DMS (ed) Pulsed magnetic resonance: NMR, ESR, and optics. Clarendon Press, Oxford, p 492   4

[191] Hatanaka H, Terao T, Hashi T (1975) J Phys Soc Japan 39: 835   51

[192] Hausser R, Noack F (1964) Z Physik 182: 93   111, 258

[193] Heer CV (1972) Statistical mechanics, kinetic theory, and stochastic processes. Academic Press, New York   104

[194] Heink W, Kärger J, Pfeifer H (1978) Chem Eng Sci 33: 1019   175

[195] Held G, Noack F, Pollak V, Melton B (1973) Z Naturforsch 28c: 59   157

[196] Hepp MA, Miller JB (1994) J Magn Reson A 111: 62   320

[197] Hills BP, Snaar JEM (1992) Mol Phys 76: 979   260

[198] Hoult DI, Richards RE (1976) J Magn Reson 24: 71   301, 319

[199] Hoult DI (1979) J Magn Reson 33: 183   204, 312

[200] Howarth MA, Lian LY, Hawkes GE, Sales KD (1986) J Magn Reson 68: 433   484, 485

[201] Hubbard PS (1962) Phys Rev 128: 650   110

[202] Hurd RE, Freeman DM (1989) Proc Natl Acad Sci USA 86: 4402   345

[203] Hyslop WB, Lauterbur PC (1991) J Magn Reson 94: 501   260

[204] Icenogle MV, Caprihan A, Fukushima E (1992) J Magn Reson 100: 376   306

[205] Jacobsen JP, Bildsøe HK, Schaumburg K (1976) J Magn Reson 23: 153   114

[206] Jänsch HJ, Arnolds H, Ebinger HD, Polenz C, Polivka B, Pietsch GJ, Preyss W, Saier V, Veith R, Fick D (1995) Phys Rev Lett 75: 120   302

[207] Jakob PM, Breitling T, Haase A (1996) In: Proceedings of the International Society for Magnetic Resonance in Medicine, New York, vol 3. Society of Magnetic Resonance, Berkeley, p 1599   357

[208] Jeener J, Broekaert P (1967) Phys. Rev 157: 232   36, 39, 429

[209] Jeener J, Meier BH, Bachmann P, Ernst RR (1979) J Chem Phys 71: 4546   222

[210] Jeffrey KR (1981) Bull Magn Reson 3: 69   113, 118

[211] Jen J (1978) J Magn Reson 30: 111   222

[212] Johnson G, Hutchison JMS, Redpath TW, Eastwood LM (1983) J Magn Reson 54: 374   238

[213] Jones GP (1966) Phys Rev 148: 332   115

[214] Joseph PM, Yuasa Y, Mukherji B, Sloviter HA (1985) Invest Radiol 20: 504   388

[215] Kanters RPF, Char BW, Addison AW (1993) J Magn Reson A 101: 23   485

[216] Karczmar GS, Twieg DB, Matson GB, Weiner MW (1989) Magn Reson Med 81: 1   21, 200

[217] Kärger J, Pfeifer H, Heink W (1988) Adv Magn Reson 12: 1   175, 186

[218] Kärger J, Pfeifer H, Vojta G (1988) Phys Rev A 37: 4514   182

[219] Kastler A (1950) J Phys Radium 11: 255   360

[220] Kay LE, McClung RED (1988) J Magn Reson 77: 258   483, 484, 492

[221] Kennedy SD, Bryant RG (1986) Biophys J 50: 669   149

[222] Kentgens APM, de Boer E, Veeman WS (1987) J Chem Phys 87: 6859   222

[223] Khazanovich TN (1963) Polymer Sci USSR 4: 727   126

[224] Kim SG, Ashe J, Hendrich K, Ellermann JM, Merkle H, Uğurbil K (1993) Science 261: 615   258

[225] Kimmich R, Noack F (1970) Z Naturforsch 25a: 1680   149

[226] Kimmich R (1971) Z Naturforsch 26b: 1168   149

[227] Kimmich R (1974) Colloid & Polymer Sci 252: 786 and 254: 918   130

[228] Kimmich R, Peters A (1975) J Magn Reson 19: 144   130

[229] Kimmich R, Peters A (1975) Chem Phys Lipids 14: 350   149

[230] Kimmich R, Voigt G (1978) Z Naturforsch 33a: 1294 and 35a: 1431 (1980)   125

[231] Kimmich R (1980) Bull Magn Reson 1: 195   138

[232] Kimmich R, Schnur G, Scheuermann A (1983) Chem Phys Lipids 32: 271   125, 149, 152

[233] Kimmich R, Nusser W, Winter F (1984) Phys Med Biol 29: 593   149

[234] Kimmich R, Winter F, Nusser W, Spohn K-H (1986) J Magn Reson 68: 263   138, 149, 150, 151

[235] Kimmich R, Hoepfel D (1987) J Magn Reson 72: 379   245, 258, 342, 357

[236] Kimmich R, Rommel E, Knüttel A (1989) J Magn Reson 81: 333   342, 343

[237] Kimmich R, Gneiting T, Kotitschke K, Schnur G (1990) Biophys J 58: 1183   158

[238] Kimmich R, Nusser W, Gneiting T (1990) Colloids and Surfaces 45: 283   149, 153, 154, 156, 157

[239] Kimmich R (1990) Makromol Chem Macromol Symp 34: 237   156

[240] Kimmich R, Unrath W, Schnur G, Rommel E (1991) J Magn Reson 91: 136   186

[241] Kimmich R, Bühler K, Knüttel A (1991) J Magn Reson 93: 256   76, 276, 357

[242] Kimmich R, Rommel E, Nickel P, Pusiol D (1992) Z Naturforsch 47a: 361   314

[243] Kimmich R, Niess J, Hafner S (1992) Chem Phys Lett 190: 503   42, 46

[244] Kimmich R, Demco D, Hafner S (1992) In: Blümich B, Kuhn W (eds) Magnetic resonance microscopy. VCH, Weinheim, p 59   319, 325, 327, 362, 382

[245] Kimmich R, Weber HW (1993) Phys Rev B 47: 11 788   134

[246] Kimmich R, Klammler F, Skirda VD, Serebrennikova IA, Maklakov AI, Fatkullin N (1993) Appl Magn Reson 4: 425   134, 152, 154, 155, 209

[247] Kimmich R, Fatkullin N, Weber HW (1994) J Non-Cryst Solids 172-174: 689   126

[248] Kimmich R, Fischer E (1994) J Magn Reson A 106: 229   176, 187, 193, 197, 223

[249] Kimmich R, Simon B, Köstler H (1995) J Magn Reson A 112: 7   200

[250] Kimmich R, Barenz J, Weis J (1995) J Magn Reson A 117: 228   143

[251] Kimmich R, Fischer E, Callaghan P, Fatkullin N (1995) J Magn Reson A 117: 53   160, 174

[252] Kimmich R, Görke U, Weis J (1995) J Trace and Microprobe Techniques 13: 285   389

[253] Kimmich R (1996) Z Naturforsch 51a: 330   314

[254] Kingsley PB (1994) J Magn Reson A (1994) 107: 14   483, 484, 485

[255] Kirchartz KR, Oertel H (1988) J Fluid Mech 192: 249   306

[256] Klafter J, Zumofen G, Blumen A (1991) J Phys A: Math Gen 24: 4835   178

[257] Klafter J, Shlesinger MF, Zumofen G (1996) Physics Today 49: 33   136

[258] Klammler F, Kimmich R (1989) J Phys E: Sci Instrum 22: 74   284, 358

[259] Klammler F, Kimmich R (1990) Phys Med Biol 35: 67   257, 258, 282, 290, 358

[260] Klinkenberg M, Blümler P, Blümich B (1996) J Magn Reson A 119: 197   388

[261] Knüttel A, Kimmich R (1988) J Magn Reson 78: 205   244

[262] Knüttel A, Rommel E, Clausen M, Kimmich R (1988) Magn Reson Med 8: 70   344

[263] Knüttel A, Kimmich R (1989) J Magn Reson 83: 335   352, 357

[264] Knüttel A, Kimmich R (1989) Magn Reson Med 10: 404   245, 345, 351

[265] Knüttel A, Kimmich R (1989) Magn Reson Med 10: 411   351, 357

[266] Knüttel A, Kimmich R, Spohn K-H (1990) J Magn Reson 86: 526   362, 383, 384

[267] Knüttel A, Spohn K-H, Kimmich R (1990) J Magn Reson 86: 542   362, 389, 390

[268] Knüttel A, Kimmich R (1990) J Magn Reson 86: 253 — 352, 357, 362

[269] Knüttel A, Kimmich R, Spohn K-H (1990) Bull Magn Reson 12: 30 — 390

[270] Knüttel A, Kimmich R, Spohn K-H Magn Reson Med 17: 470 — 362, 383

[271] Köllner R, Schweikert KH, Noack F, Zimmermann H (1993) Liq Cryst 13: 483 — 173

[272] Koenig SH, Schillinger WE (1969) J Biol Chem 244: 6520 — 149

[273] Koenig SH, Schillinger WE (1969) J Biol Chem 244: 3283 — 149

[274] Koenig SH, Brown RD III, Lindstrom TR 1981) Biophys J 34: 397 — 149

[275] Koenig SH, Brown RD III, Adams D, Emerson D, Harrison DG (1984) Inv Radiology 19: 76 — 157

[276] Koenig SH, Brown RD III, Kenworthy AK, Magid AD, Ugolini R (1993) Biophys J 64: 1178 — 149, 154

[277] Koenig SH (1994) Inv Radiology 29: 127 — 258

[278] Konrat R, Burghardt I, Bodenhausen G (1991) J Am Chem Soc 113: 9135 — 368

[279] Kose K (1991) J Magn Reson 92: 631 — 288

[280] Kose K (1992) J Magn Reson 98: 599 — 306

[281] Kose K (1994) Phys Rev Lett 72: 1467 — 253, 288

[282] Köstler H, Kimmich R (1993) J Magn Reson B 102: 177 — 362, 378, 383

[283] Köstler H, Kimmich R (1993) J Magn Reson B 102: 285 — 325, 362, 380, 383

[284] Kubo R, Tomita K (1954) J Phys Soc Japan 9: 888 — 121

[285] Kuhn W (1990) Angew Chemie 102: 1 — 298

[286] Kuhn W, Barth P, Hafner S, Simon G, Schneider H (1994) Macromolecules 27: 5773 — 317

[287] Kulagina TP, Litvinov VM, Summanen KT (1993) J Polym Sci B: Polym Phys 31: 241 — 89, 118

[288] Kumar A, Welti D, Ernst RR (1975) J Magn Reson 18: 69 — 234

[289] Kumar A, Welti D, Ernst RR (1975) Naturwissenschaften 62: 34 — 234

[290] Kumar GR, Chaddah P (1987) Cryogenics 27: 229 — 140

[291] Kuntz ID, Kauzmann W (1974) Adv Protein Chem 28: 239 — 152

[292] Kunze C, Kimmich R, Demco D (1993) J Magn Reson A 101: 277 — 325, 362, 379, 383, 391

[293] Kunze C, Kimmich R (1994) J Magn Reson B 105: 38 — 362, 383, 385, 387, 325

[294] Kunze C, Kimmich R (1994) Magn Reson Imaging 12: 805 — 391

[295] Lauterbur PC (1973) Nature 242: 190 — 234, 314, 319

[296] LeBihan D, Breton E, Lallemand D, Grenier P, Cabanis E, Laval-Jeantet   282, 288
      M (1986) Radiology 161: 401

[297] Leger L, Hervet H, Rondelez F (1981) Macromolecules 14: 1732   194

[298] Levitt MH (1991) J Chem Phys 94: 30   362, 378,
                                             392

[299] Lodge TP, Rotstein NA, Prager S (1990) Advan Chem Phys 79: 1   211

[300] Look DC, Lowe IJ (1966) J Chem Phys 44: 2995   112, 115

[301] Lowe IJ, Norberg RE (1957) Phys Rev 107: 46   30

[302] Lurie DJ, Hutchison JMS, Bell LH, Nicholson I, Bussell DM, Mallard JR   138
      (1989) J Magn Reson 84: 431

[303] Lurie DJ, Nicholson I, Mallard JR (1991) J Magn Reson 94: 197   138

[304] Luyten PR, Vold RR, Vold RL (1984) J Magn Reson 58: 484   114

[305] Luyten PR, Marien AJH, Sijtsma B, den Hollander JA (1986) J Magn   325
      Reson 67: 148

[306] Luz Z, Meiboom S (1963) J Chem Phys 39: 366   112, 222

[307] Maas WE, Laukien FH, Cory DG (1995) J Magn Reson A 113: 274   7

[308] McCall DW, Douglass DC, Anderson EW (1959) Phys Fluids 2: 87   186

[309] McConnell HM (1958) J Chem Phys 28: 430   217

[310] McCoy MA, Warren WS (1990) J Chem Phys 93: 1   7

[311] McDonald PJ, Attard JJ, Taylor DG (1987) J Magn Reson 72: 224   319

[312] Macura S, Ernst RR (1980) Mol Phys 41: 95   223, 361

[313] Madhu PK, Kumar A (1995) J Magn Reson A 114: 201   91, 240

[314] Maffei P, Mutzenhardt P, Retournand A, Diter B, Raulet R, Brondeau J,   203, 314
      Canet D (1994) J Magn Reson A 107: 40

[315] Magnus W (1954) Commun Pure Appl Math 7: 649   432

[316] Maklakov AI, Skirda VD, Fatkullin NF (1990) Self-diffusion in polymer   175
      systems. In: Cheremisinoff NP (ed) Encyclopedia of fluid mechanics,
      vol 9: Polymer flow engineering. Gulf, Houston

[317] Mallett MJD, Codd SL, Halse MR, Green TAP, Strange JH (1996) J Magn   316
      Reson A 119: 105

[318] Manassen Y, Navon G (1985) J Magn Reson 61: 363   267

[319] Manassen Y, Hamers R, Demuth JE, Castellano AJ (1989) Phys Rev Lett   323
      62: 2531

[320] Mansfield P (1965) Phys Rev A 137: 961   30

[321] Mansfield P (1971) J Phys C 4: 1444   320

[322] Mansfield P (1971) Progr NMR Spectr 8: 41   26

[323] Mansfield P, Grannell PG (1973) J Phys C 6: L422   234

[324] Mansfield P (1977) J Phys C 10: L55   252

[325] Mansfield P, Maudsley AA, Morris PG, Pykett IL (1979) J Magn Reson 33: 261 — 240

[326] Mansfield P, Morris PG (1982) NMR imaging in biomedicine. Academic Press, New York — 252

[327] Mansfield P, Bowtell RW, Blackband S (1992) J Magn Reson 99: 507 — 175

[328] Marshall AG, Verdun FR (1990) Fourier transforms in NMR, optical, and mass spectrometry. Elsevier, Amsterdam — 398

[329] Mason RH, Antich PP, Babcock EE, Gerberich JL, Nunnally RL (1989) Magn Reson Imag 7: 475 — 388

[330] Mason RH, Bansal N, Babcock EE, Nunnally RL, Antich PP (1990) Magn Reson Imag 8: 729 — 388

[331] Matsui S (1991) Chem Phys Lett 179: 187 — 50, 320

[332] Maudsley AA, Oppelt A, Ganssen A (1979) Siemens Forschung und Entwicklung 8: 326 — 274

[333] Maudsley AA, Hilal SK, Permn WH, Simon HE (1985) Magn.Reson Med 2: 218 — 274

[334] Mayo BC (1973) Chem Soc Reviews 2: 49 — 273

[335] Mefed AE, Atsarkin VA (1978) Sov Phys JETP 47: 378 — 114

[336] Mehring M (1983) Principles of high resolution NMR in solids. Springer, Berlin Heidelberg New York — 51, 420, 472

[337] Meiboom S, Gill D (1958) Rev Sci Instr 29: 688 — 252

[338] Menon RS, Ogawa S, Kim S-G, Ellermann JM, Merkle H, Tank DW, Ugurbil K (1992) Invest Radiol 27: 47 — 258

[339] Merboldt K-D, Hänicke W, Frahm J (1986) J Magn Reson 67: 336 — 283

[340] Metz KR, Boehmer JP, Bowers JL, Moore JR J Magn Reson B 103: 152 — 203, 207, 314

[341] Metzler R, Glöckle WG, Nonnenmacher TF (1994) Physica A 211: 13 — 178

[342] Millar JM, Thayer AM, Bielecki A, Zax DB, Pines A (1985) J Chem Phys 83: 934 — 142

[343] Millar JM, Thayer AM, Zimmermann H, Pines A (1986) J Magn Reson 69: 243 — 142

[344] Mills R (1973) J Phys Chem 77: 685 — 177

[345] Migchelson C, Berendson HJC (1973) J Chem Phys 59: 296 — 152

[346] Mitra PP, Sen PN, Schwartz LM, Le Doussal P (1992) Phys Rev Lett 68: 3555 — 191

[347] Mitra PP, Sen PN, Schwartz LM (1993) Phys Rev B 47: 8565 — 191

[348] Moran PR (1982) Magn Reson Imag 1: 197 — 283

[349] Morgan LO, Nolle AW (1959) J Chem Phys 31: 365 — 111

[350] Morris GA, Freeman R (1978) J Magn Reson 29: 433 — 303

[351] Morris GA, Freeman R (1979) J Am Chem Soc 101: 760 — 351, 362

[352] Morris GA (1984) In: Levy GC (ed) Topics in carbon-13 NMR spec-   360
       troscopy. Wiley, New York, p 179

[353] Morris GA, Chilvers PB (1994) J Magn Reson A 107: 236   91, 240

[354] Morris PG (1986) Nuclear magnetic resonance imaging in medicine and   238, 276
       biology. Clarendon Press, Oxford

[355] Mosher TJ, Smith MB (1990) Magn Reson Med 15: 334   306

[356] Müller H-P, Weis J, Kimmich R (1995) Phys Rev E 52: 5195   283, 284,
                                                                             286

[357] Müller L, Ernst RR (1979) Mol Phys 38: 963   362, 381,
                                                                             392

[358] Muller RN (1991) Magn Reson Med 22: 178   258

[359] Müller S, Beckmann N (1989) Magn Reson Med 12: 400   388

[360] Nakashima TT, McClung RED (1986) J Magn Reson 70: 187   483, 484

[361] Navon G, Song Y-Q, Rõõm T, Appelt S, Taylor RE, Pines A (1996) Science 360
       271: 1848

[362] Nestle N, Kimmich R (1996) Magn Reson Imag 14: 905   175, 258,
                                                                             274, 276

[363] Nestle N, Kimmich R (1996) J Phys Chem 100: 12   258, 259

[364] Nickel P, Robert H, Kimmich R, Pusiol D (1994) J Magn Reson A 111:   313
       191

[365] Nickel P, Kimmich R (1995) Mol Struct 345: 253   228

[366] Nishimura DG, Macovski A, Pauly JM (1986) IEEE Trans Med Imaging   282, 283
       MI-5: 140

[367] Noack F (1986) Progr in NMR Spectr 18: 171   138

[368] Noggle JH, Schirmer RE (1971) The nuclear Overhauser effect. Aca-   361
       demic Press, New York

[369] Nusser W, Kimmich R, Winter F (1988) J Phys Chem 92: 6808   149, 151

[370] Nusser W, Kimmich R (1990) J Phys Chem 94: 5637   149, 151

[371] Orbach R (1961) Proc Roy Soc A264: 458   86

[372] Orbach R (1961) Proc Roy Soc A264: 485   86

[373] Orbach R (1986) Science 231: 814   178, 208

[374] Ordidge RJ, Connelly A, Lohman JAB (1986) J Magn Reson 66: 283   341

[375] Otting G, Liepinsh E, Wüthrich K (1991) Science 254: 974   134

[376] Packer KJ (1969) Mol Phys 17: 355   283

[377] Packer KJ, Wright KM (1980) J Magn Reson 41: 268   361

[378] Packer KJ, Wright KM (1983) Mol Phys 50: 797   483, 484

[379] Pake GE (1948) J Chem Phys 16: 327   469

[380] Park HW, Cho CW (1986) Magn Reson Med 3: 448   267

[381] Park HW, Ro YM, Cho ZH (1988) Phys Med Biol 33: 339   270

[382] Pastawski HM, Levstein PR, Usaj G (1995) Phys Rev Lett 75: 4310   42

[383] Pastawski HM, Usaj G, Levstein PR (1996) Chem Phys Lett 261: 329   42

[384] Pincus P (1969) Solid State Comm 7: 415   126, 169

[385] Pines A, Gibby MG, Waugh JS (1972) J Chem Phys 56: 1776   362

[386] Pines A, Gibby MG, Waugh JS (1973) J Chem Phys 59: 569   362, 381

[387] Pohl DW, Schwarz SE, Irniger V (1973) Phys Rev Lett 31: 32   194

[388] Pople JA, Schneider WG, Bernstein HJ (1959) High-resolution nuclear   422, 426
magnetic resonance. McGraw-Hill, New York

[389] Posse S, Aue WP (1990) J Magn Reson 88: 473   266

[390] Powles JG, Mansfield P (1962) Phys Lett 2: 58   26

[391] Powles JG, Strange JH (1963) Proc Phys Soc 82: 6   30

[392] Pütz B, Barski D, Schulten K (1991) J Magn Reson 97: 27   260

[393] Ramsey NF (1950) Phys Rev 78: 699   422

[394] Ramsey NF (1953) Phys Rev 91: 303   426

[395] Rao MG, Gupta AK (1982) Chem Eng J 24: 181   275

[396] Raulet R, Escanyé JM, Humbert F, Canet D (1996) J Magn Reson A 119:   314
111

[397] Redfield AG (1965) Advan Magn Reson 1: 1   97

[398] Redfield AG, Kunz SD (1979) In: Opella SJ, Lu P (eds) NMR and bio-   352
chemistry. Dekker, New York, p 225

[399] Redpath TW, Norris DG, Jones RA, Hutchison JMS (1984) Phys Med   283, 284
Biol 29: 891

[400] Reinhold M, Brunner P, Ernst RR (1981) J Chem Phys 74: 184   34, 361

[401] Rhim W-K, Pines A, Waugh JS (1971) Phys Rev B3: 684   42

[402] Rhim WK, Elleman DD, Vaughan RW (1973) J Chem Phys 59: 3740   320

[403] Richardson LF (1926) Proc Roy Soc London Ser A 110: 709   178

[404] Robert H, Pusiol D, Rommel E, Kimmich R (1994) Z Naturforsch 49a: 314
35

[405] Robert H, Minuzzi A, Pusiol DJ (1996) J Magn Reson A 118: 189   315

[406] Robinson AL (1986) Science 231: 671   42

[407] Rofe CJ, Van Noort J, Back PJ, Callaghan PT (1995) J Magn Reson B   266
108: 125

[408] Roman HE (1995) Phys Rev E 51: 5422   178

[409] Roman HE, Dräger J, Bunde A, Havlin S, Stauffer D (1995) Phys Rev E   178
52: 6303

[410] Rommel E, Noack F, Meier P, Kothe G (1988) J Phys Chem 92: 2981   149, 152

[411] Rommel E, Kimmich R (1989) Magn Reson Med 12: 209   261, 357

[412] Rommel E, Kimmich R (1989) Magn Reson Med 12: 390   112, 261,
                                                                                          263, 264,
                                                                                          265, 357

[413] Rommel E, Kimmich R (1989) J Magn Reson 83: 299   325, 359

[414] Rommel E, Kimmich R (1989) Bull Magn Reson 11: 169   328

[415] Rommel E, Hafner S, Kimmich R (1990) J Magn Reson 86: 264   319

[416] Rommel E, Nickel P, Kimmich R, Pusiol D (1991) J Magn Reson 91: 630   21, 204,
                                                                           313

[417] Rommel E, Pusiol D, Nickel P, Kimmich R (1991) Meas Sci Technol 2:   204
       866

[418] Rommel E, Kimmich R, Körperich H, Kunze C, Gersonde K (1992)   261, 263,
       Magn Reson Med 24:   357

[419] Rommel E, Nickel P, Rohmer F, Kimmich R, Gonzales C, Pusiol D (1992)   223, 229
       Z Naturforsch 47a: 382

[420] Rommel E, Kimmich R, Robert H, Pusiol D (1992) Meas Sci Technol 3:   314
       446

[421] Rommel E, Kimmich R, Spülbeck M, Fatkullin N (1993) Progr Coll Pol   214
       Sci 93: 155

[422] Rose ME (1957) Elementary theory of angular momentum. John Wiley,   420, 438,
       New York   439, 459

[423] Rothwell WP, Holecek DR, Kershaw JA (1984) J Polym Sci: Polym Lett   175
       Ed 22: 241

[424] Rourke DE, Prior MJW, Morris PG, Lohman JAB (1994) J Magn Reson   240
       A 107: 203

[425] Rugar D, Yannoni CS, Sidles JA (1992) Nature 360: 563   322

[426] Rugar D, Züger O, Hoen S, Yannoni CS, Vieth H-M, Kendrick RD (1994)   322
       Science 264: 1560

[427] Ryan LM, Taylor RE, Paff AJ, Gerstein BC (1980) J Chem Phys 72: 508   320

[428] Saarinen TR, Johnson CS (1988) J Magn Reson 78: 257   176, 193,
                                                                           194

[429] Samoilenko AA, Artemov DY, Sibeldina LA (1988) JETP 47: 417   319, 325

[430] Schaefer J, Stejskal EO (1979) In: Levy GC (ed) Topics in carbon-13   434
       NMR spectroscopy. Wiley, New York, p 283

[431] Schauer G, Nusser W, Blanz M, Kimmich R (1987) J Phys E: Sci Instrum   140, 149
       20: 43

[432] Schmidt C, Wefing S, Blümich B, Spiess HW (1986) Chem Phys Lett 130:   222, 224
       84

[433] Schmidt-Rohr K, Spiess HW (1994) Multidimensional solid-state NMR   421
       and polymers. Academic Press, London

[434] Schmiedel H, Schneider H (1975) Ann Physik 32: 249   42

[435] Schneider H, Schmiedel H (1969) Phy Lett 30A: 298   42

[436] Schnur G, Kimmich R, Winter F (1986) J Magn Reson 66: 295   34, 361,
                                                                           362

[437] Schnur G, Kimmich R, Lietzenmayer R (1990) Magn Reson Med 13: 478   260, 356,
                                                                           358 388

[438] Schnur G, Kimmich R (1989) Unpublished results   260

[439] Schweiger A (1990) In: Kevan L, Bowman MK (eds) Modern pulsed and continuous-wave electron spin resonance. Wiley, New York — 222

[440] Schweikert KH, Krieg R, Noack F (1988) J Magn Reson 78: 77 — 140

[441] Sen PN, Hürlimann MD (1994) J Chem Phys 101: 5423 — 191

[442] Sepponen RE, Pohjonen JA, Sipponen JT, Tanttu JI (1985) J Comp Assist Tomography 9: 1007 — 261

[443] Shattuck MD, Behringer RP, Johnson GA, Georgiadis JG (1995) Phys Rev Lett 75: 1934 — 288

[444] Shaw D (1976) Fourier transform NMR spectroscopy. Elsevier, Amsterdam — 398

[445] Sholl CA (1981) J Phys C: Solid State Phys 14: 447 — 111, 125

[446] Shriver J (1991) J Magn Reson 94: 612 — 485

[447] Sidles JA (1991) Appl Phys Lett 58: 2854 — 322

[448] Siegle G (1968) Z Naturforsch 23a: 91 — 30, 33, 34

[449] Siegle G (1968) Z Naturforsch 23a: 1194 — 30

[450] Siegle G (1968) Z Naturforsch 23a: 556 — 33

[451] Silver MS, Joseph RI, Hoult DI (1984) J Magn Reson 59: 347 — 245

[452] Simon B, Kimmich R, Köstler H (1996) J Magn Reson A 118: 78 — 204, 205, 313

[453] Singer JR (1959) Science 130: 1652 — 283

[454] Skibbe U, Neue G (1990) Colloids Surf 45: 235 — 313

[455] Slichter CP (1990) Principles of magnetic resonance 3rd edn. Springer, Berlin Heidelberg New York — 37, 452, 459, 472

[456] Smith BA (1982) Macromolecules 15: 469 — 194

[457] Sørensen OW, Eich GW, Levitt MH, Bodenhausen G, Ernst RR (1983) Progr NMR Spectr 16: 163 — 343, 483, 484, 485

[458] Sørensen OW, Ernst RR (1983) J Magn Reson 51: 974 — 484

[459] Sørensen OW (1990) J Magn Reson 86: 435 — 392

[460] Sørensen OW (1991) J Magn Reson 93: 648 — 392

[461] Solomon I (1955) Phys Rev 99: 559 — 97

[462] Solomon I (1958) Phys Rev 110: 61 — 37

[463] Solomon I (1959) Phys Rev Lett 2: 301 — 21

[464] Sotak CH, Freeman DM, Hurd RE (1988) J Magn Reson 78: 355 — 245, 345

[465] Spiess HW (1978) In: NMR basic principles and progress, vol 15. Springer, Berlin Heidelberg New York, pp 56-214 — 113, 125, 421

[466] Spiess HW (1980) J Chem Phys 72: 6755 — 39, 493

[467] Stapf S, Kimmich R, Nieß J (1994) J Appl Phys 75: 529 — 136

[468] Stapf S, Kimmich R, Seitter R-O (1995) Phys Rev Lett 75: 2855 — 126, 136

[469] Stapf S, Kimmich R, Seitter R-O (1996) Magn Reson Imaging 14: 841   136

[470] Stefan J (1891) Ann Phys Chem 3 Folge 42: 269   275

[471] Stehling MK, Turner R, Mansfield P (1991) Science 254: 43   252

[472] Stejskal EO, Tanner JE (1965) J Chem Phys 42: 288   185

[473] Stejskal EO, Memory JD (1994) High resolution NMR in solid state.   320
       Fundamentals of CP/MAS-spectroscopy. Oxford University Press, New
       York

[474] Stilbs P (1987) Progr NMR Spectr 19: 1   186

[475] Subramanian PR, Devanathan V (1974) J Phys A: Math Nucl Gen 7:   413
       1995

[476] Sun Y, Xiong J, Lock H, Buszko ML, Haase JA, Maciel GE (1994) J Magn   320
       Reson A 110: 1

[477] Swanson SD, Quint LE, Yeung HN (1990) Magn Reson Med 15: 102   362, 388

[478] Szymanski S, Gryff-Keller AM, Binsch G (1986) J Magn Reson 68: 399   97

[479] Takegoshi K, McDowell DA (1985) Chem Phys Lett 116: 100   44, 320

[480] Takegoshi K, Ito M, Terao T (1996) Chem Phys Lett 260: 159   222

[481] Tanner JE (1970) J Chem Phys 52: 2523; erratum: J Chem Phys 57: 3586   175
       (1972)

[482] Taylor DG, Bushell MC (1985) Phys Med Biol 30: 345   290

[483] Torrey HC (1949) Phys Rev 76: 1059   91

[484] Torrey HC (1953) Phys Rev 92: 962   125

[485] Tycko R, Opella SJ (1987) J Chem Phys 86: 1761   34

[486] Tzalmona A, Armstrong RL, Menzinger M, Cross A, Lemaire C (1990)   258
       Chem Phys Lett 174: 199

[487] Tzalmona A, Armstrong RL, Menzinger M, Cross A, Lemaire C (1992)   258
       Chem Phys Lett 188: 457

[488] Ueshima Y, Yamai S, Ikehira H, Hashimoto T, Mori K, Maki T, Fakuda   388
       H, Tateno Y (1990 Magn Reson Med 15: 158

[489] van den Ven FJM, Hilbers CW (1983) J Magn Reson 54: 512   483, 484

[490] van Dijk P (1984) J Comput Assist Tomogr 8: 429   283

[491] Van Wedeen J, Meuli RA, Edelman RR, Geller SC, Frank LR, Brady TJ,   283
       Rosen BR (1985) Science 230: 946

[492] Varshalovich DA, Moskalev AN, Khersonskii VK (1988) Quantum the-   413
       ory of angular momentum. World Scientific, Singapore

[493] Veeman WS, Cory DG (1989) In: Warren WS (ed) Advances in magnetic   320
       resonance, vol 13. Academic Press, San Diego, p 43

[494] Vega AJ, Vaughan RW (1978) J Chem Phys 68: 1958   117

[495] Vega AJ, Luz Z (1987) J Chem Phys 86: 1803   493

[496] Vega S, Pines A (1977) J Chem Phys 66: 5624   493

[497] Vega S (1978) J Chem Phys 68: 5518   409

[498] Vilfan M, Kogoj M, Blinc R (1987) J Chem Phys 86: 1055 — 126

[499] Voigt G, Kimmich R (1979) Progr Coll & Polym Sci 66: 273 — 140

[500] Vold RL, Vold RR (1978) Progr NMR spectroscopy 12: 79 — 97

[501] Vold RL, Vold RR, Poupko R, Bodenhausen G (1980) J Magn Reson 38: 141 — 118

[502] Wang P-K, Slichter CP (1986) Bull Magn Reson 8: 3 — 483, 485

[503] Wangsness RK, Bloch F (1953) Phys Rev 89: 728 — 97

[504] Warren WS, Hammes SL, Bates JL (1989) J Chem Phys 91: 5895 — 7

[505] Warren WS, Richter W, Andreotti AH, Farmer BT II (1993) Science 262: 2005 — 71

[506] Waugh JS, Huber LM, Haeberlen U (1968) Phys Rev Lett 20: 180 — 320

[507] Wefing J, Spiess HW (1988) J Chem Phys 89: 1219 — 222

[508] Weigand F, Hafner S, Spiess HW (1996) J Magn Reson A 120: 201 — 318, 320

[509] Weis J, Frollo I, Budinsky L (1989) Z Naturforsch 44a: 1151 — 267, 270

[510] Weis J, Budinsky L (1990) Magn Reson Imag 8: 483 — 270

[511] Weis J, Görke U, Kimmich R (1996) Magn Reson Imag 14: 1165 — 267, 271, 272, 273, 274, 275, 292

[512] Weis J, Kimmich R, Müller H-P (1996) Magn Reson Imag 14: 319 — 288, 306

[513] Weitekamp DP, Bielecki A, Zax D, Zilm KW, Pines A (1983) Phys Rev Lett 50: 1807 — 5, 138

[514] Werbelow LG, Grant DM (1977) Adv Magn Reson 9: 189 — 97

[515] Wind RA, Creyghton JHN, Ligthelm DJ, Smidt J (1978) J Phys C: Solid State Phys 11: L223 — 325

[516] Woelk K, Rathke JW, Klingler RJ (1994) J Magn Reson A 109: 137 — 148, 204

[517] Woelk K, Gerald RE II, Klingler RJ, Rathke JW (1996) J Magn Reson A 121: 74 — 204

[518] Woessner DE (1961) J Chem Phys 34: 2057 — 186

[519] Woessner DE (1962) J Chem Phys 36: 1 — 125

[520] Woessner DE (1962) J Chem Phys 37: 647 — 125

[521] Woessner DE, Snowden BS, Meyer GH (1969) J Chem Phys 50: 719 — 125

[522] Wokaun A, Ernst RR (1977) J Chem Phys 67: 1752 — 409, 493

[523] Wolf D (1979) Spin-temperature and nuclear-spin relaxation in matter. Clarendon Press, Oxford — 84, 336, 413

[524] Wölfel W, Noack F, Stohrer M (1975) Z Naturforsch A 30: 437 — 173

[525] Wu D, Johnson CS (1995) J Magn Reson A 116: 270 — 207

[526] Wüthrich K (1986) NMR of proteins and nucleic acids. Wiley, New York — 361

[527] Yeung HN, Swanson SD (1989) J Magn Reson 83: 183 — 388

[528] Zax D, Pines A (1983) J Chem Phys 78: 6333   175

[529] Zax DB, Bielecki A, Zilm KW, Pines A (1984) Chem Phys Lett 106: 550   138

[530] Zerhouni E, Parrish D, Rodgers WJ, Yang A, Shapiro E (1988) Radiology   303, 306
       169: 59

[531] Zeuner U, Dippel T, Noack F, Müller K, Mayer C, Heaton N, Kothe G   173
       (1992) J Chem Phys 97: 3794

[532] Zhang S, Meier BH, Ernst RR (1992) Phys Rev Lett 69: 2149   42

[533] Zick K (1994) Nondestr Test Eval 11: 255   319, 325

[534] Ziherl P, Vilfan M, Žumer S (1995) Phys Rev E 52: 690   174

[535] Zimmerman JR, Brittin WE (1957) J Phys Chem 61: 1328   221

Subject Index

Springer
and the
environment

At Springer we firmly believe that an international science publisher has a special obligation to the environment, and our corporate policies consistently reflect this conviction.

We also expect our business partners – paper mills, printers, packaging manufacturers, etc. – to commit themselves to using materials and production processes that do not harm the environment. The paper in this book is made from low- or no-chlorine pulp and is acid free, in conformance with international standards for paper permanency.

 Springer